Marine Productivity: Perturbations and Resilience of Socio-ecosystems

Hubert-Jean Ceccaldi • Yves Hénocque
Yasuyuki Koike • Teruhisa Komatsu
Georges Stora
Marie-Hélène Tusseau-Vuillemin
Editors

Marine Productivity: Perturbations and Resilience of Socio-ecosystems

Proceedings of the 15th French-Japanese Oceanography Symposium

With the special cooperation of Dr Patrick Prouzet, Institut Français pour l'Exploitation de la Mer (Ifremer), Issy-les-Moulineaux, France

Springer

Editors
Hubert-Jean Ceccaldi
Directeur d'études (h) Ecole Pratique des Hautes Etudes
Institut Méditerranéen d'Océanologie (M.I.O.)
Oceanomed, Bâtiment Méditerranée
Marseille cedex 09, France

Président of S.F.J.O. (*) of France
Marseille cedex 09, France

Société franco-japonaise d'Océanographie
Marseille, France

Yasuyuki Koike
Tokyo University of Marine Science and Technology
Conseillor of S.F.J.O. (*) of Japan
Tokyo, Japan

Georges Stora
Directeur de Recherche émérite
Institut Méditerranéen d'Océanologie (M.I.O.)
Oceanomed, Bâtiment Méditerranée
Marseille cedex 09, France

Deputy Secretary-General of S.F.J.O. (*) of France
Marseille cedex 09, France

Yves Hénocque
Institut Français pour l'Exploitation de la Mer (Ifremer)
Conseiller Principal Politique maritime et Gouvernance
General-Secretary of S.F.J.O. (*) of France
Correspondant Asie-Pacifique (JAMSTEC)
Issy-les-Moulineaux cedex, France

Teruhisa Komatsu
Atmosphere and Ocean Research Institute (AORI)
University of Tokyo
President of S.F.J.O. (*) of Japan
Kashiwa-shi, Japan

Marie-Hélène Tusseau-Vuillemin
Responsable de la Direction scientifique
Institut Français pour l'Exploitation de la Mer (Ifremer)
Issy-les-Moulineaux, France

With the special cooperation of
Patrick Prouzet
Directeur de Recherche Honoraire
Direction scientifique
Institut Français pour l'Exploitation de la Mer (Ifremer)
Issy-les-Moulineaux cedex, France

(*) S.F.J.O. Société franco-japonaise d'Océanographie
S.F.J.O. of France:
c/o Institut Océanographique
Paris

S.F.J.O. of Japan:
c/o Maison franco-japonaise
Ebisu, Shibuya-ku, Tokyo

ISBN 978-3-319-13877-0 ISBN 978-3-319-13878-7 (eBook)
DOI 10.1007/978-3-319-13878-7

Library of Congress Control Number: 2015934459

Springer Cham Heidelberg New York Dordrecht London

Printed on acid-free paper

Springer International Publishing AG Switzerland is part of Springer Science+Business Media (www.springer.com)

The 15th Franco-Japanese Symposium of Oceanography "Marine Productivity, Perturbations and Resilience of Socio-Ecosystems," organized by the long-standing partners Société franco-japonaise d'Océanographie de France and Société franco-japonaise d'Océanographie du Japon, reviewed the impacts of natural (storms, typhoons, earthquakes, tsunamis, etc.) and man-made (pollution, buildings in coastal areas, aquaculture, tourism, sports, diving, etc.) perturbations inflicted on natural coastal and marine environments, and examined the resilience of the affected socio-ecosystems and ability of governance responses to face these global/local changes.

The 15th Franco-Japanese Symposium of Oceanography was supported by the two countries' highest authorities.

The book collects 43 selected papers, written by experts from numerous universities and research institutes in both countries. It addresses the needs of marine sciences researchers (natural and social sciences), decision-makers and coastal zone managers, and other stakeholders involved in coastal and marine socio-ecosystems.

French and Japanese scientists participating to the SFJO Symposium at Boulogne-sur-Mer

French and Japanese scientists participating to the SFJO Symposium at Marseille

The 15th French-Japanese Oceanography Symposium
« Marine productivity: coastal socio-ecosystems degradation and resilience »
Boulogne-sur-mer
2013 october 17-18
www.ifremer.fr/productivite_marine_2013/
SG

The 15th French-Japanese Oceanography Symposium
« Marine productivity: coastal socio-ecosystems degradation and resilience »
Marseille
2013 october 21-22
www.ifremer.fr/productivite_marine_2013/
SG

Foreword

J'ai le grand plaisir de préfacer les minutes du 15ème Colloque franco-japonais d'océanographie qui s'est tenu en octobre 2013.

En 2011, un puissant tremblement de terre a dévasté l'Est du Japon (Tohoku) et un tsunami a balayé la zone côtière de cette région entraînant le terrible accident de la centrale nucléaire de Fukushima Daiichi. Confronté à cette catastrophe, le gouvernement du Japon s'est efforcé de reconstruire ces zones le plus rapidement possible. Tout au long de ce processus, nous étions conscients que nous ne pouvions accomplir ce redressement et cette reconstruction post-catastrophe uniquement par nos propres moyens et que nous étions soutenus par nos amis du monde entier.

Les deux Sociétés franco-japonaises d'océanographie font partie des organisations qui nous ont apporté un soutien précieux et sans faille pour le redémarrage de l'aquaculture au Sanriku, une région côtière qui traverse les préfectures d'Aomori, Iwate et Miyagi. Je voudrais leur exprimer ici ma plus profonde gratitude. Ces Sociétés ont fait don à la région du Tohoku d'équipements nécessaires à la recherche dans le domaine de l'aquaculture et lui ont apporté une aide technologique. Ces équipements ont grandement contribué à la renaissance de l'aquaculture sur toute la côte du Sanriku. C'est sans nul doute grâce à la coopération de ces Sociétés franco-japonaises qui dure depuis plus de cinquante ans qu'une telle contribution a été possible.

Le Gouvernement du Japon s'est également engagé depuis 2011 dans la promotion des sciences biologiques marines et des écosystèmes marins du Tohoku. Dorénavant, nous voudrions nous concentrer sur cette expertise afin de contribuer au mieux à la reconstruction post-catastrophe. Nous espérons qu'à l'avenir, les jalons de ce premier projet donneront naissance à un pôle d'excellence international attirant les chercheurs du monde entier.

Je souhaite encourager cette relation forte qui s'inscrit dans la durée entre la France et le Japon et je formule des vœux pour que cette collaboration se renforce encore.

Je veux enfin rendre hommage aux nombreuses personnes qui ont œuvré à l'organisation de ce Colloque et leur dire que je partage leur passion pour la recherche dans le domaine de l'océanographie. Je tiens à les féliciter pour leur succès.

Ambassadeur du Japon en France Son Excellence Monsieur Yoichi Suzuki

Foreword

I am very pleased to present the minutes of the 15th French-Japanese Symposium of Oceanography, held in October 2013.

In 2011, a very powerful earthquake devastated East Japan (Tohoku), and a tsunami swept through the coastal area of this region, causing the terrible accident at Fukushima Daiichi. Faced with this disaster, the Japanese government tried to rebuild these areas as quickly as possible. Throughout this process, it was realized that the recovery and post-disaster reconstruction could not be accomplished without external support, which was readily provided by friends around the world.

The two French-Japanese Oceanography Societies are part of organizations that have provided invaluable and unwavering support for the restart of aquaculture in the Sanriku coastal region through the prefectures of Aomori, Iwate, and Miyagi.

I would like to express to these organizations my deepest gratitude. These two oceanography societies have provided financial as well as technological assistance to the Tohoku facilities for research in the field of aquaculture. These facilities have, in turn, greatly contributed to the renaissance of aquaculture along the coast of Sanriku. Undoubtedly, such contribution was only possible due to the cooperation of the French-Japanese Oceanography Societies that has lasted for more than 50 years.

The Government of Japan is also committed, since 2011, in promoting the biological marine science and marine ecosystems of Tohoku, which is now the focus in order to best contribute to the post-disaster reconstruction. It is hoped that future milestones of this first project will give rise to an international center of excellence, attracting researchers worldwide.

I want to encourage this strong relationship between France and Japan, and I hope that this collaboration will further strengthen.

Finally, I pay tribute to the many people who contributed to the organization of this conference and say that I share their passion for research in the field of oceanography. I congratulate them for their success.

Ambassador of Japan to France His Excellency Mr Yoichi Suzuki

Foreword

Née de la mer, Marseille s'est, de tous temps, tournée vers elle. L'éventail d'activités qu'elle génère ont, au fil des siècles, façonné son image et contribué à son attractivité et à son rayonnement. L'engouement touristique qu'elle suscite aujourd'hui en atteste.

Pour important qu'il soit, cet atout n'est pas le seul. Avec la présence de structures universitaires et de centres de recherches dédiées aux sciences et techniques marines en Méditerranée, Marseille occupe dans le paysage scientifique une place importante. La tenue, pour la troisième fois depuis 1985, d'un colloque Franco-Japonais d'Océanographie à Marseille, en liaison avec le Troisième Congrès International des Aires Marines Protégées, confirment l'intérêt que le monde scientifique accorde au rôle joué par notre ville en ce domaine.

Mais Marseille s'attache aussi à mettre en pratique les enseignements qui découlent des nombreux échanges scientifiques qu'elle accueille et auxquels elle participe. La mer est partie intégrante de notre cadre de vie, de notre qualité de vie, et le littoral marseillais est un espace fragile où se focalisent les enjeux et les défis liés à sa préservation et à sa valorisation.

Il s'agit de satisfaire aux besoins du présent sans compromettre, pour les générations futures, la possibilité de pouvoir répondre à leurs propres besoins. C'est dans cette perspective que la Ville de Marseille a réalisé d'importants projets: création du Parc National des Calanques – parc national périurbain terrestre et maritime le plus important d'Europe – du Parc Maritime des Iles du Frioul, de l'opération Récifs Prado et du Plan de Gestion de la Rade …

Les manifestations scientifiques d'envergure internationale consacrées au monde marin trouvent donc, à Marseille, un environnement à leur mesure. Elles ont eu toute leur place dans l'éventail des évènements culturels qui ont marqué Marseille-Provence 2013, Capitale Européenne de la Culture. Quant au 15e Colloque Franco-Japonais d'Océanographie, il s'inscrit également dans le cadre du jumelage de Marseille avec la ville-port de Kobé, avec laquelle notre ville est liée depuis plus d'un demi-siècle.

Maire de Marseille, Sénateur des Bouches-du-Rhône Jean-Claude Gaudin

Foreword

Born of the sea, Marseille has, at all times, looked upon it. The range of activities it generates have, over the centuries, shaped its image and contributed to its attractiveness and its influence. The tourist enthusiasm it generates today attests.

Although important, this advantage is not the only one. With the presence of university structures and research centers dedicated to marine science and technology in the Mediterranean, Marseille is in an important scientific landscape. Held for the third time since 1985, a new edition of the Franco-Japanese Symposium of Oceanography in Marseille, in conjunction with the Third International Congress on Marine Protected Areas, confirms the interest that the scientific community attaches to the role played by our city in this field.

But Marseille also puts into practice the lessons arising from many scientific exchanges it accepts and which it participates. The sea is an integral part of our living, our quality of life, and the Marseille coast is a fragile area where the issues and challenges related to its preservation and its valuation should be focused.

This is to meet the needs of the present without compromising, for future generations, the possibility to meet their own needs. It is in this perspective that the City of Marseille has made significant projects: e.g., creating the Calanques National Park, the largest European land and sea peri National Park; Frioul Islands Marine Park; Operation Prado Reefs; and Management Plan of the Rade.

Major international scientific events devoted to the marine world are therefore, in Marseille, their measurement environment. They had their place in the range of cultural events marked Marseille-Provence 2013, European Capital of Culture. During the 15th Symposium on Franco-Japanese Oceanography, it has also become part of the twinning of Marseille with the port city of Kobe, with which our city is linked for more than half a century.

Mayor of Marseille, Senator of Bouches-du-Rhône Jean-Claude Gaudin

Foreword

Les effets du grand tremblement de terre qui a frappé le Japon en mars 2011, et du terrible tsunami qui s'en est suivi ont renforcé, s'il en était besoin, la collaboration remarquable qui, depuis de nombreuses années, s'est établie entre les sociétés océanographiques de la France et du Japon.

Le 15e colloque franco-japonais d'Océanographie a été organisé grâce à l'appui de l'Ifremer et des instances régionales, départementales et des collectivités locales de la Région Nord Pas de Calais et de la région Provence-Alpes-Côte d'Azur. Il s'est tenu en octobre 2013, successivement à Boulogne-sur-Mer et à Marseille, sur le thème: "Productivité marine – perturbations et résilience des socio-écosystèmes" et il a revêtu une importance exceptionnelle.

Rarement solidarité scientifique, mais aussi humaine, autour des problèmes de la mer ne s'est exprimée avec une telle ampleur à travers les thèmes débattus par les spécialistes des deux pays, sous les présidences conjointes des Professeurs Hubert-Jean Ceccaldi et Teruhisa Komatsu.

Tout ce qui compte pour une mesure objective des dégâts causés aux écosystèmes marins, qu'ils soient côtiers ou de profondeur, mais aussi pour mettre en oeuvre de nouveaux programmes dévolus à la protection des aires marines ou pour améliorer les conditions de vie et de travail des pêcheurs, tous les aspects en somme d'une véritable résilience pour atteindre un nouvel équilibre favorable, ont été passés au crible et portés à la connaissance des décideurs politiques.

Ce 15e Colloque est donc bien davantage qu'une simple manifestation d'étape, dans les échanges réguliers établis entre les deux Sociétés franco-japonaises d'Océanographie. Il a valeur de vrai symbole et honore ceux qui l'ont conçu et qui y ont pris part.

Professeur François Gros

Membre de l'Institut
Ancien co-président du
Consultatif Conjoint franco-japonais
pour les Sciences et les Technologies

Foreword

The effects of the great earthquake that struck Japan in March 2011, and the terrible tsunami that followed, have reinforced, if need be, the remarkable collaboration that has over many years been established between oceanographic scientific societies from France and Japan.

The 15th French-Japanese Symposium of Oceanography was organized with the support of Ifremer (Institut Français pour l'Exploitation de la Mer) and regional, departmental and local communities in the Region Nord Pas-de-Calais and the Region Provence-Alpes-Côte d'Azur. It was held in October 2013, successively in Boulogne-sur-Mer and in Marseille, on the main theme "Marine Productivity: disturbance and resilience of socio-ecosystems" and was of exceptional importance.

Rarely scientific solidarity, but also human, around the problems of the sea has spoken to such an extent through the topics discussed by the experts of the two countries under the joint chairs of Professors Hubert-Jean Ceccaldi and Teruhisa Komatsu.

All that matter for an objective measure of damages to marine ecosystems, whether coastal or deep, but also to implement new programs vested in the protection of marine areas and to improve the conditions of life and work of fishermen, in fact all aspects of a real resilience to reach a new favorable balance, were scrutinized and brought to the attention of policy makers.

The 15th Symposium is much more than a mere expression of the stage, in regular exchanges between the two scientific societies established between France and Japan in the field of Oceanography. It has a true symbolic value and honors those who have designed and participated.

Member of the Institute Professeur François Gros
Former Co-Chairman of Franco-Japanese
Joint Advisory Council for Science and Technology

Preface

It is with great pleasure that we present this book bringing together the papers presented at the 15th Franco-Japanese Symposium of Oceanography.

Its preparation would not have been possible without the constant and very friendly collaboration between the leaders and members of the "Société Franco-Japonaise d'Océanographie au Japon" (founded in 1960) and the "Société Franco-Japonaise d'Océanographie en France" (founded in 1983).

Both Societies allow their members, mostly French and Japanese researchers, to share new initiatives in a very friendly atmosphere hence very fruitful common understanding. Networks of informal exchanges are priceless because they can establish very flexible, effective and rapid links between teams, laboratories and individuals. Eventually, they may lead to more formal and long-term collaborations.

So it is with great pleasure that we would like to express our gratitude to the organizations and individuals who provided their priceless help and continuous support for our project.

Firstly, our deepest gratitude goes to the French Research Institute for the Exploitation of the Sea (IFREMER), its Paris Headquarters led by IFREMER President François Jacq and represented by its Scientific Direction led by Marie-Hélène Tusseau-Vuillemin, and its Boulogne-sur-Mer Channel-North Sea Centre led by its Director Dominique Godefroy who, with his dedicated team, welcomed and ensured the success of the first leg of our Symposium in Boulogne-sur-Mer. The organization and the running of the latter would not have been possible without the contribution of both IFREMER Atlantic Centers in Nantes and Brest, more particularly through the efficient inputs from Catherine Coriou as regards the construction and running of the Symposium internet site and Sylvie Gros' making of beautiful promotional posters.

But above all, we want to warmly thank Dr. Patrick PROUZET for his friendly assistance, for his outstanding efficiency, and for the very important work he has done so this book is achieved. We express him our most sincere gratitude, especially for establishing remarkable functional links with the various departments of the French Institute for Exploitation of the Sea IFREMER. Here, we thank him from the bottom of our hearts.

We are also very pleased to very sincerely address our gratitude to our partners and hosts, the Provence Alpes Côte d'Azur (PACA) Regional Council, the City of Marseille, the Sasakawa Franco-Japanese Foundation, the City of Boulogne-sur-Mer, the Nord-Pas-de Calais Regional Council, the University of Aix-Marseille, the Urban Community of Boulogne-sur-mer, the Urban Community of Marseille Provence Métropole, the Chamber of Commerce and Industry of Boulogne-sur-mer for its generous material contributions, the National Center of Scientific Research (CNRS), the Pytheas Institute, the Mediterranean Institute of Oceanology, the wonderful aquarium Nausicaa in Boulogne-sur-mer, the two Poles of Competitiveness of Toulon (Mer P.A.C.A.) and in Boulogne-sur-mer (Aquimer), the Oceanographic Institute of Paris and the Academy of Sciences, Arts and Letters of Marseille.

We will not forget the hard work of the scientific committees and organization as well as executive offices in each country and the dedication of the host teams respectively in Boulogne-sur-Mer and Marseille.

We are glad also to thank the design office C'Graph in Marseille for their excellent work.

Through carefully selected topics (macro wastes, radioactive substances in food webs after Fukushima, ecological consequences of the Tohoku tsunami on ecosystems, effects of dredging at sea, pollutants in food webs, biological indicators, tropical environments, sato-umi and prud'homies, services to human communities), our Societies' main target was to look at the management of natural resources by mankind, its trajectories and adaptations, in comparing Japan's and France's respective approaches at a critical period where environmental problems are increasing at a pace and a scale never seen before.

"Our house is burning while we are looking elsewhere" ("Notre maison brûle et nous regardons ailleurs"), such was the warning of the then President of the French Republic, Jacques Chirac, at the opening of the speech he made before the plenary of the Fourth Earth Summit, on September 2, 2002, in Johannesburg, South Africa.

What shall we do, we scientists, to transfer appropriate hence useful information concerning the preservation and sustainable exploitation of coastal and marine resources?

It is difficult to convince decision makers to protect biodiversity where resources were once plentiful and therefore considered as limitless for the good of economy and employment. It is also almost impossible to convince a poacher to stop hunting rare animals when their sale is his only subsistence. Protection must not be opposed to development but should be considered together.

We are beginning to uncover the benefits that mankind withdraws from the exploitation of natural resources, services rendered by nature that no longer can be considered as free in regard to the growing damages inflicted by our activities on the environment. With this 15th French-Japanese Symposium on Oceanography, we intended to show that the functioning of natural systems were often disrupted, either by natural processes including hazards or by Man himself, while such disturbances can be to a certain extent naturally or artificially buffered thanks to the socio-ecosystem resilience, more particularly through coastal communities' capacity to adapt using and improving their ancestral practice of coastal zone management.

It is this very notion of "socio-ecosystem" and "ecosystem-based management" we wanted to illustrate a new approach, through concrete examples and development models taken from both countries. Histories, traditions, religions, customs, rules and laws rooted into the past form the settings and frameworks in which develops the exploitation of natural resources. Our aim is to contribute to bringing up new elements to light in order to make these settings and frameworks evolve along the basic rules of Ecology – the ultimate truth – which depend not only on the existence of natural ecosystems, but also on the populations who use them.

This book is intended to inform scientific communities, management structures, decision-making bodies, and administrations of both countries and others countries involved in the management of coastal ecosystems, as they will have to necessarily and concretely face the covered issues in each country-specific context. In this, our conferences' major interest goes well beyond their scientific purpose to reach the most important social and cultural aspects in the way we enjoy and manage our coastal and marine goods and services.

Marseille, France Hubert-Jean Ceccaldi
Kashiwa-shi, Japan Teruhisa Komatsu

Marine Productivity: Perturbation and Resilience of Socio-Ecosystem

Special Topics

1. Natural Perturbations
2. Anthropogenic Perturbations
3. Ecosystem Based Approach
4. People, Institutions and the Sea
5. Status and Future of Marine Protected Areas
6. Round Table
7. Short and Preliminary Communications
8. Authors Index

Committee in Charge of Organization

Boulogne-sur Mer	Marseille
Dominique Godefroy (Directeur centre Ifremer Manche- mer du Nord)	Prof. Hubert-Jean Ceccaldi (Président SFJO France)
Prof. Hubert-Jean Ceccaldi (Président SFJO France)	Prof. Teruhisa Komatsu (Président SFJO Japon)
Prof. Teruhisa Komatsu (Président SFJO Japon)	Georges Stora (CNRS/ IMO Marseille)
Prof. Yasuyuki Koike (SFJO Japon)	Prof. Yasuyuki Koike (SFJO Japon)
Patrick Prouzet (Ifremer)	Patrick Prouzet (Ifremer)
André Carpentier (Ifremer/Nouméa)	Prof. Ivan Dekeyser (OSU-UAM)
Prof. Takeshi Yamane (Univ. Kinki)	Mathias Girault (Univ. Ikashika)
Philippe Valette (D/Nausicaa)	Gérald Gregori (CNRS IMO Marseille)
Emmanuel Thouard (Ifremer/DAEI)	Didier Reault (élu chargé de la mer et du littoral ville de Marseille)
Thierry Missonier (DG Aquimer)	Prof. Bernard Quéguiner (UAM-IMO)
	Marc Tedetti (iRD, Aix-Marseille Université)

Scientific Committee

Prof. Hubert-Jean Ceccaldi (SFJO – France)	Georges Stora (CNRS/IMO Marseille)
Yves Hénocque (Ifremer-Jamstec-OPRF)	Prof. Catherine Mariojouls (Agro Paris Tech, Paris)
Patrick Prouzet (Ifremer)	Prof. Hisayuki Arakawa (TUMSAT)
Paul Marchal (Ifremer)	Prof. Denis Bailly (UBO)
Verena Trenkel (Ifremer)	Patrick Baraona (Pôle Mer PACA)
Marie-Hélène Tusseau-Vuillemin (DS/ Ifremer)	Prof. Charles-François Boudouresque (IMO-UAM)
Prof. Fabienne Petit (Université Rouen)	Prof. Ivan Dekeyser (OSU-UAM)
Prof. Robert Lafite (Université Rouen)	François Galgani (Ifremer)
Prof. J.-C. Dauvin (Université Lille)	Mathias Girault (Univ. Ikashika)
Gérald Grégori (CNRS IMO Marseille)	Guy Herrouin (Pôle Mer PACA)

Prof. Teruhisa Komatsu (SFJO Japon)	Prof. Hamano Akira (Univ. Pêches Shimonoseki)
Prof. Yasuyuki Koike (SFJO Japon)	Sylvain Pioch (Univ. Montpellier III)
Hideyuki Takahashi (FRA Hasaki)	Prof. Bernard Quéguiner (UAM-IMO)
Prof. Tsuchiya Makoto (Univ. des Ryukyu)	Prof. Bernard Salvat (EPHE Perpignan)
	Prof. Nardo Vicente (DS/Inst. Paul Ricard)
	Prof. Takeshi Yamane (Univ. Kinki)

Executive Board

Prof. Hubert-Jean Ceccaldi (SFJO France)	Yves Hénocque (Ifremer-Jamstec-OPRF)
Prof. Teruhisa Komatsu (SFJO Japon)	Georges Stora (CNRS Université Marseille)
Prof. Yasuyuki Koike (SFJO Japon)	Catherine Coriou (Ifremer – in charge of website)
Patrick Prouzet (Ifremer)	Sylvie Gros (Ifremer – in charge of web graphics)
Benoist Hitier (Ifremer)	
Bénédicte Desmytteres (Ifremer)	
Catherine Sart (Ifremer)	

Editorial Board

Prof. H.-J. Ceccaldi (Chairman SFJO France)	Georges Stora (Université Marseille)
Yves Hénocque (Ifremer- Jamstec-OPRF)	Marie – Hélène Tusseau-Vuillemin (DS/Ifremer)
Prof. Yasuyuki Koike (SFJO Japon)	With the cooperation of Patrick Prouzet (Ifremer)
Prof. Teruhisa Komatsu (Chairman SFJO Japon)	

List of Chairmen

Dr V. Trenkel	Prof. F. Petit
Prof. T. Yamane	Prof. H. Arakawa
Dr P. Marchal	Prof. T. Komatsu
Dr M. Sasano	Dr Y. Hénocque
Prof. H.-J. Ceccadi	Dr H. Takahashi
Prof. J.-C. Dauvin	Dr P. Prouzet
Prof. B. Salvat	Dr G. Grégori
Prof. I. Dekeyser	Prof. Y. Koike
Prof. C. Mariojouls	
Prof. M. Tsuchiya	Mr Y. Tanaka

Contents

Part I

Special Introductory Communication

An Example of Friendship and Cooperation Between France and Japan: Oyster Farming in Sanriku Area (Tohoku Region, Northern Japan) Before and After Tsunami – Restoration and Technical Adaptation of Culture Systems

Yasuyuki Koike

Abstract

In the 1960s, oyster culture in France was greatly perturbated by a mass mortality of cupped oyster caused by infection by the protozoan *Marteilia* sp. But oyster culture could be resumed by the timely help of export of spat from the port of Sanriku, northern Japan. After examination and identification of the pathogens, a large number of single-seed oysters were exported to France in the 1970s (Koganezawa 1984; Goto 2012).

On 11 March 2011, a huge tsunami surged against the coast of Sanriku and ravaged the fisheries and aquaculture installations included those for oyster culture.

After the disaster, with the solidarity to return assistance given in the past, the organizations of France decided to help reconstruct and reactivate the marine productivity in the coast of Sanriku. From several French cooperatives of oyster growers, equipment for oyster culture was sent to a Japanese cooperative, and from several research organizations, the contribution was sent to Japanese producers.

Independently to this contribution, in the framework of the long-term exchanges between the two "Sociétés franco-japonaises d'Océanographie" of France and of Japan, some scientifical and technological equipments such as microscopes and plankton nets were donated to Japanese research centers and fishery cooperatives.

Y. Koike (✉)
Tokyo University of Marine Science and Technology,
Conseillor of S.F.J.O. (*) of Japan, 17-16, Minami-3,
Koenji, Suginami-ku 166-0003, Tokyo, Japan
e-mail: oreillemer@ybb.ne.jp

H.-J. Ceccaldi et al. (eds.), *Marine Productivity: Perturbations and Resilience of Socio-ecosystems*,
DOI 10.1007/978-3-319-13878-7_1, © Springer International Publishing Switzerland 2015

Under the help of French colleagues, the oyster growers of Sanriku could restart the most important works, such as spat collection in the summer season, before the end of July to August in the same year as the disaster. In Miyagi Prefecture, about 75 % of oyster growers have restarted the oyster farming in the same region.

1 A Short History of Cooperations Between the Two "Sociétés Franco-Japonaises D'Océanographie" of France and Japan

Scientifical and technological relationships in the field of marine sciences between France and Japan have been facilitated by dialogues and exchanges established between the Société franco-japonaise d'Océanographie of Japan (founded in 1960 and gathering around 250 members) and the Société franco-japonaise d'Océanographie of France (founded in 1984 and gathering around 70 persons as well as public and private organizations).

These two societies are mainly associations of researchers and other specialists in marine fields who have expressed their wish to promote efficient and pleasant relationships between Japanese and French individuals as well as organizations having similar fields of activities.

For several decades, the exchanges between the members of the two societies have been very fruitful and very friendly. The exchanges reached a maximum during common meetings (colloques, symposiums) alternatively organized in France and in Japan, approximately every 2 or 3 years. It is interesting to summarize them briefly.

Each of these symposia has been concretized by special official publications:

First edition in France at Montpellier in 1983 devoted to "Aquaculture."

Second edition in Japan at Sendai in 1984 devoted to "Aquaculture" and a program of visit of fish and mollusk hatcheries.

Third edition in France at Marseille in 1985 with a special topic on "Coastal management and littoral planning," organized with the support of numerous public and private organizations: Physical oceanography and sediment dynamics, colored waters (red tides), teledetection, communications, microbiology of coastal waters, biological characteristics of coastal waters, artificial reefs, resources exploitation, and aquaculture.

Fourth edition in Japan at Shimizu in 1988 on the thema "General oceanography" presentation of communications, visits to aquaculture centers, centers of research, and universities.

Fifth edition in Japan at Hiroshima and Higashino in 1989, a series of six French-Japanese colloques, with a general topic: "Founding an algal park in Seto-nai-kai."

Sixth edition in Japan at Tokyo in 1990 on "Coastline and conflicts," held in the Maison franco-japonaise in Tokyo. Visits in Tokyo University of Marine Science and Technology (TUMSAT) and representatives of several hatcheries.

Seventh edition in Japan at Tokyo in 1990 on the "Determinism of Biological Recruitment at sea." Visits of TUMSAT and several hatcheries.

Eighth edition in France at Ifremer Nantes in 1991. Three French-Japanese symposiums were organized: "Determining factors of growth in aquaculture," "Economy and management of fisheries," and "Co-development of fisheries and recreational activities in coastal zones."

Ninth edition in Japan at Tokyo in 1991 on "Oceanic fluxes."

Tenth edition in Japan at Tokyo in 1992 with two symposiums: "Biotechnology and Environment" and "Determinism of Biological Recruitment at sea," held in "Maison franco-japonaise de Tokyo" (Nichi-Futsu Kaikan).

Eleventh edition in France at Paris in 1997 with the special topic: "Coastal zone observation and forecast in the medium and long terms"; meetings held in "Institut océanographique de Paris." Visit of the forum "SEAMER" at Paris and then a trip in Brittany to visit several hatcheries.

Twelfth edition in Japan at Tokyo in 2005 on the "Mutual new understanding for research in oceanography and fisheries, in France and in Japan." Scientific meetings at Tokyo University of Marine Science and Technology (TUMSAT) and then visits to university laboratories, research vessels, official offices, and private firms in Onjuku, Abiko, and Shimizu.

Thirteenth edition in France at Marseille in 2008 on the "Global Change: interactions mankind/marine environments." Scientific meetings held at the Faculty of Pharmacy of Marseille from the 8 to 10 September and then at the Paris Nihon Bunka Kaikan (Maison de la Culture du Japon à Paris) on 12 September with special *thema*: Oceanography, Microbiology, Bio-geo-chemistry, Coral reefs, Biodiversity, Management of littoral zones and artificial reefs, Aquaculture, and New techniques of observation in the marine environment.

Fourteenth edition in Japan at Kobe then at Tokyo in 2010 with the special topic "Toward sustainable use and management of the oceans." Meetings held at Kobe in the international exhibition hall, in the framework of the forum "Techno-Ocean." In Tokyo, other meetings on education and on fish behavior, in TUMSAT (Kaiyodai) then a conference in "Maison Franco-Japonaise" (Nichi-Futsu Kaikan). Commemoration of the 50th anniversary of the "Société franco-japonaise d'Océanographie" of Japan.

2 Transportation of Seed Oysters from Japan to France in the 1960s

The mass production of oyster seeds (spat) in Japan started in the 1920s at Mangoku-Ura bay, Ishinomaki City, near Sendai. These seed oysters were transported to all regions around the Japanese coast, and in the 1940s, they were exported to the United States by ship.

For oyster culture, in European countries, France is the top producer and consumer, producing around 120,000 tons of the cupped oyster *Crassostrea gigas* annually and an additional 1,500 tons of the flat oyster *Ostrea edulis* (Buestel et al. 2009). The amount of production is about half of that of Japan (around 250,000 tons) (Fig. 1). There has been a close relationship in

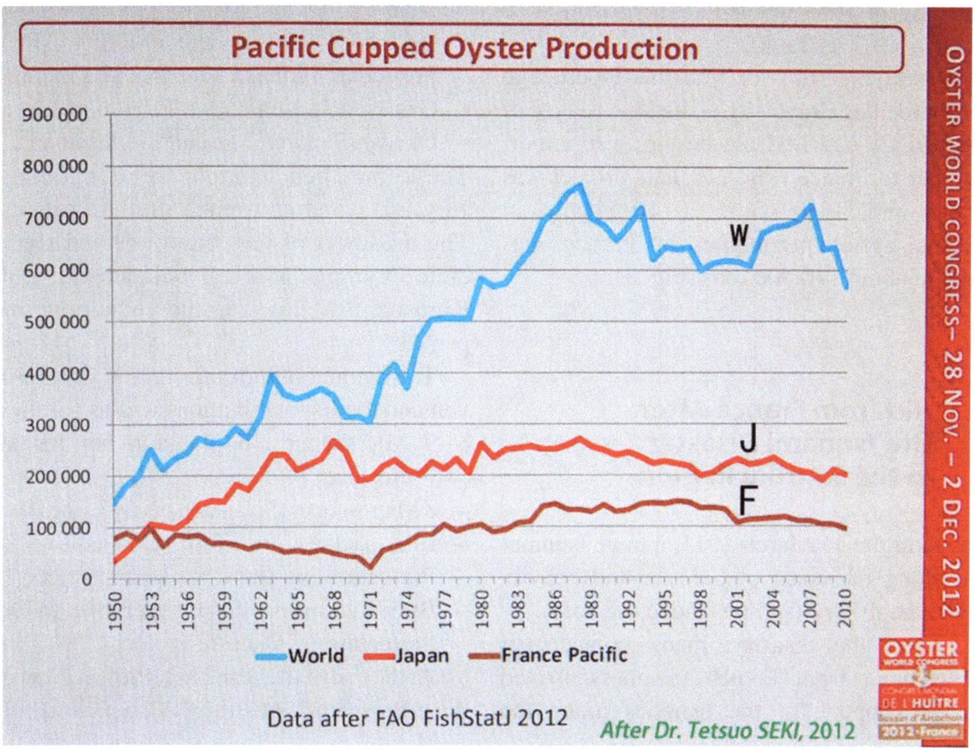

Fig. 1 Production of cupped oyster *Crassostrea gigas* in France and Japan (Adapted from Seki 2012)

the exchange of information and culture techniques between the two countries since about 40 years ago.

In the 1960s, the oyster farming in France was greatly affected by the mass mortality of *Crassostrea angulata* (called Portuguese oyster), caused by a viral disease and followed by infection from the protozoan *Marteilia* and the parasite *Bonamia*. To restore the culture industry after this problem, Dr. Trochan, the director of La Tremblade Institute (former Institut Scientifique et Technique des Pêches Maritimes), established contact with Professor Takeo Imai, of Tohoku University, asking about the possibility to export Japanese seed oysters to France.

After pathological and epizootic examinations and experiments of single oyster seeding, the project of exportation of oyster spat to France was put in place. These former examinations have been realized by the project group: Professor T. Imai (Tohoku University), Dr. Akimitsu Koganezawa (National Institute of Fisheries in Tohoku), Mr. Kunio Goto (Miyagi Prefectural Fisheries Experimental Station), and Dr. Marc Dupuis (Scientific Counselor of French Embassy in Japan).

The first trial of mass export by air took place successfully in 1969 under the authorization of both countries (Fig. 2a–k).

At that time, Professor François Doumenge from Montpellier University visited the Mayor of Ishinomaki City to find the best way to export seed oyster to France (Fig. 2l). This project was continued until 1979 (Goto 2012). Finally, in 1990, the oyster production in France was restored to about 140,000 tons (Fig. 3).

3 Aid from France After the Tsunami Disaster in the Sanriku Region

In Japan, on the 11 March 2011, a huge tsunami surged along the coast of Sanriku and heavily damaged the fishery facilities along the coast.

Just after the disaster, many contacts of encouragement from French members arrived proposing support for the fisheries along the Sanriku coast. Especially, Profs. Hubert-Jean Ceccaldi (President of Société franco-japonaise d'Océanographie) and Catherine Mariojouls (President of Association pour le Développement de l'Aquaculture) proposed to collect contributions for the fisheries. To respond to their proposals, a special committee was formed between the French and Japanese members of "Société franco-japonaise d'Océanographie."

After discussing with researchers and members of fishery cooperatives of Sanriku, equipments for technical surveys that had been lost in the tsunami were requested. So the committee decided to support them with microscopes and plankton nets to examine the oyster larvae in the first spawning season after the tsunami.

The organizations that supported this collaboration are as follows:

- Association pour le Développement de l'Aquaculture (ADA)
- Société franco-japonaise d'Océanographie-France (SFJO-F)
- La Fondation d'Entreprise Air Liquide at Paris and Teisan in Japan
- Rotary Club Saint-Jean at Marseille
- Société franco-japonaise d'Océanographie-Japan
- (SFJO-J, ex-President: Shiro Imawaki, President: Teruhisa Komatsu and members)
- Gambalo Japan Project: Region of Bretagne
- Okaeshi Project: Marennes-Oléron area

These members started the donations from July 2011 and continued until October 2012. The members of the committee greatly appreciate Olympus Medical Science Co. Ltd. and Rigosha Co. Ltd. for the reduction of their price.

The amount of the contribution was 3,302,606 yen and the list of donations was as follows:

- 9 Microscopes, 5 plankton net for Miyagi Prefecture
- 8 Microscopes, 5 plankton net for Iwate Prefecture
- Life jackets for Taro Cooperative (Iwate Prefecture) by Gambalo Japan Project
- Buoys and ropes for Cooperatives of Miyagi Prefecture by Okaeshi Project

At the end of the year 2011, Prof. C. Mariojouls, the President of the Association of the

Fig. 2 (**a–l**) From the *top* and *left* to *right*: (**a**) Dr. Takeo Imai, Tohoku University; (**b**) the Scientific Councilor of French Embassy; (**c**) preparation for exportation; (**d**) experiment for single-seed oyster; (**e**) Mr. Kunio Goto, preparing single-seed oyster; (**f**) single-seed oysters; (**g**) preparing single-seed oysters; (**h**) sorting and packing of seed oysters; (**i**) reportage by French TV crew; (**j**) sanitary and pathological check; (**k**) start to France by plane; and (**l**) Professor Doumenge with the Mayor of Ishinomaki City

Development of Aquaculture (ADA), visited the Sanriku region and related organizations and regional fishery cooperatives to encourage them. And on the beginning of February 2012, Dr. Hubert-Jean Ceccaldi, the president of SFJO of France, and Dr. Georges Stora of Centre d'Océanologie de Marseille, vice Secretary of SFJO of France, visited there also (Fig. 4a–e).

4 Exchange Information and French-Japanese Joint Seminar on Restoration of Oyster Culture in Sanriku

About one and a half years after the disaster, in September 2012, thanks to a donation of Maison Franco-Japonaise, a joint seminar was

Fig. 2 (continued)

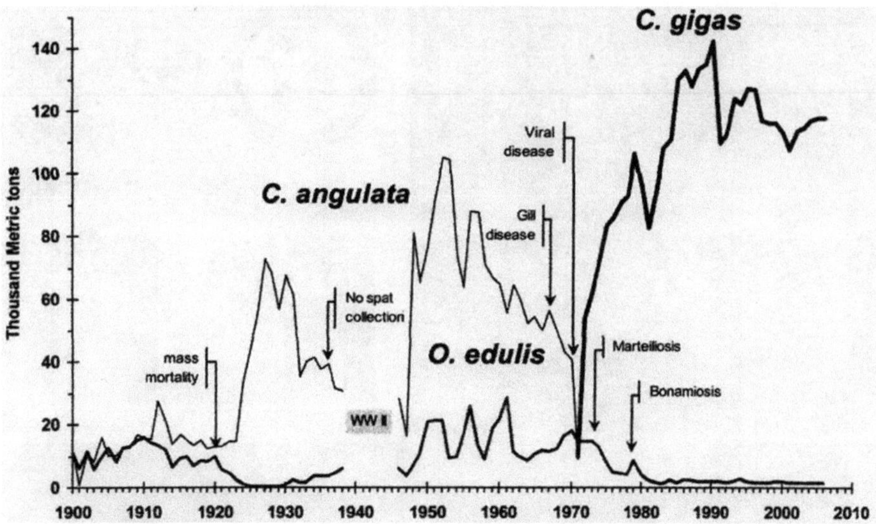

Fig. 3 Historical trend of French oyster production (Adapted from Buestel et al. 2009)

held in the Sanriku region with a French delegation. Members of the delegation were as follows;

Professor Denis Bailly (Professor of Université de Bretagne Occidentale), Prof. Catherine Mariojouls (Agro-Paris-Tech), Dr. Jean Prou (the Director of IFREMER laboratory in La Tremblade), and Mr. Olivier Laban (President of Aquitaine and Arcachon Regional Committee of Shellfish Culture). Two researchers of the Université de Bretagne Occidentale and eight oyster growers were funded by the French government.

The French delegation first visited Iwate and Miyagi Prefectures to observe the areas damaged by the tsunami, Taro and Osawa in Iwate and Shizugawa and Ishinomaki in Miyagi Prefecture. At the Iwate Fisheries Experimental Station, a ceremony of the last donation of microscopes was realized by Professor Catherine Mariojouls. Then at the Tohoku National Fisheries Research Institute at Shiogama, a joint seminar was held among the members of both the Japanese-French Oceanographic Societies and the Prefectural Fisheries Research Stations of Iwate and Miyagi and invited professors of universities and oyster growers of the regions.

After returning to Tokyo, another open seminar was carried out at the "Maison franco-japonaise" at the Ebisu quarter. The information exchanged at these seminars has been highly appreciated by both countries (Fig. 5a–e).

5 Participation to the Oyster World Congress at Arcachon and Technical Visit on Oyster Farms on the Atlantic Coast

After the joint seminar in Japan, Mr. Olivier Laban, the president of the First Oyster World Congress, proposed to the Japanese members to attend the world congress and present information on the cooperation for oyster culture between Japan and France. This congress was an epoch-marking event because the organizer was the group of oyster farmers associated with several research institutes. The date was from 28 November to 2 December 2012. About 250 participants from 27 countries attended. The members of the Japanese delegation were as follows: Dr. Tetsuo Seki (Program Officer of Agriculture, Forestry and Fisheries Technical Information Society), Mr. Kunio Goto (Adviser of the Fishery Cooperative of

Fig. 4 (**a**–**e**) From the *top* and *left* to *right*: (**a**) a part of the scientific equipment brought and given by the French delegation; (**b**) Prof. Mariojouls delivering the scientific equipment; (**c**) Dr. Mariojouls visiting Matsushima cooperative; (**d**) Dr. G. Stora, Prof. H.-J. Ceccaldi, and Prof. Y. Koike; and (**e**) news concerning the visit of Dr. Stora and Prof. Ceccaldi at Matsushima

Fig. 5 (**a–e**); (**a**) French delegation visiting the cooperative of Taro, Iwate Prefecture; (**b**) news for the visit of French delegation at Shizugawa cooperative; (**c**) visit at Mangoku-Ura bay, Ishinomaki city; (**d**) joint seminar at Tohoku National Fisheries Research Institute; and (**e**) open seminar at Maison Franco-Japonaise, Tokyo

Shiogama City), Messrs. Yoshimasa Koizumi and Tamotsu Suzuki (members of the Fishery Cooperatives of Shiogama City), and Dr. Yasuyuki Koike (Member of Maritime and Fisheries Promotion Society in Tokyo University of Marine Sciences and Technology (TUMSAT), Counselor of SFJO of Japan).

The result was very efficient. Technical information from various countries was a good reference for Japanese oyster culture. Especially, recent pathological information was very important for the seed culture of each country.

After the congress, the Japanese delegation visited several oyster farms along the Atlantic coast, Arcachon-Aquitaine, Marennes, Bretagne Sud, and the stations of Institut français pour l'Exploitation de la Mer (Ifremer).

In Marennes, Mr. Goto who examined seed oysters before exporting in 1968 was guided to the seabed in Mouillelande where the first seed from Japan was released. And he could exchange information with the oyster grower who transported the first seed that arrived in Paris to Marennes. It was just a historical

Fig. 6 (**a–d**) From the *top* and *left* to *right*: (**a**) first World Oyster Congress at Arcachon, (**b**) news for Japanese delegation visiting oyster farm at Arcachon, (**c**) news for Kunio Goto visiting the historical place where the first seed oysters from Japan were released, and (**d**) visit at Ifremer laboratory, La Tremblade

moment to recall the memories of 40 years ago (Fig. 6a–c).

After the site of culture, the delegation visited two stations of Ifremer, La Tremblade and La Trinité-sur-Mer, to exchange important pathological information about the disease and herpes of oysters with the specialists. Actually, in the last 5 years in France, there is the very important problem of disease in oysters again. The conclusion of the discussion was that seed oysters should not be exchanged from one region to another if there has been risk of pathogens being transferred. And the results of research and information about the disease must be open and readily disseminated to related countries (Fig. 6d).

6 The Restoration of Oyster Culture in Japan After the Tsunami Disaster

The time table of restoration project in 10 years by the government of Miyagi Prefecture is:

- First 3 years: The period of repair (2011–2013).
- Next 4 years: The period of regeneration (2014–2017)
- Another 3 years: The period of expansion (2018–2020)

Currently we are at the end of the first period of repair. By the statistics of the prefecture, the repair of oyster culture is going well. But we

Fig. 7 (**a**–**f**) From the *top* and *left* to *right*: (**a**) percentage of restarted oyster growers at Miyagi Prefecture in 2 years after the tsunami disaster, (**b**) reconstruction of Onagawa port, (**c**) young oyster grower returned back in family, (**d**) horizontal growing technique of oysters in use in France (Arcachon), (**e**) vertical growing technique of oysters in use in Japan (Kesennuma), and (**f**) "Fine de Claire Verte," the Marennes' speciality

must think about the 30 % of oyster growers who have abandoned their activity (Fig. 7a).

Toward the restoration, what we must do now?

Most of all, it is necessary to repair the infrastructure as well as various other things. For example, in large areas of the seabed, there are many macro debris that have not been surveyed.

Almost all fishing ports in the region were submerged by about 1 m under the level, and only 10 % of them were repaired (Fig. 7b). To repair the culture and processing facilities, more materials and funds also are needed. But in the moment as it is not completely repaired, we can think about the density of culture facilities which was too much before the tsunami. Fortunately, there are at least several young oyster growers who have returned to work with their families, and they are very positive to adopt the new methods and aspects (Fig. 7c).

After these events, such as the joint seminar in Japan, the world oyster congress, and technical visit, researchers and oyster growers of Sanriku could have important information and new ideas on the differences of oyster culture style between France and Japan. The main differences of culture style in each country are as follows.

In France, where the tidal range is very large, they can use the huge intertidal zone horizontally for the culture site. In Japan, where the tidal range is not so large, they must use the coastal water vertically by hanging culture. So there are unchanged differences in the environment of both countries (Fig. 7d, e).

The oyster growers of Sanriku started the trials to adapt the new methods and ideas from France, for example, single oyster hanging method of Etang de Thau, culture at intertidal zone, and importing French materials such as Coupelles (seed collectors) with releasing machine of spat collection. These trials should increase the value of Japanese oysters for sale under the live condition with shells like the French style. Normally, it is not so easy to change the traditional style of degustation, but after the disaster at a moment, it will be a good opportunity to change the custom (Fig. 7f).

7 Conclusion

There are numerous exchanges, some between scientists, some other ones between professionals of oyster culture, and some between the two "Sociétés Franco-Japonaises d'Océanographie." But finally, there are good exchanges between the two countries, with mutual benefits and a large part of friendships.

Acknowledgments The author appreciates the courtesy of Mrs. Kunio Goto, Minji Fukuda, and Ryo Sasaki for their agreements to use the historical pictures.

References

Buestel D, Ropert M, Prou J, Goulietquer P (2009) History, status, and future of oyster culture in France. J Shellfish Res 28(4):813–820

Goto K (2012) Historical memories of seed oyster transportation to France. Personal letter (in Japanese)

Koganezawa A (1984) History of oyster culture in Matsushima. News of Tohoku National Fisheries Research Institute 26 (in Japanese)

Seki T (2012) Oyster Industry threatening biotic causes in Japan. Oral presentation in the First Oyster World Congress in Arcachon

Part II

Communications: Natural Disturbances

Impacts of the 2011 Mega-earthquake and Tsunami on Ezo Abalone *Haliotis discus hannai* at Iwaisaki, Miyagi, Japan

Hideki Takami and Hiroshi Nakaie

Abstract

On 11 March 2011, a massive tsunami generated by a mega-earthquake hit a wide area of the Pacific coast of northeast Japan. We have analysed the effects of the tsunami on Ezo abalone *Haliotis discus hannai* population at Iwaisaki in Miyagi, Japan. Ezo abalone is among the most economically valuable fisheries resources, so assessments are needed to evaluate the effects of the disaster on this important species. Before the tsunami, algal forests dominated by the brown macroalga *Eisenia bicyclis* had developed in the survey area, where large juvenile and adult abalone >30 mm in shell length (SL) inhabited. Juvenile abalone <30 mm SL inhabited the area dominated by crustose coralline algae (CCA). After the tsunami, no apparent decrease was observed in the density of large juvenile and adult abalone inhabited in the algal forests. The impact of the tsunami was more profound in the CCA area than in the algal forests. Just before the tsunami, the intensive recruitment of zero-year-old juveniles was observed, but individuals of this year class were not detected after the event. The distribution pattern of juveniles could be a cause of the marked decrease, because most of them inhabited the CCA area where the disturbance by the massive water movement was not reduced by the effects of the macro algal forest. Since the age at first capture of abalone is at 4–5 years old, the future commercial catch may considerably decrease for at least 4–5 years after the tsunami.

1 Introduction

On 11 March 2011, the Great East Japan Earthquake occurred off the Pacific coast of the northeastern part of Honshu Island, Japan, with a moment magnitude of 9.0 at a depth of 24 km. The massive tsunami generated by this earthquake severely impacted the coast of northeastern

H. Takami (✉)
Tohoku National Fisheries Research Institute, Fisheries Research Agency, 3-27-5, Shinhama, Shiogama, Miyagi 985-0001, Japan
e-mail: htakami@affrc.go.jp

H. Nakaie
Miyagi Prefectural Government, 3-8-1 Honcho, Aoba-ku, Sendai, Miyagi 980-8570, Japan

H.-J. Ceccaldi et al. (eds.), *Marine Productivity: Perturbations and Resilience of Socio-ecosystems*,
DOI 10.1007/978-3-319-13878-7_2,

Honshu Island and resulted in the loss of nearly 19,000 people and devastated many towns in the coastal areas of this region. Fisheries are one of the most important industries in the coastal area impacted by the tsunami. Assessments are needed to evaluate the effects of the earthquake and tsunami on coastal populations of fishery resource organisms for the future fishery and stock enhancement.

On the Pacific coast of northeastern Honshu Island, which was severely disturbed by the massive tsunami, Ezo abalone *Haliotis discus hannai* is one of the most valuable fisheries resources and also play an important role in the food web of the rocky shore ecosystem (Won et al. 2011). The main objective of this study is to examine the effects of the earthquake and subsequent tsunami on the abalone populations. Data on the abalone population before the earthquake and tsunami event are essential to allow comparison and clarify the scale of impacts. Before March 2011, we regularly carried out quantitative analyses on the population of this fisheries resources at Iwaisaki, Miyagi Prefecture. To assess the impacts of the earthquake and tsunami by before and after comparisons, we carried out the survey using the same method after the earthquake and tsunami, in order to compare the results with the data collected before the event.

2 Materials and Methods

2.1 Study Site

This study was done in the rocky subtidal habitats at Iwaisaki located near the mouth of Kesennuma Bay (Fig. 1). As the site faces an open ocean, waves impact this area directly. We selected four stations of 10 m × 10 m each (named A, D, E, G) in the study site. Station A was dominated by crustose coralline alga (CCA) *Lithophyllum yessoense* and the other stations were included in kelp beds (KB) dominated by the brown macroalga *Eisenia bicyclis*. A detailed information of these stations was described in Table 1. All fisheries were closed in the study site after the earthquake at least until June 2011 when the last survey was carried out in this study.

2.2 Sample Collection

The abalone *Haliotis discus hannai* were quantitatively sampled by scuba diving in the mornings of 23 October 2009 (before the earthquake) and 15 June 2011 (after the earthquake) using 2 m×2 m quadrats. In each station, five replicate quadrats were haphazardly located, and all abalone in the quadrats were collected by hand and placed in mesh

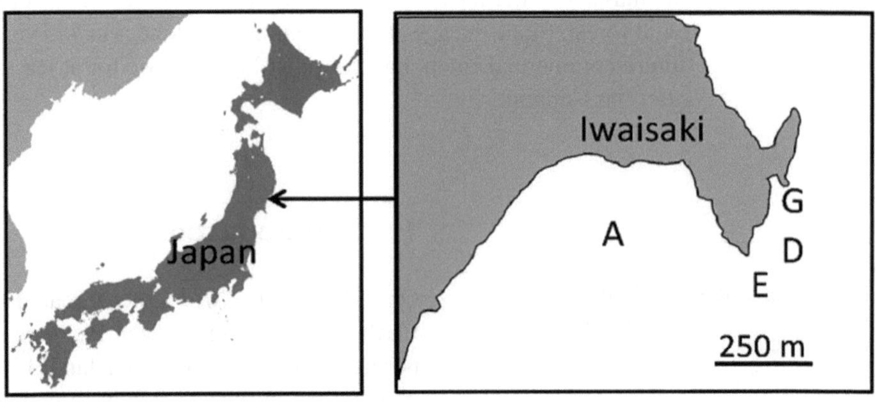

Fig. 1 Location of the study site and stations *A*, *D*, *E*, and *G*

bags before being transported to the boat. The shell length (SL) of abalone was measured to the nearest 1 mm with callipers on the boat, and measured animals were released to their original habitats.

The recruitment process during the juvenile stage of 2008–2010 year classes of abalone was also investigated with the method of Takami et al. (2013) near station A which is located in CCA. Since the distribution of juvenile abalone was too cryptic and patchy to estimate their density by the quadrat sampling, the abundance of juveniles' shell length inferior to 40 mm SL was regularly monitored by intensive visual searching from December 2009 to June 2011. In every survey, the same diver searched for juvenile abalone in every crevice and overhangs by sometimes overturning boulders and removing large *epibiota*, recording each searching time. Juvenile abalone were carefully removed from substrata and transferred to a sampling bottle with fine tweezers. The relative abundance of juvenile abalone was expressed as catch per unit effort (CPUE), accounting for the number of collected juveniles per searching time in hours. The shell length of collected juveniles was measured to the nearest 0.1 mm using a video camera system with an image analyser connected to a dissecting microscope in the laboratory.

2.3 Data Analysis

Changing in abalone densities were tested by a two-way analysis of variance (ANOVA) using date of sampling and type of algal community as fixed factors. Data were transformed in $[\log(n+1)]$ in the case of heterogeneous variation.

Table 1 Location and environmental features for each sampling station

Station	Latitude (N)	Longitude (E)	Depth (m)	Dominated algae
A	38°49.673′	141°35.995′	4–5	Crustose coralline algae
D	38°49.622′	141°36.258	4–5	Kelp
E	38°49.573′	141°36.218′	4–5	Kelp
G	38°49.717′	141°36.275′	3–4	Kelp

3 Results and Discussion

After the tsunami, the underwater visibility at the study site was much lower than that before the earthquake and tsunami due to sediment suspension. A remarkable increase in the amount of fine sediments was observed under boulders and in crevices of bedrocks. In stations D, E, and G, although the remaining holdfasts of *E. bicyclis* that lost their fronds were sometimes observed, the biomass of this alga appeared to be undamaged by the event. More severe disturbance was obvious in station A which was located in CCA. Many of the large rocks were cracked and turned over on the seafloor and in consequence bare rocks, which were not covered with any *epibiota* including crustose coralline algae, were exposed in many places.

Abalone density obtained by the quantitative quadrat survey varied significantly among stations and dates of sampling, but there was no significant date by station interaction for the abalone density by the two-way ANOVA (Table 2).

After the tsunami, station A located in CCA indicated the lowest mean densities. In the other stations located in KB, obvious changes in the abalone density were not observed (Fig. 2).

There are no clear trends in the size-frequency distributions of abalone before and after the tsunami. The abalone smaller than 30 mm in shell length was not sampled by quadrat possibly because of the relatively low density and patchy distribution of juveniles (Fig. 3). Juveniles were only detected by intensive visual searching during the recruitment process monitoring which was conducted near station A.

Table 2 Two-way ANOVA for densities of abalone *Haliotis discus hannai* between sampling dates and habitats at Iwaisaki, Japan

	df	MS	F	P
Sampling date	1	0.718	8.562	0.006
Habitats	3	2.806	11.152	0.000
Interaction	3	0.380	1.511	0.231
Residual	32	2.684		

The main spawning season of Ezo abalone at this site is from late August to early October (Nakaie and Takami 2012). In December 2009, small juveniles (4.6–12.5 mm SL) which were reproduced in the 2009 spawning season were observed as the cohort with higher CPUE than the 2008 year class. Then, the CPUE of the 2009 year class did not markedly decrease until June 2010. From December 2010 to February 2011, newly recruited juveniles, which were spawned in 2010, were detected as a single cohort. During the same period, juveniles of the 2009 year class grew to more than 25 mm SL with lower CPUE. In June 2011, 3 months after the tsunami, no individual of the 2010 year class was not observed, whereas the CPUE of the 2009 year class did not change obviously between before (February 2011) and after (June 2011) the tsunami (Fig. 4).

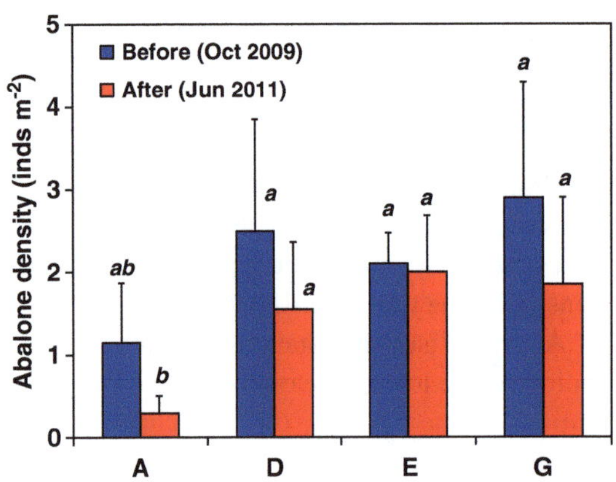

Fig. 2 Changes in density of Ezo abalone *Haliotis discus hannai* collected in stations *A*, *D*, *E*, and *G* between before and after the earthquake and tsunami

From the results of quadrat sampling and CPUE survey on juveniles, it was demonstrated that the impacts of tsunami on Ezo abalone *Haliotis discus hannai* were different among stations and their growth stages. After the tsunami, although significant difference was not marginally detected, abalone densities by quadrat sampling tend to decrease in station A which was dominated by CCA.

Newly recruited juveniles inhabit near station A was more severely affected by the tsunami than older individuals. Juveniles reproduced in 2010 were not detected after the tsunami, and such a collapse of the youngest year class has not been observed before the tsunami. Naylor and McShane (2001) suggest that the wave action caused by even usual storm disturbances is an important contributor to the mortality of new recruits in the New Zealand abalone *H. iris* because of the relatively weaker adhesion of juveniles to substrate than that of adults. Similarly, the much more serious disturbance by the tsunami event could specifically influence the survival of smaller abalone in the present study site. The distribution pattern of juvenile abalone also had significant consequences for their survival. Juveniles mainly inhabit crustose coralline algae, which were more severely disturbed than

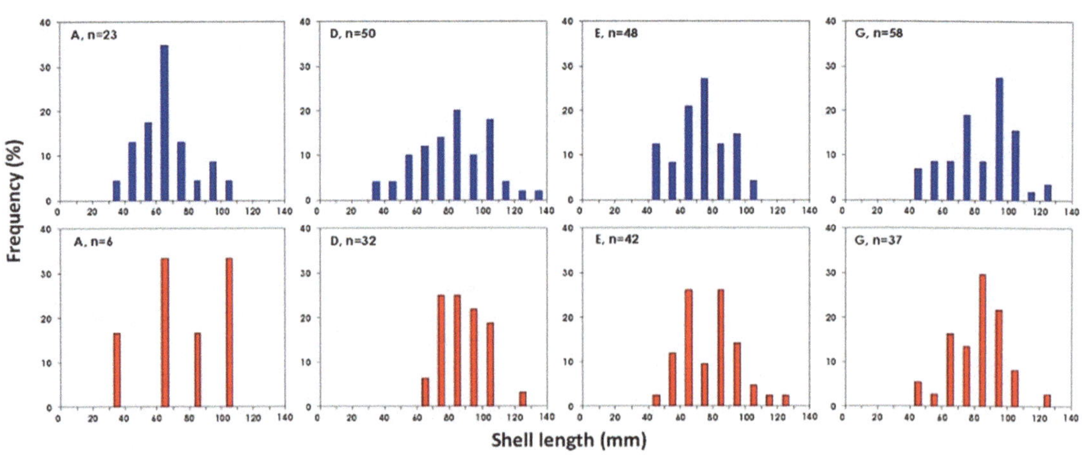

Fig. 3 Changes in size distribution of Ezo abalone *Haliotis discus hannai* collected in stations *A*, *D*, *E*, and *G* between before (October 2009, *upper row*) and after (June 2011, *lower row*) the earthquake and tsunami

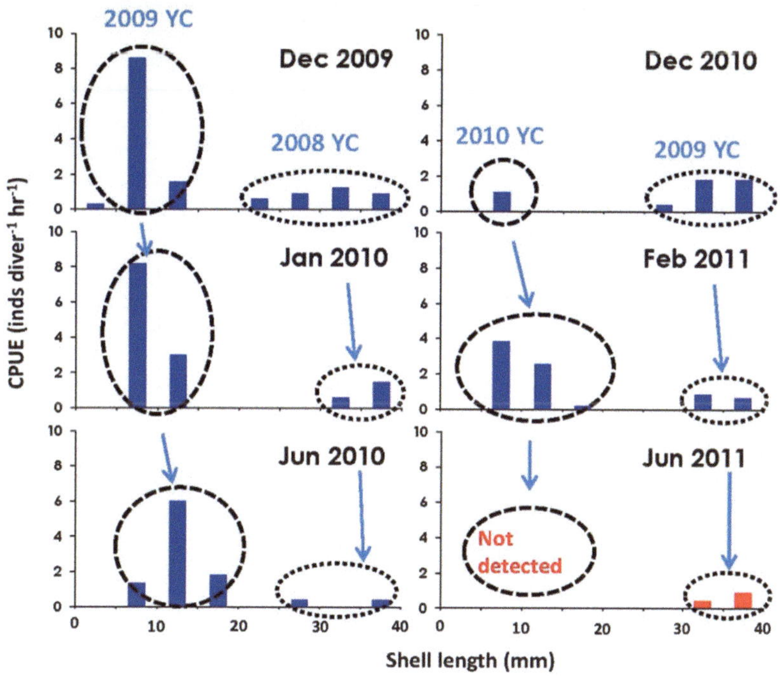

Fig. 4 Changes in CPUE for the number of collected juvenile Ezo abalone *Haliotis discus hannai* per searching time in hours with size distribution between before and after the earthquake and tsunami

in kelp beds. In contrast, the main habitat of adult abalone is kelp beds, where the current velocity of the tsunami might have been partially attenuated by the presence of the algal canopy as observed in the previous tsunami event generated by the Sumatra-Andaman earthquake in 2004 (Whanpetch et al. 2010).

Takami et al. (2013) also reported that young-of-the-year Ezo abalone (2010 year class) were seriously affected by this mega-earthquake and the tsunami at Oshika Peninsula in Miyagi which is located 53 km south from this study point. There is a possibility that the mega-earthquake and tsunami had a negative impact in a wide range in the survival of the smaller juvenile of Ezo abalone. Since the age at first capture of abalone is 4–5 years old (Sasaki 1999), the future commercial catch may considerably decrease at least after 4–5 years from the event. Continuous monitoring of the affected abalone population is needed for effective stock management to avoid collapse of these ecologically and economically important resources.

Acknowledgements We thank Ryo Sasaki, Akio Oshino, and Hiroshi Nagashima for their assistance in the field. This research was supported by the Development Projects for Application in Promoting New Policy of Agriculture, Forestry, and Fisheries, Japan (grant no. 2024).

References

Nakaie H, Takami H (2012) Comparison of recruitment processes of ezo abalone *Haliotis discus hannai* in different reproductive strategies. Nippon Suisan Gakkaishi 78:1217–1220 (in Japanese)

Naylor JR, McShane PE (2001) Mortality of post-settlement abalone *Haliotis iris* caused by conspecific adults and wave exposure. N-Z J Mar Freshw Res 35:363–369

Sasaki R (1999) Ecological studies on the recruitment structure in abalone *Haliotis discus hannai* Ino. PhD thesis, Tohoku University (in Japanese with English abstract)

Takami H, Won N-I, Kawamura T (2013) Impacts of the 2011 mega-earthquake and tsunami on abalone *Haliotis discus hannai* and sea urchin *Strongylocentrotus nudus* populations at Oshika Peninsula, Miyagi. Japan Fish Oceanogr 22:113–120

Whanpetch N, Nakaoka M, Mukai H, Suzuki T, Nojima S, Kawai T, Aryuthaka C (2010) Temporal changes in benthic communities of seagrass beds impacted by a tsunami in the Andaman Sea, Thailand. Estuar Coast Shelf Sci 87:246–252

Won N-I, Kawamura T, Takami H, Hoshikawa H, Watanabe Y (2011) Comparison of abalone (*Haliotis discus hannai*) catches in natural habitats affected by different current systems: implication of climate effects on abalone fishery. Fish Res 100:84–91

Short-Term Effects of the March 11, 2011, Tsunami on Phytoplankton Assemblages in Ofunato and Kesennuma Bays, Japan

Yutaka Okumura, Hiroshi Nakaie, Keisaku Kusaka, Tetsuroh Ishikawa, and Shinnosuke Kaga

Abstract
We investigated the changes in phytoplankton quantity and diversity after the March 11, 2011, Tohoku earthquake and tsunami in Ofunato Bay and in the mouth of Kesennuma Bay in northeastern Japan. Phytoplankton quantities increased in Ofunato Bay, and diversity changed. The phytoplankton assemblages other than diatoms increased near the water surface after the earthquake. In contrast, a clear change was not observed at Iwaisaki, at the mouth of Kesennuma Bay, during the survey. These results suggest that the influence of the tsunami on phytoplankton varied by location. In this study, phytoplankton in Ofunato Bay were affected by the tsunami, which is enclosed and was easily affected by the terrestrial environment and broken seawall. Limited changes were observed at Iwaisaki at the entrance to Kesennuma Bay.

Y. Okumura (✉)
Research Center for Coastal Fisheries & Aquaculture, Tohoku National Fisheries Research Institute, Fisheries Research Agency, 3-27-5, Shinhama, Shiogama, Miyagi 985-0001, Japan
e-mail: okumura@affrc.go.jp

H. Nakaie
Miyagi Prefectural Government, 3-8-1, Honmachi, Aoba-ku, Sendai, Miyagi 980-8570, Japan

K. Kusaka • T. Ishikawa
Kesennuma Fisheries Experiment Station, Miyagi Prefecture Fisheries Technology Institute, 47-6, Suginosawa, Akaiwa, Kesennuma, Miyagi 988-0181, Japan

S. Kaga
Iwate Fisheries Technology Center, Daisanchiwari, Heita, Kamaishi, Iwate 026-0001, Japan

H.-J. Ceccaldi et al. (eds.), *Marine Productivity: Perturbations and Resilience of Socio-ecosystems*,
DOI 10.1007/978-3-319-13878-7_3,

1 Introduction

The coastal area of northeast Japan was damaged by a tsunami after the earthquake on March 11, 2011 (Goto et al. 2011; Urabe et al. 2013). Facilities near the coast were damaged by the tsunami. More than 10,000,000 L of fuel oil flowed into Kesennuma Bay from 22 oil tanks (Sasaki 2012). In addition, a heavy oil tank was destroyed in Ofunato Bay (Sasaki 2012). As components of oil are toxic to aquatic organisms (Okumura et al. 2003), there was concern for marine pollution in the ecosystem.

More than ten sewage disposal plants on the coast stopped operations temporarily because of electrical problems caused by the tsunami (Ministry of Land, Infrastructure, Transport and Tourism 2011). If untreated wastewater flowed into the coastal area, the nitrogen and phosphate concentrations would increase in seawater. The increase in nutrients could cause eutrophication, which would deteriorate the coastal environment.

Kesennuma and Ofunato bays are aquaculture grounds for oyster, scallop, and seaweed (Hayakawa et al. 2001; Ito et al. 2007). These shellfish grow by eating suspended materials in seawater such as phytoplankton (Kusui et al. 1983). If the phytoplankton diversity changes because of an environmental disturbance, phytoplankton unsuitable for aquaculture increase in numbers, and restoration of the marine product industry may be hindered. For example, if the amounts of harmful algae increase shellfish toxin concentrations, shellfish shipments are suspended.

After the earthquake, several environmental pollution (Harino and Yatsuzaka 2012) and plankton surveys (Yamada 2012) were conducted in Kesennuma Bay (Tanaka 2012; Yokoyama and Hatakeyama 2012; Masuda 2012). High concentrations of oil in sediments (Yamamoto et al. 2012) and a high density of toxic algae (Kaga et al. 2012; Nishitani et al. 2012) were observed after the earthquake. However, comparisons with pre-earthquake levels are limited because of a lack of data. Thus, the influence of the earthquake and tsunami on total quantity and diversity of phytoplankton, which is the diet of shellfish, is unknown.

We measured the concentration of chlorophyll a (Chl *a*) as a first step to environmentally monitor the aquaculture grounds after the earthquake to determine the quantity of phytoplankton in the seawater of Kesennuma and Ofunato bays. Then, we compared the change in phytoplankton diversity from pigment concentrations before and after the tsunami because pigment profiles vary with phytoplankton taxa (Jeffrey and Vesk 1997).

2 Materials and Methods

2.1 Investigation Procedure

We took samples at Iwaisaki, at the mouth of Kesennuma Bay (Fig. 1a), approximately every month from May 2010 to February 2011 before the earthquake and from June 2011 to March 2012 after the disaster. Water temperature and salinity were measured at the sampling sites. These parameters were measured at each depth using a conductivity-temperature-depth instrument. Seawater was also collected from depths of 0, 5, 10, 15, and 20 m. We collected seawater in a water column from the surface to the sea bottom in Ofunato Bay (Fig. 1b) approximately every week from May 18 to August 17, 2011, before the earthquake and collected seawater at 2-m intervals from the surface to 22-m depth twice in June and July 2012.

2.2 Pigment and Phytoplankton Diversity Analyses

Pigment concentrations in seawater samples were determined according to a previously reported method (Okumura et al. 2012). Briefly, 200-ml seawater was filtered through Whatman GF/F glass microfiber filters (GE Healthcare UK Ltd., Buckinghamshire, UK). The phytoplankton pigments were extracted from the filter with 1-ml methanol. The pigments were analyzed by high-performance liquid chromatography (HPLC; Shimadzu, Kyoto, Japan) using the method of Zapata et al. (2000).

As phytoplankton have taxa-specific pigments (Jeffrey and Vesk 1997), the ratios of each phytoplankton taxon in seawater were calculated by

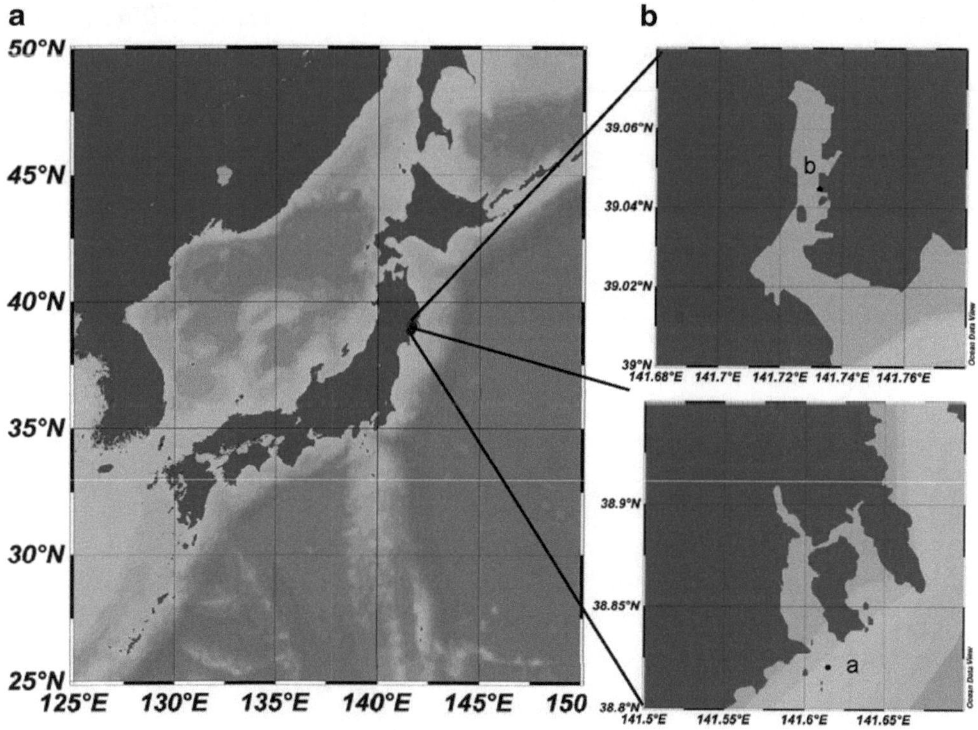

Fig. 1 (**a** and **b**) Sampling sites. (**a**) Iwaisaki, at the mouth of Kesennuma Bay near Kesennuma City, Miyagi, Prefecture. (**b**) Ofunato Bay near Ofunato City, Iwate Prefecture

chemical mass balance (Hayakari and Hanaishi 2001; Okumura et al. 2012) from eight kinds of pigments identified by HPLC. We investigated the change in phytoplankton quantities before and after the tsunami from the taxa-specific pigment or the ratio of each phytoplankton taxa.

Phytoplankton pigments can be used to determine an approximate index of phytoplankton classification, so pigment concentrations were substituted for the phytoplankton cell count data. We investigated the change in phytoplankton diversity before and after the earthquake using nonmetric multidimensional scaling (NMDS) and analysis of similarities (ANOSIM) in the vegan package (Oksanen et al. 2013) using R software ver. 2.13 (The R Project for Statistical Computing 2013).

2.3 Water Temperature and Salinity Data Analysis

The contour figures of the temporal changes in pigments, water temperature, and salinity at Iwaisaki were made with Ocean Data View software (Alfred Wegener Institute 2013).

3 Results

3.1 Pigment Concentrations Before and After the Earthquake

Chl *a* concentrations at Iwaisaki were 0.11–4.7 μg/L before the earthquake and 0.06–2.3 μg/L after the earthquake (Fig. 2a). A phytoplankton bloom was observed twice in the spring and autumn. Although Chl *a* concentrations during the spring bloom before the earthquake were higher than those after the earthquake, a major change in Chl *a* concentrations was not observed before and after the earthquake. Fucoxanthin (Fuco) concentration was high in spring, and peridinin (Per) concentration was high in autumn. The vertical distribution of Chl *a* tended to be greater near the surface (0–5 m depth) than that at

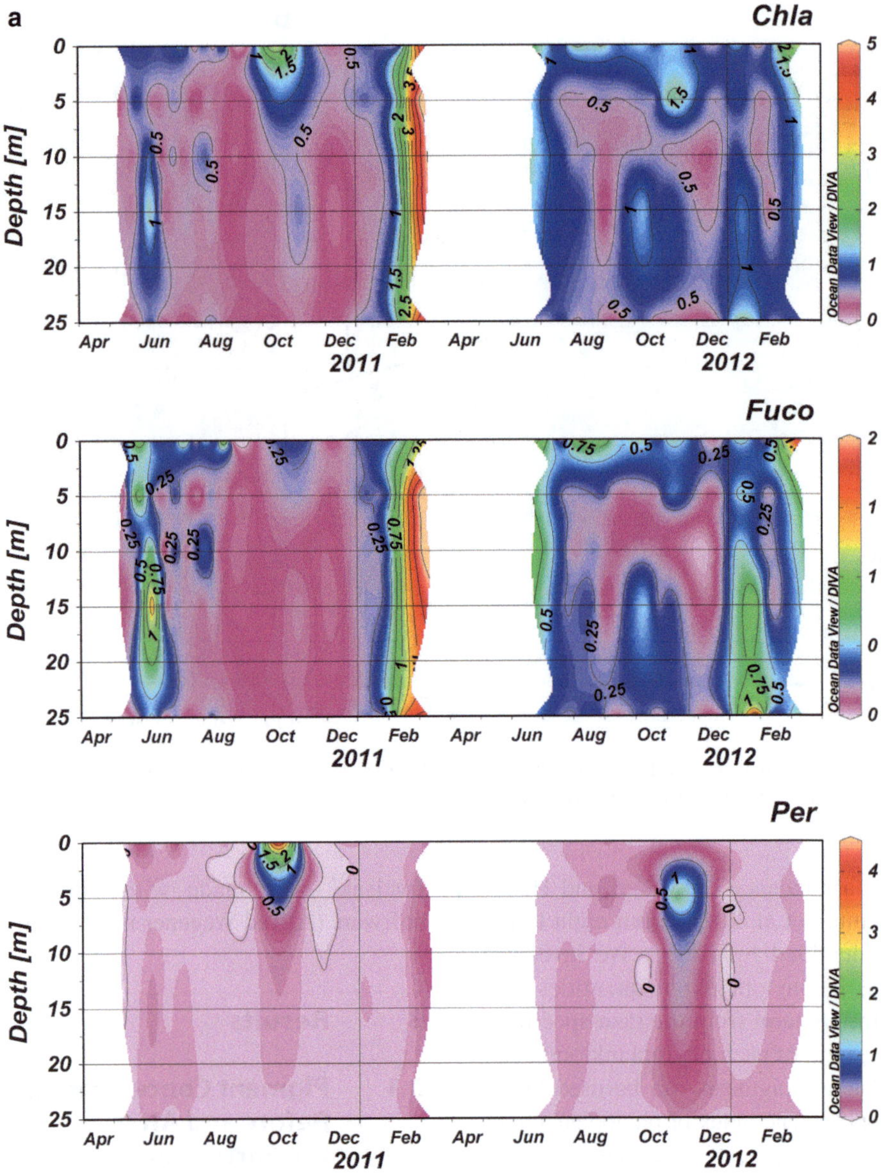

Fig. 2 Temporal changes in the Iwaisaki environment. (**a**), Chl *a* (μg/L), Fuco (μg/L), and Per (μg/L). (**b**) water temperature (°C) and salinity (PSU)

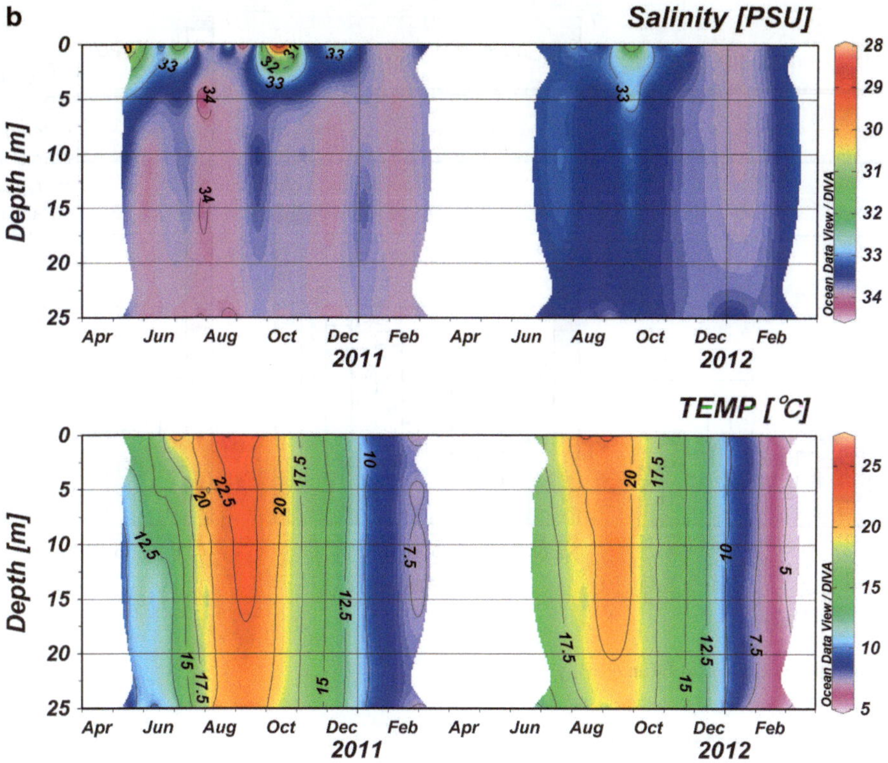

Fig. 2 (continued)

the bottom. Salinity was 28.3–34.5 PSU (Fig. 2b), and average salinity was 33.5 PSU. Water temperature was 5.1–26 °C.

The average Chl *a* concentration in the Ofunato Bay water column was about 1 μg/L before the earthquake (Fig. 3). The average concentration of Chl *a* after the earthquake was 2.59 μg/L in June and 2.42 μg/L in July. In particular, Chl *a* near the surface was about 7 μg/L. Chl *a* tended to be higher after than that before the earthquake. Although diatoms were major phytoplankton taxa during the investigation, phytoplankton assemblages other than diatoms, such as cryptophytes, increased near the water surface after the earthquake.

3.2 Comparison with a Location with Phytoplankton Diversity

Based on the NMDS plot nested in the graph, no changes in phytoplankton assemblages were observed at Iwaisaki before and after the earthquake (Fig. 4). In contrast, the plots for Ofunato Bay showed that phytoplankton locations changed after the earthquake from those before the earthquake. Phytoplankton diversity remained unchanged at Iwaisaki but changed in Ofunato Bay before and after the earthquake.

The ANOSIM statistic R and P was 0.046 and 0.004 at Iwaisaki and 0.569 and 0.001 in Ofunato Bay, respectively (Fig. 5). As the ANOSIM statistic R was near zero ($R \leq 0$), a statistically significant difference between the 2010 and 2011 groups was not observed at Iwaisaki. As the ANOSIM statistic R was >0, a statistically significant difference between groups was observed in Ofunato Bay.

4 Discussion

4.1 Particular Phytoplankton at Iwaisaki, Kesennuma Bay Mouth

Chl *a* concentrations (0.06–2.3 μg/L) in 2011–2012 after the earthquake were approximately

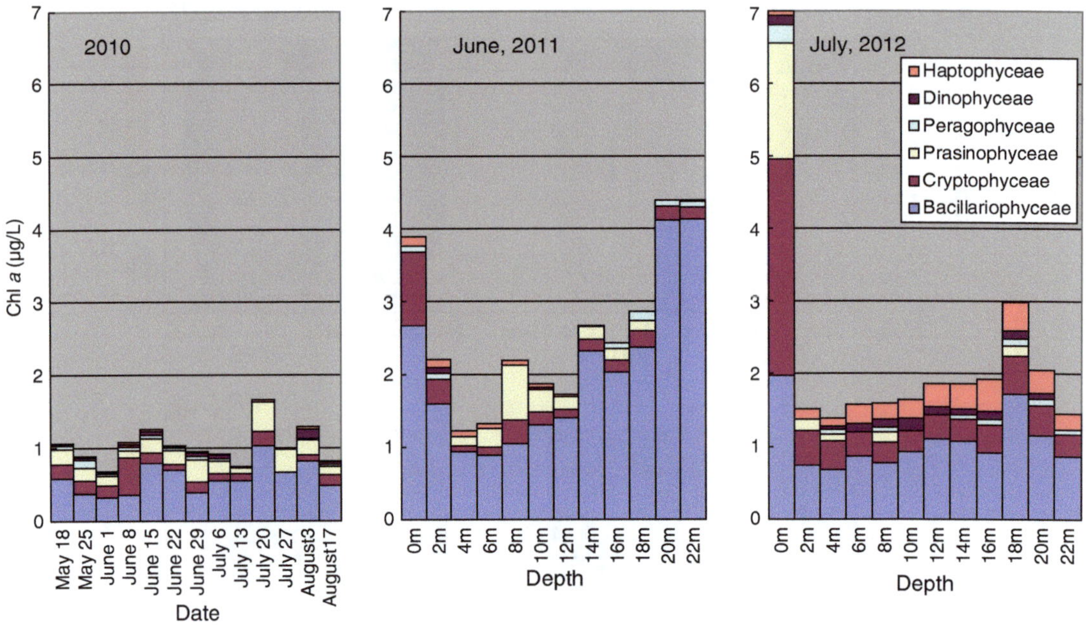

Fig. 3 Concentration of Chl *a* in Ofunato Bay

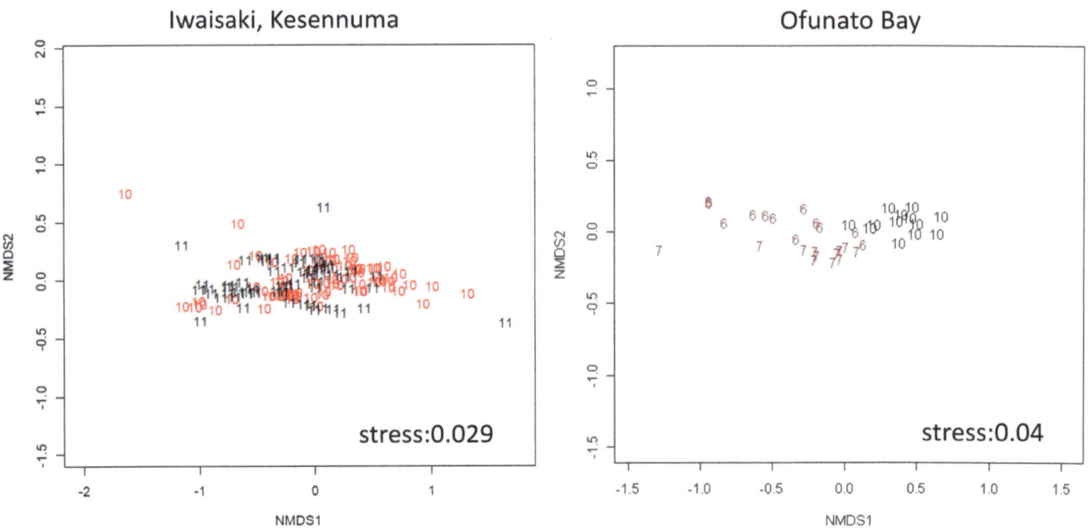

Fig. 4 Nonmetric multidimensional scaling by pigment concentration at Iwaisaki and in Ofunato Bay

equal to those from 2010 to 2011 before the earthquake (0.11–4.7 μg/L; Fig. 2a) and those from 2004 to 2005 (0.1–6.0 μg/L), which included the inner part of Kesennuma Bay (Ito et al. 2007).

The increase in Chl *a* concentration during spring agreed with the increase in fucoxanthin concentration (see Fig. 2a). The increase in Chl *a* concentration in autumn agreed with the increase in peridinin concentration. This was observed

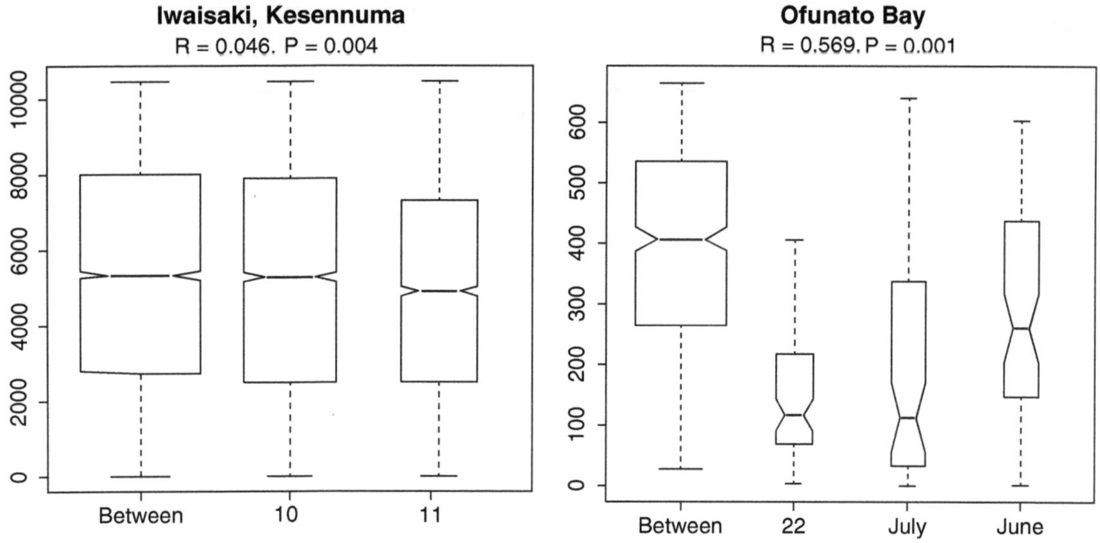

Fig. 5 Analysis of similarities by pigment concentration at Iwaisaki and in Ofunato Bay

because the major accessory pigment of diatoms is Fuco and that of dinoflagellates is Perid (Jeffrey and Vesk 1997). We concluded that the spring bloom was caused by diatoms and that the autumn bloom was due to dinoflagellates. The autumn bloom in Kesennuma Bay was thought to be mainly phytoplankton except diatoms (Itoh et al. 2006). These results agree with those of a previous report.

The quantity and diversity of phytoplankton did not change at Iwaisaki. This probably occurred because Iwaisaki is geographically slightly outside the bay mouth (Fig. 1a). The average salinity was high at 33.5 PSU (see Fig. 2b). Thus, we thought that the seawater at Iwaisaki was strongly influenced by offshore water rather than coastal water from the inner part of the bay.

4.2 Phytoplankton in Ofunato Bay

Phytoplankton quantities increased after the earthquake (see Fig. 3), although Chl *a* concentration varied with the season and year (Hayakawa et al. 2001). This may have occurred because plankton feeders such as shellfish decreased in number because of outflow of the oyster and scallop rafts, or nitrogen and phosphate runoff from the sewage disposal plants increased because the sewage disposal plants stopped operating. It is possible that the coastal area temporarily shifted to a eutrophied condition.

Amounts of phytoplankton species except diatoms increased near the surface, and phytoplankton diversity changed in Ofunato Bay after the earthquake. If nitrogen and phosphate runoff from the sewage disposal plants increased, the nitrogen, phosphorus, and silicate balance might collapse depending on location. This change in balance of nutrients could have changed the predominant phytoplankton taxa.

These results suggest that the influence of the tsunami on phytoplankton varied by location. In this study, phytoplankton in Ofunato Bay was affected by the tsunami because the bay is enclosed and was easily affected by the terrestrial environment and broken seawall, rather than at Iwaisaki, which is at the entrance of Kesennuma Bay.

Acknowledgments This work was made possible through a contribution from a project of the Fisheries Agency of Japan.

References

Goto K, Chagué-Goff C, Fujino S, Goff J, Jaffe B, Nishimura Y, Richmond B, Sugawara D, Szczuciński W, Tappin DR, Witter RC, Yulianto E (2011) New insights of tsunami hazard from the 2011 Tohoku-oki event. Mar Geol 290:46–50

Harino H, Yatsuzaka E (2012) Current status of chemical substances in Kesennuma and Moune Bay after earthquake. Aquabiology 203:531–537 (in Japanese)

Hayakari S, Hanaishi R (2001) Introduction of spreadsheet application macros for environmental data analysis and analytical examples. (II) Macro for CMB Analysis. J Jpn Soc Atmos Environ 36:39–45 (in Japanese)

Hayakawa Y, Kobayashi M, Izawa M (2001) Sedimentation flux from mariculture of oyster (*Crassostrea gigas*) in Ofunato estuary. Jpn ICES Mar Sci 58:435–444

Ito H, Chiba A, Saito K, Onodera T, Oshino A (2007) The productivity of the aquaculture ground in Kesennuma Bay. Miyagi Pref Rep Fish Sci 7:35–40 (in Japanese)

Itoh H, Fujita N, Chiba A (2006) The seasonal change of the aquaculture ground environment at the innermost in Kesennuma Bay. Miyagi Pref Rep Fish Sci 6:51–57 (in Japanese)

Jeffrey SW, Vesk M (1997) 2 Introduction to marine phytoplankton and their pigment signature. In: Jeffrey SW, Mantoura RFC, Wright SW (eds) Phytoplankton pigments in oceanography. UNESCO Publishing, Paris, pp 37–84

Kaga S, Watanabe R, Nagai S, Kamiyama T, Suzuki T (2012) Accumulation of paralytic shellfish toxins in bivalves caused by the dinoflagellate *Alexandrium tamarense* in Ofunato Bay, Iwate Prefecture, northern Pacific coast of Japan after the 2011 Tohoku-Oki earthquake. Gekkan Kaiyo 44:321–327 (in Japanese)

Kusui Y, Kimura T, Mekuchi T, Hashimoto T (1983) Growth of oyster in relation to quantities of food. Bull Hiroshima Fish Exp Stn 13:7–33 (in Japanese)

Masuda R (2012) Underwater visual census of fish assemblages in Moune bay, Kesennuma. Aquabiology 203:562–568 (in Japanese)

Ministry of Land, Infrastructure, Transport and Tourism, Japan (2011) Damage, emergency recovery, and the present status of affairs of the sewage disposal plants by tsunami. http://www.mlit.go.jp/common/000138622.pdf (in Japanese)

Nishitani G, Yamamoto M, Natsuike M, Liu D, Yoshinaga I (2012) Dynamics of phytoplankton in Kesennuma Bay and Moune Bay after the disaster 3.11. Aquabiology 203:545–555 (in Japanese)

Oksanen J, Blanchet FG, Kindt R, Legendre P, Minchin PR, O'Hara RB, Simpson GL, Solymos P, Henry M, Stevens H, Wagner H (2013) Community ecology package. http://cran.r-project.org, http://vegan.r-forge.r-project.org/

Okumura Y, Koyama J, Uno S (2003) The relationship between logPow and molecular weight of polycyclic aromatic hydrocarbons and EC50 values of marine microalgae. La Mer 41:182–191

Okumura Y, Kamiyama T, Suzuki T (2012) Application of chemical mass balance to determination of phytoplankton composition from pigment profiles in seawater. La Mer 50:69–76

Sasaki K (2012) The oil pollution with the East Japan great earthquake disaster on 11th, March 2011. Oil Pollut Inf 1:5–13 (in Japanese)

Tanaka M (2012) The aim and future perspective of "Kesennuma-Moune Investigation" on effects and recovery after the 3/11 disaster conducted in Kesennuma and Moune Bay, Miyagi Prefecture, Japan. Aquabiology 203:515–523 (in Japanese)

The Alfred Wegener Institute for Polar and Marine Research (2013) http://odv.awi.de/

The R Project for Statistical Computing (2013) http://www.r-project.org/

Urabe J, Suzuki T, Nishita T, Makino W (2013) Immediate ecological impacts of the 2011 Tohoku earthquake tsunami on intertidal flat communities. PLoS One 8:e62779

Yamada Y (2012) Seasonal changes of zooplankton community structure in Moune-Kesennuma inlet and adjacent waters. Aquabiology 203:556–561 (in Japanese)

Yamamoto M, Yokoyama K, Yoshinaga I (2012) Change of heavy metals and oil spilled in Kesennuma Bay and Moune Bay after the disaster 3.11. Aquabiology 203:538–544 (in Japanese)

Yokoyama K, Hatakeyama M (2012) Tsunami disaster in Moune bay and reverted tidal flat. Aquabiology 203:524–530 (in Japanese)

Zapata M, Rodriguez F, Garrido JL (2000) Separation of chlorophylls and carotenoids from marine phytoplankton: a new HPLC method using a reversed phase C8 column and pyridine containing mobile phases. Mar Ecol Prog Ser 195:29–45

Harmful Shell Borers, *Polydora* Species (Polychaeta: Spionidae), from Commercially Important Mollusk Shells in East Asia and Australia

Waka Sato-Okoshi, Hirokazu Abe, Kenji Okoshi, Wataru Teramoto, Jeremy Shaw, Byoung-Seol Koh, Yong-Hyun Kim, Jae-Sang Hong, and Jing-Yu Li

Abstract

Shell-boring polydorids (Polychaeta: Spionidae) are economically and ecologically important species that must be monitored owing to the risk they pose to commercially important mollusk shells. Tracking polydorid species internationally requires accurate species identification, which is based on both morphological characteristics and nuclear 18S rRNA gene sequences. Four serious shell-boring *Polydora* in East

W. Sato-Okoshi (✉) • W. Teramoto
Laboratory of Biological Oceanography, Graduate School of Agricultural Science, Tohoku University, Sendai 981-8555, Japan
e-mail: wsokoshi@bios.tohoku.ac.jp; w.teramoto@bios.tohoku.ac.jp

H. Abe
Tohoku National Fisheries Research Institute, Fisheries Research Agency, 3-27-5, Shinhama-cho, Shiogama-city, Miyagi 985-0001, Japan
e-mail: hirokazuabe@affrc.go.jp

K. Okoshi
Department of Environmental Science, Faculty of Science, Toho University, 2-2-1, Miyama, Funabashi, Chiba 274-8510, Japan
e-mail: kenji.okoshi@env.sci.toho-u.ac.jp

J. Shaw
Centre for Microscopy, Characterisation and Analysis, The University of Western Australia, 35 Stirling Highway, Crawley, WA 6009, Australia
e-mail: jeremy.shaw@uwa.edu.au

B.-S. Koh
Korea Marine Environment Management Corporation, Heagong Bldg., Samseong-dong 71, Gangnam-gu, Seoul 135-870, Republic of Korea
e-mail: bskoh@koem.or.kr

Y.-H. Kim
B&G Eco Tech Environmental Monitoring Research Institute, 140, Topri, Chukdong, Sacheon, Gyeongnan 664-811, Republic of Korea
e-mail: polychaeta@hanmail.net

J.-S. Hong
Department of Oceanography, Inha University, Incheon 402-751, Republic of Korea
e-mail: jshong@inha.ac.kr

J.-Y. Li
Laboratory of Phycology and Algal Aquaculture, Fisheries College, Ocean University of China, 5 Yushan Road, Qingdao 266003, China
e-mail: lijingyu@yahoo.co.jp

H.-J. Ceccaldi et al. (eds.), *Marine Productivity: Perturbations and Resilience of Socio-ecosystems*,
DOI 10.1007/978-3-319-13878-7_4, © Springer International Publishing Switzerland 2015

Asian and Australian aquaculture, i.e., *Polydora brevipalpa*, *P. uncinata*, *P. haswelli*, and *P. aura*, are described here. The shell-associated polydorids that have been reported from these areas were summarized and reviewed at the same time.

1 Introduction

Spionid polychaetes are distributed in a wide variety of habitats from shallow water to deepwater depths in all oceans. They are commonly the most abundant polychaete groups in terms of biomass and number of species in the coastal benthic environment. Species of *Polydora* and related genera (Polychaeta, Spionidae) are the so-called polydorids, and they are found in a wide variety of substrata, ranging from soft mud to hard calcareous materials (Blake 1996; Sato-Okoshi 1999, 2000). Polydorids are well known for their ability to bore into various calcareous substrates, e.g., coralline algae, corals, and mollusk shells (Blake and Evans 1973; Blake 1996; Sato-Okoshi 1999). Although they excavate their burrows for use as a habitat, some species have been frequently reported as harmful invaders from the viewpoint of aquaculture, as they often damage the commercially important mollusk shells by inducing abnormal shell formation, decreasing their commercial value, reducing their growth rate and meat yield, and causing heavy mortality (Sato-Okoshi et al. 1990, 2008; Okoshi and Sato-Okoshi 1996; Handley and Bergquist 1997; Mortensen et al. 2000; Lleonart et al. 2003; Simon et al. 2006). In recent years, the heavily infested abalone shells were observed to be dead in tank-cultured system and it was forced to stop cultivating in several prefectures in Japan (personal communication). Measures of preventing or controlling polydorid infestation in commercially important mollusk shells have been suggested (Handley and Bergquist 1997; Diggles et al. 2002; Simon et al. 2010), but the problem remains unresolved.

An expansion of molluscan aquaculture has resulted in a global distribution of certain commercially important mollusk shells (Cohen and Carlton 1998). Consequently, the polydorid species that associate with these mollusk shells, such as borers and crevice inhabitants, have also spread by accompanying commercially important host shells (Bailey-Brock 2000; Radashevsky and Olivares 2005; Simon et al. 2006). Those species are not only a source of concern economically (Radashevsky and Olivares 2005; Simon et al. 2006) but also pose a threat ecologically (Cohen and Carlton 1998). Ecological disturbances caused by invasive organisms (including polychaetes) associated with these mollusks are currently seen as a major cause of diversity loss worldwide (Mack et al. 2000; Miura 2007).

The polydorids include several species groups that are morphologically indistinguishable, with an insufficient number of distinct characters to enable species level identification (Radashevsky and Pankova 2006; Sato-Okoshi and Abe 2013). At the same time, some species possess a high degree of intraspecific variation, particularly with respect to pigmentation patterns (Sato-Okoshi and Abe 2013; Teramoto et al. 2013). Recently, molecular sequence analyses were conducted in some shell-associated *Polydora* species (Sato-Okoshi and Abe 2012, 2013; Teramoto et al. 2013), and the results demonstrated that the nuclear 18S rRNA gene analysis would be an effective and useful tool to accurately discriminate between these problematic species.

The present study first summarizes the polydorid species to understand the background of the relevant species associated with commercially important mollusk shells in East Asia and Australia. Secondly, the study focuses on harmful *Polydora* species known to be responsible for heavy host shell damage and with the potential to become an invasive species as a result of accidental transport on host shells. Morphological, ecological, and molecular biological characteristics of these high-risk *Polydora* species are described here in order to better trace and monitor the threat they pose to mollusk fisheries and ocean ecologies.

2 Distribution and Composition of Polydorid Species Associated with Mollusk Shells

There are numerous polydorids that have been reported as being associated with commercially important mollusk shells, i.e., oysters, scallops, and abalone, in Japan, South Korea, China, and Australia (Table 1). A total of 28 species from five genera have been reported from these waters since 1970, during which time the commercial transportation of mollusk shells has increased. The percentage composition of the number of species by genus was 16 species (57 %) in *Polydora*, seven species (25 %) in *Dipolydora*, three species (11 %) in *Boccardia*, and one species (3 %) in each *Boccardiella* and *Pseudopolydora*, respectively. The results show that over half of the target species belong to the genus *Polydora*. Among these, estimates of the shell infestation condition by *Polydora brevipalpa* and *P. uncinata* from Japan (Imajima and Sato 1984; Sato-Okoshi et al. 1990; Sato-Okoshi 1994, 1998; Sato-Okoshi and Abe 2012); *P. haswelli*, *P. aura*, and *P. uncinata* from South Korea (Sato-Okoshi et al. 2012); *P. brevipalpa* from China (Sato-Okoshi et al. 2013); and *P. uncinata*, *P. hoplura*, and *Boccardia knoxi* from Australia (Lleonart et al. 2003; Sato-Okoshi et al. 2008; Sato-Okoshi and Abe 2012) revealed that these polydorid species should be considered particularly dangerous. To date, *Polydora uncinata* is the only species revealed to be harmful across all these countries, with the exception of China. This species was reported to be especially harmful in abalone shells, i.e., land-based tank-cultured abalone.

3 Morphological, Biological, and Ecological Characteristics of the Four Harmful *Polydora* Species

Detailed characteristics of the four most harmful species from the 28 polydorid species are described below. These include *Polydora brevipalpa*, *P. uncinata*, *P. haswelli*, and *P. aura*. Photographs of living specimens (Fig. 1) and the infestation condition of scallop, oyster, and abalone shells (Fig. 2) are shown.

3.1 *Polydora brevipalpa* (Fig. 1a, b)

3.1.1 Morphological Characteristics

Large-sized worms. Conspicuous black bands appear on palps. The anterior end of the prostomium is rounded, and there is no notochaetae in the 5th chaetiger. The color of the pygidium is light tan, white, black, or partially black.

3.1.2 Biological and Ecological Characteristics

The larvae showed planktotrophic development. The species is distributed in cold waters. It is very common in scallop shells particularly abundant in the left valves of wild and sown cultured scallops in the Okhotsk Sea (Fig. 2a) (Mori et al. 1985). Recently, the species was first extracted from the shells of wild and sown cultured abalone, other than scallop, in north China (Sato-Okoshi et al. 2013). They reproduce egg capsules repeatedly, except during seasons of high temperature, and showed high reproductive capacity. The larvae settle at the periphery of the left valve of the scallop during drift ice period and their life span is approximately 2.5 years (Sato-Okoshi 1994). Severe infestation by the species was observed not only in Japanese scallops but also in Chinese sown cultured scallops (Sato-Okoshi et al. 2013).

3.2 *Polydora uncinata* (Fig. 1c, d)

3.2.1 Morphological Characteristics

Large-sized worms. Conspicuous black bands appear on palps and black pigmentation presents on both dorsal and ventral sides of anterior chaetigers. Rarely no conspicuous black pigmentation is observed on the palps and body. Short occipital tentacle is present. There are special recurved notochaetae in posterior chaetigers.

Table 1 Polydorid species associated with commercially important mollusk shells in East Asia (Japan, South Korea, China) and Australia

Country	Host shell		Associated spionid species: Boring	Associated spionid species: Non-boring	References
Japan	Scallop	*Patinopecten yessoensis*^{S}	***Polydora brevipalpa***, *P. websteri, P. curiosa, Dipolydora alborectalis, D. concharum, D. bidentata*		Sato-Okoshi (1999) and Sato-Okoshi and Abe (2012)
		Patinopecten yessoensis^{W}	***Polydora brevipalpa***, *P. websteri, Dipolydora alborectalis, D. concharum, D. bidentata*		Sato-Okoshi (1999)
		Patinopecten yessoensis^{C}	***Polydora brevipalpa***, *P. onagawaensis*		Sato-Okoshi (1999) and Teramoto et al. (2013)
	Oyster	*Crassostrea gigas*^{W}	*Polydora onagawaensis, P. uncinata, P. calcarea, P. websteri, P. haswelli, Dipolydora bidentata, D. concharum*	*Polydora cornuta, Boccardiella hamata, Boccardia proboscidea, B. pseudonatrix, Pseudopolydora antennata*	Sato-Okoshi (1999, 2000), Sato-Okoshi and Abe (2013), Present paper
		Crassostrea gigas^{C}	*Polydora onagawaensis, P. uncinata, P. websteri, P. aura, P. haswelli, P. curiosa, Dipolydora bidentata, D. concharum, D. giardi*	*Dipolydora socialis, Boccardiella hamata*	Sato-Okoshi (1999, 2000), Sato-Okoshi and Abe (2013), Present paper
	Abalone	*Haliotis discus hannai*^{W}	*Polydora calcarea, P. websteri, P.* sp., *Dipolydora armata, D. giardi*		Sato-Okoshi (1999), Present paper
		Haliotis discus hannai^{S}	*Polydora* sp.		Present paper
		Haliotis discus hannai^{C*}	***Polydora uncinata***		Present paper
		Haliotis discus discus^{C*}	***Polydora uncinata***, *Dipolydora armata*		Present paper
		Haliotis diversicolor aquatilis^{W}	*Polydora ciliata, P. flava orientalis, P. websteri, Dipolydora giardi, D. armata*		Kojima and Imajima (1982) and Sato-Okoshi (1999)
		Haliotis diversicolor aquatilis^{C*}	***Polydora uncinata***, *Dipolydora armata*		Present paper
South Korea	Scallop	*Chlamys farreri*^{W}	*Polydora haswelli, Dipolydora alborectalis*	*Polydora limicola*	Sato-Okoshi et al. (2012)
	Oyster	*Crassostrea gigas*^{C}	*Polydora haswelli, P. aura,* ***P. uncinata***		Sato-Okoshi et al. (2012)
		Crassostrea gigas^{W}		*Boccardiella hamata*	Sato-Okoshi et al. (2012)
	Abalone	*Haliotis discus discus*^{W}	***Polydora haswelli, P. aura***		Sato-Okoshi et al. (2012)
		Haliotis discus discus^{C}	***Polydora haswelli, P. aura***		Sato-Okoshi et al. (2012)

China	Scallop	*Patinopecten yessoensis*[S]	***Polydora brevipalpa***, *P. onagawaensis*		Sato-Okoshi et al. (2013)
		Patinopecten yessoensis[C]	*Polydora onagawaensis*		Sato-Okoshi et al. (2013)
		Chlamys farreri[W]	*Polydora onagawaensis*		Sato-Okoshi et al. (2013)
		Chlamys farreri[C]	*Polydora onagawaensis*		Sato-Okoshi et al. (2013)
	Oyster	*Crassostrea gigas*[W]	*Polydora onagawaensis*	*Boccardiella hamata*	Sato-Okoshi et al. (2013)
		Crassostrea gigas[C]	*Polydora onagawaensis, P. websteri*		Sato-Okoshi et al. (2013)
	Abalone	*Haliotis discus hannai*[W]	*Polydora onagawaensis, P. brevipalpa*		Sato-Okoshi et al. (2013)
		Haliotis discus hannai[S]	*Polydora onagawaensis, P. brevipalpa*		Sato-Okoshi et al. (2013)
Australia	Scallop	*Chlamys australis*[W]	*Polydora uncinata*		Sato-Okoshi et al. (2008)
		Pecten alba	*Polydora latispinosa*		Blake and Kudenov (1978)
	Oyster	*Crassostrea gigas*[W]	*Polydora calcarea*	*Boccardia proboscidea*	Sato-Okoshi and Abe (2013), Present paper
		Crassostrea gigas[C]	*Polydora websteri, P. hoplura, Dipolydora giardi*	*Boccardia proboscidea, B. pseudonatrix*	Blake and Kudenov (1978), Present paper
		Saccostrea commercialis[C]	*Polydora websteri* (cited as *P. haswelli*)	*Boccardia pseudonatrix* (cited as *B. knoxi*)	Blake and Kudenov (1978), Sato-Okoshi et al. (2008)
		?	*Polydora haswelli, P. latispinosa*	*Boccardia chilensis*	Blake and Kudenov (1978)
	Abalone	*Haliotis roei*[W]	***Polydora uncinata***, *Dipolydora giardi, D. armata, D. aciculata*	*Boccardia proboscidea*	Sato-Okoshi et al. (2008)
		Haliotis roei[C*]	***Polydora uncinata***, *Dipolydora armata* (cited as *P. armata*)		Blake and Kudenov (1978), Sato-Okoshi et al. (2008)
		Haliotis rubra[C]	*Polydora hoplura*	***Boccardia knoxi***	Lleonart et al. (2003)
		Haliotis laevigata[C*, C]	***Polydora hoplura, P. uncinata***		Lleonart et al. (2003) and Sato-Okoshi et al. (2008)
		Haliotis conicopora[W]	*Dipolydora armata, D. aciculata*		Sato-Okoshi et al. (2008)
		Haliotis denov	*Polydora woodwicki, Dipolydora armata* (cited as *P. armata*)		Blake and Kudenov (1978)

Bolds show harmful species for the host shells
W wild, *C* culture, *C** Land-based tank culture, *S* sown culture

Fig. 1 Pictures of live individuals extracted from the shells. *Polydora brevipalpa* (**a**), anterior chaetigers with palps with black bars and a white pygidium, both protruding from their burrows in scallop shell (From Sato-Okoshi and Abe 2012). (**b**) A complete individual extracted from its burrow (From Sato-Okoshi and Abe 2012). *Polydora uncinata* (**c**), anterior chaetigers with palps with inconspicuous black bars (From Sato-Okoshi and Abe 2012). (**d**) Posterior chaetigers with special hooks on the notopodia (From Sato-Okoshi and Abe 2012). *Polydora haswelli* (**e**), distinct black pigmentation along the prostomium, on peristomium, and on anterior chaetigers and palp with black bars (From Sato-Okoshi et al. 2012). *Polydora aura* (**f**), a complete individual with orange in color (From Sato-Okoshi and Abe 2012)

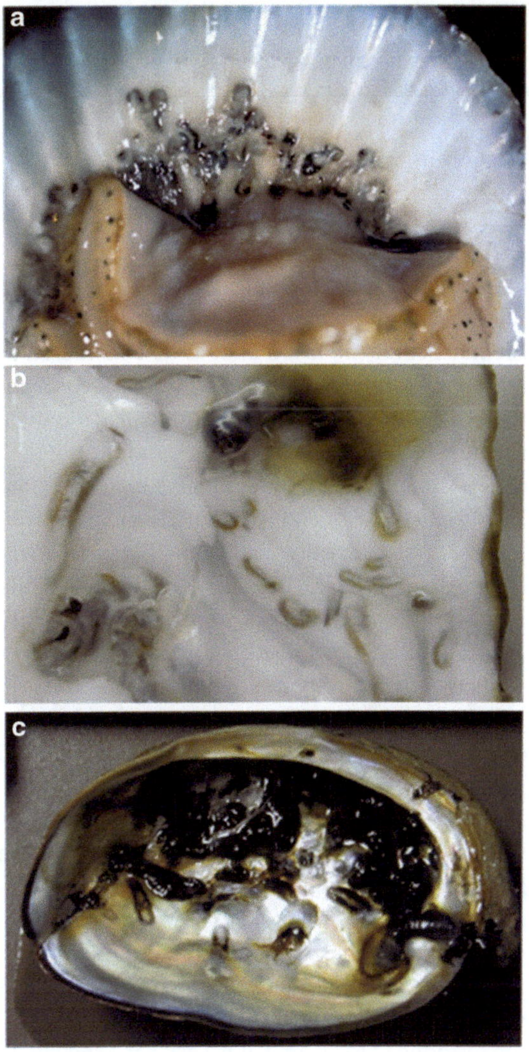

Fig. 2 Pictures of shells infested by *Polydora* species. (**a**) Inner surface of the left valve of scallop shell showing heavy infestation by *P. brevipalpa* after removing the mantle (From Sato-Okoshi and Abe 2012). (**b**) Inner surface of the oyster shell showing heavy infestation by *P. uncinata* and *P. haswelli* (From Sato-Okoshi and Okoshi 2014). (**c**) Inner surface of the abalone shell showing heavy infestation by *P. uncinata* (From Sato-Okoshi et al. 2012)

3.2.2 Biological and Ecological Characteristics

Larvae showed direct development and adelphophagy with many nurse eggs within egg capsules. Their hatched-out stage was ca 17-chaetiger larvae and showed no or very short planktonic larval stage. One female seemed to reproduce egg capsules repeatedly and reproduction capacity was suggested to be very high. Life span was predicted to be more than 2 years in the tank-cultured environment. The species was distributed widely in Asian and Australian waters, and it was extracted commonly from both wild and cultured mollusk shells. It is noteworthy to cite that a high number of these species have been suddenly observed in suspended cultured oyster shells in South Korea (Fig. 2b), which is speculated to be the result of host oysters being introduced from different waters (Sekino et al. 2003; Sato-Okoshi et al. 2012). In addition to Japan and Australia, heavy infestation of this species has been reported on land-based tank-cultured abalone in Chile. It is also speculated that this infestation was the result of host abalone being transported from Japan to Chile accompanying the *P. uncinata* population (Radashevsky and Olivares 2005). This species is regarded as one of the most dangerous polydorids, especially in the case of enclosed aquaculture systems such as abalone land-based tank culture (Fig. 2c).

3.3 *Polydora haswelli* (Fig. 1e)

3.3.1 Morphological Characteristics

Medium- to large-sized worms. Conspicuous black bands appear on palps and black pigmentation presents on both dorsal and ventral sides of anterior chaetigers.

3.3.2 Biological and Ecological Characteristics

Larvae revealed planktotrophic development. Little is currently known about other characteristics, e.g., life history or reproductive characteristics. Although the species exhibits medium sizes in Japan, it reaches larger sizes and is very common in South Korean mollusk shells (Fig. 2b) (Sato-Okoshi et al. 2012). The species causes harmful infestations in many commercially important shells, but appears limited to South Korea (Sato-Okoshi et al. 2012).

3.4 *Polydora aura* (Fig. 1f)

3.4.1 Morphological Characteristics

Medium- to large-sized worms. The body and palps are light orange but some worms without conspicuous color. A short inconspicuous occipital tentacle is present. Special notochaetae are present in the posterior chaetigers which exhibit tight cylindrical bundles of short needles.

3.4.2 Biological and Ecological Characteristics

All the larvae in the capsule developed simultaneously and revealed planktotrophic development. Little is currently known about other characteristics, i.e., life history or reproductive characteristics. As for *P. haswelli*, this species was very large in size and common in South Korean mollusk shells (Sato-Okoshi et al. 2012). The species caused severe infestations in many commercially important shells, but again appears limited to South Korea (Sato-Okoshi et al. 2012).

4 Phylogenetic Analysis of 18S rRNA Gene Sequences

Misidentification and taxonomic confusion of *Polydora* species often arise because of a lack of efficient morphological key characteristics and a high degree of intraspecific variation. In recent years, an increasing number of taxonomic studies are relying on molecular methods for accurate species identification. 18S rRNA gene sequences of seven *Polydora* species (*P. onagawaensis*, *P. calcarea*, *P. brevipalpa*, *P. websteri*, *P. haswelli*, *P. aura*, and *P. uncinata*) have been analyzed (Sato-Okoshi and Abe 2012, 2013; Teramoto et al. 2013). The results of these studies showed that the nucleotide sequence of the 18S rRNA gene was completely identical within a species even between separate populations and distinct between different species (Fig. 3). Therefore, 18S rRNA gene analysis is an effective tool for the identification of these *Polydora* species, including the four species described above.

5 Discussion and Conclusion

As little data exists on the species characteristics of *Boccardia knoxi*, a species reported to be a severe shell-boring parasite of abalone shells by Lleonart et al. (2003), it should be considered as the focus of future studies. Additionally, discriminating between the morphologically identical species *P. hoplura* and *P. uncinata*, for which only *P. uncinata* has undergone molecular analysis (Sato-Okoshi and Abe 2012), remains problematic. As such, only data from *P. uncinata* will be presented here. The key morphologically identifying characteristics of these four species are summarized in Table 2.

The *Polydora* species differed according to locality and in their host shells. The species considered to be the most damaging to East Asian and Australian aquaculture were *P. brevipalpa*, *P. uncinata*, *P. haswelli*, and *P. aura*. It is now possible to speculate that some of these species have been transported all over the world with commercially important host shells over a prolonged period of time, making it difficult or impossible to trace their origins. It has been demonstrated that once invasive nonindigenous species successfully are introduced to an area, it is very hard and costly to eradicate them (Miura 2007). Although it is difficult to prevent polydorid infestation after they succeed in boring into the shells, one of the most effective ways to avoid their influences may be preventing their early settlement on shells applying their life history traits. However, the harmful species tended to be continuous breeders and have high reproductive ability and possess long life span, so as a result, they show long settling periods. Moreover, once *P. uncinata* is introduced to the land-based tank-cultured system, the species has the capacity to increase its population rapidly according to its direct development.

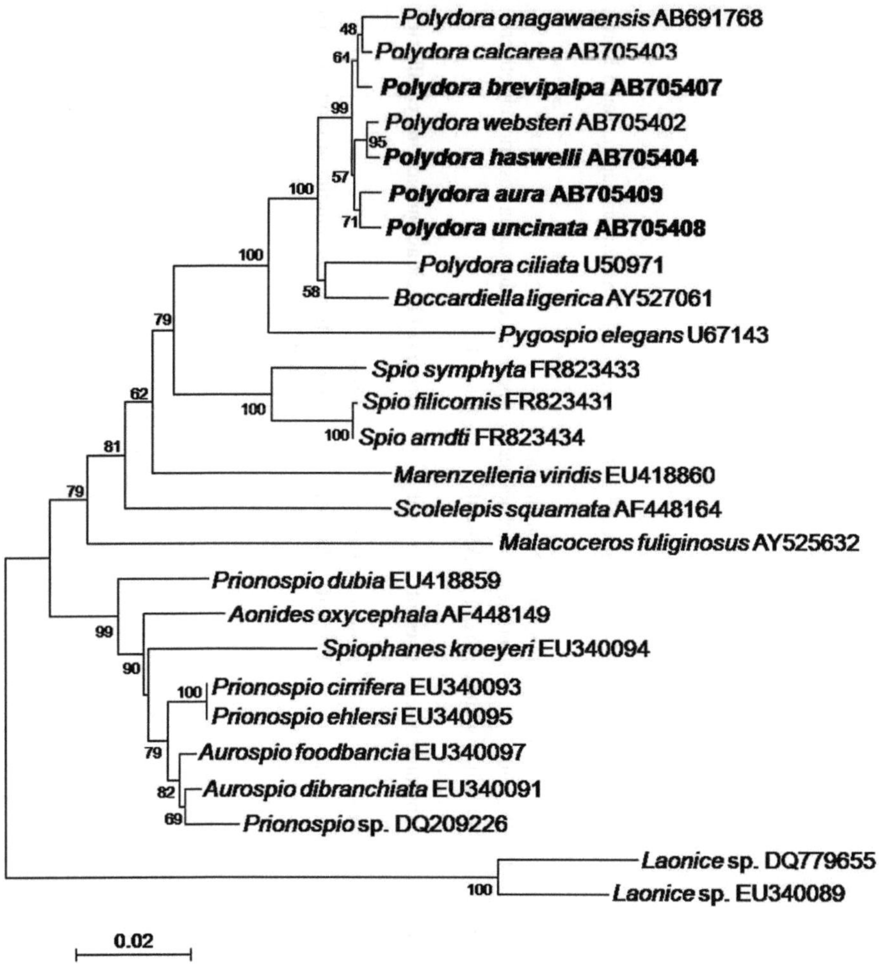

Fig. 3 A neighbor-joining tree inferred from the nuclear 18S rRNA gene sequences of spionid polychaetes. The 18S rRNA gene sequences of the four harmful *Polydora* species are highlighted in *boldface type*. Bootstrap values of >50 % as a percentage of 1,000 bootstrap replicates are given at the respective nodes. The *scale bar* represents the number of substitutions per site

The transportation of live animals therefore requires great care and suitable precautions to maintain sustainable aquaculture and biological diversity. Monitoring studies are indeed necessary to avoid the further dispersal of species already known to be harmful in existing cultured and natural environments. It is needless to say that not only these polydorid polychaetes but other as yet unrecognized and potentially invasive animals and plants may have been unintentionally transported as a result of the international trade in economically important mollusk species.

Table 2 Comparison of morphological characteristics among four harmful shell borers, *Polydora brevipalpa*, *P. uncinata*, *P. haswelli*, and *P. aura* from Japan, South Korea, northeast China, and Australia

		Pigmentation				Occipital	Notochaetae	
Polydora species	Body color	Palps	Prostomium	Chaetigers 1–4	Pygidium	tentacle	on chaetiger 5	Special notochaetae in posterior chaetigers
Polydora brevipalpa	Tan	**Black bars**	Absent	**Absent**	Absent or white or black	Absent	**Absent**	Absent
Polydora uncinata	Tan	**Black bars or absent**	Black or absent	Black or absent	Absent or black	**Present**	Present	**Recurved hook accompanying capillaries**
Polydora haswelli	Tan	**Black bars**	Black or absent	Black or absent	Absent	Absent	Present	Absent
Polydora aura	**Orange or tan**	Absent	Absent	Absent	Absent	**Present**	**Absent**	**Tight cylindrical bundles of short needles accompanying capillaries**

Bold entries show distinguishable characters of the species

References

Bailey-Brock JH (2000) A new record of the polychaete *Boccardia proboscidea* (family Spionidae), imported to Hawaii with oysters. Pac Sci 54:27–30

Blake JA (1996) Family Spionidae Grube, 1850. In: Blake JA, Hilbig B, Scott PH (eds) Taxonomic atlas of the benthic fauna of the Santa Maria Basin and Western Santa Barbara Channel, vol 6, The Annelida, Part 3. Santa Barbara Museum of Natural History, Santa Barbara, pp 81–224

Blake JA, Evans JD (1973) *Polydora* and related genera (Polychaeta: Spionidae) as borers in mollusk shells and other calcareous substrates. Veliger 15:235–249

Blake JA, Kudenov JD (1978) The Spionidae (Polychaeta) from southeastern Australia and adjacent areas with a revision of the genera. Mem Natl Mus Victoria 39:171–280

Cohen AN, Carlton JT (1998) Accelerating invasion rate in a highly invaded estuary. Science 279:555–558

Diggles BK, Hine PM, Handly SJ, Boustead NC (2002) A handbook of diseases of importance to aquaculture in New Zealand. NIWA Sci Tech Ser 49:1–200

Handley SJ, Bergquist PR (1997) Spionid polychaete infestations of intertidal pacific oysters *Crassostrea gigas* (Thunberg), Mahurangi Harbour, northern New Zealand. Aquaculture 153:191–205

Imajima M, Sato W (1984) A new species of *Polydora* (Polychaeta, Spionidae) collected from Abashiri Bay, Hokkaido. Bull Natl Sci Mus Ser A 10:57–62

Kojima H, Imajima M (1982) Burrowing polychaetes in the shells of abalone *Haliotis diversicolor aquatilis* chiefly on the species of Polydora. Bull Jpn Soc Sci Fish 48:31–35

Lleonart M, Handlinger J, Powell M (2003) Spionid mudworm infestation of farmed abalone (*Haliotis* spp.). Aquaculture 221:85–96

Mack RN, Simberloff D, Lonsdale WM, Evans H, Clout M, Bazzaz FA (2000) Biotic invations: causes, epidemiology, global consequences, and control. Ecol Appl 10:689–710

Miura O (2007) Molecular genetic approaches to elucidate the ecological and evolutionary issues associated with biological invasions. Ecol Res 22:876–883

Mori K, Sato W, Nomura T, Imajima M (1985) Infestation of the Japanese Scallop *Patinopecten yessoensis* by the boring polychaetes, *Polydora*, on the Okhotsk Sea Coast of Hokkaido, especially in Abashiri Waters. Bull Jpn Soc Sci Fish 51:371–380 (in Japanese with English abstract)

Mortensen S, van der Meeren T, Fosshagen A, Hernar I, HarkestadL TL, Bergh O (2000) Mortality of scallop spat in cultivation, infested with tube dwelling bristle worms, *Polydora* sp. Aquac Int 8:267–271

Okoshi K, Sato-Okoshi W (1996) Biomineralization in molluscan aquaculture– growth and disease. Bull de l'Inst Oceanogr Monaco 14:151–169

Radashevsky VI, Olivares C (2005) *Polydora uncinata* (Polychaeta: Spionidae) in Chile: an accidental transportation across the Pacific. Biol Invasions 7(3):489–496

Radashevsky VI, Pankova VV (2006) The morphology of two sibling sympatric *Polydora* species (Polychaeta: Spionidae) from the Sea of Japan. J Mar Biol Assoc UK 86:245–252

Sato-Okoshi W (1994) Life history of the polychaete, *Polydora variegata*, that bores into the shells of scallops in northern Japan. Mem Mus Nat d'Hist Natl 162:549–558

Sato-Okoshi W (1998) Three new species of polydorids (Polychaeta, Spionidae) from Japan. Species Diversity 3:277–288

Sato-Okoshi W (1999) Polydorid species (Polychaeta, Spionidae) in Japan, with descriptions of morphology, ecology and burrow structure. 1 Boring species. J Mar Biol Assoc UK 79:831–848

Sato-Okoshi W (2000) Polydorid species (Polychaeta, Spionidae) in Japan, with descriptions of morphology, ecology and burrow structure. 2 Non-boring species. J Mar Biol Assoc UK 80:443–456

Sato-Okoshi W, Abe H (2012) Morphological and molecular sequence analysis of the harmful shell boring species of *Polydora* (Polychaeta: Spionidae) from Japan and Australia. Aquaculture 368–369:40–47

Sato-Okoshi W, Abe H (2013) Morphology and molecular analysis of the 18S rRNA gene of oyster shell borers, *Polydora* species (Polychaeta: Spionidae), from Japan and Australia. J Mar Biol Assoc UK 93:1279–1286

Sato-Okoshi W, Okoshi K (2014) Spionid polychaetes expand their distribution by accompanying oyster shells during transportation. In: Turner JP (ed) Oysters: biology, consumption and ecological importance. Nova Science Publishers, New York

Sato-Okoshi W, Sugawara Y, Nomura T (1990) Reproduction on the boring polychaete *Polydora variegata* inhabiting scallops in Abashiri Bay, North Japan. Mar Biol 104:61–66

Sato-Okoshi W, Okoshi K, Shaw J (2008) Polydorid species (Polychaeta, Spionidae) in southwestern Australian waters with special reference to *Polydora uncinata* and *Boccardia knoxi*. J Mar Biol Assoc UK 88:491–501

Sato-Okoshi W, Okoshi K, Koh B-S, Kim Y-H, Hong J-S (2012) Polydorid species (Polychaeta, Spionidae) associated with commercially important mollusc shells in Korean waters. Aquaculture 350–353:82–90

Sato-Okoshi W, Okoshi K, Abe H, Jing-Yu L (2013) Polydorid species (Polychaeta, Spionidae) associated with commercially important mollusk shells from eastern China. Aquaculture 406–407:153–159

Sekino M, Hamaguchi M, Aranishi F, Okoshi K (2003) Development of novel microsatellite DNA markers from the Pacific oyster *Crassostrea gigas*. Mar Biotechnol 5:227–233

Simon CA, Ludford A, Wynne S (2006) Spionid polychaetes infesting cultured abalone, *Haliotis midae*, in South Africa. Afr J Mar Sci 28:167–171

Simon CA, Bentley MG, Caldwell GS (2010) 2, 4-Decadienal: exploring a novel approach for the control of polychaete pests on cultured abalone. Aquaculture 310:52–60

Teramoto W, Sato-Okoshi W, Abe H, Nishitani G, Endo Y (2013) Morphology, 18S rRNA gene sequence, and life history of a new *Polydora* species (Polychaeta, Spionidae) from northeastern Japan. Aquat Biol 18:31–45

Impact of the 2011 Tsunami on Seagrass and Seaweed Beds in Otsuchi Bay, Sanriku Coast, Japan

Teruhisa Komatsu, Takayoshi Ohtaki, Shingo Sakamoto, Shuhei Sawayama, Yasuaki Hamana, Michiharu Shibata, Kohji Shibata, and Shuji Sasa

Abstract

The Sanriku Coast facing the Pacific Ocean is an important area for fishing and aquaculture. The Sanriku Coast was hit hard by the catastrophic tsunami event produced by the Great East Japan Earthquake on 11 March 2011. The tsunami destroyed not only infrastructures of coastal fisheries and aquacultures but also coastal ecosystems, such as the seagrass and seaweed beds vital for sustainable fisheries. This study surveyed the seagrass and seaweed beds in Otsuchi Bay, Iwate prefecture in June 2011 and found that the seaweed beds around the mouth and middle of the bay were not impacted by the tsunami, but that the seagrass beds in the bay head had been destroyed. However, observations verified the growth of seagrass seedlings germinated from seeds produced in previous years. This response of the seagrass demonstrates its resilience to huge tsunami events occurring in an interval of several decades. Huge seawall construction plan against tsunami threatens their recovery. The recovery program in Sanriku Coast must be established on a holistic approach including social and ecological aspects between land and the sea because coastal areas are ecotone and also "sociotone" between them.

1 Introduction

The Sanriku Coast is a highly fertile coastal zone where waters from the Oyashio cold water current encounter the Kuroshio warm water current (Fig. 1) (Kawai 1972). Geomorphologically, the Sanriku Coast comprises semi-enclosed and ria bays that protect raft and fishery facilities from rough sea conditions. Numerous aquaculture facilities are deployed in the ria bays (Komatsu et al. 2002), formerly belonging to the Rikuchu National

T. Komatsu (✉) • T. Ohtaki • S. Sakamoto
S. Sawayama • Y. Hamana • S. Sasa
Atmosphere and Ocean Research Institute (AORI), University of Tokyo, President of S.F.J.O. (*) of Japan, 5-1-5, Kashiwanoha, 277-8564 Kashiwa-shi, Japan
e-mail: komatsu@aori.u-tokyo.ac.jp

M. Shibata • K. Shibata
Toyo Corporation, 1-1-6, Yaesu, Chuoku, Tokyo 103-8234, Japan

H.-J. Ceccaldi et al. (eds.), *Marine Productivity: Perturbations and Resilience of Socio-ecosystems*, DOI 10.1007/978-3-319-13878-7_5, © Springer International Publishing Switzerland 2015

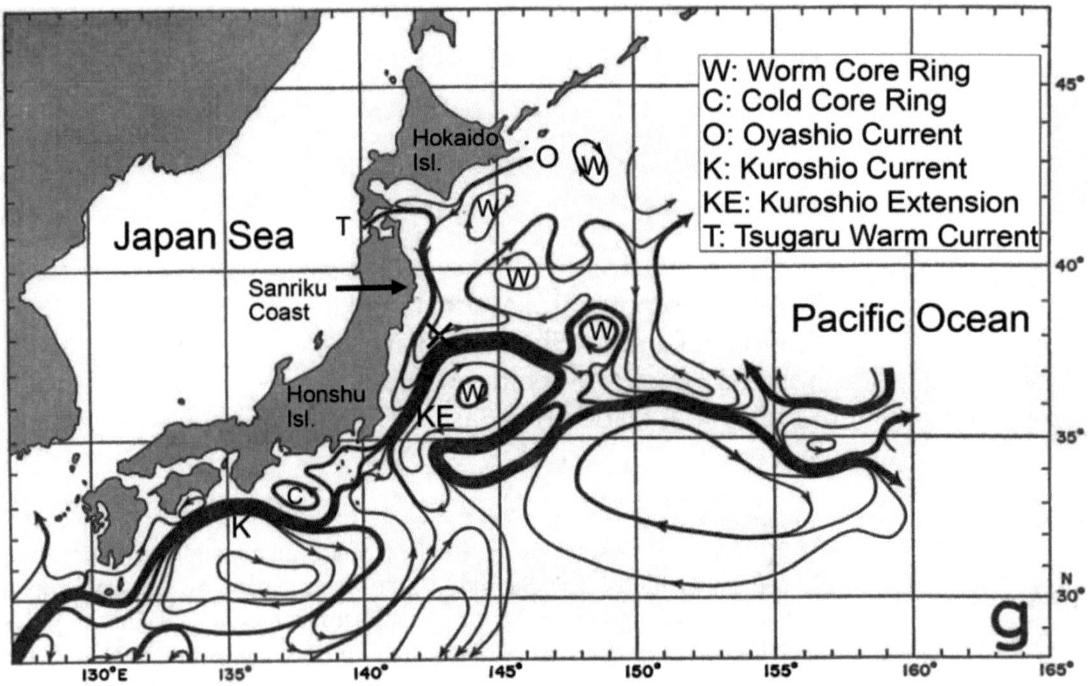

Fig. 1 Map showing the current systems off the Sanriku Coast and the epicentre of the Great East Japan Earthquake, indicated by a *closed cross* in the northwestern Pacific Ocean, based on Kawai (1972)

Park due to the attractive landform of the bays and now part of the Sanriku Recovery National Park. The majority of the aquaculture operating on the Sanriku Coast comprises of seaweeds, oysters, ascidians and scallops, which do not require artificial feed input. Shells and ascidians are filter feeders that depend on particulate organic matter (POM) in the seawater (Dame 1993; Petersen 2007). POM is supplied from rivers and the ocean (Wada et al. 1987; Kawamiya et al. 1996). Ocean-derived POM consists of phytoplankton, zooplankton and other particulate materials. In the Sanriku ria bays, the seaweed and seagrass beds produce POM not only from their own biomass but also from epiphytic organisms (e.g. Oshima et al. 1999). The seaweed beds are distributed mainly in the mouth and middle of the bays, and the seagrass beds are distributed mainly at the head of the bays, because the seaweed grows on rock beds, and most of the seagrass grows on sand beds.

Seagrass and seaweed beds play an ecologically important role as substrates for flora and fauna, including epiphytic and benthic organisms, as well as their key role in fisheries (Fortes 1996; Coles et al. 1993). Seagrass beds contribute to the marine environment by stabilising bottom sediments and maintaining coastal water quality and clarity (Ward et al. 1984; Jeudy de Grissac and Boudouresque 1985; Komatsu and Yamano 2000; Komatsu et al. 2004). Seaweed and seagrass beds contribute to forming marine environments by influencing the spatial and temporal distributions of water flow (Komatsu and Murakami 1994), water temperature (Komatsu et al. 1982, 1985, 1994), downward illumination via shading by their canopies (Komatsu 1989; Komatsu et al. 1990), pH distribution (Komatsu and Kawai 1986) and dissolved oxygen content by their respiration and assimilation (Komatsu 1989; Komatsu et al. 1990). Many commercially important species spawn in seagrass and seaweed beds (e.g. molluscs, sea urchins, balaos, cuttlefish), and larvae and juveniles use the beds as nursery grounds (Arasaki and Arasaki 1978).

Thus, seagrass and seaweed beds support biodiversity and are important habitats for marine

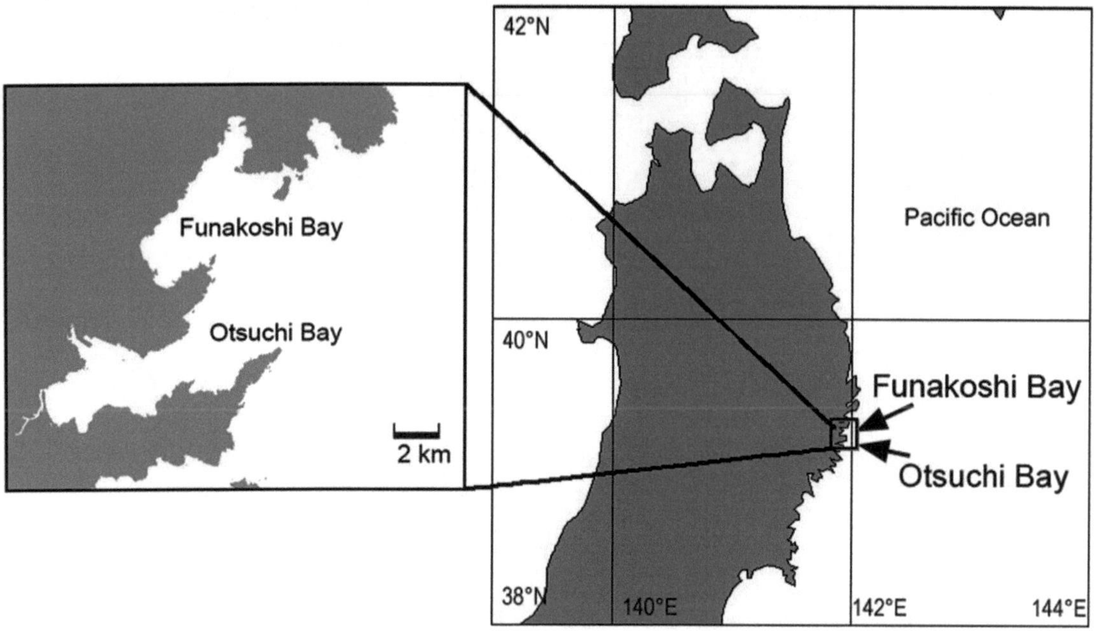

Fig. 2 Map showing the Otsuchi Bay and Funakoshi Bay, Iwate prefecture and Sanriku Coast, Japan

animals and hence, contribute significantly to coastal fisheries (Komatsu and Tastukawa 1998). Recently, it has been reported that floating seaweeds play a vital role as spawning and nursery grounds of fishes and associated animals (e.g. Safran 1990; Safran and Omori 1990; Thiel and Gutow 2005; Komatsu et al. 2009; Abé et al. 2013).

In Japan, the most important floating seaweeds are *Sargassum horneri* (Turner) C. Agardh and other *Sargassum* species (Komatsu et al. 2007) and are seasonally distributed in offshore waters (Komatsu et al. 2014; Mizuno et al. 2014) and around the Japanese coastline (Yoshida 1963). Japanese saury (*Cololabis saira*) and some flying fish species spawn on the floating seaweeds, and Japanese saury fisheries is one of the most important fisheries along the Sanriku Coast in autumn because the saury migrates south from waters off Hokkaido Island to those off the Coast. Thus, seaweed beds consisting of *Sargassum* species including *S. horneri* are highly important in supporting fisheries on this coast.

Zostera caulescens Miki, the world's tallest seagrass, reaching a length of 7 m (Aioi et al. 1996; 1998), grows extensively along the Sanriku Coast and in Funakoshi Bay (Fig. 2) to a water depth of ~ 20 m (Tastukawa et al. 1996; Sultana and Komatsu 2002). Komatsu et al. (2003) mapped a bed of *Z. caulescens* with a narrow multi-beam sonar system in Otsuchi Bay (Fig. 2), where the seagrass beds were distributed on a shallow bottom at a depth of <7 m. Furthermore, a considerable amount of data has been collected on the seagrass and seaweed beds in Otsuchi Bay and Funakoshi Bay because the marine research station belonging to the Atmosphere and Ocean Research Institute of the University of Tokyo is located in Otsuchi Bay.

On 11 March 2011, the Great East Japan Earthquake occurred off the Sanriku Coast, causing a massive tsunami (Department of International Affairs of Japan Science and Technology Agency 2011) that struck the coastline and caused extensive human casualties and damage to cars, houses, ports, boats, roads and the Fukushima first nuclear power plant. The widespread damage to fisheries infrastructures, such as fishing boats, fish markets, freezing store houses, fuel stands, ice-producing facilities,

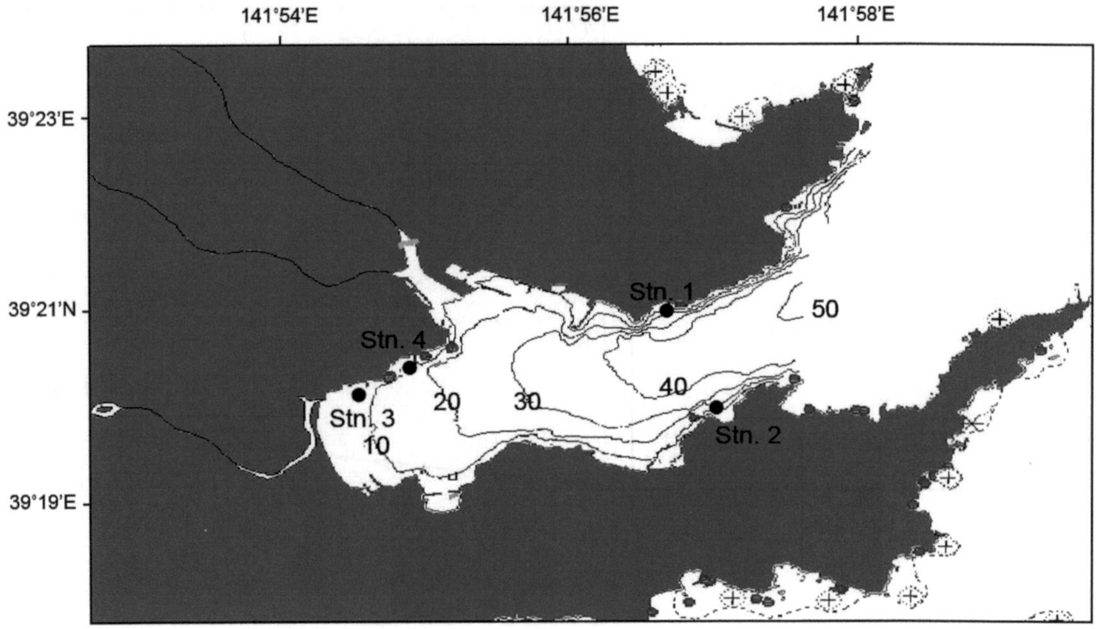

Fig. 3 Map showing the seafloor topography and stations in Otsuchi Bay

roads, bridges and dikes, was immediately evident. However, it was difficult to observe damage to the seagrass and seaweed beds, which serve as "natural infrastructures" for fisheries. Therefore, we opted to survey the seagrass and seaweed beds in Otsuchi Bay after the tsunami.

In this article, we examine the status of seagrass and seaweed beds before and after the 2011 tsunami to provide information on these beds, their role in coastal fisheries and their resilience to natural catastrophic disasters occurring on the Sanriku Coast.

2 Materials and Methods

2.1 Study Site

As our study site, we selected Otsuchi Bay on the Sanriku Coast, which represents a typical ria bay for this coastline, facing Pacific Ocean. The mouth of the bay and the longitudinal axis are 1 and 5 km in length, respectively (Fig. 3). Bottom depths along the axis are 30–50 m deep in the mouth and centre of the bay. Because the bay mouth is narrow, the inside of the bay is protected from waves and currents. Numerous aquaculture facilities are located in the bay. The inundation height of the tsunami was estimated to be 10–14 m near the fishing port located in the bay (Yagi 2012).

2.2 Field Survey

We were unable to conduct any field survey shortly after the tsunami, which occurred on 11 March 2011 because most of the boats had been lost in the tsunami and no hotels were open. Thus, the field survey was conducted from 19 to 23 June 2011, 3 months after the tsunami. An underwater camera (QI Co., Model FM 4100) was deployed at stations where seagrass and seaweed beds had been distributed before the tsunami (Fig. 3), and the positions of the stations were recorded by GPS.

3 Results

3.1 Seaweed Beds

At station 1, seaweed beds consisting of *Laminaria* sp. and *S. horneri* were observed in 2009. In June 2011, we observed *S. horneri* with

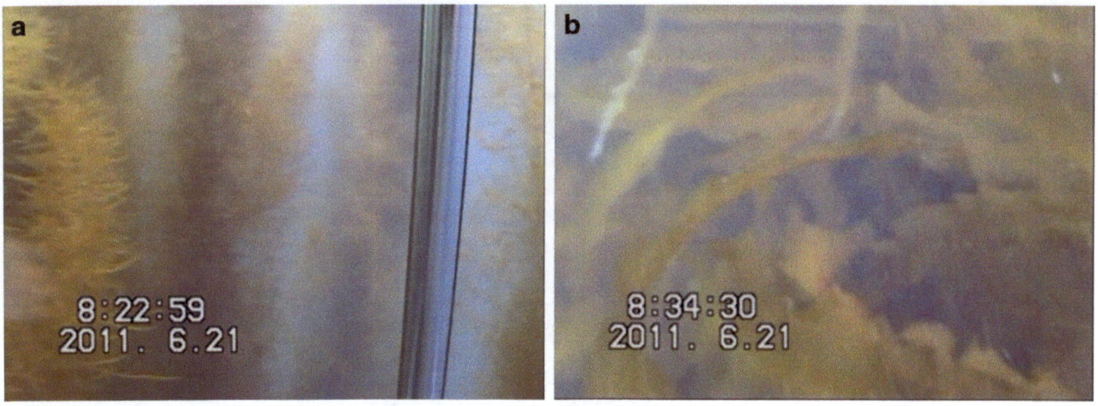

Fig. 4 Photographs of *Sargassum horneri* (**a**) and *Laminaria* spp. (**b**) at Station 1, taken in June 2011

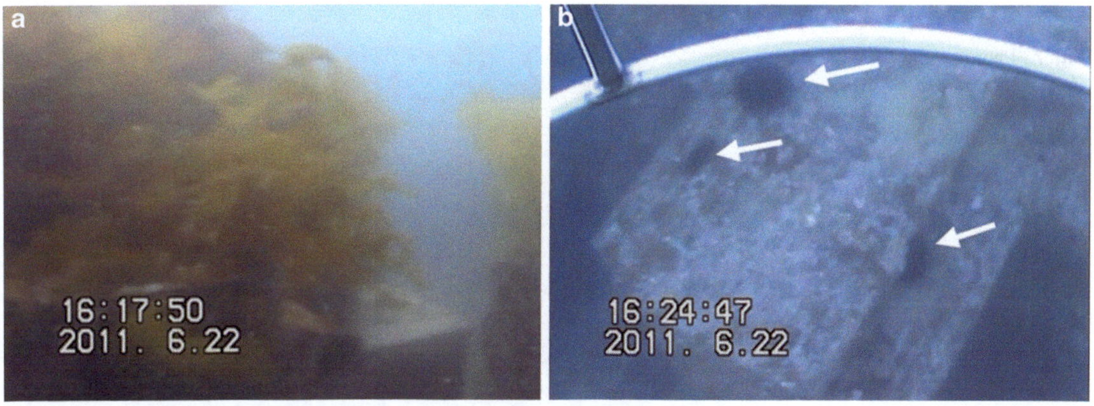

Fig. 5 Photographs of *Undaria pinnatifida* (**a**) and an artificial reef with sea urchins (white arrows) (**b**) at Station 2, taken in June 2011

stipes 7 m long (Fig. 4a). *Laminaria* species with a luxuriant canopy covered the seafloor (Fig. 4b).

At station 2, thriving *Undaria pinnatifida* (Harvey) Suringar was observed on artificial reefs built for abalone (Fig. 5a). Several sea urchins were attached to these artificial reefs (Fig. 5b).

3.2 Seagrass Beds

In the middle and south of the bay head around station 3, we found no flowering or vegetative shoots of seagrasses, but did observe seagrass seedlings (Fig. 6). The seedlings were identified as *Zostera caulescens* Miki and *Zostera marina* Linnea. Around Station 4, we found flowering shoots of *Zostera caespitosa* Miki (Fig. 7).

4 Discussion

4.1 Impact of the Tsunami on Seaweed Beds

Sargassum horneri is an annual (Umezaki 1984) and matures in summer on the Sanriku Coast. *Undaria pinnatifida* matures in late spring to early summer on the Coast (Miyagi Prefecture 2014). According to the information from the Fisheries Agency of Japan (2014), *Laminaria religiosa* Miyabe is distributed in Otsuchi Bay and is biennial (Yoshida 1989) and matures in summer. If the tsunami had destroyed these seaweed species, they would not have been thriving only 3 months later. When rocks have been turned over, their surfaces are white because the bottom surface that has been in contact with the

Fig. 6 Photograph of *Zostera* spp. seedlings (*white arrows*) at Station 3, taken in June 2011

Fig. 7 Photograph of *Zostera caespitosa* flowering shoots at Station 4, taken in June 2011

seafloor is neither covered with seaweed nor are any animals attached. Because phytal organisms cannot grow on the bottom surface of an artificial reef, a clean surface would suggest that the artificial reef had been overturned. However, we observed no clean surfaces in the rocks and artificial reefs and, hence, no evidence for their rolling by the tsunami. Thus, we concluded that the tsunami had no significant impact on the seaweeds around the coast of the mouth and middle of the bay. The reasons for this are as follows: (1) the rocks or rock plates that acted as substrates for seaweeds were sufficiently heavy to remain unmoved by the tsunami; (2) the seaweeds were sufficiently small and flexible in March before the onset of summer, their maturation season, to not be removed by the tsunami; (3) since the seaweed beds were distributed close to the deep and steep seafloor around the coast of the middle and mouth of the ria bays, the tsunami had not yet formed high waves due to the considerable water depth, which is supported by the observation of one of the students staying at the International Centre for Coastal Oceanography of the University of Tokyo located on the coast near the centre of Otsuchi Bay, who saw that the sea level rose and fell slowly during the tsunami (Kokubu, personal communication)—the shallower the water depth, the higher the wave formed (Japan

Weather Association 2014); (4) since the Otsuchi Bay has a U-shape horizontal profile, the wave height in the bay head was higher than that in the bay mouth (Japan Weather Association 2014); and (5) the tsunami waves did not break, and the speed of water movement was lower than that required to remove small seaweeds from the substrates.

4.2 Impact of the Tsunami on Seagrass Beds and Their Responses

Elevated waves from the tsunami struck the bay head where the water was very shallow (Japan Weather Association 2014). The resulting powerful spilling waves and undertows transported seagrasses growing on the sandy seafloor. We discovered solitary seagrass seedlings without rhizomes linking to other flowering or vegetative shoots when we removed seedling samples in October 2011, which indicated that the seedlings had germinated from seeds. Shabaka and Komatsu (2010), using bimonthly sampling, found that *Z. caulescens* flowered between June and August in Funakoshi Bay, which lies near Otsuchi Bay. It is hypothesised that *Z. marina* also flowers in June and August, similar to *Z. caulescens*. Hence, we assume that the seeds had been produced before 2011 and had remained buried until the tsunami. Yamaki et al. (2006) found that 90 % of seagrass seeds that had been preserved for 1 year germinated in a fresh water medium. Thus, it is possible that seeds produced in 2010, and even before 2010, were capable of germination in 2011. Seeds resuspended by the tsunami that then fell on the sand beds would have been able to germinate.

4.3 Germination of Seagrass Seeds

Field and laboratory studies of seagrass germination have focused primarily on salinity, temperature, light, scarification and, more recently, the sediments in which the seeds germinate (e.g. oxygen [oxygen-reduction profiles or Eh]), as critical factors influencing germination processes (Orth et al. 2000). A massive tsunami does not change the salinity or water temperature of the seeds' environment, but does alter the light, sediments and scarification. Some studies have reported that light is not important for the germination of *Z. marina* (e.g. Moore et al. 1993). While the latter study demonstrated more rapid germination of *Z. marina* seeds in sediments with no oxygen, mixing by a tsunami oxygenates sediments. Harrison (1991) reported that scarification of the seed coat resulted in increased germination rates for *Z. marina*, and it is possible that collision of seagrass seeds with sediments during the tsunami caused scarification of the seed coat and promoted the seeds' germination.

4.4 Influences of Environments on Seed Germination and Seedling Growth of Seagrasses

It was reported that the lowest water temperature in which *Z. marina* seeds germinate was <5 °C, and the optimal water temperature was between 5 and 10 °C in the Chesapeake Bay, USA (Orth and Moore 1983) and Odawa Bay, Japan (Kawasaki et al. 1986). In Otsuchi Bay, mean water temperatures in March, April, May and June at a depth of 1 m from 2000 to 2009 were 7.1, 9.1, 12.3 and 14.8 °C, respectively (Michida et al. 2010). Abe et al. (2008) examined germination rates using culture experiments at water temperatures from 5 to 25 °C at intervals of 5 °C over a period of 30 d following seed storage for 0–60 d at 0 °C at a depth of 1.5 cm in sand. They revealed that the optimal water temperature for seed germination was in the range of 10–15 °C (the highest germination rate occurred at 10 °C) with a PAR [photosynthetic active radiation] of 50 μmol regardless of storage duration. 50 % of the seeds were germinated at 10 °C within 2 months. The germinated seeds grew to seedlings approximately 10 cm long within 2–3 weeks in water temperatures of 5–20 °C. Thus, water temperatures in Otsuchi Bay might be favourable for seeds to germinate and grow to longer than 10 cm, which correlates with our observation that seedlings of *Zostera* species were present in mid-June. Germinated seeds require light, although some studies have reported that light had no influence on the germination of *Z. marina* seeds. Because the tsunami removed flowering and vegetative shoots, the lack of shading from these absent shoots enabled seedlings to receive light, as has been reported previously (Olesen and Sand-Jensen 1994; Abe et al. 2004).

4.5 Resilience of Seagrasses to Disturbances

In general, seagrasses undergo asexual reproduction through production of shoots on the root systems in subtidal zones that provide a stable environment (Philipps et al. 1983). However, seedlings are also produced after disappearance of seagrass beds caused by storm (Aioi and Komatsu 1996) and tsunami events. It is believed that seagrasses have two reproductive strategies: asexual in a stable environment and sexual, when a sudden loss of seagrass beds has occurred. The huge tsunami hit Sanriku Coast in 1896 (off Sanriku with moment magnitude, Mw 8.0–8.1), 1933 (off Sanriku with Mw 8.4), 1960 (off Chile with Mw 9.5) (Iwate Prefecture 2014; Kanamori 1977; Omachi et al. 2003) and 2011 (off Sanriku with Mw 9.0) (Takeuchi and Chavoshian 2011). The last one in 2011 was caused by the great earthquake which tends to occur at an interval of 700 years along this coastline facing Japan trench (Satake 2011). *Zostera* species on the Sanriku Coast invest energy to produce seeds as an insurance against the disappearance of the mother plants; this provides the seagrass beds with resilience against catastrophic natural disasters, such as huge tsunamis. Thus, it is recommended that recovery operations after such events (e.g. restoration of ports and removal of debris (fragments of dykes) in shallow waters) incorporate measures to avoid damage to recovering seagrass beds and do not take place before regrowth of broad seagrass beds near the sites of such operations because of the fundamental role of these beds in fisheries.

4.6 Socio-ecosystem Resilience

Coastal areas in Otsuchi Bay and Sanriku Coast depend on fisheries and aquacultures. Restoration of the areas from the tsunami disaster needs recovery of coastal fisheries and aquacultures, which are based on the coastal ecosystem. In Sanriku Coast, damaged seagrass beds started to regenerate on the beds where their mother plants had been distributed. On the other hand, the Ministry of Land and Transport with Miyagi and Iwate Prefectures is constructing huge seawalls with a wall height of 10–15 m and a base width of 43–63 m from the against the huge tsunami along the coast in the places of broken seawalls to protect people and their properties without environmental assessments (Anon 2013) with a far-fetched reason that construction of huge seawalls is a recovery work for sea-walls usually with about 3–4 m high before the tsunami. If fortification of the coast by the huge seawalls cuts continuity of land and the sea and material flow between them, it destroys an ecotone including seagrass beds by direct and indirect ways due to construction. It is needed to remember that society based on the fisheries and aquacultures utilising coastal ecosystems exists with the nature that has original resilience against natural disasters. Thus, the recovery program from the tsunami must be established on holistic approach including social and ecological aspects of land and the sea based on the ecosystem-based concept that secures material flows between land and the sea from the sustainable point of view because the coastal area is an ecotone of land and the sea and also a "sociotone" between land and the sea.

Acknowledgements We thank Mr. Masataka Kurosawa and Mr. Masaaki Hirano, staffs of the International Centre for Coastal Oceanography, University of Tokyo, for their assistance during this study. We are also grateful to Mr. Kazuhiro Hantani of Toyo Corp. for his encouragement to carry out our research. This study was supported by grants from the Marine Alliance Organization of the University of Tokyo, the Environment Research and Technology Development Funds (S9) and (S13) of the Japanese Ministry of the Environment and the Tohoku Ecosystem-Associated Marine Sciences (TEAMS) research program of the Ministry of Education, Culture, Sports, Science and Technology (MEXT).

References

Abe M, Hashimoto N, Kurashima A, Maegawa M (2004) Structure and seasonal change of the *Zostera marina* population on the coast of Matsunase, Mie Prefecture, central Japan. Nippon Suisan Gakk 70:523–529 (in Japanese with English abstract and figure captions)

Abe M, Kurashima A, Maegawa M (2008) Temperature requirements for seed germination and seedling growth of *Zostera marina* from central Japan. Fish Sci 74:589–593

Abé H, Komatsu T, Kokubu Y, Natheer A, Rothäusler EA, Shishido H, Yoshizawa S, Ajisaka T (2013) Invertebrate fauna associated with floating *Sargassum horneri* (Fucales: Sargassaceae) in the East China Sea. Species Diver 18:75–85

Aioi K, Komatsu T (1996) A preliminary report on the recovery process of *Zostera marina* in Otsuchi Bay, Northern Japan after disturbance by large atmospheric depressions. In: Kuo J, Phillips RC, Walker DI, Kirkman H (eds) Seagrass biology: proceedings of an international workshop, Rottnest Island, Western Australia, 25–29 January 1996. The University of Western Australia, Nedlands, pp 319–322

Aioi K, Komatsu T, Morita K (1996) Giant seagrass, *Zostera caulescens*, discovered in Funakoshi Bay, Iwate Prefecture. Bull Jpn Soc Fish Oceanogr 60:7–10 (in Japanese with English Abstract and figure captions)

Aioi K, Komatsu T, Morita K (1998) The world's longest seagrass, *Zostera caulescens* from northeastern Japan. Aquat Bot 61:87–93

Anon (2013) Is a plan for constructing seawalls appropriate? The Nature Conservation Society of Japan, Tokyo, 2 pp (in Japanese). http://www.nacsj.or.jp/katsudo/kaiho/pdf/No534-p2-3.pdf

Arasaki M, Arasaki T (1978) Compilation of stories on algae. Tokai University Press, Tokyo, 228 pp (in Japanese)

Coles RG, Lee Long WJ, Watson RA, Derbyshire KJ (1993) Distribution of seagrasses, and their fish and penaeid prawn communities, in Cairns Harbour, a tropical estuary, northern Queensland, Australia. Aust J Mar Freshwat Res 44:193–210

Dame RF (1993) The role of bivalve filter feeder material fluxes in estuarine ecosystems. Nato ASI Series 33:245–269

Department of International Affairs of Japan Science and Technology Agency (2011) The Great East Japan Earthquake information from official websites. Japan Science and Technology Agency, Tokyo, p 67

Fisheries Agency of Japan (2014) Dominant seaweeds in Otsuchi Bay. http://www.jfa.maff.go.jp/j/kikaku/wpaper/h23_h/trend/1/zoom_f016.html. Accessed 14 July 2014

Fortes MD (1996) Seagrass research in Southeast Asia: where does it lead to? In: Kuo J, Phillips RC, Walker DI, Kirkman H (eds) Seagrass biology: proceedings of an international workshop, Rottnest Island, Western Australia, 25–29 January 1996. The University of Western Australia, Nedlands, pp 359–366

Harrison PG (1991) Mechanisms of seed dormancy in an annual population of *Zostera manna* (eelgrass) from the Netherlands. Can J Bot 69:1972–1976

Iwate Prefecture (2014) History of tsunami hit in Iwate (in Japanese) http://www.bunka.pref.iwate.jp/shizen/umi/data/kiso03.html. Accessed 10 Aug 2014

Japan Weather Association (2014) Basic knowledge of tsunami. Japan Weather Association, Tokyo, pp 1–10 (in Japanese) http://www.jwa.or.jp/news/docs/津波の基礎知識.pdf. Accessed 14 July 2014

Jeudy de Grissac A, Boudouresque C-F (1985) Rôle des herbiers de Phanérogames marines dans les mouvements des sédiments côtiers: les herbiers à *Posidonia oceanica*. In: Ceccaldi HJ, Champalbert G (eds) Actes du colloque pluridisciplinaire franco-japonais d'Océanographie Fascicule 1. Société Franco-japonaise d'Océanographie, Marseille, pp 143–151

Kanamori H (1977) The energy release in great earthquakes. J Geophys Res 82(20):2981–2987

Kawai H (1972) Hydrography of the Kuroshio Extension. In: Stommel H, Yoshida K (eds) Kuroshio – its physical aspects. University of Tokyo Press, Tokyo, pp 235–354

Kawamiya M, Kishi MJ, Ahmed MDK, Sugimoto R (1996) Causes and consequences of phytoplankton blooms in Otsuchi Bay, Japan. Cont Shelf Res 16:1683–1698

Kawasaki Y, Iituka T, Goto H, Terawaki T, Shimo S (1986) Effects of temperature on *Zostera marina* L. I. Seed germination and seedling development. Report of Central Institute of Electric Power U485028, Abiko, pp 1–22

Komatsu T (1985) Temporal fluctuations of water temperature in a *Sargassum* forest. J Oceanogr Soc Jpn 41:235–243

Komatsu T (1989) Day-night reversion in the horizontal distributions of dissolved oxygen content and pH in a *Sargassum* forest. J Oceanogr Soc Jpn 45:106–115

Komatsu T, Kawai H (1986) Diurnal changes of pH distributions and the cascading of shore water in a *Sargassum* forest. J Oceanogr Soc Jpn 42:447–458

Komatsu T, Murakami S (1994) Influence of a *Sargassum* forest on the spatial distribution of water flow. Fish Oceanogr 3:256–266

Komatsu T, Tatsukawa K (1998) Mapping of *Zostera marina* L. beds in Ajino Bay, Seto Inland Sea, Japan, by using echosounder and global positioning systems. J Rech Océanogr 23:39–46

Komatsu T, Yamano H (2000) Influence of seagrass vegetation on bottom topography and sediment distribution on a small spatial scale in the Dravuni Island Lagoon, Fiji. Biol Mar Medit 7:243–246

Komatsu T, Ariyama H, Nakahara H, Sakamoto W (1982) Spatial and temporal distributions of water temperature in a *Sargassum* forest. J Oceanogr Soc Jpn 38:63–72

Komatsu T, Kawai H, Sakamoto W (1990) Influence of *Sargassum* forests on marine environment. Bull Coast Oceanogr 27:115–126 (in Japanese with English abstract)

Komatsu T, Murakami S, Kawai H (1994) Some features of jump of water temperature in a *Sargassum* forest. J Oceanogr 52:109–124

Komatsu T, Takahashi M, Ishida K, Suzuki T, Hiraisi T, Tameishi H (2002) Mapping of aquaculture facilities in Yamada Bay in Sanriku Coast, Japan, by IKONOS satellite imagery. Fish Sci 68(suppl I):584–587

Komatsu T, Igarashi C, Tatsukawa K, Matsuoka Y, Harada S, Sultana S (2003) Use of multi-beam sonar to map seagrass beds in Otsuchi Bay, on the Sanriku Coast of Japan. Aquat Living Resour 16(3):223–230

Komatsu T, Umezawa Y, Nakaoka M, Supanwanid C, Kanamoto Z (2004) Water flow and sediment in *Enhalus acoroides* and other seagrass beds in the Andaman Sea, off Khao Bae Na, Thailand. Coast Mar Sci 29:63–68

Komatsu T, Tatsukawa K, Filippi JB, Sagawa T, Matsunaga D, Mikami A, Ishida K, Ajisaka T, Tanaka K, Aoki M, Wang WD, Liu HF, Zhang SD, Zhou MD, Sugimoto T (2007) Distribution of drifting seaweeds in eastern East China Sea. J Mar Syst 67:245–252

Komatsu T, Matsunaga D, Mikami A, Sagawa T, Boisnier E, Tatsukawa K, Aoki M, Ajisaka T, Uwai S, Tanaka K, Ishida K, Tanoue H, Sugimoto T (2009) Abundance of drifting seaweeds in eastern East China Sea. J Appl Phycol 20:801–809

Komatsu T, Mizuno S, Natheer A, Kantachonpoo A, Tanaka K, Morimoto A, Hsiao S-H, Rothäusler EA, Shishidou H, Aoki M, Ajisaka T (2014) Unusual distribution of floating seaweeds in the East China Sea in the early spring of 2012. J Appl Phycol 26:1169–1179

Michida Y, Sato K, Otsuki M, Kurosawa M, Morita K, Hirano M, Yamazaki T (2010) Observation results of hydrography and weather in Otsuchi Bay in 2009. Res Rep Cent Int Coast Oceanogr 35:57–63 (in Japanese)

Miyagi Prefecture (2014) Seaweeds. Traditional fishing gears and methods (in Japanese). http://www.pref.miyagi.jp/soshiki/mtsc/dentokaiso.html. Accessed 14 July 2014

Mizuno S, Ajisaka T, Lahbib S, Kokubu Y, Alabsi NM, Komatsu T (2014) Spatial distributions of floating seaweeds in the East China Sea from late winter to early spring. J Appl Phycol 26:1159–1167

Moore KA, Orth RJ, Nowak JF (1993) Environmental regulation of seed germination in *Zostera marina* L. (eelgrass) in Chesapeake Bay: effects of light, oxygen and sediment burial. Aquat Bot 45:79–91

Olesen B, Sand-Jensen K (1994) Patch dynamics of eelgrass *Zostera marina*. Mar Ecol Prog Ser 106:147–156

Omachi T, Tobata M, Inoue S (2003) Analytical verification of causes of tsunami earthquake generation. Annu J Coast Eng 50:331–335 (in Japanese)

Orth RJ, Moore KA (1983) Seed germination and seedling growth of *Zostera marina* L. (eelgrass) in the Chesapeake Bay. Aquat Bot 15:117–131

Orth RJ, Harwell MC, Bailey EM, Bartholomew A, Jawad JT, Lombana AV, Moore KA, Rhode JM, Woods HE (2000) A review of issues in seagrass seed dormancy and germination: implications for conservation and restoration. Mar Ecol Prog Ser 200:277–288

Oshima Y, Kishi MJ, Sugimoto T (1999) Evaluation of the nutrient budget in a seagrass bed. Ecol Model 115:19–33

Petersen JK (2007) Ascidian suspension feeding. J Exp Mar Biol Ecol 342:127–137

Philipps R, Grant WS, McRoy CP (1983) Reproductive strategies of eelgrass (*Zostera marina* L.). Aquat Bot 16:1–20

Safran P (1990) Drifting seaweed and associated ichthyofauna: floating nursery in Tohoku waters. La Mer 28(4):225–239

Safran P, Omori M (1990) Some ecological observations on fishes associated with drifting seaweed off Tohoku coast, Japan. Mar Biol 105:395–402

Satake K (2011) Supercycles of great earthquakes along Japan Trench. Bull Coord Comm Earthq Predict 86:3–15 (in Japanese). http://cais.gsi.go.jp/YOCHIREN/report/kaihou86/03_15.pdf. Accessed 14 July 2014

Shabaka S, Komatsu T (2010) Phenology and morphology of the marine phanerogam *Zostera caulescens* Miki in Funakoshi Bay, northeast Honshu Island, Japan. Coast Mar Sci 34:13–23

Sultana S, Komatsu T (2002) Preliminary study on shoot density and biomass of seagrass, *Zostera caulescens*, in Funakoshi Bay off the Sanriku Coast, Japan. Otsuchi Mar Sci 27:23–27

Takeuchi K, Chavoshian A (2011) Quick report of Great Eastern Japan Earthquake and tsunami. International Centre for Water Hazard and Risk Management under the auspices of UNESCO (ICHARM), Tsukuba, 13 pp. http://www.iwra.org/doc/Tsunami_report_Kuniyoshi_Takeuchi_ICHARM.pdf

Tastukawa K, Komatsu T, Aioi K, Morita K (1996) Distribution of seagrasses off Kirikiri in Funakoshi Bay, Iwate Prefecture, Japan. Otsuchi Mar Res Cent Rep 21:489–494 (in Japanese)

Thiel M, Gutow L (2005) The ecology of rafting in the marine environment. 1. the floating substrata. Oceanogr Mar Biol 42:181–263

Umezaki I (1984) Ecological studies of *Sargassum horneri* (Turner) C. Agardh in Obama Bay, Japan Sea. Bull Jpn Soc Sci Fish 50:1193–1200

Wada E, Minagawa M, Mizutani H, Tsuji T, Imaizumi R, Karasawa K (1987) Biogeochemical studies on the transport of organic matter along the Otsuchi River watershed, Japan. Estuar Coast Shelf Sci 25:321–333

Ward LG, Kemp WM, Boyton WR (1984) The influence of waves and seagrass communities on suspended particulates in an estuarine embayment. Mar Geol 59:85–103

Yagi H (2012) Fishing port of Otsuchi. In: Institute for Fisheries Engineering, Alpha Hydraulic Engineering Consultants Co., Ltd. and Fukken Co., Ltd. (eds) Activity report collection of Fisheries Research Agency for reconstruction from the earthquake disaster, No. 1. Research report on damages of earthquake and tsunami caused by the East Japan Great Earthquake, No. 1. Fisheries Research Agency, Yokohama, pp 31–42 (in Japanese). http://www.fra.affrc.go.jp/eq/repo_res/report01.pdf

Yamaki K, Ogawa H, Yoshikawa H, Nanba N (2006) Effects of salinity and temperature on germination activation of eelgrass seeds. Aquac Sci 54:347–351 (in Japanese with English abstract)

Yoshida T (1963) Studies on the distribution and drift of the floating seaweed. Bull Tohoku Reg Fish Res Lab 23:141–186 (in Japanese with English abstract)

Yoshida T (1989) Marine algae of Japan. Uchida Rokakuho Publication, Tokyo, 1252 pp (in Japanese)

Impact of Repeating Massive Earthquakes on Intertidal Mollusk Community in Japan

Kenji Okoshi

Abstract

History repeats itself. In these 3,500 years, massive earthquake like the Pacific coast of Tohoku earthquake in 2011 has occurred at least seven times. Major earthquakes with big tsunami have occurred in 500-year cycles along the Pacific coast of northern part of Japan. The tsunami caused a lot of damage to the Tohoku region including marine life. During the earthquake, wavelike movement observed on the surface of the tidal flats. Liquefaction occurred just after the earthquake before tsunami. Bivalves distributed inside the sediment were dug out to surface with a jet of water. Not only the tsunami but also the liquefaction brought quick damage in the coastal marine life. The earthquake caused sudden land subsidence of about 1 m in maximum where intertidal zone became subtidal zone. It may take at least several 10 years to recover the coseismic land subsidence. We have to clarify not only short-term effect which may explain an initial recovery of intertidal and subtidal animals but also a long-term effect which may explain continuous changes in mollusk population with land subsidence. Construction of tsunami seawalls has begun without environmental assessment. Huge seawalls break the continuity of nature to land from the sea. We have to clarify the impact on the coastal ecosystem of artificial structures.

1 Introduction

At least seven magnitude 9-class earthquakes have occurred along the Pacific coast from Hokkaido to Tohoku over the past 3,500 years (source: *Sankei Shimbun*, February 3, 2012), generating huge tsunami that inundated the coastline after analyzing volcanic ash and sand layers in

K. Okoshi (✉)
Department of Environmental Science,
Faculty of Science, Toho University,
2-2-1, Miyama, Funabashi, Chiba 274-8510, Japan
e-mail: kenji.okoshi@env.sci.toho-u.ac.jp

H.-J. Ceccaldi et al. (eds.), *Marine Productivity: Perturbations and Resilience of Socio-ecosystems*,
DOI 10.1007/978-3-319-13878-7_6,

Fig. 1 Imaginary picture of the tsunami that occurred in the Jomon period in Japan illustrated by Kaiki Okoshi

trenches excavated at lowland. Because the marine species of diatoms were collected only from sand layers, indicating that these layers were originated from marine event sediments, brought by tsunamis from the ocean (Dawson et al. 1996). Major earthquakes like the Pacific coast of Tohoku earthquake in 2011 have occurred in 500-year cycles along the Pacific coast of northern part of Japan. The tsunami caused a lot of damage to the Tohoku region including marine life.

It is considered that a magnitude 9-class earthquake had occurred along the Pacific coast of Japan during the Jomon period (Fig. 1). The Jomon period is the time in Japanese prehistory from about 14,000 BC to about 300 BC, when Japan was inhabited by a Neolithic culture. We have a clear evidence of a big earthquake in 869 from the ancient documents and geological analysis. The 869 Jogan Sanriku earthquake and associated tsunami struck the area around Sendai in the northern part of Japan on 9 July 869 (Abe et al. 1990; Minoura et al. 2001).

The earthquake had an estimated magnitude of 8.6 on the surface wave magnitude scale. The geological analysis indicated that the tsunami caused widespread flooding of the Sendai plain, with sand deposits being found up to 4 km from the coast. It is written in ancient documents that the tsunami reached in front of the local government office in Tagajyo (Fig. 2).

In this way, major earthquake with massive tsunami has struck repeatedly in northern Japan. There has been no report about the effects of the magnitude 9-class earthquake and tsunami to the intertidal organisms including mollusks. This earthquake in 2011 was a magnitude 9-class earthquake for the first time happened since the industrial revolution in Japan. Fire of the ship and corruption of the oil tank took place here and there. Combustion products and fuel have flowed into the sea. This earthquake was also the first earthquake of large quantities of chemicals that flows into the sea.

In this chapter, the author would like to focus on the rapid impact of the earthquake for mollusks distributed in intertidal sandy shore and also on the long-term effect of the earthquake including land subsidence and artificial structures constructed after the earthquake.

2 Rapid Impact by Liquefaction and Tsunami

Three minutes before the earthquake of 2011, the water was clear to the bottom at low tide in Yatsu tidal flats in Tokyo Bay. During the earthquake,

Fig. 2 Imaginary picture of the Jogan-Sanriku Tsunami in 869 illustrated by Kaiki Okoshi

Fig. 3 Yatsu tidal flat in Tokyo Bay. (**a** and **b**) 14:43, 11th of March 2011, 3 min before the earthquake. The water was clear to the *bottom* at low tide. (**c** and **d**) 22 min after the earthquake. Clear water changed to muddy water (From Okoshi 2012b)

wavelike movement observed on the surface of the tidal flats. Liquefaction occurred just after the earthquake before tsunami. Liquefaction is a phenomenon in which the strength and stiffness of a soil are reduced by earthquake shaking. The water spurted from the cracks, which looked like a jet of water. I observed many fissures moving and a jet of water from liquefaction coming to the surface around the Yatsu tidal flats (Okoshi 2012a, b, 2013; Okoshi et al. 2014) (Fig. 3).

Twenty minutes after the earthquake, clear water changed to muddy water. Sanbanze tidal flats are connected with Yatsu tidal flats through two channels.

Fig. 4 (**a**) Sand boils that erupted in Sanbanze tidal flat in Tokyo Bay during the earthquake (From Okoshi 2013) (**b**) The lawn of the baseball ground was divided into pieces and covered with *Ulva* sp. (*arrow*) after the earthquake

The muddy water was flowing to the Sanbanze tidal flats through the sewage outlet and two channels. There were many sand boils that erupted in tidal flats during the earthquake (Okoshi 2013) (Fig. 4).

It is thought that this was caused by ground liquefaction. Many cracks caused by the earthquake were also observed in the baseball ground connected to the beach. The lawn of the baseball ground was divided into pieces and covered with *Ulva* sp. after the tsunami (Fig. 4). Many broken shells of trough shell *Mactra veneriformis* were observed along the coastline. There were a few broken shells in Manila clam *Ruditapes philippinarum* in the same area. There is a possibility that the difference in thickness of the shell is influenced in this matter.

Bivalves like sand gaper *Mya arenaria oonogai* normally distributed in deep were dug out to surface with a jet of water in Gamo tidal flats in Miyagi and Matsukawa-ura Inlet where 50 km north from the Fukushima nuclear power plant in Fukushima, northern Japan (Okoshi 2012a, b, 2013; Okoshi et al. 2014).

Some species of infaunal bivalves were also rushed by the tsunami in Matsukawa-ura Inlet. Matsukawa-ura Inlet was attacked by tsunami reached over 9 m. Strong tsunami stream dug up the bottom sand with many organisms living inside the substratum (Fig. 5).

Fig. 5 Bivalves damaged in Matsukawa-ura tidal flat from the 2011 earthquake, where buoyancy caused by soil liquefaction pushed up many underground organisms including *Mya arenaria oonogai* (photo taken the 8th of April 2011)

In this way, not only tsunami but also the liquefaction brought quick damage in the coastal marine life. The infaunal bivalves dug from the tidal ground were not killed instantly. Some adult specimens of deep burrowing bivalves *Macoma contaculata* and *Mya arenaria oonogai* had been survived on the tidal surface in Matsukawa-ura Inlet, Fukushima, by the end of June 2011 (Okoshi 2013).

3 Long-Term Effect by Land Subsidence

The earthquake caused sudden land subsidence of about 1 m in maximum around the Oshika Peninsula, Miyagi, where intertidal zone became subtidal zone. Mangoku-ura Inlet near Oshika Peninsula was occurred land subsidence of ca. 80 cm. Therefore, intertidal zone became subtidal zone and part of the land became intertidal zone. Although many mollusks were damaged by land subsidence and tsunami, plural recruitment of juvenile was observed after July 2011. Spat of Pacific oyster *Crassostrea gigas* was observed to attach to new hard substances after summer. Because the area of larval attachment became higher after the summer in 2011, the distribution area of the adult and juvenile was no longer overlap. The snails *Littorina brevicula* and *Batillaria cumingi* that can move itself were found at shores that were land before the earthquake (Okoshi 2012a, b). Also during low tide, the part of newly formed tidal flat has changed to a tide pool. The tide pools and their periphery were subject to the influx of soil and landform changed by typhoons, waves, and so on. Tide pools change their environment by sea and land natural phenomena, so what creatures advance to there from sea? The great influential typhoon came in September 2011, and soils were influx into the tide pool from the mountain surrounded, and the environment was dramatically changed. *B. cumingi* and *Assiminea spp.* were found in the tide pool extensively after the typhoon. Many juveniles of Venus clam *Cyclina sinensis* and some of Manila clam *R. philippinarum* were also found in and around tide pools (Okoshi 2013). These results showed that larval recruitment occurred in newly formed intertidal zone after the earthquake. In the future, they have possibility to reproduce in the tide pool.

It may take at least several 10 years to recover the coseismic land subsidence. What creatures will advance to new environment and will disappear? We have to clarify not only short-term effect which may explain an initial recovery of intertidal and subtidal animals but also a long-term effect which may explain continuous changes in population with land subsidence.

4 Effects of Man-Made Structures

Almost all tidal flats in Mangoku-ura Inlet have lost as a result of land subsidence after the earthquake. The mountain sediment has been putting into sea from October 2013 to construct artificial tidal flats for Manila clam culture (Fig. 6). Silt is included in a certain percentage of the mountain sediment, turbidity sometimes occurred in the sea when low pressure came. We need to clarify whether or not there is a new recruitment of juvenile after the summer 2014.

The magnitude 7.7 earthquake occurred in 1993 in the Sea of Japan near the Okushiri Island in southwest-off Hokkaido, northern Japan (Satake and Tanioka 1995). The tsunami reached several minutes after the earthquake. The tsunami was 32 m high in maximum. The whole island subsided by 5–80 cm. The tsunami inundated large parts of urban area in Okushiri Island especially in Aonae area, despite its artificial wall of tsunami defenses. The destructive power of this tsunami led to an overhaul of the sea defenses

Fig. 6 (**a** and **b**) Artificial tidal flats made by mountain sediment in Mangoku-ura Inlet; (**c**) Turbidity occurs when stir the newly-introduced mountain sediment

Fig. 7 Coastline of Okushiri Island, Hokkaido Japan. (**a**) Natural coast. Coastal vegetation is continuous until the beach. (**b**) Coast which are divided by the tsunami seawalls with some vegetation on the shore. (**c**) Huge seawalls. There is no vegetation at all on the shore

involving the construction of higher tsunami seawalls. Huge seawalls break the continuity of nature to land from the sea (Fig. 7). Twenty years after the Okushiri earthquake, the construction of tsunami seawalls has begun in the Pacific coast of Tohoku district without environmental assessment. The height of the tsunami seawalls will be exceeding 14 m in maximum.

References

Abe H, Sugeno Y, Chigama A (1990) Estimation of the height of the Sanriku Jogan 11 Earthquake-Tsunami (A.D. 869) in the Sendai Plain. Zishin 43:513–525 (in Japanese with English abstract)

Dawson S, Smith DE, Ruffman A, Shi S (1996) The diatom biostratigraphy of tsunami sediments: examples from recent and middle holocene events. Phys Chem Earth 21:87–92

Minoura K, Imamura F, Sugawara D, Kono Y, Iwashita T (2001) The 869 Jogan tsunami deposit and recurrence interval of large-scale tsunami on the Pacific coast of northeast Japan. J Nat Disaster Sci 23:83–88
Okoshi K (2012a) How do the 3.11 tsunami affect the intertidal animals? Jpn J Benthol 66:117–119 (in Japanese)
Okoshi K (2012b) Effect of earthquake and tsunami to the coastal marine life. In: Iwatsuki K, Domoto A (eds) A catastrophic disaster and biodiversity. Biodiversity Network, Tohoku, pp 20–25 (in Japanese)
Okoshi K (2013) Changes in the distribution and abundance of intertidal bivalves in sandy shore one and a half years after the Pacific coast of Tohoku Earthquake. J Jpn Soc Water Environ 36:44–48 (in Japanese)
Okoshi K, Suzuki M, Maruyama Y, Shinohara W (2014) Traces of the earthquake and tsunami observed in shells. Chikyu Mon 36:42–46 (in Japanese)
Satake K, Tanioka Y (1995) Tsunami generation of the 1993 Hokkaido Nansei-Oki earthquake. Pure Appl Geophys 144:803–821

Rising to the Challenge of Reconstructing the Coastal Fisheries Environment After the Massive Tsunami: The National "Tohoku Ecosystem-Associated Marine Sciences (TEAMS)" Project

Toshiki Nakano

Abstract

On 11 March 2011, the fifth most powerful earthquake in recorded history occurred off Japan in the northeastern (Tohoku) Pacific coastal area of Japan known as the Sanriku Coast. Now known as the Great East Japan Earthquake (or Tohoku-Oki Earthquake), it was a moment magnitude 9.0 undersea megathrust-type earthquake with its epicenter approximately 70 km east of the Oshika Peninsula in Miyagi Prefecture. Approximately 20,000 people were confirmed as dead or are still missing. In addition, the tsunami-related nuclear accidents at the Fukushima Daiichi nuclear power plant in Fukushima Prefecture resulted in the release of large amounts of radioactive substances into the environment. The impact of the earthquake and tsunami on the Sanriku area and the subsequent processes of transition over time are yet to be determined. Tohoku Ecosystem-Associated Marine Sciences (TEAMS) is a national project to observe marine ecosystem change after the Great East Japan Earthquake in Tohoku region. TEAMS is composed of research groups at three institutions in Japan. Tohoku University is a representative institution of TEAMS and conducts studies on the environmental change processes in the fishing environment. The Atmosphere and Ocean Research Institute (AORI) of the University of Tokyo is performing "Studies on the mechanisms of marine ecosystem change." AORI will evaluate the processes of reconstruction in the Sanriku sea area. The Japan Agency for Marine-Earth Science and Technology (JAMSTEC) is charged with "Elucidation of the mechanisms of ecosystem change on the seafloor around the coastal area." JAMSTEC also manages

T. Nakano (✉)
Marine Biochemistry Laboratory, Graduate School of Agricultural Science, Tohoku University, Aoba-ku, Sendai 981-8555, Japan
e-mail: nakanot@bios.tohoku.ac.jp

H.-J. Ceccaldi et al. (eds.), *Marine Productivity: Perturbations and Resilience of Socio-ecosystems*, DOI 10.1007/978-3-319-13878-7_7,

data for TEAMS. TEAMS has been constructing a database of research information that will be accessible to the fishing industry, local residents, and researchers. The main objective of TEAMS is to facilitate reconstruction of the coastal environment and fisheries in the Tohoku area.

1 Introduction

On 11 March 2011, the fifth most powerful earthquake in recorded history occurred off the northeastern (Tohoku) Pacific coastal area of Japan known as the Sanriku Coast. Now known as the Great East Japan Earthquake (or Tohoku-Oki Earthquake), it was a moment magnitude (Mw) 9.0 undersea megathrust-type earthquake with its epicenter approximately 70 km east of the Oshika Peninsula in Miyagi Prefecture. The focal region inferred from the distribution of aftershocks stretches approximately 400–500 km long by 200 km wide offshore (Sato et al. 2011). This earthquake caused several huge tsunamis that which reached heights of up to 40 m at Miyako in Iwate Prefecture and traveled up to 10 km inland at Sendai in Miyagi Prefecture. These tsunamis seriously damaged towns, villages, ports, and boats in the coastal areas. Approximately 20,000 people were confirmed as dead or are still missing. This earthquake and the ensuing tsunami are one of the greatest natural disasters ever experienced in Japan (Kahoku Shimpo 2011; Ozawa et al. 2011; Sagiya 2011; Suzuki 2011; Tsuji 2011).

In addition, the tsunami-related nuclear accidents at the Fukushima Daiichi nuclear power plant (Fukushima Daiichi NPP) in Fukushima Prefecture resulted in the release of large amounts of radioactive substances into the environment, including highly contaminated cooling water that was released into the ocean. An estimated 3.5–27 petabecquerels of radioactive substances were released into the Pacific (Kahoku Shimpo 2011; Yoshida and Kanda 2012; Normile 2013b; Motojima 2014). Although the radioactive level in marine epipelagic fish has been decreasing after the nuclear accident, the radioactive level in marine demersal and freshwater fish has been increasing (Yokota and Kikkawa 2013). Levels of radioactive contaminants in fish and shellfish have mostly been below the under minimum detection limits (Sakurai et al. 2012).

The northwest Pacific is one of the major fishing areas in the world, and the Sanriku Coast is recognized as one of the largest seafood-producing areas in Japan. Many communities in Tohoku coastal areas rely on marine products and fisheries industries. Accordingly, it is important to reconstruct the coastal fisheries industry in these disaster-stricken areas, such as Sanriku. However, the effects of the disaster caused by the earthquake and tsunamis on marine ecosystems are unknown, including the effects of large amounts of debris, subsidence, contamination caused by oil spills and radioactive substances release, and the loss of seaweed beds. The impact of the earthquake and tsunami on the Sanriku area and the subsequent processes of transition over time are yet to be determined, although such information is vital to effectively restore the fishing industry.

Furthermore, other recent disasters, such as the northern Sumatra (Indian Ocean) earthquake and tsunami in 2004 and super typhoon Haiyan (Yolanda) around Philippine island in 2013, were also known to cause serious damage to living environments and marine ecosystems. Accordingly, it is important to record and determine the effects of natural disasters on ecosystems and the subsequent processes of transition for restoration and optimal use of ecosystem services.

Various research projects have been underway to understand the mechanism of occurrence of this massive earthquake and to reconstruct disaster-stricken areas in Japan. Tohoku Ecosystem-Associated Marine Sciences (TEAMS) is a national project to observe marine ecosystem change after the Great East Japan

Earthquake in the Tohoku region. The project was launched in January 2012 by the Japan's Ministry of Education, Culture, Sports, Science, and Technology (MEXT) (Urushihara 2013; Hara 2013; Normile 2013a; TEAMS HP 2013). This commentary will briefly describe the aims and activities of TEAMS, particularly the project being conducted by Tohoku University.

2 General Features of TEAMS

As shown in Fig. 1, TEAMS primarily comprises research groups at three institutions in Japan: Tohoku University, the Atmosphere and Ocean Research Institute (AORI) of the University of Tokyo, and the Japan Agency for Marine-Earth Science and Technology (JAMSTEC).

Tohoku University is undertaking "Elucidation of the environmental change process in the fishing environment."

AORI is performing "Studies on the mechanisms of marine ecosystem change," which includes (1) development of advanced monitoring systems, (2) disturbances and recovering process ecosystems, (3) changes in material circulation processes, (4) influx of environmental pollutants, (5) integrative model of coastal ecosystems, (6) effects of forests and rivers on the marine environment, and (7) physical changes to coastal ecosystems (Seike et al. 2013). AORI will evaluate the processes of reconstruction in the Sanriku sea area.

JAMSTEC is charged with "Elucidation of the mechanisms of ecosystem change on the seafloor around the coastal area," which includes (1) present situations and processes of debris decomposition, (2) surveying distributions of species and analyzing population structures, (3) realized distributions and materials spread by tsunamis and turbidity currents, (4) long-term monitoring of fishery resources in the deep sea, (5) evaluating the effects of toxic substances on nutrition stages, and (6) habitat mapping for deep-sea ecosystems (Oguri et al. 2013). JAMSTEC also manages data for TEAMS.

TEAMS has been constructing a database of research information that will be accessible to the fishing industry, local residents, and researchers.

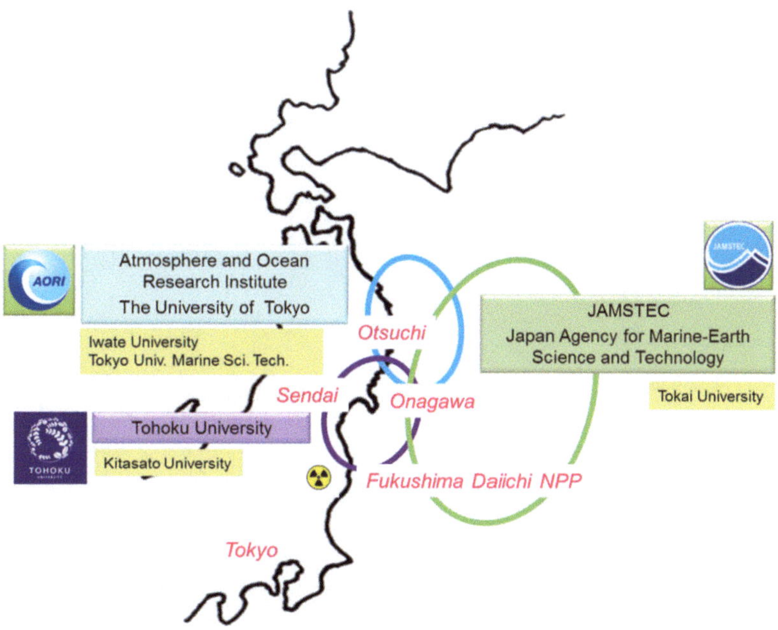

Fig. 1 The research sea areas near Sanriku Coast managed by Tohoku Ecosystem-Associated Marine Sciences (TEAMS)

3 Studies on the Elucidation of the Environmental Change Process in the Fishing Environment by Tohoku University

Tohoku University is a representative institution of TEAMS and conducts studies on the environmental change processes in the fishing environment. The main fields where Tohoku University has been observing and researching are shown in Fig. 2. The research subjects for the "Study on ecological succession of fishery grounds in Miyagi Prefecture" are the following (Urushihara 2013; Hara 2013; TEAMS HP 2013).

3.1 Environmental Monitoring in Coastal Areas

After the massive tsunami in 2011, the environments of the coastal sea areas in Miyagi Prefecture were dramatically changed. An automated rise- and fall-type or floating-type multifunction buoy has been developed to observe current directions and speeds, water temperature, salinity, turbidity, and the amounts of chlorophyll a (Fig. 3). This is a real-time monitoring system for use in the sea. Periodic water quality changes, such as water temperature, dissolved oxygen concentrations, coliform bacteria counts, nitrite-nitrogen levels, and silica, have been measured by a research vessel. These data should provide useful information for understanding the environment of the fisheries in the coastal sea areas in Miyagi Prefecture.

After the tsunamis, silt and mud sediments were found in Onagawa Bay in Miyagi Prefecture. A decrease in the number and species of invertebrates, which may have been due to the turbulence caused by the tsunami, was observed in Onagawa Bay.

3.2 Ecosystems and Genetic Research

The tsunami and subsequent subsidence changed the sea environment. The dynamics and recovery processes of seaweed bed communities, such as brown algae *arame* kelp *Eisenia bicyclis* beds, can be used as markers for assessing of the damage caused by the tsunami and for conserving the coastal ecosystem.

The recruitment of brown algae, abalone, and sea urchins was reduced due to silt and mud sedimentation and subsequent subsidence in the rocky subtidal areas in Shidugawa Bay in Miyagi Prefecture. The genetic diversity of benthic invertebrates and fish has also been monitored.

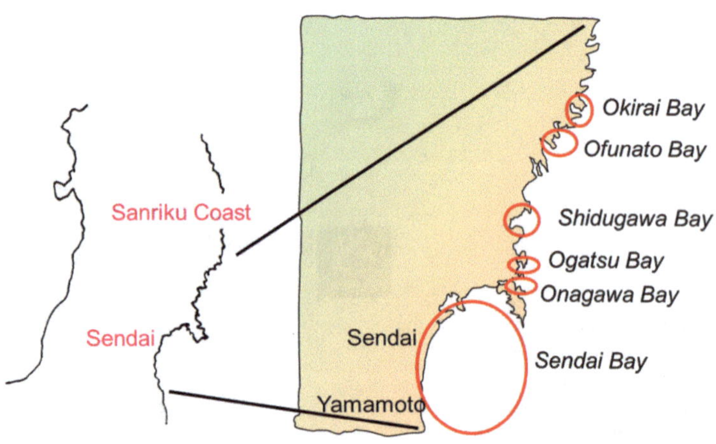

Fig. 2 The main fields where Tohoku University has been observing and researching

Fig. 3 Real-time monitoring system with multifunction buoy to observe sea conditions

3.3 Coastal Fishery Resources and Tidal Flat Fauna

The tsunami subsidence and debris have obstructed the recovery of ecosystems and fisheries activity. The fishery resources, including benthos, planktons, and fauna in the coastal shallow sea areas, tidal flats, and lower rivers in Miyagi Prefecture, have been monitored to understand the recovering processes of the coastal ecosystems.

Large amount of debris are still found in the coastal zone near Yamamoto in Miyagi Prefecture. Accordingly, it will be hard to catch bivalves, such as the surf clam *Pseudocardium sachalinensis*, from the sea bottom in these areas.

3.4 Feature of Aquacultural Environment and Innovation of Aquaculture

The massive tsunami caused extensive damage to aquacultural facilities for fish and shellfish culture. The efficient production of aquacultural species, such as oysters and scallops, has been studied for rebuilding aquacultural cultivation systems.

A single seed experiment with oysters with a floating upweller system, "FLUPSY," has been used for intermediate culture in Ogatsu Bay in Miyagi Prefecture and Yamada Bay in Iwate Prefecture. High growth and survival rates were observed in spat oyster cultures by this FLUPSY.

The changes in the features of bottom materials, such as the levels of chemicals, in Onagawa Bay in Miyagi Prefecture have been measured. The levels of chemical contaminants, such as extracts from the bottom sediment with organic solvents (n-hexane or methyl alcohol) in the entrance and the middle areas of Onagawa Bay, have decreased. However, the changing patterns of the levels of these extracts from the bottom sediment from Onagawa Bay appear to be variable. The several extracts from the bottom sediment of Onagawa Bay with organic solvents showed an inhibitory effect for cultured mammalian cell. Several chemical compounds, such as polycyclic aromatic hydrocarbons (PAHs), have been detected in the bottom sediment from Onagawa Bay. But the origins and levels of PAHs are still unknown and further studies should be needed (Yamaguchi et al. 2014).

3.5 Coastal Environment and Marine Resources in the Southern Part of Iwate Prefecture

Periodic water quality and biological feature changes have been observed in Okirai Bay and Ofunato Bay in southern Iwate Prefecture. The genetic structures of the Yezo abalone *Nordotis discus hannai*, Japanese common sea cucumber *Apostichopus japonicus*, and northern sea urchin *Strongylocentrotus nudus*, from the perspective of natural population genetics, have also been determined.

4 Conclusions and Perspectives

From a scientific standpoint, TEAMS will contribute to reconstructing the regional fisheries industries by determining revealing the mechanisms of ecosystem recovery based on long-term surveys and research. The biological, chemical, and physical data, which have been archived to build databases available to the public, are expected to provide significant information for facilitating the reconstruction of the coastal environment and fisheries in the Tohoku area and for predicting when the next disaster will occur in the world.

Further investigations (now in progress) are required to determine the effects of massive tsunamis on marine ecosystems and life in general.

Acknowledgments The author is grateful to Profs. A. Kijima and M. Hara at Tohoku Univ., Japan, for the valuable suggestion and offer of materials to prepare this paper and the presentation at the 15th CFJO. The author also wishes to thank Dr. H.-J. Ceccaldi, Prof. Emeritus, at Ecole Pratique des Hautes Etudes, France, for giving an opportunity to present our research topics during the 15th CFJO symposium. The author would like to dedicate this commentary to the late Dr. T. Nomura, Prof. Emeritus, at Tohoku Univ., Japan, who was decorated with officier of the Palmes Academiques in 2013 and introduced the author to Prof. H.-J. Ceccaldi. TEAMS project has been supported by MEXT, Japan.

References

Hara M (2013) Introduction of Tohoku ecosystem-associated marine science (TEAMS) project. In: Book of abstract, The 10th international symposium on integrated field science, Sendai, Japan, March 23, p 4

Kahoku Shimpo (2011) The Great East Japan Earthquake and Tsunami. Kahoku Shimpo Pub. Co., Sendai

Motojima I (2014) Nuclear accident in Fukushima. J Jpn Sci 49:6–12

Normile D (2013a) Taking the pulse of a ravaged ocean. Science 340:546

Normile D (2013b) The Pacific swallows Fukushima's fallout. Science 340:547

Oguri K, Kawamura K, Sakaguchi A, Toyofuku T, Kasaya T, Murayama M, Fujikura K, Glud RN, Kitazato H (2013) Hadal disturbance in the Japan Trench induced by the 2011 Tohoku-Oki earthquake. Sci Rep 3:1915. doi:10.1038/srep01915

Ozawa S, Nishimura T, Saito H, Kobayashi T, Tobita M, Imakiire T (2011) Coseismic and postseismic slip of the 2011 magnitude-9 Tohoku-Oki earthquake. Nature 475:373–376

Sagiya T (2011) The great east Japan earthquake. J Jpn Sci 46:1340–1346

Sakurai Y, Wada T, Katayama S (2012) Beyond the terrible disaster of the great east Japan earthquake. PICES Press 20:10–12

Sato M, Ishikawa T, Ujihara N, Yoshida S, Fujita M, Mochizuki M, Asada A (2011) Displacement above the hypocenter of the 2011 Tohoku-Oki earthquake. Science 332:1395

Seike K, Shirai K, Kogure Y (2013) Disturbance of shallow marine soft-bottom environments and megabenthos assemblages by a huge tsunami induced by the 2011 M 9.0 Tohoku-Oki Earthquake. PLoS ONE 8:e65417. doi:10.1371/journal.pone.0065417

Suzuki Y (2011) Unexpected problems in the great east Japan earthquake. J Jpn Sci 46:1347–1353

TEAMS project (2013) Available: http://www.agri.tohoku.ac.jp/teams/english/index.Html

Tsuji Y (2011) Tsunami and the great east Japan earthquake. J Jpn Sci 46:1354–1360

Urushihara J (2013) Development of new sustainable fishing industries in disaster stricken area, where huge tsunami occurred. Sci Window 4:20–23

Yamaguchi T, Nakano T, Yamauchi A, Kaneko K, Seki A, Sasaki M (2014) The changes in the features of bottom materials in Pacific coastal area of Miyagi Prefecture. In: Book of abstract, 3rd annual conference TEAMS, Sendai, Japan, May 17, p 6

Yokota M, Kikkawa T (2013) Radioactive substances detected in marine products. Foods Food Ingredients J 218:216–223

Yoshida N, Kanda J (2012) Tracking the Fukushima radionuclides. Science 336:1115–1116

Part III

Communications: Anthropogenic Disturbances

Distribution of Radioactive Material in Marine Ecosystems Off the Fukushima Coast: Radioactive Cesium Levels in Fukushima Marine Organisms

Hisayuki Arakawa, Tadashi Tokai, Yoshinori Miyamoto, Seiji Akiyama, Keiichi Uchida, Akira Matsumoto, Miho Narita, Hiroshi Myouse, Yukio Agatsuma, Satoshi Katayama, Masaru Aoki, Ikuo Matsumoto, and Naoto Hirakawa

Abstract

As a result of the earthquake and tsunami of March 2011, a large quantity of radioactive material was emitted into the sea from the Fukushima Daiichi Nuclear Power Plant (FNPP). Seawater, sediments, and many marine organisms with high radiation levels have been reported from coastal areas in Fukushima Prefecture. Currently, fishing in Fukushima Prefecture has been banned. This projected study will examine the following issues: (1) diffusion of radioactive material via the food chain, (2) diffusion via migration/movement of various organisms, and, as a long-term project, (3) the biological half-life of selected radionuclide cesium within the fish body. Sampling of marine organisms was carried out between November 2011 and May 2013 by scuba diving, gillnetting, and seine netting. A rocky shore and a sandy beach at Yotsukura (35 km south of FNPP; depth, 0.5–1 m), Ena (50 km south of FNPP; depth, 5–6 m), and Souma (50 km north of FNPP; depth, 3–7 m) were selected as sampling locations. Sampled species included 15 seaweeds, 25 invertebrates, and 43 fish species. Concentrations of radioactive cesium ($^{134+137}Cs$) were measured using a germanium semiconductor detector. In addition, an

H. Arakawa (✉) • T. Tokai • Y. Miyamoto
S. Akiyama • K. Uchida • A. Matsumoto
M. Narita • H. Myouse
Tokyo University of Marine Science and Technology, 5-7, Konan-4, Minato, Tokyo 108-8477, Japan
e-mail: arakawa@kaiyodai.ac.jp

Y. Agatsuma • S. Katayama • M. Aoki
Tohoku University, 1-1, Amamiya-machi, Tsutsumidori, Aoba-ku, Sendai 981-8555, Japan

I. Matsumoto • N. Hirakawa
Fukushima Prefectural Fisheries Experimental Station, 13-2, Shimokajiro, Onahama, Iwaki, Fukushima 970-0316, Japan

H.-J. Ceccaldi et al. (eds.), *Marine Productivity: Perturbations and Resilience of Socio-ecosystems*, DOI 10.1007/978-3-319-13878-7_8, © Springer International Publishing Switzerland 2015

ultrasonic pinger was used to investigate the ranging behavior of the rockfish *Sebastes cheni* in an area contaminated with high concentrations of cesium. Immediately after the accident at FNPP, very high levels of radioactive cesium were recorded in coastal marine organisms, but these decreased with time. At present, the concentrations found in demersal coastal fish are higher than in pelagic species. The ranging behavior of rockfish, contaminated at comparatively high levels, was very limited, with the fish remaining within a small territorial area and thus limiting the diffusion of radioactive cesium. It is 3 years since the Fukushima accident; however, continuous monitoring of radioactive cesium concentrations in benthic and coastal fish is still required to implement suitable management policies.

1 Introduction

On March 11, 2011, a massive earthquake and tsunami occurred in northeastern Japan. The Fukushima Daiichi Nuclear Power Plant (FNPP) lost power as a result of the tsunami, and a subsequent explosion within the building released a large quantity of radioactive material into the atmosphere and sea (Honda et al. 2012; Wakeford 2011; Akahane et al. 2012; Chino et al. 2012; Aoyama et al. 2013). According to previous reports (TEPCO 2012; Estournel et al. 2012), the total amount of radioactive ^{137}Cs released into the ocean was 3.6×10^{15} Bq.

Radioactive Cs levels exceeding the Japanese limit for human consumption (100 Bq/kg) were detected in various fish species caught in coastal Fukushima (Watabe 2012). Until September 2013, fishing in Fukushima Prefecture was banned, while the Fisheries Agency of Japan and authorities in Fukushima Prefecture measured radioactive Cs concentrations in marine organisms after the accident. These analytical data have been published by Wada et al. (2013). Concentrations of radioactive Cs in octopus species (*Enteroctopus dofleini* and *Octopus conispadiceus*) and gastropods (*Buccinum isaotakii*) decreased rapidly. Two hundred days after the accident, radioactive Cs was not detected in these organisms. Concentrations of radioactive Cs in anchovy larvae (*Engraulis japonicus*), the bivalve *Pseudocardium sachalinense*, and the seaweed Eisenia *bicyclis* decreased with time. However, concentrations of radioactive Cs in coastal demersal fish, such as *Microstomus achne*, *Sebastes cheni*, and *Okamejei kenojei*, decreased slowly, with levels in excess of 100 Bq/kg recorded in 2012.

The reasons for these differences in the rate of decrease in radioactive cesium among the sampled species may include local or large migrations, nutritional physiology, digestive processes, and mode of excretion. To clarify how radioactive material is dispersed within the marine ecosystem and how levels in organisms may vary in the future, cesium concentrations near FNPP were monitored.

This ongoing project involved the assessment and measurement of (1) diffusion via the food web, (2) diffusion via migration/ranging, and (3) a long-term study of the biological half-life of radioactive Cs within the fish body (Fig. 1). This report contains a brief outline and some preliminary key results from the study.

2 Materials and Method

This study measured the spatial and temporal changes in radioactive Cs concentrations in coastal organisms near the damaged FNPP. Sampling locations were selected at Yotsukura (35 km south of FNPP; depth, 0.5–1 m), Ena (50 km south of FNPP; depth, 5–6 m), and Souma (50 km north of FNPP; depth, 5–6 m) (Fig. 2).

Immediately after the accident, released radioactive material flowed south due to the

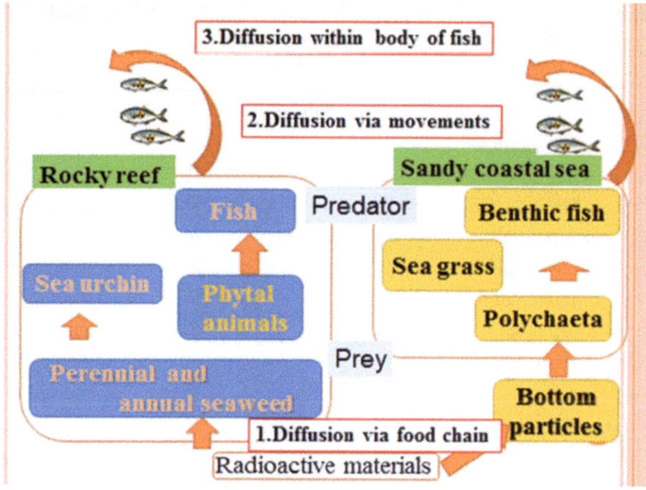

Fig. 1 Outline of the projected study

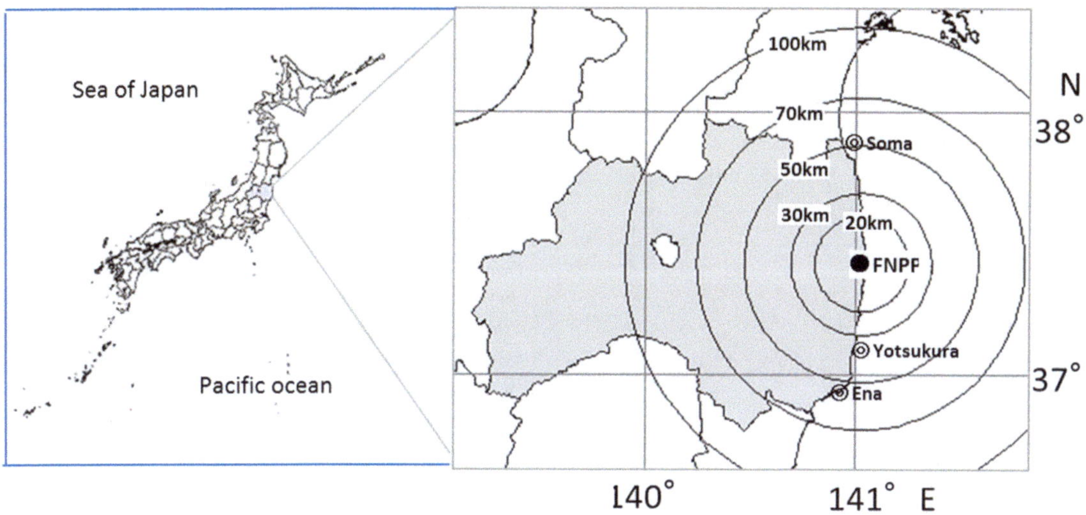

Fig. 2 Location of the sampling stations

coastal current (Honda et al. 2012); therefore, bottom sediments and fish species in the area south of FNPP (i.e., off Yotsukura) were most contaminated (Otosaka and Kobayashi 2012; Wada et al. 2013).

In general, rocky and sandy ecosystems are populated by different species; therefore, in this study, the sampling locations were divided into a rocky reef and sandy area, and Cs levels in sampled species recorded. The observation period extended from September 2012 to September 2013.

Organisms in both the reef and sand ecosystems were sampled using gillnetting and seine netting, respectively. Seaweeds and invertebrates were collected by scuba diving. Sampled species included 15 seaweeds, 25 invertebrates, and 43 fish species (Tables 1, 2, and 3).

Radioactive Cs concentrations and stable isotope ratios in each organism were measured to assess contamination and the prey–predator relationship effect. Levels of radioactive Cs were measured using a germanium semiconductor detector (GEM20-70, Seiko EG&G Co. Ltd., Tokyo, Japan). Cs concentration is shown as the sum of ^{134}Cs and ^{137}Cs.

Diffusion of radioactive Cs via the movement/migration of some organisms was also examined using a stationary-type biotelemetry system. Biotelemetry involves installing a receiver in a

Table 1 Summary of sampled fish species. Symbols P, R, and S in the Habitat column indicate pelagic substrate, rock substrate, and sand substrate, respectively

Species	Cs detection	Habitat
Engraulis japonicus		P
Seriola quinqueradiata		P
Zeus faber	*	P
Conger japonicus	*	R
Ditrema temmincki temmincki	*	R
Hexagrammos otakii	*	R
Lateolabrax japonicus	*	R
Oplegnathus punctatus		R
Pagrus major	*	R
Physiculus maximowiczi	*	R
Sebastes cheni	*	R
Sebastes inermis	*	R
Sebastes pachycephalus pachycephalus	*	R
Sebastes vulpes	*	R
Sebastiscus marmoratus	*	R
Takifugu poecilonotus (it seems that there is two Takifugu poecilonotus)	*	R
Takifugu poecilonotus	*	R
Takifugu rubripes	*	R
Takifugu snyderi	*	R
Chelidonichthys spinosus	*	S
Cynoglossus joyneri	*	S
Dasyatis akajei	*	S
Dasyatis matsubarai	*	S
Kareius bicoloratus	*	S
Lepidotrigla microptera	*	S
Microstomus achne	*	S
Mustelus manazo	*	S
Nibea mitsukurii	*	S
Occella iburia	*	S
Okamejei kenojei	*	S
Paralichthys olivaceus	*	S
Paraplagusia japonica	*	S
Pennahia argentata	*	S
Platycephalus sp. 2	*	S
Pleuronectes yokohamae	*	S
Scyliorhinus torazame	*	S
Squatina japonica	*	S
Suggrundus meerdervoortii	*	S
Thamnaconus modestus	*	S
Triakis scyllium	*	S
Upeneus japonicus	*	S

Table 2 Summary of sampled invertebrate species

Species	Cs detection	Classification
Charonia lampas sauliae	*	Mollusca
Enteroctopus dofleini		Mollusca
Euprymna morsei		Mollusca
Loliolus japonica	*	Mollusca
Neptunea arthritica	*	Mollusca
Neptunea kuroshio	*	Mollusca
Octopus ocellatus	*	Mollusca
Octopus vulgaris		Mollusca
Pleurobranchaea maculata		Mollusca
Sepia andreana	*	Mollusca
Tugali gigas		Mollusca
Crangon uritai	*	Crustacea
Metapenaeopsis dalei	*	Crustacea
Oratosquilla oratoria		Crustacea
Pagurus gracilipes	*	Crustacea
Pagurus ochotensis	*	Crustacea
Panulirus japonicus	*	Crustacea
Paradorippe granulata	*	Crustacea
Philyra syndactyla	*	Crustacea
Portunus trituberculatus	*	Crustacea
Trachysalambria curvirostris	*	Crustacea
Asterias amurensis	*	Echinodermata
Astrocladus coniferus	*	Echinodermata
Distolasterias nipon		Echinodermata
Echinocardium cordatum	*	Echinodermata
Glyptocidaris crenularis	*	Echinodermata
Luidia quinaria	*	Echinodermata
Patiria pectinifera		Echinodermata
Strongylocentrotus nudus	*	Echinodermata
Synaptidae sp.		Echinodermata
Actiniaria sp.	*	Cnidaria
Cavernularia obesa		Cnidaria
Rhizostomeae sp.		Cnidaria
Halocynthia roretzi	*	Chordata
Demospongiae sp.	*	Porifera
Polychaeta spp.	*	Annelida

sea area, receiving ultrasonic signals from a transmitter implanted in a fish, and monitoring its location and movement from the signal.

The selected fish species, rockfish (*Sebastes cheni*; $n=4$; body length, 22.3–26.0 cm) and greenling (*Hexagrammos otakii*; $n=2$; body length, 27.0–32.5 cm), were monitored in an area where high levels of radioactive Cs were detected. One system was set up in a rocky area off Ena. Four rockfish and two greenlings were caught in this area and an ultrasonic transmitter (pinger) implanted in each fish, which were then released in the same place (Fig. 3). The observation period lasted from November 2012 to May 2013.

Table 3 Summary of sampled seaweed and sea grass species

Species	Cs detection	Classification
Phyllospadix iwatensis	*	Sea grass
Eisenia bicyclis	*	Phaeophyceae
Myelophycus simplex		Phaeophyceae
Saccharina japonica	*	Phaeophyceae
Sargassum horneri	*	Phaeophyceae
Sargassum fulvellum	*	Phaeophyceae
Sargassum micracanthum	*	Phaeophyceae
Sargassum miyabei		Phaeophyceae
Sargassum siliquastrum	*	Phaeophyceae
Undaria pinnatifida	*	Phaeophyceae
Gelidium elegans	*	Rhodophyceae
Gloiopeltis furcata	*	Rhodophyceae
Grateloupia turuturu		Rhodophyceae
Marginisporum aberrans	*	Rhodophyceae
Pachymeniopsis elliptica	*	Rhodophyceae

3 Results and Discussion

The duration of the planned study period is 2012–2014 and this paper reports the results to May 2013.

3.1 Diffusion via the Food Web

Radioactive Cs concentrations in fish caught during 2012 and 2013 are reported. Radioactive Cs was detected in all fish, except pelagic or surface-dwelling species. Average radioactive Cs values of fish from the sandy substrate were in the range of 92–212 Bq/kg (wet weight, ww) in Yotsukura and 58–109 Bq/kg ww in Ena, while average values from the rocky reef were in the range of

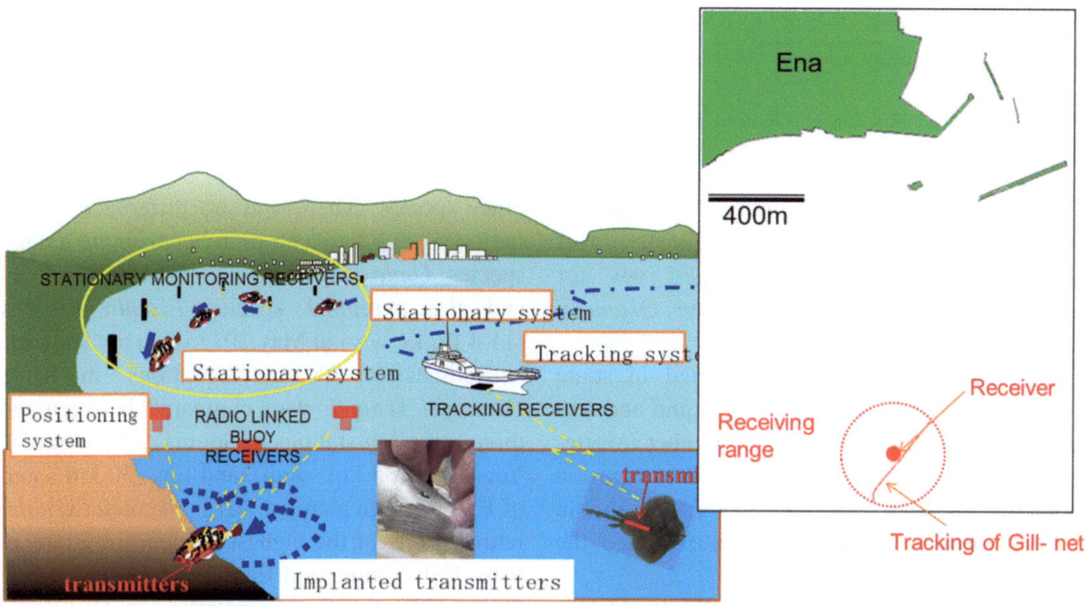

Fig. 3 Biotelemetry system off Ena

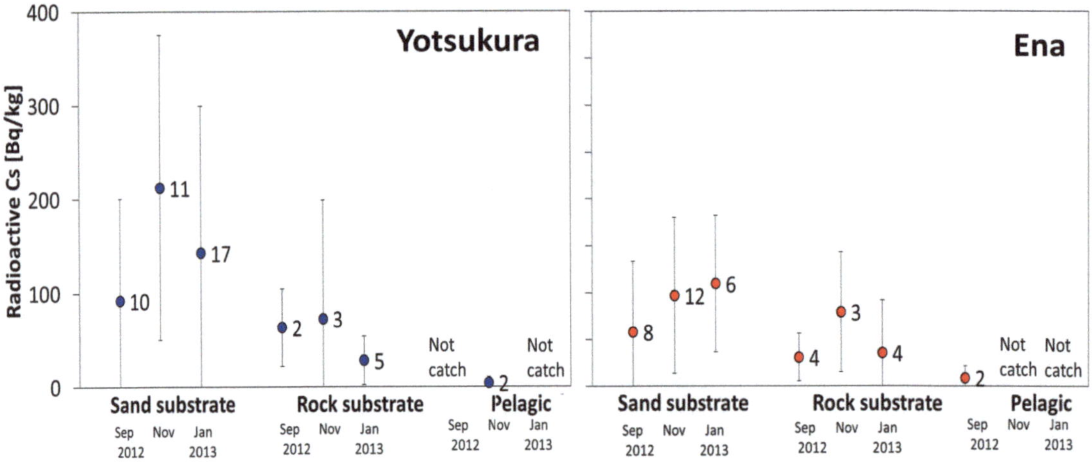

Fig. 4 Radioactive Cs concentration in fish species at the Yotsukura and Ena sampling stations off the Fukushima coast. Numerals in the figures indicate sample species numbers

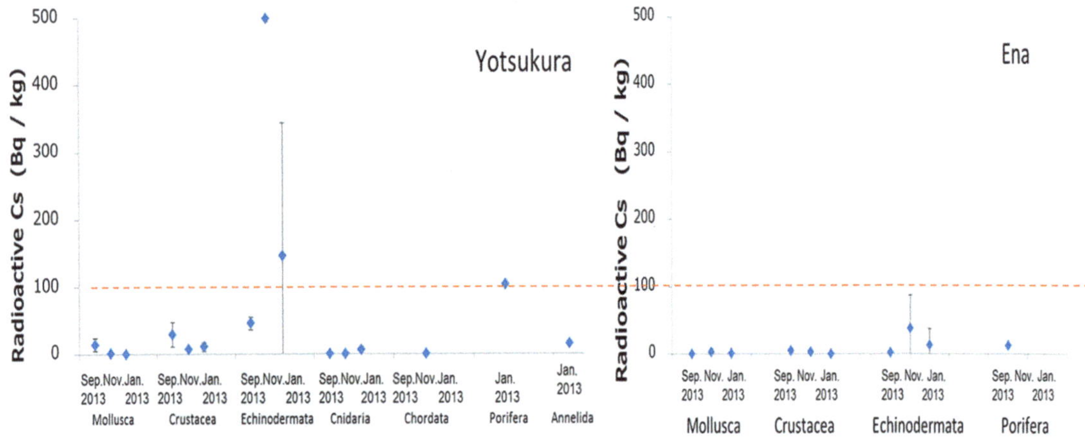

Fig. 5 Radioactive Cs concentration in invertebrates at the Yotsukura and Ena sampling stations off the Fukushima coast

28–73 Bq/kg ww in Yotsukura and 30–79 Bq/kg ww in Ena. Levels were frequently over the Japanese limit of 100 Bq/kg ww for rockfish *Sebastes cheni* and skate *Okamejei kenojei* (Fig. 4).

Radioactive Cs was not detected in some invertebrates such as octopus and squid and was generally detected at low levels in other invertebrates, with the exception of two species, a sponge *Porifera* sp. and a sea urchin *Echinocardium cordatum*, which exceeded the limit of 100 Bq/kg ww (Fig. 5).

Radioactive Cs levels in each species of seaweed and sea grass decreased with time. Concentrations were higher in red algae than in other species. There was no difference in the radioactive Cs concentration between the annual species *Undaria pinnatifida* (10.4 Bq/kg ww) and the perennial species *Eisenia bicyclis* (11.4 Bq/kg ww) in May 2012 (data not shown).

Radioactive Cs concentrations in skate *Okamejei kenojei* were compared between the three sampling stations of Souma (50 km north), Yotsukura (35 km south), and Ena (50 km south of FNPP) on September 2012. *Okamejei kenojei* returned one of the highest Cs concentrations in fish. With Souma and Ena at the same distance from FNPP, the Cs concentration in *O. kenojei* from the south (108.6 ± 14.5 Bq/kg ww) was higher than in *O. kenojei* from the north

(37.9 ± 11.5 Bq/kg ww). In a southerly direction, a concentration of 426.6 ± 163.5 Bq/kg ww recorded at the Yotsukura sampling station and closer to FNPP was higher than at the Ena station, 15 km further south. The distribution of radioactive Cs levels in fish was similar to the Cs levels in sediments (Otosaka and Kobayashi 2012). It is believed that this distribution is because of the influence of littoral southern currents at the accident site.

The prey–predator relationship effect was examined using a carbon and nitrogen stable isotope ratio analysis at the sandy sampling station at Souma. The carbon/nitrogen (CN) isotope ratio of fish was in the range of −18 to −15‰ for ^{13}C and 12–16‰ for ^{15}N. Generally, CN isotope ratios are in the range of −32 to −20‰ for particulate terrestrial organic matter, −24 to −18‰ for phytoplankton, −20 to −10‰ for benthic diatom, and −27 to −8‰ for seaweed (Fry and Sherr 1984; Ogawa and Ogura 1997). Thus, it was concluded that the food of fish in the sandy sampling area of Souma consisted of benthic diatoms living on the surface seabed.

This result indicates that the main source of primary production for fish in this sea area is benthic microalgae.

3.2 Diffusion via Movement/Migration of Organisms

The number of signals received from the transmitter implanted in fish decreased with decreasing water temperature. However, *S. cheni* were continuously recorded using the ultrasonic pinger within the same area during the fall, winter, and spring. Based on previous reports (Nemoto and Ishida 2006), it was concluded that migration from shallow to deep water occurred with seasonal changes in water temperature. Nevertheless, it was confirmed that *S. cheni* did not migrate but remained in the same area and that rockfish were contaminated by the radioactive Cs within this littoral area.

4 Conclusions

Regarding diffusion via the food web, radioactive Cs concentrations in all species decreased with time. However, many coastal demersal species had elevated radioactive Cs concentrations, especially fish. The food source of many species in this area originates from organic matter, principally microalgae. The study of diffusion via movement/migration showed that rockfish, which had high Cs levels, tended to remain within a small territorial area.

A study of diffusion within the body of fish is now being undertaken. It is necessary to observe radioactive Cs concentrations in marine organisms over a prolonged period of time to clarify the distribution and variation in Cs levels, including the biological and ecological half-life in selected marine organisms.

The Fukushima coast was decimated by the tsunami and subsequent nuclear power plant accident. The tsunami washed massive quantities of sand and rubble into the sea. According to the Fukushima Prefectural Fisheries Experimental Station, the sea urchin population was reduced significantly (unpublished data); however, the influence on other organisms has not been clarified.

Numerous coastal species were contaminated by radioactive material, including Cs, from the accident at FNPP (Wada et al. 2013). However, it is believed that Cs contamination had almost no influence on ecology or physiology function.

Currently, fishing for 41 commercial species is banned in waters near the Fukushima site; therefore, there is no fishing pressure and, in effect, it is virtually a marine protected area. Iwasaki et al. (2013) has predicted via a numerical study that flatfish resources will increase, aided by the fishing ban, though it will be several years before marine productivity of the Fukushima coast returns to normal. Thus, for future marine resource management strategies, the effects of the “fishing ban” on littoral areas need to be examined in detail.

Acknowledgments This research was supported by the Environment Research and Technology Development Fund (4ZD-1201) of the Ministry of the Environment, Japan. The authors appreciate the assistance from the Aquamarine Fukushima aquarium and the Fukushima Prefectural Fishermen's Cooperative Association.

References

Akahane K, Yonai S, Miyahara S, Yasuda H, Iwaoka K, Matsumoto M, Fukumura A, Akashi M (2012) The Fukushima Nuclear Power Plant accident and exposures in the environment. Environmentalist 32:136–143

Aoyama M, Tsumune D, Hamajima Y (2013) Distribution of 137Cs and 134Cs in the North Pacific Ocean: impacts of the TEPCO Fukushima-Daiichi NPP accident. J Radioanal Nucl Chem 296:535–539

Chino M, Terada H, Katata G, Nagai H, Nakayama H, Yamazawa H, Hirao S, Ohara T, Takigawa M, Hayami H, Aoyama M (2012) Resulting from the Fukushima Daiichi Nuclear Power Plant accident. Natl Inst Radiol Sci 252:127–135

Estournel C, Ulses C, Marsaleix P, Duhaut T, Michaud H, Auclair F, Bosc E, Osvath I, Bocquet M, Winiarek V, Nguyen C, Lyard F (2012) Assessment of the amount of cesium-137 released into the Pacific Ocean after the Fukushima accident and analysis of its dispersion in Japanese coastal waters. J Geophys Res 117:c11014-c11014-13

Fry B, Sherr EB (1984) δ13C measurements as indicators of carbon flow in marine and freshwater ecosystems. Contrib Mar Sci 27:13–47

Honda MC, Aono T, Aoyama M, Hamajima Y, Kitamura M, Masumoto Y, Miyazawa Y, Takigawa M, Saino T (2012) Dispersion of artificial caesium- 134 and -137 in the western North Pacific one month after the Fukushima accident. Geochem J 46:e1–e9

Iwasaki T, Hirakawa N, Sohtome T (2013) Evaluation of the voluntary restraint of Fukushima's coastal fishery on four flatfish resources. Bull Fukushima Pref Fish Exp Stat 16:31–43

Nemoto Y, Ishida T (2006) Ecology and stock status of black rockfish, *Sebastes inermis*, along the coast of Fukushima. Bull Fukushima Pref Fish Exp Stat 13:63–76

Ogawa N, Ogura N (1997) Dynamics of particulate organic matter in the Tamagawa Estuary and Inner Tokyo Bay. Estuar Coast Shelf Sci 44:263–273

Otosaka S, Kobayashi T (2012) Sedimentation and remobilization of radiocesium in the coastal area of Ibaraki, 70 km south of the Fukushima Dai-ichi Nuclear Power Plant. Environ Monit Assess. doi:10.1007/s10661-012-2956-7

TEPCO (Tokyo Electric Power Co.) (2012) TEPCO News Press Releases. http://www.tepco.co.jp/cc/press/2012/1204619_1834.html. 27 Oct 2013

Wada T, Nemoto Y, Shimamura S, Fujita T, Mizuno T, Sohtome T, Kamiyama K, Morita T, Igarashi S (2013) Effects of the nuclear disaster on marine products in Fukushima. J Environ Radioact 124:246–254

Wakeford R (2011) And now, Fukushima. J Radiol Prot 31:167–176

Watabe T (2012) Potential radiation effects on marine organisms due to the accident in Fukushima Dai-ichi Nuclear Power Plant. Mar Organ 34: 231–243

Fate of Antibiotics and Antibiotic-Resistant Fecal Bacteria in Water and Sediments from the Contamination Source to the Estuary: Impact and/or Resilience? Resilience to Contamination by Antibiotics

Fabienne Petit, Erick Denamur, Olivier Clermont, Roland Leclercq, Julien Deloffre, Vincent Cattoir, Kenny Oberlé, Hélène Budzinski, and Thierry Berthe

Abstract

The aim of this study based on a multidisciplinary research program (FLASH) (FLASH (FLuxes of Antibiotic and Antibiotic-resistant bacteria and the corresponding genes in Surface Hydrosystems) a research project founded by CNRS EC2CO/GIP Seine-Aval), associating chemists, hydrologists, and clinical and environmental microbiologists, was to monitor the impact of antibiotic prescription in human and veterinary practices on water and sediment contamination by antibiotic and *Escherichia coli* (antibiotic resistance, integrons) and *Enterococci* (diversity, antibiotic resistance, and the corresponding genes) along a medical center–WWTP–river continuum. For this purpose, a multi-residue chemical methodology was developed in order to detect low levels of 34 antibiotics. In the medical center, the main prescribed antibiotic (amoxicillin) was weakly found in effluents. Along the continuum, contamination of water by antibiotics decreased from 160 $\mu g.L^{-1}$ (cefotaxime) in hospital effluents to 1 $ng.L^{-1}$ (ofloxacin) in the river. These concentrations were too low to exert a selective pressure ($mg.L^{-1}$) on antibiotic-resistant bacteria.

F. Petit (✉) • J. Deloffre • K. Oberlé • T. Berthe
UMR CNRS M2C, SFR SCALE,
Normandie Université, Rouen, France
e-mail: fabienne.petit@univ-rouen.fr

E. Denamur • O. Clermont
INSERM U1137 IAME, Université Paris Diderot,
Sorbonne Paris Cité, France

R. Leclercq • V. Cattoir
EA4655, CHU de Caen, Université de Caen Basse,
Normandie Caen, France

H. Budzinski
UMR CNRS 5805 EPOC-LPTC,
Université de Bordeaux, Talence, France

H.-J. Ceccaldi et al. (eds.), *Marine Productivity: Perturbations and Resilience of Socio-ecosystems*,
DOI 10.1007/978-3-319-13878-7_9,

In the same samples, occurrence of antibiotic-resistant *E. coli* and those harboring a class 1 integrons was significantly (p-value < 0.001) decreased along the continuum. Among *Enterococcus* populations, *E. faecium* was mainly isolated (from 89 to 98 %). All *E. faecium* isolates from medical center effluents were multiple antibiotic resistant, containing *erm*(B) and *mef*(A) genes, and belonged to the hospital-adapted clonal complex 17 (CC17). The relative proportion of CC17 decreased in favor of other subpopulations, less resistant to antibiotics. In water, only persistent compounds were found (quinolones, macrolides, sulfonamides), but they did not correspond to the major resistances in *E. coli* and *Enterococcus* (penicillins, tetracyclines).

1 Introduction

1.1 Why Does the Ecosystem Matter for Human Health in the Emergence of Antibiotic Resistance?

One of the major challenges of the next decades will be the assessment of how ecosystem changes could affect human health. Thus, the assessment of the microbiological vulnerability and the resilience of the aquatic environment to contamination by fecal germs, which in industrialized countries, is accompanied by pharmaceutical contamination, will be a major health concern (Millennium Ecosystem Assessment, http://milleniumassessment.org).

Among pharmaceuticals, antibiotics have a unique character: with emerging contaminants of aquatic environments, their intensive use in human and animal medicine is also responsible for the increase of resistance to antibiotics in bacteria (Aarestrup 2005; Seveno et al. 2002). To date, the concentrations of antibiotics in water reported in the bibliography are dependent on the analytical methods and methods of sampling. However, the maximal values (in the order of a hundred μg L^{-1}) are observed in effluent from hospitals or from pharmaceutical factories. In surface water, lower concentrations (in the order of ten μg L^{-1}) are detected immediately downstream from the discharge of wastewater treatment plants (WWTPs), mainly due to the dilution in the receiving environment and the degradation of the antibiotic compounds (Kummerer 2009; Tamtam et al. 2008).

The emergence of bacterial resistance has been recognized as a major problem in public health by the World Health Organization (WHO 2001). Since the 1950s and the beginning of the large-scale use of antibiotics, pathogenic bacteria, initially sensitive to antibiotics, have very rapidly acquired resistance mechanisms, some of which were acquired from environmental bacteria. As an example, the pandemic of the extended-spectrum beta-lactamases (ESBLs), whose therapeutic consequences are dramatic, is consecutive to a genetic transfer from environmental bacteria (e.g., CTX-M from *Kluyvera* spp., Poirel et al. 2002).

1.2 Bacterial Antibiotic Resistance: A Global Ecological Process

Bacterial resistance to antibiotics is a complex ecological phenomenon which should be understood, by considering the circulation of microorganisms and the corresponding resistance genes within the four major ecosystems: humans, animals, soil, and water (Martinez 2008, 2009; Nwosu 2001; Wellington et al. 2013; Fig. 1).

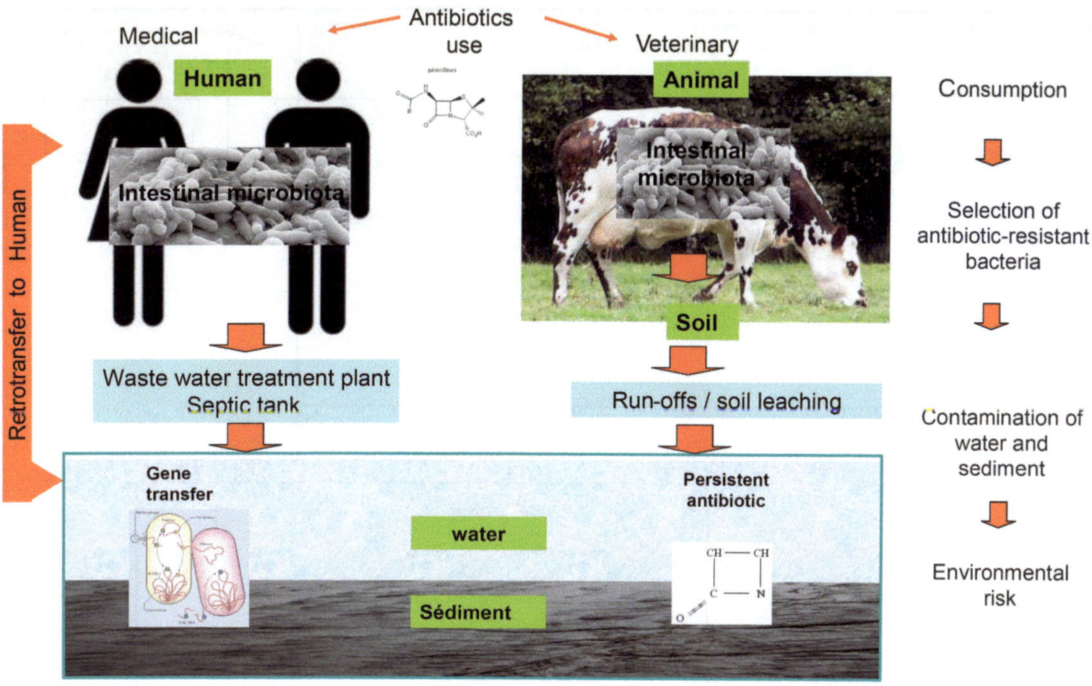

Fig. 1 Circulation of antibiotic-resistant microorganisms and antibiotics within the four main ecosystems: Humans and animals under antibiotic treatment, soil, and water

Prescribed in human and veterinary medicine depending on the legislation in force[1], antibiotics or their metabolites are discharged into the environment via urine and feces. The emergence of bacteria resistant to antibiotics is the consequence of the selective pressure exerted on the intestinal microbiota of humans and animals treated with antibiotics. Consequently, in parallel with contamination by antibiotics, fecal antibiotic-resistant bacteria, mainly *Escherichia coli* (E. coli) and *Enterococcus* spp., are released into the water environment (Goni-Urriza et al. 2000; Koczura et al. 2011; Laroche et al. 2009; Leclercq et al. 2013; Oberlé et al. 2012; Watkinson et al. 2007). The most important sources of contamination are the treated effluent from WWTP, the direct discharges in the countries which do not treat their wastewaters, the soil leaching, and the runoffs, to which can be added the uncontrolled waste from pharmaceutical factories and from fish farming (Jiménez et al. 2012; Watkinson et al. 2009).

[1] The prescription of antibiotics as an additive in animal food is forbidden by EU regulations (rule n° 1831/2003).

1.3 The Seine Estuary: One of the Most Anthropized Estuaries in Europe

Chemical and microbiological contamination of coastal areas and estuaries reflects the anthropic pressure exerted on their catchments. On the northwest European continental shelf, the Seine estuary is the outlet of a drainage basin of 79,000 km² where 30 % of the French population, 40 % of the economic activity, and 30 % of the national agricultural activity are concentrated. The microbial quality of the Seine estuary waters is poor and is mainly dominated in high-flow

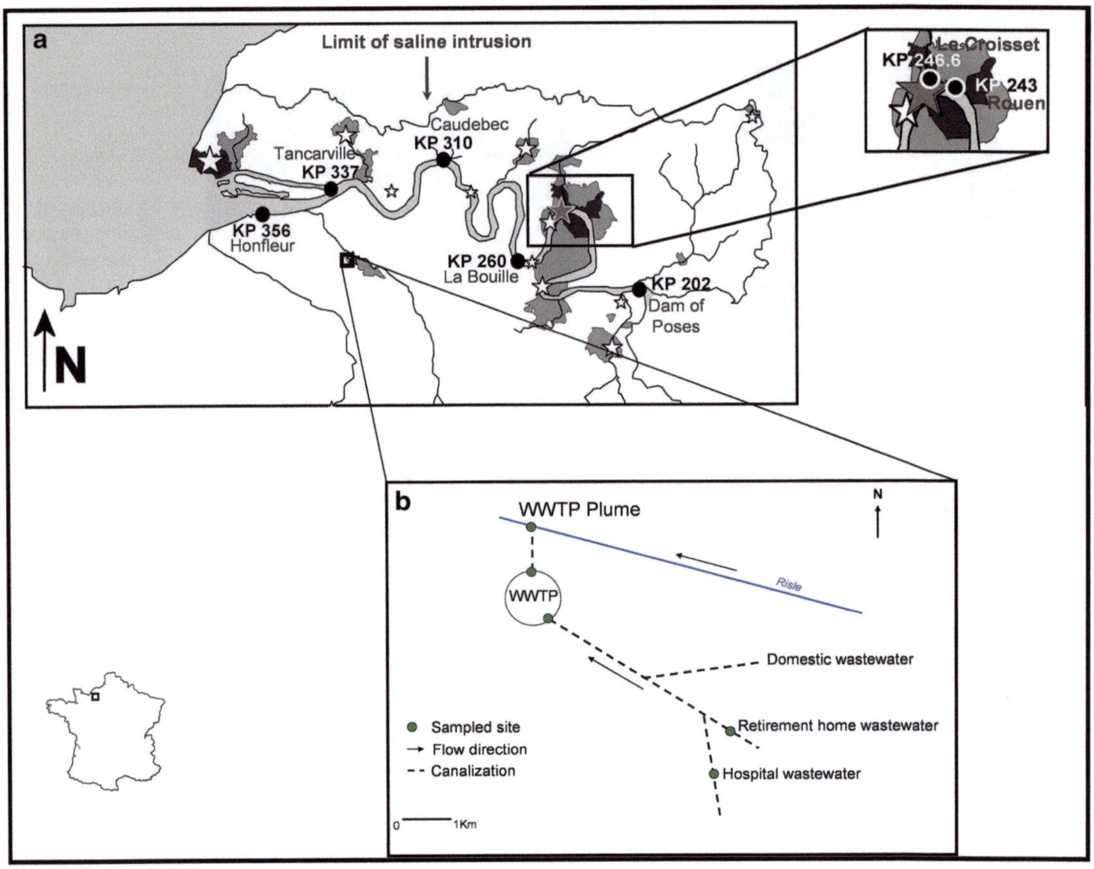

Fig. 2 Map of the Seine estuary, (**a**) one of the most anthropized estuary from Europe with a population density stated in inhabitants.km^2 (*black*, 3,000–6,000 inh. km^2; *gray*, 500–3,000 inh.km^2; *white*, <500 inh.km^2; INSEE dataset), mainly contaminated by WWTP (*gray star*) and the upstream input from Poses; Kp, kilometric point (kp 0 at Pont Marie in Paris); and of the studied zone (**b**) where the medical center (hospital of 87 beds and retirement home of 180 beds)–WWTP–river continuum (4 km) is located, for which the receiving river corresponds to the tributary (Risle) which flows in the mouth of the Seine estuary

periods by the upstream inputs, the WWTP which treats the wastewater from Paris and its suburbs (about five million inhabitants) being located 120 km upstream the Dam of Poses (Fig. 2). However, in a period of low flow, intra-estuary supplies are predominant with a contamination from tributaries close to 76 % (Garcia-Armisen et al. 2005). In the mouth of the estuary, where the microbiological quality of the water influences that of the mussels from the coastal zone, there is a permanent contamination by fecal bacteria mainly due to the input of the Risle tributary (Touron et al. 2007).

The goal of this study was to have a better understanding of the fate of antibiotics, antibiotic-resistant bacteria, and the corresponding genes, from the human reservoirs to the estuary. However, the multiplicity of intra-estuary supplies combined with the hydrosedimentary dynamic makes the studies difficult to interpret in this aquatic environment. So, to better analyze the data observed in the seine estuary, in particular to analyze more finely the relationship that exists between the consumption of antibiotics and the contamination of water, a small-scale survey has been conducted on a medical center–WWTP–river continuum, whose receiving water is a tributary of the Seine estuary (Risle) which flows in the mouth of the Seine estuary. Thus, the aim of the multidisciplinary research program (FLASH) – associating chemists, hydrologists, and clinical and environmental microbiologists – was to determine to what extent the hospital effluent has an ecological impact on the downstream aquatic environment. For this purpose, the fate of *Escherichia coli* (antibiotic resistance, gene content integrons; 342 strains) and *Enterococci* (antibiotic resistance, clonal complex 17, gene content *erm*B, *mef*A; 235 strains) was analyzed in water along a medical center–WWTP–river–estuary continuum, during a high epidemiologic period. A multiresidue chemical methodology was developed in order to detect low levels of 34 antibiotics in water. To link the occurrence of antibiotic-resistant bacteria in water and antibiotic prescription, we use the data collection from the hospital and the antibiotic sales information.

2 Results

2.1 The Fate of Antibiotics: From the Source of Contamination to the Estuary

2.1.1 Antibiotic Contamination of the Seine Estuary

The concentrations of antibiotics observed in the Seine estuary well illustrate this quantitative and qualitative decrease from the sources to the receiving environment (Tables 1 and 2). Thus, the treated effluents from the WTTP and a tributary of the Seine estuary (Robec) – impacted by effluents from a hospital – are the two major sources of intra-estuary contamination by antibiotics, with about twenty different molecules and a maximal concentration of 482 ng.L^{-1} observed for clarithromycin (macrolide family) (Table 1). Antibiotics released by WWTP effluent are rapidly diluted in the estuary, by at least a factor of 10 depending on the molecules, as shown in concentrations measured in the dense urban area (Rouen), especially on the site closest to the WWTP discharge (Table 2). Only nine antibiotics were detected, at concentrations ranging in ng.L^{-1}. With the exception of sulfamethoxazole (sulfonamide family), this decrease continues for all the antibiotics along the estuary especially in Caudebec (limit of the salinity) and then in the mouth of the estuary. The antibiotics predominantly found are those that are most stable in the aquatic environment: fluoroquinolones, sulfonamides, and macrolides.

2.1.2 Relationship Between Antibiotic Use and Antibiotic Contamination Along a Medical Center–WWTP–River Continuum

However, the multiplicity of intra-estuary supplies combined with the hydrosedimentary dynamic does not allow a clear understanding of the fate of antibiotics in this aquatic environment. A finer scale analysis of the relationship between antibiotic use and the contamination of water by antibiotics has been therefore carried out along a medical center (hospital of 87 beds and retirement home of 180 beds)–WWTP–river continuum (4 km). The receiving river (Risle) was the last tributary (Risle) which flows in the mouth of the Seine estuary (Fig. 2).

Whereas penicillins (amoxicillin) are the most prescribed antibiotics in the hospital and in the town's pharmacies, the antibiotics mainly found in the effluent from the hospital are the molecules that are most persistent in water: the quinolones/fluoroquinolones (ofloxacin, 68 μg L^{-1}; pipemidic acid, 59 μg L^{-1}) and the macrolides, to which can be added cephalosporins (160 μg L^{-1}) (Tables 3 and 4). All the antibiotics prescribed at the hospital and in the town's pharmacies are found in the raw effluent of the WWTP, but the concentrations observed are low (in the order of ng.L^{-1}) and decrease further in the receiving water where only the quinolones, the sulfonamides, and the macrolides are detected at concentrations in the order of ng.L^{-1}. These results highlight that antibiotic contamination of waters is mainly consecutive to the medical prescriptions (type of molecule and epidemic period), but also to the stability of antibiotics and/or their metabolites once released in the water.

Table 1 Major sources of contamination of the Seine estuary by antibiotics: the tributary (Robec) contaminated by effluent of a hospital (1,250-bed capacity) which flows at Kp 242 (urban zone of Rouen) and the two most important WWTPs located in the urban zone, Rouen (Kp 247; treatment capacity, 550,000 inhabitants) and Elbeuf (Kp 220; treatment capacity, 100,000 inhabitants) (Garcia-Armisen et al. 2005; Touron et al. 2007)

Antibiotics ng.L^{-1}	Robec tributary	WWTP Elbeuf	WWTP Rouen
β-lactamines			
Amoxicillin	235		41
Ampicillin	–	–	–
Penicillin G	–	–	–
Penicillin V	–	–	–
Oxacillin	12	–	–
Cloxacillin	40	–	–
Dicloxacillin	–	–	–
Cephalosporins			
Cephalexin	–	–	–
Cefotaxime	368	31	12
Cefpodoxime	–	–	–
Ceftiofur	–	–	–
Cefuroxime	–	–	–
Quinolones/fluoroquinolones			
Ciprofloxacin	114	130	265
Enrofloxacin	–	–	–
Marbofloxacin	–	–	–
Norfloxacin	–	22	165
Ofloxacin	62	77	86
Pipemidic acid	–	–	148
Oxolinic acid	–	–	–
Flumequine	–	161	138
Tetracyclines			
Tetracycline	–	–	9
Oxytetracycline	–	–	–
Chlortetracycline	–	–	–
Doxycycline	–	–	17
Sulfonamides			
Sulfadiazine	–	342	2
Sulfadimethoxine	–	129	0
Sulfamerazine	–	–	–
Sulfamethazine	–	105	–
Sulfamethizole	–	10	4
Sulfamethoxazole	5	276	200
Sulfanilamide	–	137	67
Sulfapyridine	–	235	396
Sulfathiazole	–	–	–
Macrolides			
Azithromycin		208	248
Clarithromycin	3	89	482
Erythromycin	3	126	295
Josamycin	–	11	21
Roxithromycin	–	73	239
Spiramycin	–	39	50

A chemical analysis methodology, based on solid phase extraction coupled with liquid chromatography–tandem mass spectrometry, was developed in order to detect 34 molecules of different antibiotics (Oberlé et al. 2012)

Table 2 Antibiotic concentration in water along the Seine estuary (July 2011; see Fig. 1)

Antibiotic ng.L^{-1}	Poses	Rouen	Le croisset	Les docks	La bouille	Caudebec	Tancarville	Berville	Honfleur
β-lactamase									
Amoxicillin	–	–	–	–	–	–	–	–	–
Ampicillin	–	–	–	–	–	–	–	–	–
Penicillin G	–	–	–	–	–	–	–	–	–
Penicillin V	–	–	–	–	–	–	–	–	–
Oxacillin	–	–	–	–	–	–	–	–	–
Cloxacillin	–	–	–	–	–	–	–	–	–
Dicloxacillin	–	–	–	–	–	–	–	–	–
Cephalosporins									
Cephalexin	–	–	–	–	–	–	–	–	–
Cefotaxime	–	–	–	–	–	–	–	–	–
Cefpodoxime	–	–	–	–	–	–	–	–	–
Ceftiofur	–	–	–	–	–	–	–	–	–
Cefuroxime	–	–	–	–	–	–	–	–	–
Quinolones/fluoroquinolones									
Ciprofloxacin	–	–	–	–	–	–	–	–	–
Enrofloxacin	–	–	–	–	–	–	–	–	–
Marbofloxacin	–	–	–	–	–	–	–	–	–
Norfloxacin	–	–	–	–	–	–	–	–	–
Ofloxacin	3	2	2	2	1		1	1	1
Pipemidic acid	–	–	–	–	–	–	–	–	–
Oxolinic acid	–	–	–	–	–	–	–	–	–
Flumequine	9	9	9	8	13	7	3	6	
Tetracyclines									
Tetracycline	–	–	–	–	–	–	–	–	–
Oxytetracycline	–	–	–	–	–	–	–	–	–
Chlortetracycline	–	–	–	–	–	–	–	–	–
Doxycycline	–	–	–	–	–	–	–	–	–
Sulfonamides									
Sulfadiazine	–	–	–	–	–	–	–	–	–
Sulfadimethoxine	1	0	1	0	0		0		
Sulfamerazine	–	–	–	–	–	–	–	–	–
Sulfamethazine	–	–	–	–	–	–	–	–	–
Sulfamethizole	–	–	–	–	–	–	–	–	–
Sulfamethoxazole	24	5	24	16	21	51	22	56	12
Sulfanilamide	–	–	–	–	–	–	–	–	–
Sulfapyridine	12		6	5	5	16	7	14	2
Sulfathiazole	–	–	–	–	–	–	–	–	–
Macrolides									
Azithromycin	–	–	–	–	–	–	–	–	–
Clarithromycin	8	9	11	7	11	1	0	1	2
Erythromycin	20	19	24	18	20	9	2	4	3
Josamycin	–	–	–	–	–	–	–	–	–
Roxithromycin	2	3	4	2	3	1		0	
Spiramycin	5	2	3	2	2	1	0	1	1

Table 3 Water contamination by antibiotics and fecal antibiotic-resistant bacteria, *Escherichia coli* and *Enterococcus*, along the continuum medical center–WWTP (collecting the wastewaters from the medical center and from a community of 9,058 inhabitants)–river (Risle), during the winter period and a high epidemic period (December 2009)

	Medical center														
	Hospital			Retirement home			WWTP influent			WWTP effluent			River (Risle)		
Antibiotic	Concentration (µg.L^{-1})	Resistant bacteria (%)		Concentration (µg.L^{-1})	Resistant bacteria (%)		Concentration (µg.L^{-1})	Resistant bacteria (%)		Concentration (µg.L^{-1})	Resistant bacteria (%)		Concentration (µg.L^{-1})	Resistant bacteria (%)	
Ampicillin	0.0	10	100[b]	0.7	38[a]	87.5[b]	0.0	6.0[a]	19.1[a]	0.0	10.0[a]	19.0[b]	0.0	16[a]	4.0[b]
Other β-lactams	161	–	–	0.0	–	–	0.0	–	–	0.0	–	–	0.0	–	–
Ciprofloxacin	73	8	75[b]	0.0	42[a]	100.0[b]	0.0	16[a]	15.6[b]	0.04	4.0[a]	17.0[b]	0.0	4.0[a]	5.0[b]
Other quinolones	68	–	–	59	–	–	0.4	–	–	0.1	–	–	0.001	–	–
Tetracycline	0.0	16[a]	0.0[b]	0.0	38[a]	21.0[b]	0.08	34[a]	28.0[b]	0.01	20[a]	26.0[b]	0.0	18[a]	24.0[b]
Other tetracyclines	1.0	–	–	0.0	–	–	0.07	–	–	0.0	–	–	0.0	–	–
Erythromycin	0.4	–	72[b]	0.0	–	87.0[b]	0.4	–	77.0[b]	0.05	–	80.0[b]	0.002	–	72.0[b]
Other macrolides	7.5	–	–	21.9	–	–	3.1	–	–	1.6	–	–	0.005	–	–
Cephalosporin	160	2[a]	–	0.0	6[a]	–	0.62	6.0[a]	–	0.04	2.0[a]	–	0.000	2.0[a]	–
Sulfonamides	0.0	2[a]	–	0.006	24[a]	–	0.924	16[a]	–	0.715	8.0[a]	–	0.005	6.0[a]	–

The medical center is composed of a hospital (87 beds, residence time from 4 to 28 days) and a retirement home (180 beds, residence time of an average of 10 years) (Leclercq et al. 2013; Oberlé et al. 2012)

[a]*E. coli*

[b]*Enterococcus*

– Resistance not tested

Table 4 Antibiotic consumption

Antibiotic (g)	Hospital[a]	Retirement home[a]	Community[b]
Glycopeptide			
Vancomycin	0	0	NA
β-lactams			
Amoxicillin	1,930	544	258,430
Cloxacillin	0	16	736
Ampicillin	0	0	0
Tetracyclines			
Tetracycline	0	0	0
Doxycycline	1.1	0	1,562
Fluoroquinolones			
Ciprofloxacin	115	0	980
Macrolides			
Erythromycin	75	20	NA

NA not available (From Leclercq et al. 2013)
[a]Antibiotic consumption 1 month before sampling
[b]Antibiotic consumption 6 months before sampling based on the sales of antibiotics for all the pharmacies in the studied zone

2.2 The Fate of Antibiotic-Resistant Fecal Bacteria: From the Source of Contamination to the Estuary

2.2.1 Occurrence of Antibiotic-Resistant *Escherichia coli* and *Enterococcus* in the Seine Estuary

In parallel with the contamination by antibiotics, a permanent high level of antibiotic-resistant *E. coli* has been observed in the Seine estuary (Laroche et al. 2009; Table 5). A high proportion of these strains are multiresistant to antibiotics (60.5–80 %), part of which (11.1 %) harboring an integron[2] (Table 6). The class 1 integrons are found among the *E. coli* strains all along the estuary, whereas only the strains isolated in the urban zone (Rouen) harbor class 2 integrons.

[2] Integrons are genetic supports responsible for the capture of antibiotic resistance genes. They play a major role in the dissemination of antibiotic resistance genes, mainly among Gram-negative bacteria.

2.2.2 Relationship Between Antibiotic Use and Antibiotic Contamination in Water Environment

Water contamination by antibiotic-resistant bacteria was also investigated along the medical center–WWTP–river continuum. The results show that the percentages of strains of *E. coli* resistant to at least one antibiotic, and multiple-resistant strains (from three to eight resistances), are greater in the effluent of the retirement home than in the effluent of hospital (Table 3 and Fig. 3). Thus, the long residence time of the patients enables to observe the long-term consequences of antibiotic treatment in the emergence of antibiotic-resistant bacteria which is mainly due to the selection pressure exerted on the intestinal microbiota of patients.

The epidemiologic strains from the medical center, carrying specific genetic supports (integrons, a gene for resistance to *erm* gene encoding resistance to macrolides) decreased along the *continuum* in favor of strains better adapted to the environment (Figs. 3, 4, and 5). Thus, *E. coli* resistant to at least one antibiotic and those carrying the class 1 integron decreased significantly along the continuum (p-value<0.001) from the retirement home effluents to the river (Table 3 and Fig. 3, Oberlé et al. 2012).

Among the *Enterococci* population, mainly identified as *Enterococcus faecium*, all isolates from the medical center are multiple antibiotic resistant containing *erm*(B) and *mef*(A) and belonged to the hospital-adapted clonal complex 17(CC17) (Leclercq et al. 2013). It decreases in the treated effluent from the wastewater treatment plant (19 and 17 %) to reach 4 and 5 % in the waters of the Risle. Interestingly, while the proportion of *Enterococcus faecium* resistant to erythromycin is greater than 70 % along the *continuum*, only the hospital isolates exhibit a high level of resistance. The prevalence of the *erm*(B) gene reaches 75 % of the hospital isolates and only 6.7 % of the strain isolates in the river (Fig. 4). Similarly, the occurrence of the clonal complex "CC17", sequence of DNA present among the epidemic hospital strains (Fig. 5;

Table 5 Proportion of *E. coli* isolates (%) resistant to at least one antibiotic in water, along the Seine estuary (January 2006, $N=279$; Laroche et al. 2009)

	Stations	Number of strains of *E. coli* tested	% of antibiotic-resistant isolates[a] min-max	Geom.mean
Upstream input	Poses	113	16.0–50.0	30.2
Urban zone	Rouen	77	36.8–39.6	38.2
	Le Croisset	104	24.5–94.7	56.6
	La Bouille	92	31.0–68.7	49.6
Limit of salinity	Caudebec	94	13.3–58.3	35.7
Mouth of estuary	Tancarville–Honfleur	174	44.2–60.0	50.1

Table 6 Occurrence of *E. coli* strains carrying an integron (*intI1, intI2*, or *intI3*) in water sampled along the Seine estuary (January 2006, $N=279$, Laroche et al. 2009)

	Stations	Nb. of *E. coli* isolated	% resistant isolates[a]	% of isolates with integron *intI1*	*intI2*	*intI1 + intI2*	*intI3*	Total
Upstream input	Poses	16	25.0	6.2	0	0	0	6.2
Urban zone	Rouen	58	39.6	8.6	5.2	0	0	13.8
	Le Croisset	22	54.5	9.1	4.5	0	0	13.6
	La Bouille	35	42.8	2.7	0	0	0	2.7
Limit of salinity	Caudebec	53	45.3	3.8	0	0	0	5.7
Mouth of the estuary	Tancarville–Honfleur	95	44.2	14.7	0	1.1	0	15.8

[a]% of *E. coli* isolates resistant to one or more antibiotics

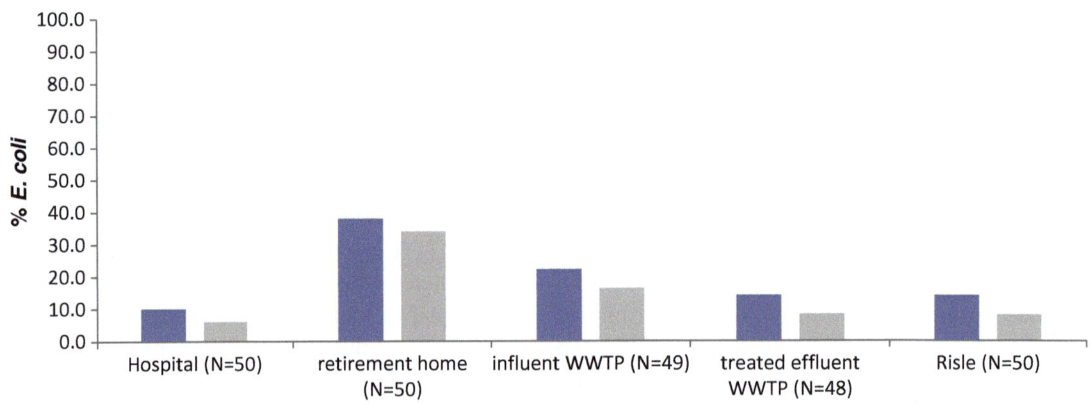

Fig. 3 Occurrence of multiple-resistant strains of *E. coli* along the medical center (*hospital* and retirement home)–wastewater treatment plant (WWTP)–river (Risle) continuum: percentage of isolates of *E. coli* resistant to at least three antibiotics ■, percentage of isolates of *E. coli* carrying class 1 integron ■ (Oberlé et al. 2012)

Leavis et al. 2006), is maximal among the strains of *Enterococcus faecium* isolated in the medical center effluents and then decreases along the continuum. This study shows a preferential disappearance of epidemiologic strains of *E. coli* and *Enterococcus faecium* in favor of other subpopulations of these bacterial species less resistant to antibiotic and probably better adapted to the aquatic environment (Berthe et al. 2013; Phan and Ferenci 2013; Ratajczak et al. 2010).

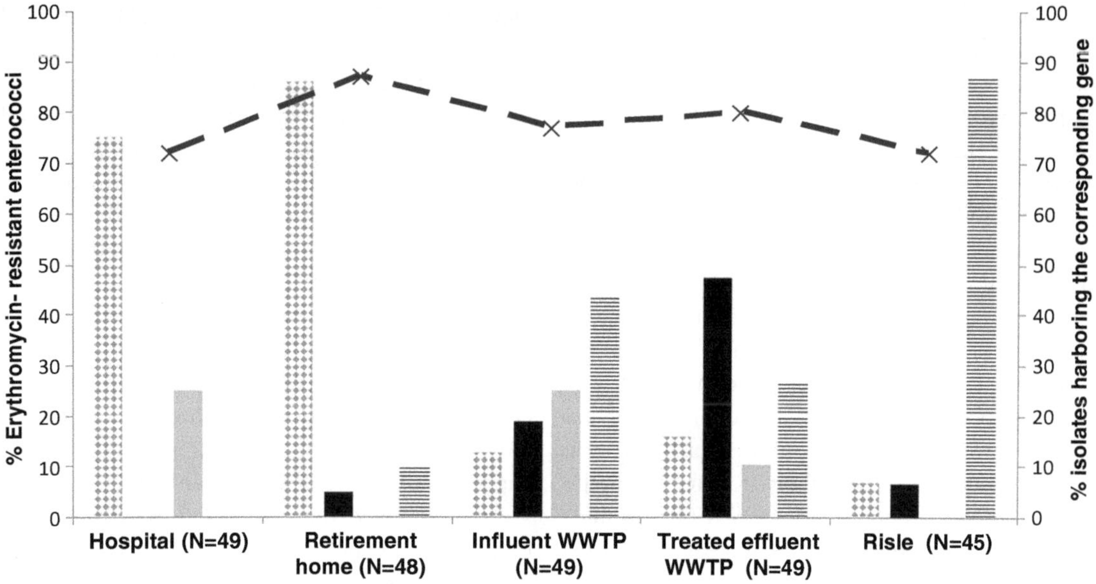

Fig. 4 Occurrence of *Enterococcus* strains resistant to erythromycin, along the medical center (hospital and retirement home)–wastewater treatment plant (WWTP)–river (Risle) continuum: (- - -), percentage of *Enterococcus* strains resistant to erythromycin (macrolide family), percentage of strains of *Enterococcus* resistant to erythromycin carrying corresponding genes *erm*B ▦, *mef*A ■, and *erm*B+*mef*A ▩ and unknown genes ▤ (Leclercq et al. 2013)

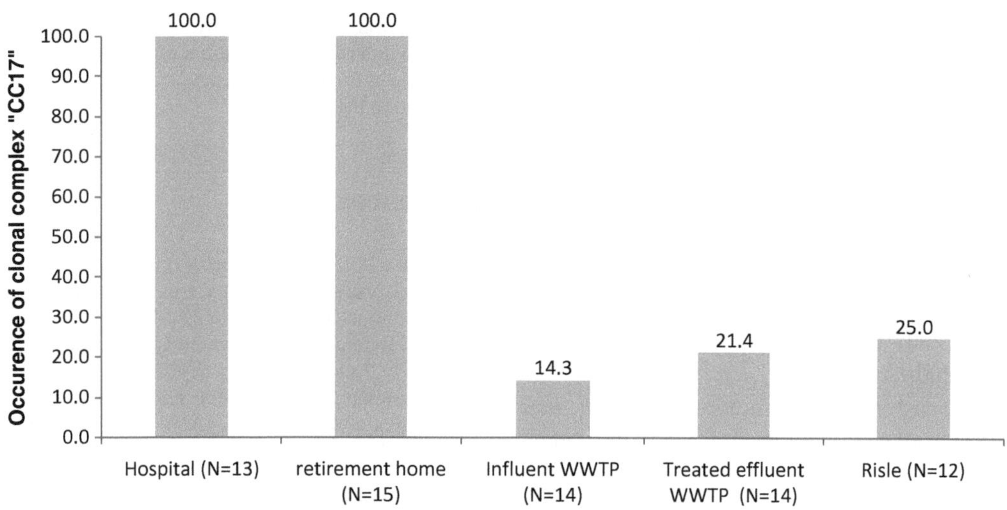

Fig. 5 Occurrence of epidemiologic strains of *Enterococcus*, carrying the DNA sequence marker CC17, within the population of *Enterococcus faecium*, along the medical center (hospital and retirement home)–wastewater treatment plant (WWTP)–river (Risle) continuum (Leclercq et al. 2013)

3 Conclusion

In the next decades, the estuaries will be exposed to a great chemical and microbial contamination, linked to the increase of the demography and human activity on their watershed. The study carried out in the Seine estuary and at small scale of a hospital–WWTP–river continuum shows not only the vulnerability but also the resilience of this environment to the contamination by antibiotics and antibiotic-resistant bacteria.

Thus, both the concentrations of antibiotic and the occurrence of antibiotic-resistant bacteria greatly decrease during their transfer from the main source of contamination (WWTP, hospital) to the estuary. The substantial decrease of antibiotic concentration along this continuum is mainly due to the degradation or elimination by WWTP treatments and the subsequent dilution in the receiving environment. Whatever the antibiotic, the concentrations observed are lower than the minimal inhibitory concentrations (MIC ranging from 0.032 to 256 μg mL^{-1}) responsible for the selection of antibiotic-resistant bacteria demonstrating the resilience of this water environment to the contamination by antibiotics (Oberlé et al. 2012; Leclercq et al. 2013). However, possible effects on bacterial physiology have been described for subinhibitory or sublethal concentrations (0.9x CMI to 0.25x CMI) (Davies et al. 2006; Kohanski et al. 2010). Indeed, the most stable molecules, such as the quinolones, the macrolides, and the sulfonamides, which persist longer in water could be accumulated in the environment such as the biofilms of periphytons or estuary mudflats.

The occurrence of antibiotic-resistant *E. coli* and *Enterococcus* strains released in waters results from the selective pressure exerted on the intestinal microbiota of human under antibiotic treatment. In water environment, no simple relationship exists between the antibiotic detected, the antibiotic use, and the antibiotic-resistant phenotypes of fecal bacteria. Indeed, only the most stable molecules are detectable in water, and the bacteria can harbor integrons that confer resistance to several antibiotics (Laroche et al. 2009; Oberlé et al. 2012). However, the abundance of antibiotic-resistant fecal bacteria decreases during their transfer from the source to the estuary, mainly due to a more important decay of hospital strains, that harbor gene implicated in the spread of antibiotic resistance (integrons, *erm* gene), in favor of strains less resistant to antibiotic and probably better adapted to the environment. These results also underline the resilience capacity of this aquatic environment (Berthe et al. 2013; Leclercq et al. 2013).

However, the estuarine sediments chronically exposed to multiple chemical contaminants, including antibiotics, to which are added supplies of antibiotic-resistant bacteria, are vulnerable environments. Indeed, this area could be a hot spot zone favorable to the transfer of antibiotic-resistance genes within the microbial communities.

Acknowledgments These studies were supported by the project FLASH (GIP Seine-Aval/EC2CO CNRS), by the SFR SCALE, and by research grants from the Haute-Normandie Regional Council (France) for the PhD of Kenny Oberlé, Mehdy Ratajczak, and Emilie Laroche. We thank Michel Simon, Caroline Bance, and Michel Auzou for the excellent technical assistance. We thank also Michel Leroux, Aurélie Lamy, Yvon Goarvot, Sophie Coté, and downtown pharmacists for antibiotic consumption data and the people that let us access to the medical center and to the WWTP.

References

Aarestrup FM (2005) Veterinary drug usage and antimicrobial resistance in bacteria of animal origin. Basic Clin Pharmacol Toxicol 96:271–81

Berthe T, Ratajczak M, Clermont O, Denamur E, Petit F (2013) Evidence for co-existence of distinct *Escherichia coli* population regarding their survival in aquatic environment. Appl Environ Microbiol 79:4684–4693

Davies J, Spiegelman GB, Yim G (2006) The world of subinhibitory antibiotic concentrations. Curr Opin Microbiol 9(5):445–453

Garcia-Armisen T, Touron A, Petit F, Servais P (2005) Sources of faecal contamination in the Seine estuary (France). Estuar Coasts 28(4):627–633

Goni-Urriza M, Capdepuy M, Arpin C, Raymond N, Caumette P, Quentin C (2000) Impact of an urban effluent on antibiotic resistance of riverine Enterobacteriaceae and *Aeromonas* spp. Appl Environ Microbiol 66:125–32

Jiménez C, Narvaez JF, Penuela GA (2012) Pharmaceutical products in the environment: sources, effects and risks. Vitae 19(1):92–108

Koczura R, Mokracka J, Jabłońska L, Gozdecka E, Kubek M, Kaznowski A (2011) Antimicrobial resistance of integron-harboring *Escherichia coli* isolates from clinical samples, wastewater treatment plant and river water. Sci Total Environ 14(1):680–685

Kohanski MA, DePristo MA, Collins JJ (2010) Sublethal antibiotic treatment leads to multidrug resistance via radical-induced mutagenesis. Mol Cell 37(3):311–320

Kümmerer K (2009) Antibiotics in the aquatic environment-A review-Part I. Chemosphere 75(4):417–434

Laroche E, Pawlak B, Berthe T, Skurnik D, Petit F (2009) Occurrence of antibiotic resistance and class 1, 2 and 3 integrons in *Escherichia coli* isolated from a densely populated estuary (Seine, France). FEMS Microbiol Ecol 68(1):118–130

Leavis HL, Willems RJL, Top J, Bonten MJM (2006) High-level ciprofloxacin resistance from point mutations in *gyrA* and *parC* confined to global hospital-adapted clonal lineage CC17 of *Enterococcus faecium*. J Clin Microbiol 44(3):1059–1064

Leclercq R, Oberlé K, Galopin S, Cattoir V, Budzinski H, Petit F (2013) Changes in enterococcal populations and related antibiotic resistance along a medical center-wastewater treatment plant-river continuum. Appl Environ Microbiol 79:2428–2434

Martinez JL (2008) Antibiotics and antibiotic resistance genes in natural environments. Science 321:365–7

Martinez JL (2009) The role of natural environments in the evolution of resistance traits in pathogenic bacteria. Proc R Soc B Biol Sci 276(1667):2521–2530

Nwosu VC (2001) Antibiotic resistance with particular reference to soil microorganisms. Res Microbiol 152:421–30

Oberlé K, Capdeville MJ, Berthe T, Budzinski H, Petit F (2012) Evidence for a complex relationship between antibiotics and antibiotic-resistant *Escherichia coli*: from medical center patients to a receiving environment. Environ Sci Technol 46:1859–68

Phan K, Ferenci T (2013) A design- constraint trade-off underpins the diversity in ecologically important traits in species *Escherichia coli*. ISME J 7:2034–2043

Poirel L, Kämpfer P, Nordmann P (2002) Chromosome-encoded Ambler class A beta-lactamase of *Kluyvera georgiana*, a probable progenitor of a subgroup of CTX-M extended-spectrum beta-lactamases. Antimicrob Agents Chemother 46:4038–40

Ratajczak M, Laroche E, Berthe T, Clermont O, Pawlak B, Denamur E, Petit F (2010) Influence of hydrological conditions on the *Escherichia coli* population structure in the water of a creek on a rural watershed. BMC Microbiol 10:222

Séveno N, Kallifidas D, Smalla K, Van Elsas J, Collard J, Karagouni A, Wellington E (2002) Occurrence and reservoirs of antibiotic resistance genes in the environment. Rev Med Microbiol 13:15

Tamtam F, Mercier F, Le Bot B, Eurin J, Tuc DQ, Clément M, Chevreuil M (2008) Occurrence and fate of antibiotics in the Seine River in various hydrological conditions. Sci Total Environ 393(1): 84–95

Touron A, Berthe T, Gargala G, Fournier M, Ratajczak M, Servais P, Petit F (2007) Assessment of faecal contamination and the relationship between pathogens and faecal bacterial indicators in an estuarine environment (Seine, France). Mar Pollut Bull 54(9):1441–1450

Watkinson AJ, Micalizzi GB, Graham GM, Bates JB, Costanzo SD (2007) Antibiotic resistant Escherichia coli in wastewaters, surface waters and oysters from an urban riverine system. Appl Environ Microbiol 73(17):5667–5670

Watkinson AJ, Murby EJ, Kolpin DW, Costanzo SD (2009) The occurrence of antibiotics in an urban watershed: from wastewater to drinking water. Sci Total Environ 407(8):2711–2723

Wellington E, Boxall AB, Cross P, Feil EJ, Gaze WH, Hawkey PM, Johnson-Rollings AS, Jones DL, Lee NM, Thomas CM, Williams AP (2013) The role of the natural environment in the emergent of antibiotic resistance in gram negative bacteria. Lancet Infect Dis 13:155–65

World Health Organization (2001) WHO global strategy for containment of antimicrobial resistance. http://apps.who.int//iris/handle/10665/66860

Marine Litter Within the European Marine Strategy Framework Directive

François Galgani

Abstract

One of the most evident changes in the last half-century is the ubiquity and abundance of litter in the marine environment. The EU Marine Strategy Framework Directive (MSFD, 2008/56/EC) established a framework within which EU Member States shall take action to achieve the good environmental status (GES) of their marine waters by 2020. Amongst the 11 MSFD qualitative descriptors, Descriptor 10 (D 10) concerns marine litter. Here we report the general strategy in support to the implementation of MSFD for marine litter. It includes various aspects such as scientific background, monitoring strategies and protocols, definition of good environmental status and targets and support to management policies.

A technical group on marine litter (DG Environment/TSG ML) was created to provide scientific and technical background for the implementation of MSFD requirements and support to Member States with regard to D 10. The work of TSG ML focuses on the specification of monitoring methods through the development of monitoring protocols for litter in the different marine compartments, including microplastics and litter in biota. Further consideration is also being given to the identification of sources of marine litter and a better understanding of the harm caused by marine litter.

1 Introduction

Pressures and demands on marine resources are often excessive and action must be taken in order to minimize impacts on the marine environment.

The European Commission developed the Marine Strategy Framework Directive (MSFD) for the protection and sustainable use of marine ecosystems. It establishes a framework within which Member States must take action to achieve or maintain good environmental status (GES) for the marine environment by 2020.

Member States have to take six procedural steps until 2016 to develop a marine strategy for their waters: (1) an initial assessment of the

F. Galgani (✉)
IFREMER/LER/PAC, Bastia, France
e-mail: Francois.galgani@ifremer.fr

H.-J. Ceccaldi et al. (eds.), *Marine Productivity: Perturbations and Resilience of Socio-ecosystems*,
DOI 10.1007/978-3-319-13878-7_10, © Springer International Publishing Switzerland 2015

current environmental status, (2) the determination of good environmental status, (3) the establishment of environmental targets and associated indicators, (4) the establishment and implementation of a monitoring programme, (5) the development of a programme of measures and (6) the entry into operation of this programme of measures.

Periodic assessments of the state of the marine environment, monitoring and the formulation of environmental targets are perceived as part of the continuous management process. Accordingly, provisions are made for the modification of adopted marine strategies and measures.

Of the 11 descriptors listed in Annex I of the MSFD for determining GES, Descriptor 10 has been defined as "Properties and quantities of marine litter do not cause harm to the coastal and marine environment". Commission Decision 2010/477/EU identifies the following criteria and associated four indicators for Descriptor 10:

1. Trends in the amount of litter washed ashore and/or deposited on coastlines, including analysis of its composition, spatial distribution and, where possible, source (indicator 10.1.1)
2. Trends in the amount of litter in the water column (including floating at the surface) and deposited on the seafloor, including analysis of its composition, spatial distribution and, where possible, source (indicator 10.1.2)
3. Trends in the amount, distribution and, where possible, composition of microparticles (in particular microplastics) (indicator 10.1.3)
4. Trends in the amount and composition of litter ingested by marine animals (e.g. stomach analysis) (indicator 10.2.1)

Human pressures on the oceans have increased substantially in recent decades and coastal and marine human activities generate considerable quantities of waste. Much of this litter will persist in the sea for years, decades or even centuries. The occurrence of litter has been demonstrated worldwide, in oceanic gyres, on shorelines, in sediments and in the deep sea. Europe is on one of the main affected area. To follow these specific issues, the European Marine Directors and DG environment established a technical group under the Working Group for the implementation of MSFD Descriptor 10.

The group defined marine litter as any persistent, manufactured or processed solid material discarded, disposed of or abandoned in the marine and coastal environment (Galgani et al. 2011). The group identified and presented 15 options/protocols for the monitoring of litter in the different marine compartments and the biological impact of the ingested litter or microlitter. It furthermore contains considerations about sources, GES, objectives, environmental targets and research needs, as well as general monitoring strategies and the understanding of harm.[1]

2 Marine Litter

Current knowledge of the quantities of litter in European seas, the degradation and fate of litter in the marine environment and its potentially harmful biological, physical and chemical impacts on marine life and habitats is insufficient. Encounters with marine litter were reported for 663 species (CBD 2012). Over half of the reported species (about 370) were associated with entanglement in and ingestion of marine debris, representing an increase of more than 40 % since the last review in 1997.

The entanglement of species in marine litter has frequently been described as a serious mortality factor leading to potential losses in biodiversity (Macfadyen et al. 2009). Entanglement in marine debris including lost (ghost) nets has been reported for pinniped species, cetaceans, all 7 species of marine turtles and more than 56 species of marine and coastal birds (Katsanevakis et al. 2007). Many species of fish have also been reported to ingest marine litter (Katsanevakis et al. 2007; CBD 2012).

An emerging area of concern is the accumulation of microplastic fragments in the water column and in sediments (Thompson et al. 2004). Pieces of common polymers of less than 5 mm have been recorded worldwide (Barnes

[1] (See http://publications.jrc.ec.europa.eu/repository/handle/111111111/22826).

et al. 2009). Plastics are biologically inert and probably stay in the marine environment for long periods. If ingested, plastics release chemicals (mainly phthalates and bisphenol A) but also sorb hydrophobic pollutants (including PCBs and DDT). More research is however needed to establish the full environmental relevance and potential impact of these microparticles, notably on distribution, transport, degradation/weathering processes and sorption/release mechanisms.

Other known impacts of marine litter include alteration and degradation of benthic habitats (Katsanevakis et al. 2007) but also assist invasions of alien species, including of algae associated with red tides (Barnes 2002).

From a socio-economic perspective, harm can include the cost of degradation of ecosystem goods and services, including the reduction in recreational, aesthetic or educational values of an area such as beaches, as well as risks to human health such as the threat of floating objects to navigation. Economic harm includes significant impact by direct costs and loss of income due to marine litter and affects a range of maritime sectors (including aquaculture, agriculture, fisheries, shipping and leisure boating), power generation and industrial use, local authorities and tourism. Levels of economic "harm" may run into millions of euro annually even at subregional scales (Mouat et al. 2010). Marine litter is also a serious offence to the visual and aesthetic sensitivities of visitors to beaches. Furthermore, sanitary, sewage-related and medical waste may cause injuries and/or be a risk to human health (Ivar do Sul and Costa 2007).

There is no common understanding of what exactly constitutes "harm" from marine litter or how it can be assessed with respect to the implementation of the MSFD. Research must consider this and develop robust approaches for assessing harm measures to reduce pollution from litter need to target different sources. Identifying the source of many litter items is however a complex task because of point and diffuse sources and possible long-distance travels before final destinations. Factors such as ocean current patterns, winds and tides and the proximity to urban centres, industrial and recreational areas, shipping lanes and fishing grounds influence the types, nature and amount of litter that is found in the open sea or collected along beaches, waterways or underwater. Sources of marine litter can be characterized as either land based or sea based, depending on how the litter enters marine waters. These broad categories can be further broken down into sources such as recreational litter, shipping or fishing litter, urban sources, etc. Some items can be attributed to certain sources. Such so-called use categories provide information for setting targets and reduction measures, as they can easily be linked to sources.

For the Mediterranean Sea, PNUE/PAM/MEDPOL (UNEP 2009) reported that most of the marine litter comes from land-based rather than sea-based sources (e.g. ships). Litter enters the sea mainly from the shoreline and results from recreational activities. It is composed mainly of plastics, aluminium and glass. Recordings of floating litter confirmed the overwhelming presence of plastics in the Mediterranean Sea (83 % of the observed marine litter items). The situation in the North Sea is different. The large diversity of items found on the coasts indicates that in the Northeast Atlantic, maritime activities in the form of shipping, fishing and offshore installations are the predominant sources together with coastal recreational and tourism activities (OSPAR 2009). A considerable proportion of litter enters the North Sea through transport by wind, currents and rivers and via the English Channel. Upcoming work will lead to better understand of waste pathways depending on the type of litter. Mapping the sources and their quantities remains a necessary step in order to plan effective measures (Fig. 1).

3 Monitoring

In Europe, the existing different, but compatible, methods for monitoring need to be adapted and harmonized to take account of regional differences, e.g. in the type of coastline or prevailing currents in offshore areas. Methodologies for source assessment are mostly based on the identification and reporting of collected/observed marine litter. Due to the difference in

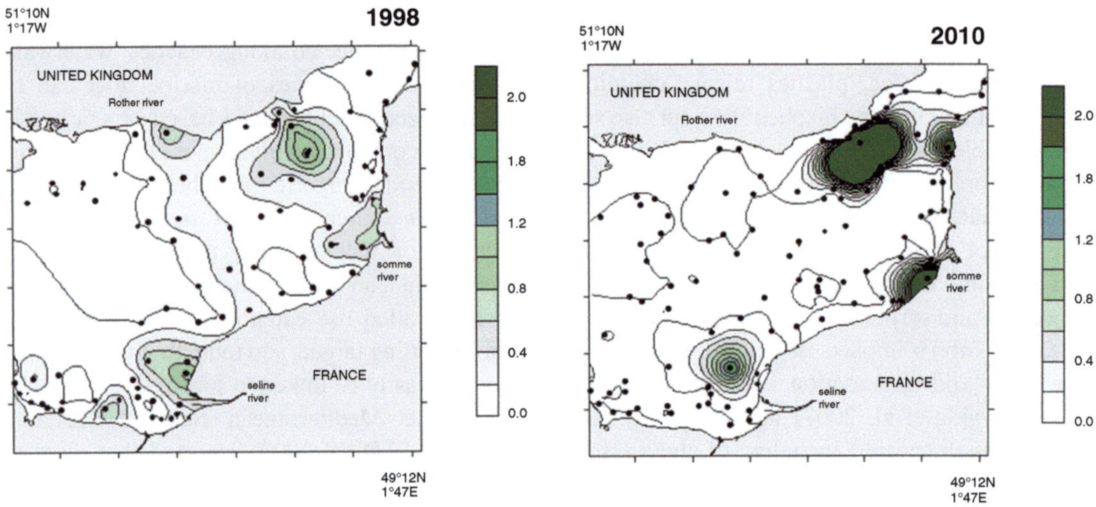

Fig. 1 Increase in litter densities in the French/English channel between 1998 and 2010 (numbers of items per hectare) (After Galgani et al., 5th International Marine Debris Conference, Honolulu, Hawaii, 2011)

the monitoring approaches, the possibilities for identifying the nature (category) of objects vary between the different environmental compartments. It has been recommended (Galgani et al. 2013) that the categories for reporting should be compatible between different survey types (beaches, sea surface, sea floor) so that outcomes are comparable. Reporting of marine litter for source attribution needs still further development as the efficiency of measures targeting specific litter sources requires the distinction of the nature/categories of litter in the different environmental compartments.

When planning monitoring schemes, consideration should be given to adequate spatial and temporal scales. Beach-litter surveys, seafloor monitoring on continental shelves and socio-economic studies can be readily applied at a European scale (Galgani et al. 2010), but priority should be given to the monitoring of marine areas that are most affected by litter. Methodological protocols in Europe are currently available for the assessment of certain types and occurrences of litter on coastlines (OSPAR 2009). These standards should be adjusted to European needs and harmonized for an extension to other regions. Pilot projects have also indicated that litter on the seafloor could be measured alongside routine biological trawling surveys (e.g. International Bottom Trawl Surveys in the OSPAR area, Mediterranean International Trawl surveys in the MEDPOL area) and could include an evaluation of sources.

In the framework of the OSPAR Convention, amounts of plastics in fulmar (*Fulmarus glacialis*) stomachs are used to assess temporal trends, local differences and compliance with a set target for acceptable pressure in the North Sea (OSPAR 2009). Such monitoring could be extended to other marine regions using region-specific indicator species, such as turtles for the Mediterranean Sea. Further work will have to focus on developing common monitoring protocols, facilitating the implementation of fit for purpose monitoring programmes, evaluating new monitoring tools, estimating the costs and developing standardized litter categories.

The evaluation of waste flows between the different compartments of the marine environment is a necessary step and goal for understanding the mechanisms of transport, fluxes and potential impacts. Finally, understanding the transport mechanisms will help to clarify transformation and provide a better description of the spatial distribution of marine litter (Fig. 2).

The gaps in knowledge are a constraint in identifying targeted and effective measures to reduce litter pollution.

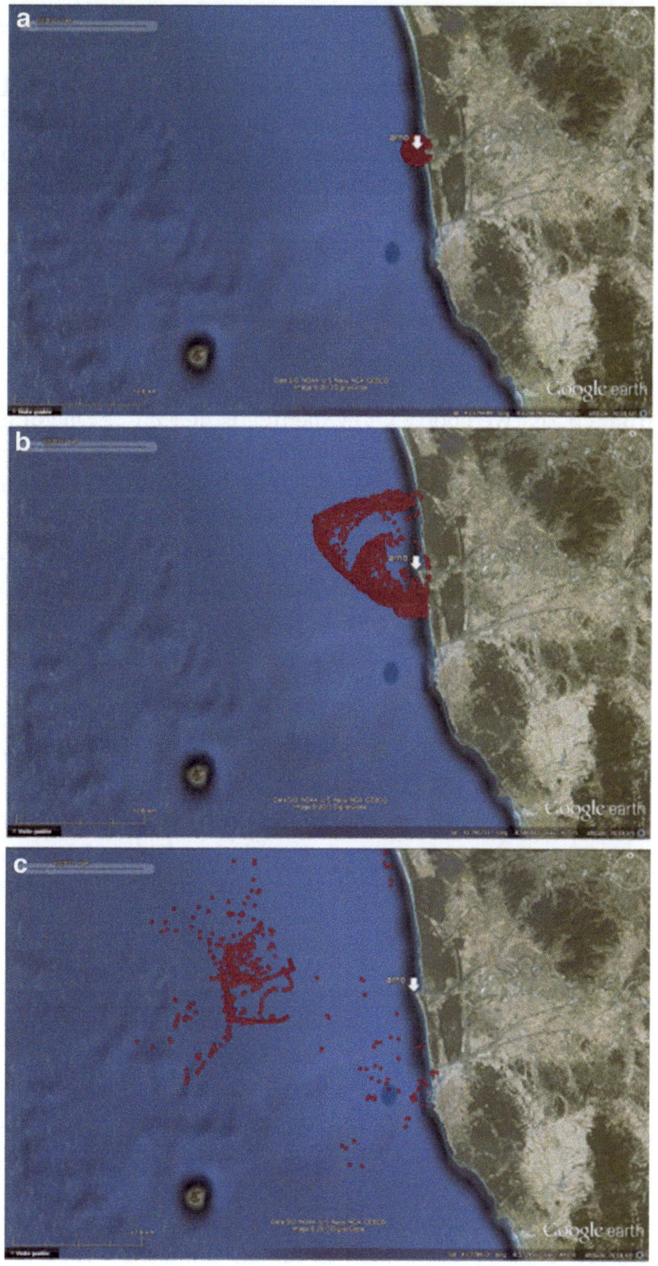

Fig. 2 Simulated transport of particles originating from the Arno estuary (Tuscany, Italy) from day 0 (**a**) to day 2 (**b**) and day 6 (**c**) [Ichthyop/Mars (IFREMER) simulation model (Courtesy from O. Gerigny, IFREMER/Bastia])

The abundance of litter at sea can be estimated either by direct observations of large debris items (e.g. submersible remote observation vehicles (ROVs) for monitoring litter on the seabed or ship-based and aerial observations for debris floating at the sea surface) or by large-scale imagery application and net trawls (for smaller items). Net-based evaluation surveys are the most widespread and adequate methods to date (Goldberg 1994), similar to the methodology for monitoring the benthic species.

The interpretation of trends is difficult because the fate of plastics at depth is not well investigated, and the accumulation of plastics on the seabed had begun long before specific scientific investigations started in the 1990s. Of the areas investigated to date along the European coasts (Galgani et al. 2000), Mediterranean sites tend to show the greatest densities of litter accumulation. Debris, mainly plastic, that reaches the seabed may have been transported a considerable distance from its source, only sinking to the ground when weighed down by fouling. The consequence is an accumulation of plastic debris in bays and canyons rather than in the open sea (Galgani et al. 2000; Katsanevakis et al. 2007).

However, due to large-scale residual ocean circulation patterns, some accumulation zones in the Atlantic Ocean and the Mediterranean Sea have very high debris densities despite being far from coasts. We poorly know the trends in accumulation of debris at sea, but available data indicate considerable variability. Abundances slightly decreased in the Gulf of Lion (France) during a 15-year period (1994–2009). However, in some areas around Greece, the abundance of debris at depth has increased over a period of 8 years (Katsanevakis 2008). Debris is progressively fragmented in the marine environment (Thompson et al. 2004) to microparticles. Concern is about the accumulation of microscopic pieces of plastic ("microplastic") due to their high prevalence at sea and the slow rate of their chemical and biological degradation. This also includes the spillage of pre-production (resin pellets) plastics, granules. At most locations, current quantities appear to be relatively low. However, plastic microparticles have been reported in quantities exceeding 100,000 items/km^2 in the North Sea. Similar quantities of debris have been reported in the northwest Mediterranean Sea (Collignon et al. 2012) where 115,000 items/km^2 were calculated, giving an extrapolated total of 250 billion items in the whole basin.

In a number of reports, the Ecological Quality Objective (EcoQO) for litter in fulmar stomachs in the OSPAR framework proved able to provide valuable information on the temporal changes in, and the spatial distribution of, the abundance of marine litter, on the differences between trends in industrial and user plastics, and on the sources of marine litter (Van Franeker et al. 2011). The EcoQO currently applies to the North Sea but can be adapted to apply in most areas of the Northeast Atlantic. Pilot studies for biomonitoring of litter should also consider other species, especially marine turtles that are regularly stranded in the Mediterranean region and which often contain fatal quantities of ingested litter. Fish, zooplankton species, shellfish and seals may be considered in the future as generally applicable target species for most European seas.

4 Consideration of the Marine Strategy Framework Directive

One of the key challenges for EU Member States in implementing the MSFD is to determine "good environmental status" because the term has different meanings in the EU marine regions or subregions and is therefore open to interpretation. More than one indicator will be required to assess GES in relation to the different compartments of the marine environment and the different aspects of litter pollution. Metrics are not yet available for evaluating some of the biological impacts that litter may have. In their absence, the thresholds may be replaced by trends in pressure-related indicators, such as the amount of litter on the seafloor or on beaches, to provide proxies for evaluating progress towards GES.

As stated above, "harm" caused by marine litter can be divided into three general categories: (1) social harm, i.e. loss in aesthetic value and public health; (2) economic harm, such as the cost to tourism, damage to vessels (net and ropes in propellers), fishing gear and facilities cleaning costs; and (3) ecological harm, e.g. mortality of, or sublethal effects on, animals through entanglement in fishing gears or harm resulting from ingestion of litter, including the uptake of microparticles. GES is regarded as achieved when litter and its degradation products present in and entering EU marine waters (1) do not cause harm to marine life and habitats, (2) do not pose direct or indirect risks to human health and (3) do not lead to negative socio-economic impacts.

Descriptor 10 is particularly related to human health and to socio-economic interests. The use of trend indicators as listed in the Commission Decision (10.1.1; 10.1.2; 10.1.3), aimed to observe and assess trends in litter occurrence in the different marine compartments, will help to predict both health and socio-economic consequences.

It is not generally feasible for assessments to provide information on the extent of harm at the population, community or ecosystem level, and it is thus essential to consider harm at the level of the individual organism. Estimates of the number of individuals affected are likely to offer the most feasible and representative conclusions about biological impacts. Following the example of the OSPAR EcoQ for plastic litter items in fulmar stomachs in the North Sea region, additional indicator species must be found for the other EU marine regions (such as sea turtles for the Mediterranean Sea), and additional indicators on ecological impacts of litter (e.g. on entanglements) may be required by reference to Commission Decision 2010/477/EU.

Reaching GES may be understood as a continuous reduction of inputs with the aim of reducing the total amount of marine litter by 2020 to a level that does not cause harm. Activities to remove litter that has already entered the marine environment will assist in reaching this goal, but some important points have to be considered. One of the difficulties in target setting for some marine regions is the lack of data for developing a baseline. Moreover, any assessment of marine litter should consider short-term variations caused by meteorological and/or hydrodynamic events and natural fluctuations, which could influence our ability to detect underlying trends. Given the variability of litter data, which is influenced greatly by season, weather conditions and water currents, a 5-year running mean is considered appropriate for providing a baseline in terms of an average level of pollution. However, the reduction in litter inputs may not lead to a measurable reduction of total litter levels in the marine environment in the short term. This is due to the persistence of some materials and the long degradation processes for certain types of litter (plastics, metal, glass and rubber). Timescales of observations should therefore be adapted to ensure multiannual frequency of surveying.

Beached litter surveys can be applied to the European spatial scale while deep-seafloor monitoring, restricted to a few areas, is more relevant at smaller scales and over longer periods. It would not be reasonable to argue that the ultimate goal of the MSFD should be 0 % litter in the marine environment. Targets for the different compartments of the marine environment need to be set by Member States on the basis of the initial level of pollution within the area considered. Increasing knowledge of the amount and dynamics of litter in the marine environment will help to determine whether targets need to be defined at the regional level in addition to targets set by individual EU Member States.

For litter on beaches and at sea (floating and on the seafloor) and for microplastics, it is proposed that the reduction goal recommended by TSG ML is adopted as a first step. This goal is to achieve a general measurable and statistically significant reduction in beach litter until 2020. Such trend-based target maybe appropriate until the evidence supports other procedures.

For microplastics, not currently measured on a regular basis, sufficient monitoring should be carried out and a baseline established before defining a precise target. Based on such monitoring, a potential target for the significant decrease of microparticles by 2020 could be formulated and the occurrence of microparticles in sediments should also be considered.

Recent studies on industrial plastics found in beached fulmars in the North Sea showed that reductions in abundance of specific marine litter items in the order of 50 % per decade are a feasible target if adequate measures are taken. OSPAR defined its more reasonable target for ecological pressure concerning litter in the North Sea that, 10 % of Northern Fulmars should be allowed to have 0.1 g plastic in their stomach. This OSPAR EcoQO cannot be directly transferred to other marine areas where fulmars do not occur. Appropriate indicator species still need to be established and it may be more suitable, after a reference value will have been established, to describe GES in relation to the ingestion of litter in terms of a trend, e.g. x% annual reduction in the quantity of ingested litter.

Harmonization is necessary for common and comparable monitoring approaches and for recommendations and guidelines to assess GES at regional, national and European scales. Research will need to incorporate the improvement of knowledge concerning impacts of litter on marine life, degradation processes of litter at sea, the study of litter-related microparticles, the study of chemicals associated with litter, the factors influencing the distribution and densities of litter at sea (human factors, hydrodynamics, geomorphology, etc.), the comparability of monitoring methods and the determination of thresholds for GES. The assessment and monitoring of socio-economic harm will also need to be addressed, and research will have to consider novel methods and automated monitoring devices and finally the rationalization of monitoring. Because monitoring is to be started in 2014, optimization of monitoring (standards/baselines, data management/quality assurance) and the extension of monitoring protocols to all MSFD regions/subregions are the most critical gap.

References

Barnes DKA (2002) Invasions by marine life on plastic debris. Nature 416:808–809. doi:10.1038/416808a

Barnes DKA, Galgani F, Thompson RC, Barlaz M (2009) Accumulation and fragmentation of plastic debris in global environments. Philos Trans R Soc Lond B Biol Sci 364:1985–1998

CBD (Convention on Biological Diversity) (2012) Impacts of marine debris on biodiversity: current status and potential solutions, Technical Series No. 67. CBD (Secretariat of the Convention on Biological Diversity and the Scientific and Technical Advisory Panel GEF), Montreal, 61 pp

Collignon A, Hecq JH, Galgani F, Voisin P, Goffard A (2012) Neustonic microplastics and zooplankton in the western Mediterranean sea. Mar Pollut Bull 64:861–864

Galgani F, Leaute JP, Moguedet P, Souplet A, Verin Y, Carpentier A, Goraguer H et al (2000) Litter on the sea floor along European coasts. Mar Pollut Bull 40:516–527

Galgani F, Fleet D, van Franeker J, Katsavenakis S, Maes T, Mouat J, Oosterbaan L et al (2010) Marine Strategy Framework Directive Task Group 10 report marine litter, JRC scientific and technical report, ICES/JRC/IFREMER Joint Report (No. 31210–2009/ 2010), Ed. by N Zampoukas, 57 pp

Galgani F, Hanke G, Werner S, Piha H (2011) MSFDGES technical subgroup on Marine Litter. Technical recommendations for the implementation of MSFD requirements. JRC Scientific and Technical Report, EUR 25009 EN – 2011, 93 pp

Galgani F, Hanke G, Werner S, Oosterbaan L, Nilsson P, Fleet D, Kinsey S, Thompson R, Van Franeker J, Vlachogianni T, Scoullos M, Mira Veiga J, Palatinus A, Matiddi M, Maes T, Korpinen S, Budziak A, Leslie H, Gago J, Liebezeit G (2013) Monitoring guidance for marine litter in European seas. MSFD GES Technical Subgroup on Marine Litter (TSG-ML). DRAFT REPORT, 120p

Goldberg E (1994) Diamonds and plastics are forever? Mar Pollut Bull 28:466

Ivar do Sul JA, Costa MF (2007) Marine debris review for Latin America and the Wider Caribbean Region: from the 1970s until now, and where do we go from here? Mar Pollut Bull 54:1087–1104

Katsanevakis S (2008) Marine debris, a growing problem: sources, distribution, composition, and impacts. In: Hofer TN (ed) Marine pollution: new research. Nova Science Publishers, New York, pp 53–100

Katsanevakis S, Verriopoulos G, Nikolaidou A, Thessalou-Legaki M (2007) Effect of marine pollution with litter on the benthic megafauna of coastal soft bottoms. Mar Pollut Bull 54:771–778

Macfadyen G, Huntington T, Cappell R (2009) Abandoned, lost or otherwise discarded fishing gear, UNEP Regional seas reports and studies 185 and FAO fisheries and aquaculture technical paper 523. UNEP/FAO, Rome, 115 pp

Mouat J, Llozano R, Bateson H (2010) Economic impacts of Q17, Q18 Marine Litter. Kimo report. Ed. by Kimo, 100 pp. http://www.kimo.org

OSPAR (2009) Marine litter in the North-East Atlantic Region: assessment and priorities for response. London, UK, 127 pp. PNUE/PAM/MEDPOL. 2009. Results of the assessment of the status of marine litter in the Mediterranean. Meeting of MED POL Focal Points No. 334, 91 pp

Thompson RC, Olsen Y, Mitchell RP, Davis A, Rowland SJ, John AWG, McGonigle D et al (2004) Lost at sea: where is all the plastic? Science 304:838. doi:10.1126/science.1094559

UNEP (2009) Marine litter: a global challenge. UNEP, Nairobi, 232 pp

Van Franeker JA, Blaize C, Danielsen J, Fairclough K, Gollan J, Guse N, Hansen P-L et al (2011) Monitoring plastic ingestion by the northern fulmar Fulmarus glacialis in the North Sea. Environ Pollut 159:2609–2615. doi:10.1016/j.env pol.2011.06.008

Distribution of the Marine Debris on Seafloor from the Primary Report of Five Cruises After the Great East Japan Earthquake 2011

Takehisa Yamakita, Hiroyuki Yamamoto, Yuka Yokoyama, Izumi Sakamoto, Shinji Tsuchida, Dhugal Lindsay, Yoshihiro Fujiwara, Masaru Kawato, Takafumi Kasaya, and Hiroshi Kitazato

Abstract

The Tohoku Ecosystem-Associated Marine Science (TEAMS) project started from 2011 with the aim of revealing the impacts of the 2011 tsunami and to assess the present state of the marine ecosystem in affected areas over time during the next decade. TEAMS collected various data such as bathymetry, geography, oceanography, benthos, and tsunami debris not only from our project but also from the archived data at our respective institutes and in public databases. Integration of data and creation of habitat maps are expected to contribute to our understanding of the effects of large tsunamis, present the status of marine biodiversity in the area, and lead to the identification of key areas for ensuring ecosystem resilience. Here we focus on information of marine debris identified from deep-sea images and sonar data. We conducted data integration and collated information in a GIS database.

Our preliminary results suggest that the impact of the tsunami debris in deep-sea areas (ca. 200–500 m) was not as severe as in coastal environments. Accumulations of tsunami debris were found mainly in submarine canyons. Although different types of debris occur so far in mixed aggregations, differences in the movement and decomposition patterns of these different materials are expected to lead to both positive and negative effects on the environment in different ways.

T. Yamakita (✉) • H. Yamamoto • S. Tsuchida
D. Lindsay • Y. Fujiwara • M. Kawato • T. Kasaya
H. Kitazato
Institute of Biogeosciences, Japan Agency for Marine-Earth Science and Technology, 2-15, Natsushima-cho, Yokosuka-city, Kanagawa 237-0061, Japan
e-mail: yamakitat@jamstec.go.jp

Y. Yokoyama • I. Sakamoto
Department of Marine Mineral Resources, School of Marine Science and Technology, Tokai University, 3-20-1 Orido, Shimizu-ku, Shizuoka 424-8610, Japan

H.-J. Ceccaldi et al. (eds.), *Marine Productivity: Perturbations and Resilience of Socio-ecosystems*, DOI 10.1007/978-3-319-13878-7_11,

1 Introduction

The Great East Japan Earthquake 2011 (2011 Tohoku-Oki Earthquake) occurred on the 11th of March 2011. After the earthquake, tsunami waves washed over the Pacific side of the Tohoku area – the northern part of the main island of Japan. Tsunamis destroyed many coastal towns, even those protected by seawalls of 10 m in height, and damaged and disabled a nuclear power plant as well. The Japanese government directed efforts to help the area recover from earthquake and tsunami damage, and several programs were conducted in terrestrial areas. During this time, surveys of marine ecosystems were conducted separately.

An initiative called the Tohoku Ecosystem-Associated Marine Science (TEAMS) project was launched at the end of the fiscal year 2011 with the support of the Ministry of Education, Culture, Sports, Science and Technology. The TEAMS project is required to contribute to the following in the ocean areas off Tohoku: (a) elucidate the effects of earthquakes and tsunamis; (b) with a view to assessing recovery, reveal the dynamics of organisms and their environment; (c) suggest fisheries that could be sustainably exploited; and (d) assess the safety of such products in terms of anthropogenic chemical compound content. As part of this project, multidisciplinary surveys of the environmental conditions and state of the ecosystem have been conducted from coastal areas to the continental slope, where fisheries are conducted, in order to assess the effect of earthquakes and tsunamis.

Previous research suggested heavy damage by the earthquake and tsunami in coastal and shallow ocean areas. Large amounts of sedimentation (Goto et al. 2012), increases in marine debris (Shibata et al. 2013), decrease in ground level/seafloor depth (Ozawa et al. 2011), and changes in the distribution of benthic organisms (Seike et al. 2013; Urabe et al. 2013) are considered to be the main changes in the ecosystem. Although extensive research and cleanup activities have been conducted in shallow waters, research results from deep-sea areas are scarce. A few reports, some sedimentation studies, suggested difficulty in identifying the effects of the earthquake and the tsunami event in the upper part of the deep sea (such as shallower than 750 m) because of bioturbation (Arai et al. 2013). Nonetheless, marine debris is expected to have either positive or negative effects in deep-sea areas (Goto 2012; Ramirez-Llodra et al. 2011).

In the present paper, we introduce preliminary results on the distribution of debris in the datasets collected from deep-sea surveys by TEAMS-related cruises and from archived data from other studies. We focus in particular on the spatial distribution of marine debris and the types of marine debris to assess the positive or negative effects of marine debris originating from the earthquake on ecosystems in the upper part of the deep sea.

2 Methods

2.1 Study Area

The Tohoku offing is located on the northern Pacific side of the main island of Japan and is well-known as a frontal area where the warm, northward-flowing Kuroshio current encounters the cold, southward-flowing Oyashio current. Since 2011, we have surveyed the deep-sea area using a side-scan sonar, deep-tow submergible cameras, and an ROV (remotely operated vehicle) system. The target region for the TEAMS surveys encompassed the area from the coast to the continental slope, and this study focused on the offshore region of Miyagi and Iwate Prefectures. The survey areas and sites are plotted in Fig. 1.

2.2 Data Collection

We examined the data from five different surveys to evaluate quantitatively the occurrence of marine debris in each survey area or along each survey transect. Details of each survey are as follows: towed camera surveys by the 4KC (4,000 m camera deep-tow camera) and YKDT (Yokosuka deep-tow camera) were conducted from May 16

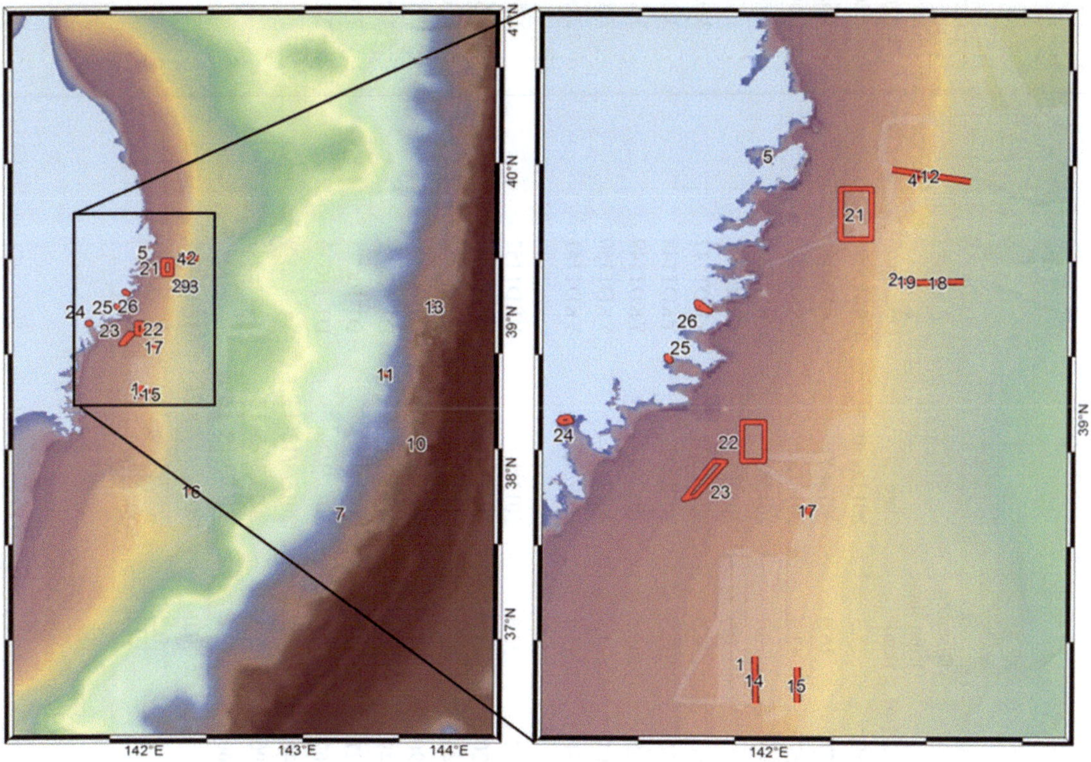

Fig. 1 Location of study sites for research on species and/or marine debris. Site numbers correspond with those in Table 1

to May 20 and July 31 to August 7, 2012. The 4KC was equipped with a Super-HARP (high-gain avalanche rushing photoconductor) video camera recording in NTSC (National Television System Committee specification) format for downward-facing seafloor observations, and this allowed observation of an area of approximately 3×3 m at an altitude of 3 m above the seafloor. The YKDT was equipped with a similar camera configuration but with a less light-sensitive CCD (charge-coupled device) camera and also recording in NTSC format. An ROV (remotely operated vehicle) survey was carried out during June 30 to July 5 in 2013. The ROV *Crambon* was equipped with a pan-tiltable, forward-facing, high-definition (1080i) video camera with a 67° view field angle and with video recorded onto a hard disk cartridge (AJA WZZ-KiPro2400) at ProRes 4:2:2HQ resolution via HD-SDI and saved in QuickTime format (.mov). Video and other data were time-stamped and recorded synchronously using an Ethernet-enabled logging/playback system based on that developed by Lindsay et al. (2012). We assumed 2 m as the observation width for the calculation of the approximate number of debris items per km^2. Side-scan sonar (EdgeTech 4200MP, 400/100 kHz) was used to obtain topographic data and marine debris occurrence information. In the case of coastal areas, another side-scan sonar system (System 3000, Klein Inc., 500/100 kHz) was deployed from a fishing boat.

3 Results

The deep-tow camera survey from May 16 to May 20, 2012, revealed a broad distribution of the brittle star *Ophiura sarsii* (Lütken 1855). It was especially dense during transect DT-6C but not during DT-7C (nos. 14 and 15 in Table 1 and Fig. 1). During transects DT-6C to DT-9C, several small items of debris, such as splinters of

Table 1 Survey point information

Cruise no. / platform	ID	Site name	Latitude	Longitude	Depth [m]	Distance from shore [km]	Date	Dive no.	Survey area [km^2]	Survey length [m]
KK-13-1	**1**	Off Minamisanriku-cho	38.56540	141.95063	**289**	27.3	2013/6/30	CRM005		267
ROV (*Crambon*)	**2**	Off Kamaishi (canyon)	39.24870	142.21763	**477**	16.9	2013/7/1	CRM006		622
	4	Off Otsuchi (canyon)	39.42470	142.25160	**457**	13.4	2013/7/2	CRM008		1,660
	5	Yamada Bay	39.46550	141.99539	46	0.7	2013/7/3	CRM009		42
	6	Yamada Bay	39.46580	141.99550	46	0.6	2013/7/3	CRM010		53
YK12_12	7	Bacterial mat site, trench	37.73170	143.28459	3,503	126.6	2012/7/31	YKDT147		3,056
Deep-tow (YKDT)	9	39 N site, Japan Trench	39.10900	143.88912	5,221	132.3	2012/8/2	YKDT148		3,002
	10	Crack site, Japan Trench	38.19850	143.78230	5,665	147.5	2012/8/3	YKDT149		3,816
	11	Animal site, Japan Trench	38.65600	143.57869	3,101	121.4	2012/8/4	YKDT150		2,808
	12	Submarine canyon off Miyako	39.43170	142.28046	**551**	11.1	2012/8/5	YKDT151		12,008
	13	39 N site, Japan Trench	39.10420	143.89455	5,152	133.3	2012/8/7	YKDT152		1,596
MR12_E2	**14**	Off Oshika Peninsula	38.53640	141.97525	**320**	27.6	2012/3/16	DT-6C		8,618
Deep-tow (4KC)	**15**	Off Oshika Peninsula	38.52760	142.04983	**440**	32.7	2012/3/17	DT-7C		6,333
	16	Off Sendai Bay	37.88120	142.31117	**890**	57.5	2012/3/18	DT-8C		3,830
	17	Off Kesennuma	38.83690	142.06783	**349**	20.5	2012/3/19	DT-9C		1,080
	18	Off Kamaishi (canyon)	39.24240	142.29475	**727**	19.5	2012/3/20	DT-10C		7,167
	19	Off Kamaishi (canyon)	39.24390	142.24025	**567**	17.6	2012/3/20	DT-11C		2,257
Chartered ship	**21**	Off Iwate	39.36259	142.14869	**240**	7.2	2013/3/12–13		46.0	
Side-scan sonar	**22**	Off Miyagi	38.95898	141.96978	**190**	9.8	2013/3/15–17		26.0	
	23	Off Miyagi	38.89315	141.88641	**160**	12.2	2013/3/15–17		12.0	
Chartered ship	24	Hirota Bay	39.10483	141.82088	15	0.0	2012/7/31–8/2, 8/6–11, 9/3–8		1.5	
Side-scan sonar	25	Okirai Bay	38.99597	141.64145	15	0.2	2012/7/26–30, 9/9–13		0.3	
	26	Tōni Bay	39.19732	141.87830	20	0.3	2012/7/23–25, 9/13–16		1.4	

Depths in bold-type letters were our main study depths

Fig. 2 Accumulated marine debris in a submarine valley (Deep Tow 10C image)

Fig. 3 Cans observed near the trench

wood and plastic bags, were found. During transects DT-10C and DT-11C, near a submarine valley off Iwate, many differently sized items of debris were observed clumped together as shown in Fig. 2.

During the cruise from July 31 to August 7, 2012, marine debris were found at 3000 m depth in the Japan Trench area, more than 200 km from land, using the YKDT deep-tow camera (nos. 7–13 in Table 1). A fishing net, cans, plastic bags, and sunken wood were scattered over an area of seafloor (Fig. 3). The origin of these debris could not be determined as to whether it was related to the tsunami event or not. At a shallower study site (depth 551 m, no. 12 in Table 2), we discovered high abundances of invertebrates and fish (such as *Ophiura sarsii*, sea stars, Anguilliformes sp., and *Sebastolobus macrochir* (Günther 1877)) in a submarine canyon area where much debris had accumulated. An extremely large number of the anthomedusan jellyfish *Euphysa japonica* were observed near the seafloor at some stations, but the identity and substrate habitat preferences of its benthic polyp stages have not yet been determined (Lindsay et al. 2008), so an analysis of correlations with debris distributions was not possible at the present time.

During the ROV survey in 2013, high densities of brittle stars were observed at stations 1 and 2 (Table 1 and 2) but not as many occurred at station 4. At station 2, the body sizes of brittle stars were significantly different from those occurring at the other stations. At station 4, we found drag marks made by trawl nets and a lower diversity and population size of macrofauna.

Although the area surveyed by the ROV *Crambon* was limited, detailed information on debris could be collected (Fig. 4). The number of debris items was obviously higher within the bay area (no. 5; the Yamada Bay) on the July of 4th. Eight debris items were observed in an approximately 50×50 m area.

Surveys by side-scan sonar were able to investigate broader areas of the deep-sea floor, although the resolution of acoustic images was lower than that of the video images obtained during the ROV survey (nos. 21 to 23 in Table 2). The acoustic survey off Miyagi found a large debris item of approximately 10–15 m in length (no. 23). In coastal areas, numerous small debris items, mostly less than 1 m in length, were able to be distinguished in the acoustic images (Fig. 5; nos. 24–26 in Table 2).

As shown in Tables 1 and 2, the amount of debris decreased substantially with increasing water depth and/or distance from the coastline. The average number of debris items at the ten main, video-surveyed study sites (1, 2, 4, 12, 14–19) at depths from 289 to 890 m and at distances from 11 to 57 km from the coastline, was 5,195 (standard deviation [SD]=5,133) per km^2 (Tables 1 and 2). It was 1,116 per km^2 (SD=961) at the five sites (7, 9–11, 13) in deeper areas near the trench (>1,000 m depth), which were over 100 km from the coastline. In canyon areas (2, 4, 12, 18, 19), the average number of debris items per km^2 was 6,761 (SD=3,528), and this was higher than at other areas of similar depth that we studied (mean=3,629, SD=6,381). Of the observed debris categories, "plastics" were most numerous (113), followed by "wood and paper" (85). A large subset of the debris (over 237 items) were unclassifiable.

4 Discussion

Our preliminary results suggest that the direct impact of tsunami-derived debris in deep-sea areas off the Sanriku region (ca. 200–800 m depth) is not as severe as in coastal environments (Seike et al. 2013; Urabe et al. 2013). Accumulations of tsunami-derived debris were found mainly within submarine canyons in deep-sea areas, while in coastal areas the debris was scattered across the seafloor. The transportation power of tsunamis, turbidity currents, and the geomorphological features of the seafloor appear to greatly affect the distribution pattern of marine debris (Arai et al. 2013).

Our observations also showed that different kinds of debris, such as wood, plastics, and construction materials, still occur in mixed aggregations. These debris items accumulate sessile species that need hard substrates on which to attach. This is assumed to be because of the lack of such surfaces on the largely muddy deep-sea floor, as was observed in the canyon area. Different materials are expected to move and decompose in different ways, and the composition of the debris aggregations is therefore expected to change over time due to transportation and decomposition. For example, plastics can be harmful when accidentally ingested (Ramirez-Llodra et al. 2011). Flustering of plastics is also suspected to contribute to the accumulation of harmful substances in organisms, where micro-plastics are known to be a problem (Andrady 2011). In the case of wood or mixtures of wood and fabric materials, such debris attracts

Table 2 Properties of marine debris items extracted by the primary analysis

ID	Site name	Dive no.	No. of debris items	Debris per km^2	Types of debris (%): Wood, paper	Cans, bottles	Plastics, packages	Steel, concrete	Fabrics, cloth	Nets, cables	Others
1	Off Minamisanriku-cho	CRM005	8	15,009	50	13	0	13	13	0	13
2	Off Kamaishi (canyon)	CRM006	7	5,627	29	14	29	0	0	0	29
4	Off Otsuchi (canyon)	CRM008	9	2,710	78	0	11	0	0	0	11
5	Yamada Bay	CRM009	2	954	0	0	0	0	0	100	0
6	Yamada Bay	CRM010	6	2,267	17	17	0	33	0	17	17
7	Bacterial mat site	YKDT147	24	2,618	21	4	17	8	0	0	50
9	39 N site, Japan Trench	YKDT148	9	999	0	0	33	0	0	11	56
10	Crack site, Japan Trench	YKDT149	0	0	–						
11	Animal site, Japan Trench	YKDT150	6	712	50	17	0	0	0	0	33
12	Submarine canyon off Miyako	YKDT151	226	6,274	15	12	17	1	2	8	46
13	39 N site, Japan Trench	YKDT152	6	1,253	33	0	17	0	0	0	50
14	Off Oshika peninsula	DT-6C	31	1,199	29	10	0	23	0	3	35
15	Off Oshika Peninsula	DT-7C	26	1,368	31	0	12	8	8	0	42
16	Off Sendai Bay	DT-8C	3	261	0	0	33	0	0	0	67
17	Off Kesennuma	DT-9C	1	309	0	0	0	0	0	0	100
18	Off Kamaishi (canyon)	DT-10C	146	6,790	6	20	22	4	4	3	41
19	Off Kamaishi (canyon)	DT-11C	84	12,406	2	19	33	4	15	0	26
21	Off Iwate		0	0	–						
22	Off Miyagi		0	0	–						
23	Off Miyagi		2	0	Not distinguishable						
24	Hirota Bay		1,148	765	Not distinguishable						
25	Okirai Bay		506	1,687	Not distinguishable						
26	Tōni Bay		530	379	Not distinguishable						

IDs of bold-type letters are our main focus

Fig. 4 Wood observed by the ROV

Fig. 5 Marine debris observed by the side-scan sonar

species that can decompose cellulose materials, such as members of the wood-boring mussel subfamily Xylophagidae, which were retrieved in a sample from this area. In the case of nets or wire debris, a negative effect on fisheries has already been reported by Goto (2012). For sustainable fisheries, as a first step the number of ghost fishing nets and plastics of ingestible sizes should be reduced. The next step should be to track the stage/status of flustering in plastics and determine the distribution of various marine debris materials.

Acknowledgments We would like to thank all members of cruises for their kind help before, after, and during surveys. Especially, Yasuo Furushima, Katsunori Fujikura, Shuro Yoshikawa, and Haruka Shibata helped us to access data from each cruise. We also would like to thank two reviewers for their helpful comments. This research was partially supported by the Tohoku Ecosystem-Associated Marine Sciences (TEAMS) research program funded by the Ministry of Education, Culture, Sports, Science and Technology (MEXT).

References

Andrady A (2011) Microplastics in the marine environment. Mar Pollut Bull 62:1596–1605

Arai K, Naruse H, Miura R, Kawamura K, Hino R, Ito Y, Inazu D, Yokokawa M, Izumi N, Murayama M, Kasaya T (2013) Tsunami-generated turbidity current of the 2011 Tohoku-Oki earthquake. Geology 41:1195–1198

Goto T (2012) Assessment of gear condition and ghost-fishing ability in bottom gillnets left in the coastal waters of Iwate, Pacific coast of northern Japan, by the Tohoku earthquake and tsunami disaster in March 2011. Nippon Suisan Gakkaishi 78:1187–1189

Goto K, Chagué-Goff C, Goff J, Jaffe B (2012) The future of tsunami research following the 2011 Tohoku-oki event. Sediment Geol 282:1–13

Lindsay DJ, Pagès F, Corbera J, Miyake H, Hunt JC, Ichikawa T, Segawa K, Yoshida H (2008) The anthomedusan fauna of the Japan Trench: preliminary results from *in situ* surveys with manned and unmanned vehicles. J Mar Biol Assoc U K 88(8):1519–1539

Lindsay DJ, Yoshida H, Uemura K, Yamamoto H, Ishibashi S, Nishikawa J, Reimer JD, Fitzpatrick R, Fujikura K, Maruyama T (2012) The untethered remotely-operated vehicle PICASSO-1 and its deployment from chartered dive vessels for deep sea surveys off Okinawa, Japan, and Osprey Reef, Coral Sea, Australia. Mar Technol Soc J 46(4):20–32

Ozawa S, Nishimura T, Suito H, Kobayashi T (2011) Coseismic and postseismic slip of the 2011 magnitude-9 Tohoku-Oki earthquake. Nature 475:373–376

Ramirez-Llodra E, Tyler P, Baker M (2011) Man and the last great wilderness: human impact on the deep sea. PLoS ONE 6:e22588

Seike K, Shirai K, Kogure Y (2013) Disturbance of shallow marine soft-bottom environments and megabenthos assemblages by a huge tsunami induced by the 2011 M9. 0 Tohoku-Oki. PLoS ONE 8:e65417

Shibata H, Furushima Y, Fujikura K, Miyake H, MR12-E02 and YK12-12 Members A (2013) Distribution of Tsunami debris at submarine canyon off Sanriku. In: Abstract B. Blue Earth 2013 BE13:181

Urabe J, Suzuki T, Nishita T, Makino W (2013) Immediate ecological impacts of the 2011 Tohoku earthquake tsunami on intertidal flat communities. PLoS ONE 8:e62779

Oithona davisae, the Most Predominant Copepod in Tokyo Bay, a Highly Eutrophic Embayment: Why Are They So Predominant?

Yuji Tanaka and Tatsuro Akiba

Abstract

Oithona davisae is a planktonic cyclopoid copepod species, known to be extremely abundant and often predominates in coastal marine embayment. To understand why this animal is so successful, we have been looking into its swimming ability to escape from predators. Here we show how *O. davisae* escapes from the moon-jelly *Aurelia aurita*, which devours zooplankton and occurs in a huge number in embayments including Tokyo Bay. Direct observations revealed that *O. davisae* is agile enough to escape from the moon-jelly's *ephyra* larvae, which appear much more numerously than adult moon-jelly. This agility reducing the predation mortality may be crucial for *O. davisae* to predominate in this bay and somewhere else that are full of predators. Direct observations by the use of video cameras suggested that *O. davisae* may be recognized as a genius in escaping from predators, comparing with some other planktonic animals such as *Acartia* (larger copepod), barnacle *cypris*, decapod zoeas, etc., being less agile than *O. davisae*.

1 Introduction

Among various factors, anthropogenic loading and global warming are considered to be causing significant changes in the coastal marine environment. We need to holistically understand mechanisms in such changes to predict how the environment would be in the future. Prerequisite is detailed knowledge on biological processes as well as physical and chemical ones in response to the environmental changes. Planktonic copepods are important in the aquatic ecosystems because of their abundance and diversity; they are considered to bridge the primary producers (phytoplankton) and higher consumers (such as larger zooplankton and fish). (Nishida 1985; Tsuda and Nemoto 1988; Itoh and Aoki 2010).

Among them, a cyclopoid copepod *Oithona davisae* (Fig. 1) is one of the most predominant species in the zooplankton community of coastal

Y. Tanaka (✉)
Tokyo University of Marine Science and Technology,
4-5-7 Konan, Minato, Tokyo 108-8477, Japan
e-mail: ytanaka@kaiyodai.ac.jp

T. Akiba
National Institute of Advanced Industrial Science
and Technology, 1-1-1 Higashi,
Tsukuba, Ibaraki 305-8522, Japan
e-mail: ta-akiba@aist.go.jp

H.-J. Ceccaldi et al. (eds.), *Marine Productivity: Perturbations and Resilience of Socio-ecosystems*,
DOI 10.1007/978-3-319-13878-7_12,

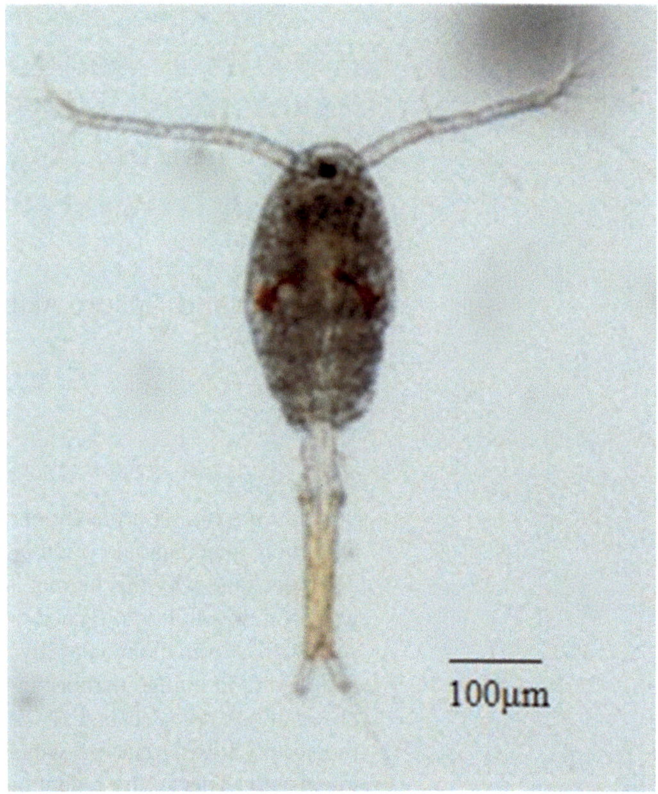

Fig. 1 Adult of *Oithona davisae*, of which prosome length is ca. 400 μm

waters including embayments such as Tokyo Bay (Ferrari and Orsi 1984; Anakubo and Murano 1991; Itoh et al. 2011), one of the most eutrophic embayment in the world.

This bay is also known to be inhabited by abundant moon-jelly *Aurelia aurita*, having planktonic stages being voracious to smaller zooplanktons such as copepods (Omori et al. 1995). It is intriguing that *O. davisae* remains so predominant in spite of the abundant existence of the "copepod eater" *A. aurita*. In this context, we may speculate that certain reasons should exist for the coexistence of such abundant *A. aurita* and *O. davisae* as its prey.

In the present paper, we will (a) show some direct observations on the feeding behavior of moon-jelly's *ephyra* in laboratory conditions; (b) discuss the special feature of the behavior of *O. davisae*, able to escape easily from the *ephyra*; and (c) speculate the reason why *O. davisae* is predominant in Tokyo Bay today.

2 Materials and Method

2.1 *Ephyra* as Predator

In order to perform direct observations of predator-prey interactions between moon-jelly's *ephyra* and zooplankton, after the method previously established (Ishii et al. 2004), from mature moon-jelly *Aurelia aurita* captured from the innermost part of Tokyo Bay, planula larvae were obtained first, and then through the polyps, after strobilation, the *ephyra* larvae were liberated (Fig. 2).

2.2 Zooplankton as Prey

As preys of the *ephyra*, zooplankton such as *Oithona davisae*, *Acartia omorii*, *cypris* larvae of barnacles, zoea larvae of decapods, and adults of a small species of hydromedusae *Rathkea*

Fig. 2 *Ephyras* of moon-jelly *Aurelia aurita*. These were newly liberated from strobila being cultured in the laboratory. The whole diameter is ca. 3 mm

octopunctata were captured from the innermost part of Tokyo Bay and kept in a healthy condition. In addition, nauplii of *Artemia* were obtained from canned cysts.

2.3 Video Recording

A flat-drum chamber (100 mm diameter, 25 mm thick) made of Plexiglas filled with filtered seawater was used to perform direct observations on the feeding of *ephyra* on each kind of zooplankton. Holding the flat surface upright, while slowly rotating the chamber (1 revolution per minute) containing the *ephyra* and a limited number of one kind of zooplankton, recordings were done with a Sony Handycam video (30 frames per second).

Transparent cells of 40 mm cube and also cells for absorbance meter (10 mm × 10 mm and 50 mm tall) filled with filtered seawater and standing still were also used to make closer view of the prey and predators' behavior. Video images were analyzed frame by frame afterwards.

3 Results

3.1 *Oithona*

Live adult of *Oithona davisae* did not show any special behavior while the *ephyras* were not in the vicinity. However, once the *ephyras* were came near, they easily escaped. As a consequence of such behavior, almost no *Oithona* was captured by *ephyra* (Fig. 3).

3.2 *Acartia*

Acartia which is larger in size than *Oithona* is expected to swim faster and thus thought to be able to escape better than *Oithona*. However, unlike the case of *Oithona*, *Acartia* did not always succeed to escape, but sometimes were in contact with *ephyra* and eventually eaten (Fig. 4).

3.3 *Cypris*

Cypris larvae of barnacles were very often into contact with *ephyra*, but never eaten.

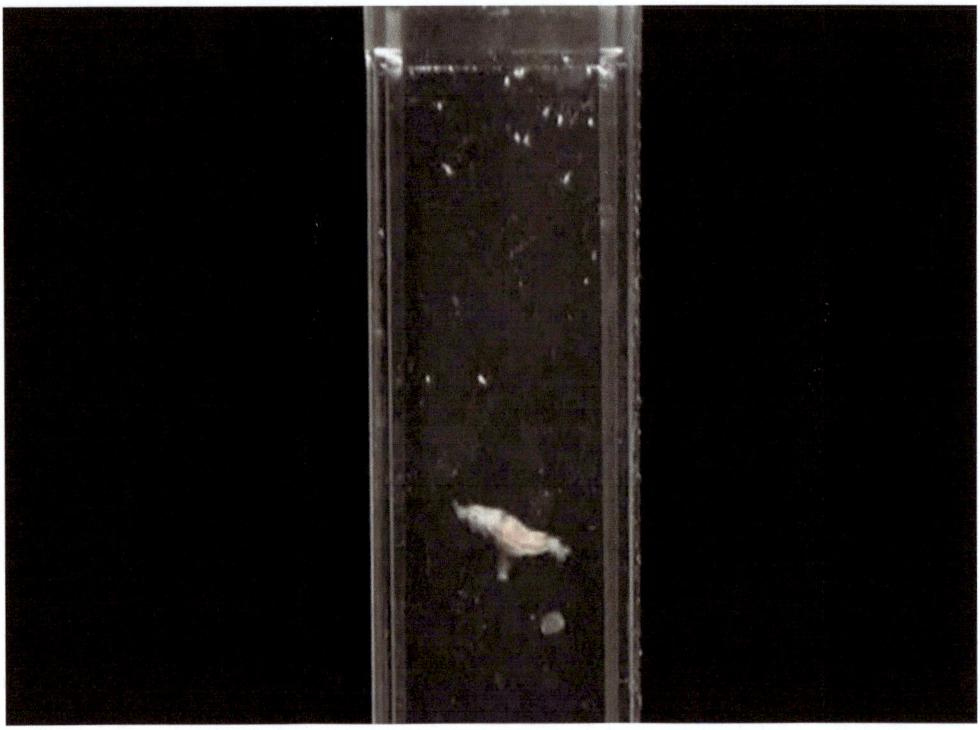

Fig. 3 Side view of the 10 mm absorbance cell containing an *ephyra* and ten *Oithona*. After this clip, the *ephyra* swam upward near the surface where *Oithona* is swimming. Nevertheless, not a single *Oithona* was in contact with *ephyra*. All the individuals of *Oithona* easily escaped from the *ephyra*

Fig. 4 Side view of the 40 mm cubic chamber containing an *ephyra* and a few *Acartia*. After this clip, the *ephyra* swam upward near the surface where *Acartia* were located. Unlike the case of *Oithona*, *Acartia* did not always succeed to escape, but sometimes were in contact with *ephyra* and eventually eaten

Fig. 5 Side view of the rotating flat drum chamber containing the *ephyra* and *Artemia* larvae. The *ephyra* in the top-middle of the image just captured two *Artemia*

After the contact with *ephyra, cypris* stopped swimming and made the body curled to sink. They restarted swimming after a delay ranging from several seconds and 2 min after the contact.

3.4 Zoea

Zoea larvae of decapods were often in contact with *ephyra*, but never eaten. It seems they were not seriously influenced by the *ephyral* venomous stings, probably having harder shell to avoid stings of the jellyfish tentacles.

3.5 *Rathkea*

In the vicinity of *ephyra*, the small hydromedusa *Rathkea octopunctata* seemed to try escaping. A single hit of *ephyra* seemed to be was enough to paralyze and capture *Rathkea*.

3.6 *Artemia*

Nauplius larvae of *Artemia* did not seem to avoid *ephyra* and were easily captured (Fig. 5). Once in contact with *ephyra*, the nauplii quickly got paralyzed probably by the *ephyra* venom. Additional observations using *Artemia* eggs as prey showed that the eggs were not captured by the *ephyra* (Fig. 5).

4 Discussion

We observed that *O. davisae*, being different from other zooplanktons, are not easily eaten by the *ephyra* of moon-jelly. In other words, they are capable of avoiding *ephyral* predation very well, suggesting that not all kinds of zooplankton are passively eaten by every stage of the moon-jelly. This observation may give a clue to understand why *O. davisae* are predominant in Tokyo Bay today. Our speculation is as follows:

- The recent predominance of *O. davisae* in Tokyo Bay may partly be explained by the increase in the abundance of moon-jelly due to the heavy increase of artificial constructions on the coast offering the substrata of the asexually growing polyps of the moon-jelly (Ishii and Katsukoshi 2010), directly followed by the *ephyra* and medusae.
- In addition, the increase in the occurrence of oxygen depletion in the lower layer of inner Tokyo Bay (Ando et al. 2005) may cause the general decrease of species diversity of planktons except for the moon-jelly that is less vulnerable in such oxygen-depleted environment (Ishii et al. 2004).
- Increase in *ephyras* of moon-jelly due possibly to the above reasons, which voraciously eat zooplanktons except for *O. davisae*, may eventually lead to enhance the predominance of *O. davisae*.
- Moreover, not spawning but carrying the eggs, being different from many other copepods, reproduction of *O. davisae* may not seriously be affected by the development of oxygen depletion in the bottom layer (Uye 1994).

It has been pointed out that the predominance of *O. davisae*, with a very small size (even adults >0.5 mm) and thus grazing smaller phytoplankton such as dinoflagellates but not diatoms, leads to leftovers and excess growth of diatoms that sink to the aphotic bottom (7). The increase of organic materials in the aphotic bottom deteriorates the oxygen depletion due to excess decomposition. This possible process of enhancement of oxygen depletion may also cause the increase of moon-jelly and predominance of *O. davisae* again. These processes may lead to some loss of species diversity in Tokyo Bay. To avoid this spiral, reduction of nutrient input and mitigation of artificial construction of the coastline are possible measures. Resilience of the Tokyo Bay ecosystem may not well function without reviving the natural coastline or without reducing the nutrient loading. To examine if the above speculation is valid, we are working on more quantitative studies on the prey-predator relationship between zooplanktons including *O. davisae* and the moon-jelly *A. aurita* (not only the *ephyra* stage but also the polyp and medusa stages).

References

Anakubo T, Murano M (1991) Seasonal variation of zooplankton in Tokyo Bay. J Tokyo Univ Fish 78(2):145–165

Ando H, Kashiwagi N, Ninomiya K, Ogura H, Kawai T (2005) Changes in the state of water pollution in Tokyo Bay since 1980 – trend analysis of water quality using monitoring data obtained by Local Governments. In: Annual report of the Tokyo Metropolitan Research Institute for Environmental Protection, Tokyo Metropolitan Research Institute for Environmental Protection, Tokyo, pp 141–150 (in Japanese)

Ferrari FD, Orsi J (1984) *Oithona davisae*, new species, and *Limnoithona sinensis* (Burckhardt, 1912) (Copepoda: Oithonidae) from the Sacramento-San Joaquin Estuary, California. J Crustac Biol 14(1):106–126

Ishii H, Katsukoshi K (2010) Seasonal and vertical distribution of *Aurelia aurita* polyps on a pylon in the innermost part of Tokyo Bay. J Oceanogr 66(3):329–336

Ishii H, Kojima S, Tanaka Y (2004) Survivorship and production of *Aurelia aurita ephyrae* in the innermost part of Tokyo Bay, Japan. Plankton Biol Ecol 51(1):26–35

Itoh H, Aoki N (2010) Temporal and spatial distribution of planktonic copepods in Tokyo Bay: seasonal occurrence in the innermost part of the bay, in the early 1990s. Bull Plankton Soc Jpn 57(2):94–104 (in Japanese)

Itoh H et al (2011) Vertical distribution of planktonic copepods in Tokyo Bay in summer. Plankton Benthos Res 6(2):129–134

Nishida S (1985) Taxonomy and distribution of the family Oithonidae (Copepoda, Cyclopoida) in the Pacific and Indian Oceans. Bull Ocean Res Inst Univ Tokyo 20:1–167

Omori M, Ishii H, Fujinaga A (1995) Life history strategy of *Aurelia aurita* (Cnidaria, Scyphomedusae) and its impact on the zooplankton community of Tokyo Bay. ICES J Mar Sci J du Conseil 52(3–4):597–603

Tsuda A, Nemoto T (1988) Feeding of copepods on natural suspended particles in Tokyo Bay. J Oceanogr 44(5):217–227

Uye S (1994) Replacement of large copepods by small ones with eutrophication of embayments: cause and consequence. Hydrobiologia 292/293:513–519

Monitoring of a Potential Harmful Algal Species in the Berre Lagoon by Automated *In Situ* Flow Cytometry

Mathilde Dugenne, Mélilotus Thyssen, Nicole Garcia, Nicolas Mayot, Guillaume Bernard, and Gérald Grégori

Abstract

The vicinity of urban activity and industry (petrochemistry) around the Berre lagoon (southeast of France) has induced the degradation of its ecosystem, characterized by a permanent eutrophic state. In particular, a power plant has discharged substantial inputs of enriched freshwater in the lagoon since 1966. Due to these high nutrient inputs and also to regeneration rates, several species of phytoplankton regularly bloom in the lagoon at spring, summer, or autumn. Peaks of phytoplanktonic biomass (>150 μg Chl*a*/dm^3) are generally followed by intense heterotrophic activities leading to O_2 consumption with hypoxic or anoxic episodes. The study of phytoplankton dynamics is thus of primary importance.

Within the framework of an ecological survey, an automated platform of *in situ* instruments was set up in a laboratory located on the "Berre l'Etang" harbor to assess biological and hydrological features of the lagoon at short time scale during October 2011. The phytoplanktonic community was characterized by a Cytosense autonomous flow cytometer (Cytobuoy, Netherlands) operating at high frequency (hourly sampling) in order to detect sudden changes of species compositions and abundances. In parallel, hydrological sensors have measured several physicochemical variables of the water directly pumped from the lagoon (1.5 m depth) to the field laboratory.

In October 2011, among the various phytoplanktonic species optically resolved by the Cytosense, the dinoflagellate *Akashiwo sanguinea* has

M. Dugenne • M. Thyssen
N. Garcia • G. Grégori (✉)
Université Sud Toulon Var, IRD, CNRS,
Aix-Marseille University, Marseille, France

Mediterranean Institute of Oceanography,
Marseille, France
e-mail: gerald.gregori@mio.osupytheas.fr

N. Mayot • G. Bernard
GIPREB Syndicat Mixte, Berre-l'Étang, France

H.-J. Ceccaldi et al. (eds.), *Marine Productivity: Perturbations and Resilience of Socio-ecosystems*,
DOI 10.1007/978-3-319-13878-7_13, © Springer International Publishing Switzerland 2015

been detected. It is known for causing fish and seabird death when its abundance is high and for having a potential toxicity on human consumed shellfish. On October 6, nutrient concentration combined with a weak hydrodynamic state triggered a large development of this species. Concentration reached up to 450 cells/cm^3, leading to a fast increase in chlorophyll concentration (>20 µg Chl*a*/dm^3). However, these dinoflagellates were quickly affected after a sudden wind (mistral, 330°–360°) event with mean speed exceeding 20 m/s: on October 7, 2011, *A. sanguinea* abundance dropped down below 20 cells/cm^3.

These results, out of reach of more conventional methods with manual sampling, underline the power of *in situ* automated monitoring to follow in near real time the dynamics of phytoplankton in general and also to target some particular species of interest such as harmful algae.

1 Introduction

Phytoplanktonic species play a key role for marine productivity. These primary producers fuel the entire trophic network and modulate biogeochemical cycles (N, C, P, Si, S, Fe). When environmental conditions favor their bloom, some species can be responsible for massive death of organisms, commercial species (Smayda 1997). These events, known as harmful algal blooms (HABs), represent a major risk for the marine fauna, from filter feeders to fishes, but also marine mammals and seabirds, as they can be subject to toxic or nontoxic (anoxia, gill clogging, physiological changes) effects leading to their death (Zigone and Enevoldsen 2000). HABs are caused by numerous species of dinoflagellates, cyanobacteria, diatoms, prymnesiophytes, and raphidophytes which are harmful in many ways. Thus, they require an accurate monitoring to detect them as early as possible.

Phytoplankton analyses are based on their optical properties via microscopy or fluorimetry, both considered as traditional methods used to detect algal proliferations (Anderson et al. 2001). In France, the national program REPHY (REseau de surveillance du PHYtoplancton et des PHY-cotoxines) conducted by IFREMER (Institut Français de Recherche pour l'Exploitation de la MER) is in charge of phytoplankton surveys along the French shore. It aims at enhancing our understanding about factors controlling phytoplanktonic blooms. By means of early warnings, this network also intends for preventing poisoned shellfish consumption by triggering some particular toxic species. Observations are based on microscopic counts which are time consuming and require expertise. The sampling frequency is, typically, from one per month to up to one per week during risky periods (in May/June for most of the sites).

Located in the south of France, the Berre lagoon is one of the largest marine lagoons in Europe and around the Mediterranean Sea. Because it has received a large amount of nutrients over the years, the lagoon used to regularly host algal blooms characterized by the dominance of few species, with extreme biomass (Malkassian 2012). Since 1995, this site benefits from a special attention relying on European (Directive Cadre sur l'Eau (DCE)), regional (Réseau de Suivi Lagunaire (RSL)), and local (Groupement d'Interêt Public pour la Réhabilitation de l'Etang de Berre (GIPREB)) frameworks. In general, lagoons are both highly dynamic and sensitive to human activities because of their intermediate state between land and sea. Besides a long-term monthly monitoring of phytoplankton in the Berre lagoon, it appears that the establishment of the annual budget of primary production is still incomplete. This is likely due to important seasonal changes (with, for instance, sporadic events such as natural flooding or strong wind events) (Gouze et al. 2008).

Even nowadays, little is known about the phytoplankton dynamics at fine scale and about the variables controlling it. In shallow coastal ecosystems, phytoplankton abundances are influenced, at different time scales, by grazing, local inputs (nutrients), and turbulent mixing, as described by Cloern (1996). However, conventional sampling methods are not suitable to assess their changes induced by sudden sporadic events such as wind episodes, freshwater discharges, or spates. There is thus a need for devices able to run automated and high-frequency analysis independently of the weather conditions and of the presence of a person. Such instruments would be very valuable to detect blooms at an early stage and provide a warning about a potential harmful algal bloom. This obviously requires analyses performed at a taxonomic level.

In this context, we report in this study a high-frequency *in situ* monitoring of phytoplankton in the Berre lagoon by means of a Cytosense automated flow cytometer (Cytobuoy b.v., Netherlands). This project has been supported by CNRS EC2CO, FEDER, and the Council of Provence Alpes Côte d'Azur Region. Abundances were measured every hour during October 2011 for species ranging from pico- to small micro-phytoplankton, some of which identified as harmful such as *Akashiwo sanguinea*.

2 The Berre Lagoon

The Berre lagoon receives freshwater inputs from both three natural rivers and from the artificial channel of a hydroelectric power station (Electricité de France (EDF)) built on the north-eastern shore. The freshwater from the Durance River is thus derived and discharged in the lagoon, bringing with it massive amounts of sediments and organic matter. The lagoon also receives Mediterranean seawater through the Caronte channel. It is subjected to multiple natural (frequent wind events, freshwater discharges, and spates) and anthropic forcing (industries, inhabitations). Due to the proximity of several small cities and from the Marseille metropolis (about 800,000 inhabitants), a pressuring instatement of urban, industrial, and agricultural exploitations has emerged all around the lagoon shore. Since 1966, the main disturbing infrastructure has been the power plant, using and discharging freshwater of the Durance River for electricity supply (by turbines) into the Berre lagoon, with a drastic impact on salinity (by important dilution). All these human activities have generated wastes and inputs that were drained in the lagoon and enriched it, particularly in nitrogen and phosphorus. In 2011, the power plant was responsible for 50 % and 15 % of total nitrogen and phosphorus inputs, respectively. These high inputs combined with the fast regeneration rates of the microbial community (Gouze et al. 2008) have fueled the phytoplanktonic species, and as a result, the water mass has become ultimately eutrophic. For many years, blooms associated with extreme autotrophic biomass (>150 µg Chl*a*/dm^3) and colored waters have been observed until a regulation of the power plant discharges was imposed to EDF in 1994 and 2005 (Malkassian 2012). The salinity gradient between surface and bottom of the lagoon maintains a quasi-permanent stratification of the water column (Nérini et al. 2001) and prevents reoxygenation of the deep layer after the oxygen depletion that follows blooms. Hypoxia or in extreme case anoxia, light attenuation, and toxin production during these blooms have perturbed the entire ecosystem, which remains eutrophic (Fig. 1) and in a "bad" ecological state according to RSL (Réseau de Suivi Lagunaire) evaluation criteria (Mayot et al. 2013).

3 The Experimental Setup

In order to detect sudden changes in composition and abundances of the phytoplanktonic community in the Berre lagoon, the Cytosense (Cytobuoy, Netherlands) flow cytometer was placed in a laboratory nearby the "Berre l'Etang" harbor (Fig. 2).

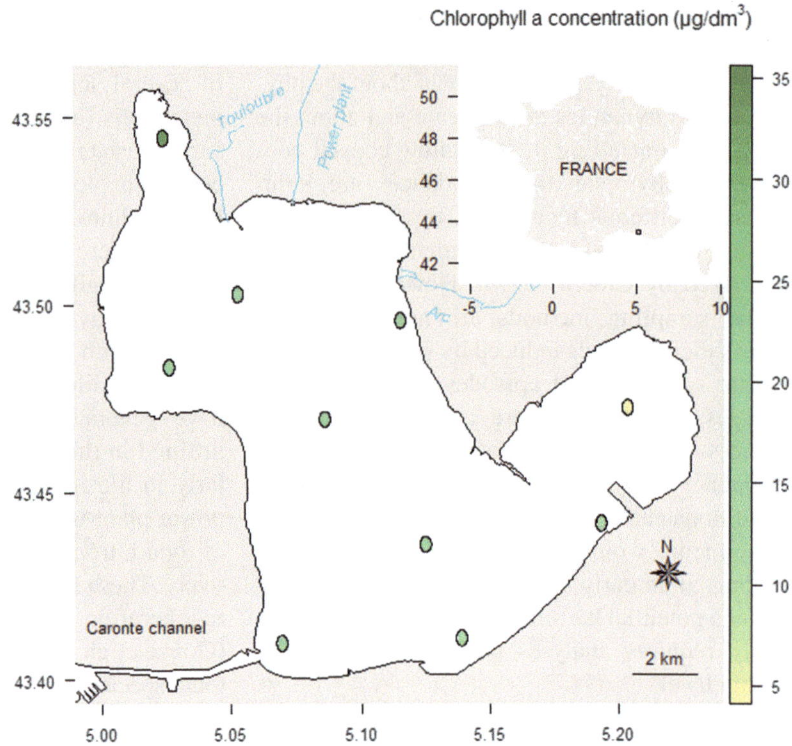

Fig. 1 Surface chlorophyll *a* concentration measured at the 10 hydrological stations in the Berre lagoon for the sampling performed in September 2011

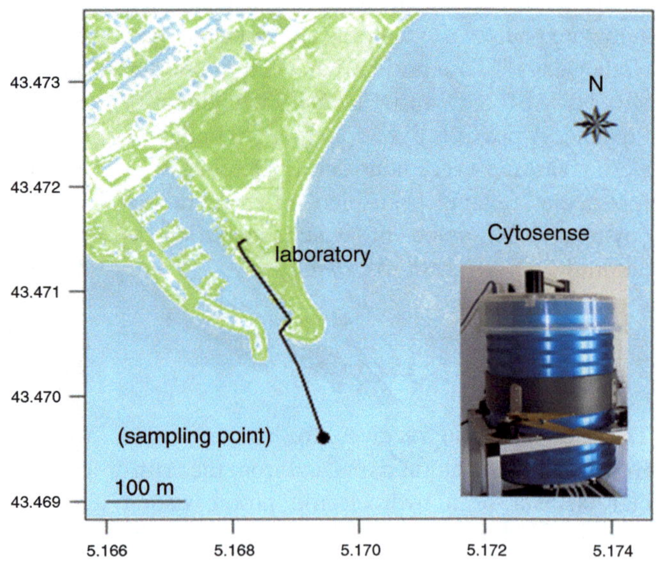

Fig. 2 Location of the sampling point (*black circle*) and the 250 m pipe (*black line*) until the laboratory, on the "Berre l'Etang" harbor. Flow cytometry analyses were automatically run every hour by the Cytosense flow cytometer. When needed, the cytometer could be remotely operated thanks to an Internet connection

This instrument is designed to operate autonomous *in situ* monitoring of a wide number of phytoplanktonic species in the range of size between 1 and 800 µm (Dubelaar et al. 1999). Analyses are made at the single cell level and at high frequency (up to once every 10 min) (Dubelaar and Gerritzen 2000; Peeters et al. 1989). This instrument has been used to reveal the temporal and spatial phytoplankton variability *in situ* (Thyssen et al. 2008). Autotrophic cells are triggered on the basis of their chlorophyll *a* content, which naturally emits red fluorescence. When the cells pass through the light source excitation (a 488 nm laser beam), five optical profiles (two diffusion intensities (forward angle and sideward light scatter) and three fluorescence intensities (red, orange, and yellow fluorescences)) are recorded for each single particle (cell). Cells sharing similar optical properties are therefore grouped together when plotted in 2D projections of the numerous variables collected by the flow cytometer (2 light scatter and three fluorescences, with for each one the area under the curve, the peak and length of the curve, etc.). In addition, an "image-in-flow" device mounted in the Cytosense takes pictures of cells of interest after their passage through the 488 nm laser beam.

In parallel to flow cytometry analyses, hydrological variables were also measured by automated sensors (ISUS for nitrate concentration and Hydrolab probe for temperature, salinity, turbidity, chlorophyll *a* content, pH) in the water pumped from the sampling point (43°28′10.57 N, 5°10′9.91 E, 2.5 m depth) to the field laboratory. Water was carried out to the laboratory through a 250 m hose (50 mm inner diameter) at a flow rate of 30 dm^3/min during 17 min before each analysis (in order to completely flush the entire hose). Each hour, temperature, salinity, turbidity, pH, nitrate, and chlorophyll *a* concentrations were measured by the various sensors installed in an 80 dm^3 tank receiving the pumped water. A dedicated volume of 1 dm^3, set up between the pipe and the 80 dm^3 tank, was used for the flow cytometry analyses performed by the Cytosense. An *in situ* HOBO® Pendant® Temperature/Light Data Logger has been installed next to the pipe inlet in order to measure the incident light intensity at the very sampling depth. A strainer brass was set *in situ* at the end of the hose to prevent biofouling and the passage of larger particles (several millimeters long) which may clog the hose and/or the instruments.

4 Results

4.1 Phytoplankton Clusters Defined by Flow Cytometry

During the October 2011 monitoring, up to 12 separate clusters of photoautotrophic cells (arbitrary labeled C1 to C12) have been resolved thanks to their optical properties recorded by the scanning flow cytometer (Fig. 3). Clusters showed a proportional relation between light scatter (related to cell size) and red fluorescence (related to chlorophyll a pigment content) intensities.

The length of the cells has been determined both from the pictures taken by the image-in-flow and from calibration microspheres (Polysciences) of various sizes (from 1 to 20 μm in diameter). It covered a large range of sizes, from 0.9±0.1 μm (for C12) to 56.4±12.2 μm (for C1) (Table 1). The largest cells pictured by the image-in-flow camera correspond to clusters C1 to C4 and belong to the microphytoplankton class. As revealed by the pictures, they were monospecific and composed of *Akashiwo sanguinea* (Hirasaka) for C1, *Prorocentrum micans* (Ehrenberg) for C2, *Scrippsiella* sp. (Balech) for C3, and *Gymnodinium* sp. (Stein) for C4.

4.2 Dynamics of *Akashiwo sanguinea* and of the Environmental Variables During the Sampling Period

The chlorophyll *a* concentration of the overall phytoplanktonic community strongly varied in the Berre lagoon during the sampling period, with *maxima* correlated to high nitrate concentrations ($n=215$, correlation coefficient 0.53, $p<0.001$). Proliferation of microphytoplanktonic cells, including *A. sanguinea*, has led to several peaks of chlorophyll *a* concentration which reached up to 21.81 μg/dm^3 on October 5 at 05:30 pm. Such high biomass have not been detected by the monthly monitoring carried out in the Berre lagoon; actually, in October the lowest concentration has been measured (Fig. 4).

During the sampling period, one major forcing that occurred in the lagoon was a turbulent mixing induced by a strong mistral event (wind from 330° to 360° from October 6 to 10) (Fig. 5).

The mean temperature of the water dropped down by 5.2 °C (from 23.2±0.2 °C to 18.0±0.3 °C between October 6 and 7). Nitrates and micro- and nanophytoplankton concentrations decreased and *Akashiwo sanguinea* abundance dropped

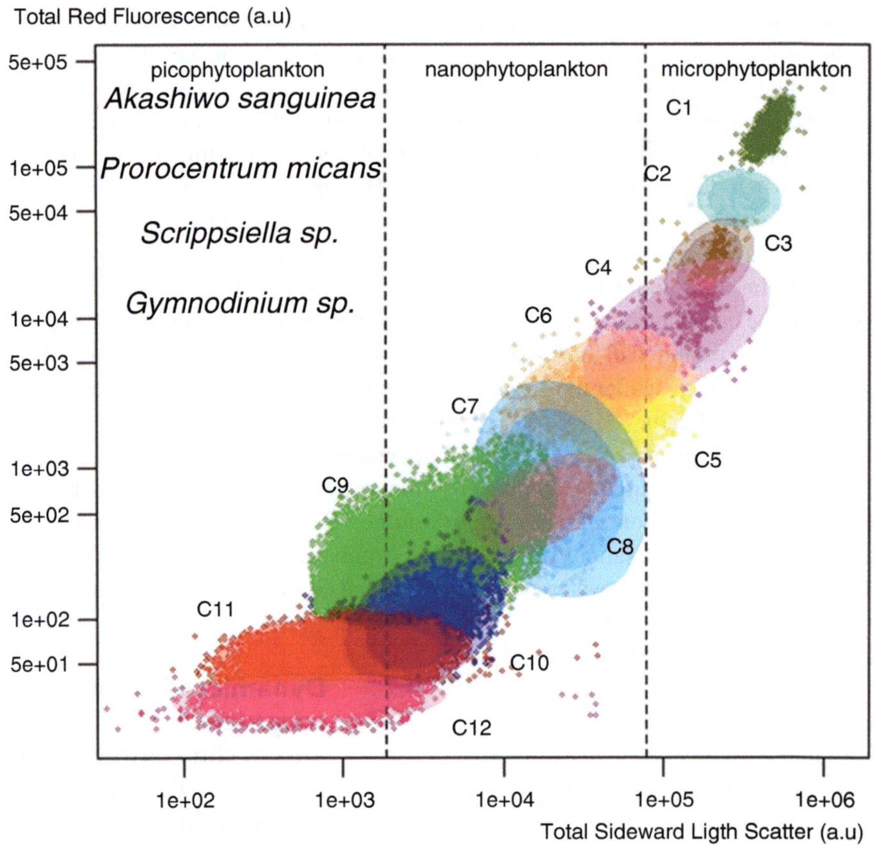

Fig. 3 Cluster description on a red fluorescence (total red fluorescence, arbitrary unit)/light scatter area under the curve (total sideward light scatter, arbitrary unit) cytogram with 0.8 and 0.9 confidence ellipses

Table 1 Mean lengths (µm) of cells composing the various cytometric clusters pictured by the image-in-flow camera

Cluster	C1	C2	C3	C4	C5	C6
	A. sanguinea	*P. micans*	*Scrippsiella* sp.	*Gymnodinium* sp.		
Mean length ± standard deviation	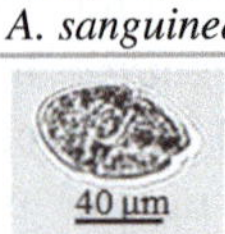	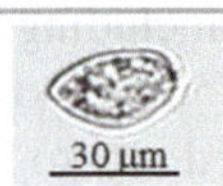	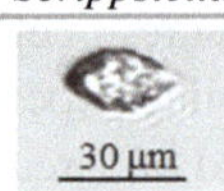			
	56.4 ± 12.2	37.7 ± 10.5	23.3 ± 8.8	20.1 ± 13.5	13.1 ± 4.2	13.2 ± 4.3
	C7	C8	C9	C10	C11	C12
	13.9 ± 5.6	8.6 ± 1.3	3.5 ± 0.5	4.1 ± 1.0	1.1 ± 0.1	0.9 ± 0.1

from 445.5 cells/cm^3 on October 6, 11:00 pm, to 12.2 cells/cm^3 on October 7, 12:00 pm (Fig. 6).

Water turbidity which rose to 15.9 NTU dropped down to 2.02 ± 0.77 NTU as the biggest autotrophic cells decreased in abundance. The maximum light intensity measured *in situ* increased from 9,300 lux to 14,467 lux, while there was no significant difference of the solar irradiation.

5 Discussion

Since the instatement of the EDF power plant on the northern shore of its basin, the Berre lagoon has been subject to recurrent algal blooms in spring, summer, and autumn. In October, these algal proliferations are caused by the persistence of summer-like temperatures combined with the

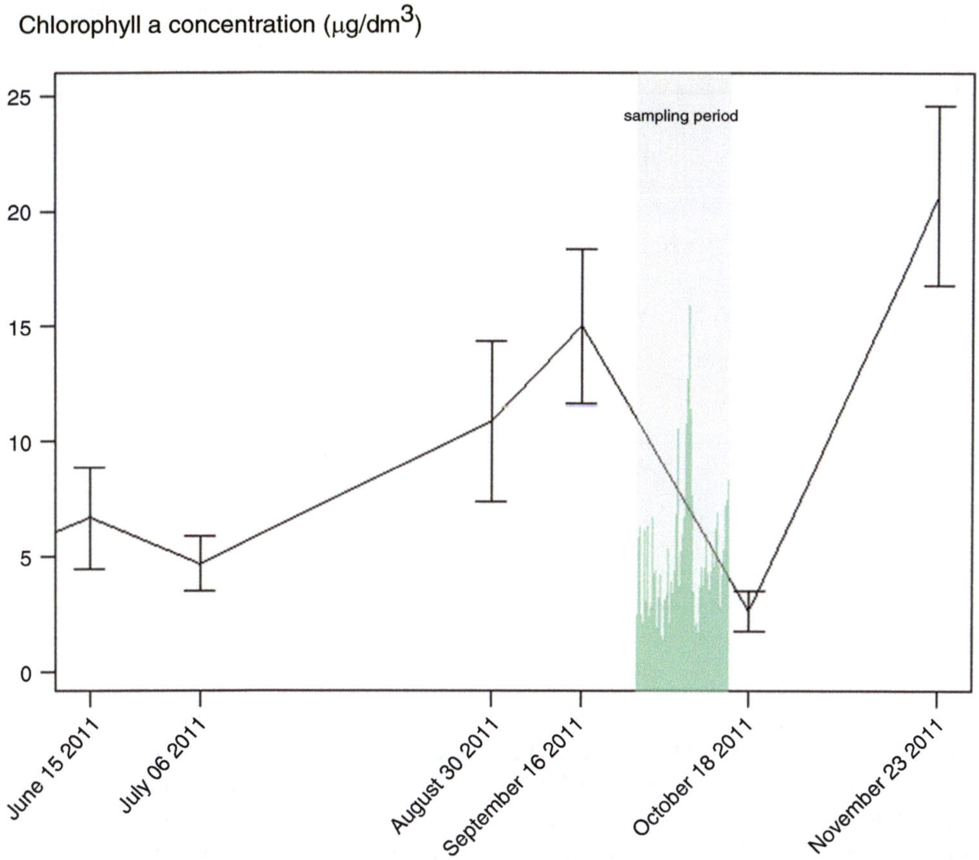

Fig. 4 Chlorophyll *a* concentration measured during the monthly monitoring of the lagoon (*black line*, with the standard deviation, $n=20$) and during the hourly sampling period from October 5 to 11 (in *green*)

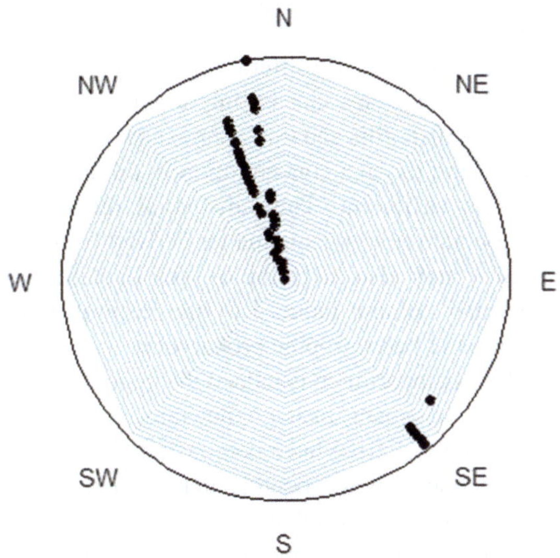

Fig. 5 Distribution of the wind direction with mean speed exceeding 20 m/s during the sampling period

resumption of nutrient supply generated by the power plant activities (Minas 1976). During the sampling period, freshwater discharges reached 150.10^6 m^3 for a volume of the lagoon estimated at 980.10^6 m^3. Mean temperature was 21.4 ± 2.1 °C.

Few dinoflagellate and diatom species like *Prorocentrum minimum* and *Cyclotella* sp. have then always been dominant for both new and regenerated production (Raimbault et al. 2013). Historically, abundances of *P. minimum* could reach 40,000 cells/cm^3 in the lagoon. Before the water policy management, colored waters were observed with concentration of up to 660 cells/cm^3, and fish and mollusk mortality was induced by toxins close to the paralytic shellfish poisoning (PSP) with concentration of 5,184 cells/cm^3 (Belin and Berthome 1988). However, since the first regulation of the power plant discharges

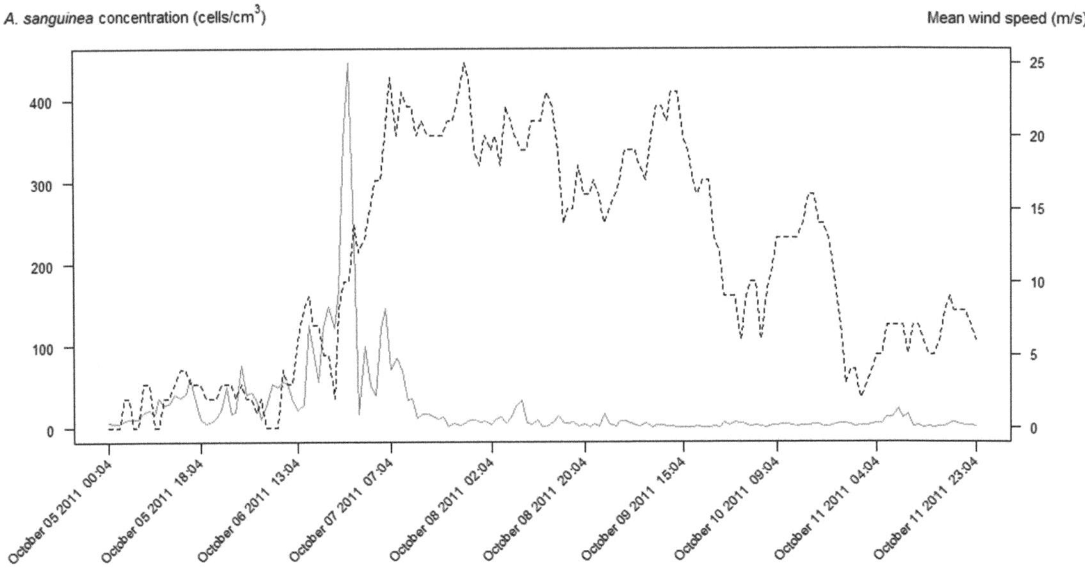

Fig. 6 Dynamics of *A. sanguinea* (*green line*) and mean wind speed (*dashed line*) before, during, and after a strong mistral event

in 1994, the abundance of this species decreased and more eurythermal and euryhaline as well as marine species like *Scrippsiella sp.* were able to develop (Malkassian 2012). *Gymnodinium sp.*, *Akashiwo sanguinea*, *Prorocentrum micans*, and several other species of nanoflagellates are also present in the lagoon (identified by microscopy by the monthly monitoring supervised by the GIPREB). Like *A. sanguinea*, able to grow at temperatures between 10 and 30 °C and salinities between 10 and 40 (Matsubara et al. 2007), these species are well adapted to the important hydrological variations of the lagoon (Nérini et al. 2001).

In the literature, *A. sanguinea* has been referred as a potentially toxic species (Tindall et al. 1984), harmful for fishes, marine birds, mollusks, and human after consumption of contaminated shellfish and corals (Cardwell et al. 1979; Botes et al. 2003; Jessup et al. 2009; Vazquez et al. 2011). However, no biotoxin have been isolated yet and evidences of a direct mortality mode have only been reported recently in case of seabird's stranding with no current toxins involved (Jessup et al. 2009). For Du et al. (2011), a massive red tide dominated by *A. sanguinea* in concentration at only 400 cells/cm^3 was responsible for the mortality of several seabird species. The saponification of their feathers' oil caused by the massive production of surfactant amino acids (MAAs) has led to hypothermia and ultimately to the death of the birds.

In addition to the direct mortality caused by secondary metabolites, harmful algal bloom species can also be indirectly noxious through the exceptional concentration and biomass they reach, especially for benthic species needing oxygen and light. Proliferation of some phytoplanktonic species can be important enough to clog shellfish gills and thus asphyxiate organisms even if the oxygen concentration in the environment is sufficient. The substantial release of phytoplanktonic organic matter derived from the blooms may also favor the oxygen depletion in the water column induced by the respiration of the aerobic heterotrophic prokaryotes.

This compartment is indeed the major responsible for the organic matter remineralization. Manté and Michez (2010) have demonstrated that species richness of the benthic macrofauna of the Berre lagoon is lower than in other Mediterranean lagoons due to frequent anoxia. The thermic or haline stratification of the water column prevents the reoxygenation of the deep layers by gas exchange with the surface, leading to hypoxia or anoxia in extreme cases. The variation of nitrate concentration and salinity during the sampling period with the strong wind (mistral) event clearly demonstrates that the Berre lagoon was in a stratified state in early October. On October 18, vertical profiles of nutrient concentrations showed differences between surface and bottom (GIPREB 2011). The mistral event has contributed to the mixing of the water mass and has favored its reoxygenation.

Due to their capacity to migrate along the water column (i.e., nycthemeral migration), flagellate species including *A. sanguinea* are known to proliferate in worldwide stratified waters (Shipe et al. 2008; Du et al. 2011). Hence, they contribute for a great part to the regenerated production based on nutrient remineralization by the microbial loop in deeper layers. When photosynthesis becomes light limited, flagellates can swim downward the column water to find inorganic regenerated form of nitrogen such as ammonium while nitrates are not consumed.

In October 2011, peaks of phytoplanktonic biomass due to dinoflagellate proliferation were correlated to higher nitrate concentrations. When the chlorophyll *a* content of the lagoon was the highest (21.81 $\mu g/dm^3$), *A. sanguinea* measured concentration reached up to 445 $cells/cm^3$. A recent study conducted on nutrient uptakes and regeneration rate in the water column suggests that pelagic regeneration of ammonium can largely encompass the demand for primary producers' development (Gouze et al. 2008).

As emphasized by Anderson (2009), harmful phytoplanktonic outbreaks are increasing over the years. The first invoked cause is intensification of industrialization and pollution. But global warming is also believed to trigger algal blooms (Moore et al. 2008). Increase of surface temperature is indeed coherent with dinoflagellate proliferation in stratified waters. Moreover, higher exposure to UV due to the depletion of ozone is a factor of control of the biosynthesis of photoprotective MAAs (Litchman et al. 2002).

Climate change could thus enhance the development of harmful dinoflagellate species as well as their deadliness. However, this very study and particularly the effects of the mistral event point out their consequence may not be clear concerning these harmful algal blooms and even display opposite responses. According to the Mermex group (2011), the stratification of water mass could be balanced by more frequent northern wind event like mistral in the Mediterranean Sea.

In this study, the occurrence of a mistral event characterized by a mean speed exceeding 20 m/s was phased with the disappearance of micro- and nanophytoplankton cells. However, wind and storms have been considered as factors of resuspension and dispersal leading to the increase of harmful algal blooms (Anderson 1989; Kremp 2001). Resuspension and dispersal of a high concentrated inoculum of resistant cysts could indeed conduct to massive and sudden proliferation if environmental conditions favor their germination.

6 Conclusion and Perspective

Algal bloom monitoring is a major preoccupation as proliferations of harmful species can induce great economic losses and sanitary issues if they are not anticipated. The Berre lagoon is a brackish water pool known to host regular phytoplanktonic blooms involving harmful and red tides species. As lagoons represent important sources of socio-economic incomes, their management constitutes a major concern for today's society. Since 2000, the European Water Framework Directive policy order regular surveys to state about the eutrophication of these ecosystems, aiming at a coordinate management that could lead to the rehabilitation of hundreds of sites, including the Berre lagoon. This study demonstrates that the deployment of an autonomous flow cytometer in this area constitutes a complementary method to the monthly monitoring of the phytoplanktonic community performed for many years now. During one month, the cytometer has characterized at high frequency the phytoplanktonic assemblage, resolving several clusters ranging from pico- to microphytoplankton. Pictures taken by the instrument helped to identify the largest ones, composed of red tide dinoflagellate species. This strategy combined to hydrological variables records has evidenced a highly dynamic community capable of responding very quickly to sporadic environmental forcing. A fast increase of dinoflagellates' abundance, including the harmful species *Akashiwo sanguinea*, was initiated by the stratification of the water column for which they present physiological advantages compared to other phytoplankton species. The potential outbreak of a harmful algal bloom in the lagoon was however disturbed by the mistral episode with a mean speed exceeding 20 m/s. The possibility of monitoring a great range of phytoplankton, at high frequency and automatically (independently of the weather conditions), with the associated environmental conditions, opens new ways to better understand the circumstances under which harmful algal blooms can arise or disappear. And since they constitute a main effect of eutrophication, autonomous flow cytometry obviously appears appropriate to follow the rehabilitation of the Berre lagoon which is expected after the discharge restrictions imposed to the EDF power plant. The characterization of the resilience of this ecosystem will be eventually determinant for the restoration of certain economical activities, explaining why the interest of such instrumentation does not only stand for fundamental researchers in the field of marine microbiology but also for people, networks, and stakeholders directly or indirectly concerned by eutrophication and the quality of the aquatic environment.

References

Anderson DM (1989) Toxic algal blooms and red tides: a global perspective. In: Red tides: biology, environmental science and toxicology. Elsevier Science Publishing Co., pp 11–16

Anderson DM (2009) Approaches to monitoring, control and management of harmful algal blooms (HABs). Ocean Coast Manag 52:342–347

Anderson DM, Andersen P, Bricelj VM, Cullen JJ, Rensel JE (2001) Monitoring and management strategies for harmful algal blooms in coastal waters, APEC #201-MR-01.1, Asia Pacific economic program, Singapore, and intergovernmental oceanographic commission technical series no. 59, Paris

Belin C, Berthome J-P (1988) Bilan des perturbations phytoplanctoniques observées sur les côtes françaises en 1986. Rapport IFREMER, pp 1–184

Botes L, Smit AJ, Cook PA (2003) The potential threat of algal blooms to the abalone (*Haliotis midae*) mariculture industry situated around the South African coast. Harmful Algae 2:247–259

Cardwell RD, Olsen S, Carr MI, Sanborn EW (1979) Causes of oyster mortality in South Puget Sound. NOAA Technical Memorandum, ERL MESA-39, Environmental Research Laboratories, Marine Ecosystems Analysis Program, Boulder

Cloern JE (1996) Phytoplankton bloom dynamics in coastal ecosystems: a review with some general lessons from sustained investigation of San Francisco Bay, California. Rev Geophys 34:127–168

Du X, Peterson W, McCulloch A, Liu G (2011) An unusual bloom of the dinoflagellate *Akashiwo sanguinea* off the central Oregon, USA, coast in autumn 2009. Harmful Algae 10:784–793

Dubelaar GBJ, Gerritzen PL (2000) CytoBuoy: a step forward towards using flow cytometry in operational oceanography. Sci Mar 64:255–265

Dubelaar GBJ, Gerritzen PL, Beeker AER, Jonker RR, Tangen K (1999) Design and first results of CytoBuoy: a wireless flow cytometer for *in situ* analysis of marine and fresh waters. Cytometry 37:247–254

GIPREB (2011) Bilan du suivi écologique 2011. http://www.etangdeberre.org/upload/file/maxi/rapport-suivi-2011_250.pdf

Gouze E, Raimbault P, Garcia N, Picon P (2008) Nutrient dynamics and primary production in the eutrophic Berre lagoon (Mediterranean, France). Transit Waters Bull 2:17–40

Group M (2011) Marine ecosystems' responses to climatic and anthropogenic forcings in the Mediterranean. Prog Oceanogr 91:97–166

Jessup DA, Miller MA, Ryan J-P et al (2009) Mass stranding of marine birds caused by a surfactant-producing red tide. PLoS ONE 4:e4550

Kremp A (2001) Effects of cyst resuspension on germination and seeding of two bloom- forming dinoflagellates in the Baltic Sea. Mar Ecol Prog Ser 216:57–66

Litchman E, Neale P, Banaszak A (2002) Increased sensitivity to ultraviolet radiation in nitrogen-limited dinoflagellates: photoprotection and repair. Limnol Oceanogr 47:86–94

Malkassian A (2012) Méthodes d'analyse fonctionnelle et multivariée appliquées à l'étude du fonctionnement écologique des assemblages phytoplanctoniques de l'étang de Berre. Thesis. Université de la Méditerranée

Manté C, Michez N (2010) Expertise pour le traitement et l'analyse statistique des données issues du suivi des indicateurs biologiques dans l'étang de Berre. Rapport GIPREB, pp 1–82

Matsubara T, Nagasoe S, Yamasaki Y, Shikata T, Shimasaki Y, Oshima Y, Honjo T (2007) Effects of temperature, salinity and irradiance on the growth of the dinoflagellates *Akashiwo sanguinea*. J Exp Mar Biol Ecol 342:226–230

Mayot N, Gouze E, Malet N (2013) Bilan des apports à l'étang de Berre et état des lieux de l'eutrophisation. Les actes des rencontres LAGUN'R, pp 128–141

Minas M (1976) Production organique primaire dans un milieu saumâtre eutrophe (Etang de Berre). Effets d'une forte dilution (dérivation des eaux de la Durance). Mar Biol 35:13–29

Moore SK, Trainer VL, Mantua NJ, Parker MS, Laws EA, Backer LC, Fleming LE (2008) Impacts of climate variability and future climate change on harmful algal blooms and human health. Environ Health 7:1–12

Nérini D, Manté C, Durbec J-P, Garica F (2001) Une méthode statistique de détermination de séquences caractéristiques dans une série temporelle de plusieurs variables. Application à la physico-chimie des eaux de l'étang de Berre. Earth Planet Sci 332:457–464

Peeters JCH, Dubelaar GBJ, Ringelberg J, Visser JWM (1989) Optical plankton analyser: a flow cytometer for plankton analysis, I: design considerations. Cytometry 10:522–528

Raimbault P, Beker B, Garica N, Fornier M, Gouze E (2013) L'étang de Berre, cycle de la matière, production primaire et communautés phytoplanctoniques. Les actes des rencontres LAGUN'R, pp 143–157

Shipe RF, Leinweber A, Gruber N (2008) Abiotic controls of potentially harmful algal blooms in Santa Monica Bay, California. Cont Shelf Res 28:2584–2593

Smayda TJ (1997) What is a bloom? Limnol Oceanogr 42:1132–1136

Thyssen M, Tarran GA, Zubkov MV, Holland RJ, Gregori G, Burkill PH, Denis M (2008) The emergence of automated high-frequency flow cytometry: revealing temporal and spatial phytoplankton variability. J Plankton Res 30:333–343

Tindall DR, Dickey RW, Carlson RD, Morey-Gaines G (1984) Ciguatoxigenic dinoflagellates from the Caribbean Sea. Seafood Toxin 21:225–240

Vazquez EJ, Lizarraga I, Schmidt CJ, Tapia A, Cortes D, Sandoval F, Tapia A, Guzman JJ (2011) Impact of harmful algal blooms on wild and cultured animals in the Gulf of California. J Environ Biol 32:413–423

Zigone A, Enevoldsen HO (2000) The diversity of harmful algal blooms: a challenge for science and management. Ocean Coast Manag 43:725–748

When Complexity Rimes with Sanity: Loss of Fractal and Multifractal Behavioural Complexity as an Indicator of Sublethal Contaminations in Zooplankton

Laurent Seuront

Abstract
Natural and anthropogenic sources of contamination such as plankton toxins and hydrocarbons are nearly ubiquitous in the marine environment. Specifically, they are a pernicious threat especially at low concentration as nonlethal effects on the plankton propagate through the food chain and accumulate in the tissues of top predators, ultimately putting human health at risk. In this contribution, I first describe how the complexity observed in the spatial and temporal patterns of copepod swimming behaviour can be objectively quantified using a series of 'behavioural stress indexes' based on fractal and multifractal analyses of copepod swimming behaviour and swimming sequences. These indexes are suggested as a potential tool to critically assess behavioural responses to natural and anthropogenic forcing in the marine environment.

1 Introduction

As observed across the whole spectrum of social and natural sciences, behavioural data are inherently very complex (Seuront 2010a, b), a priori lacking of any spatial pattern or temporal structure (Fig. 1). This complexity is believed to be biologically adaptive as it avoids restricting the functional response of an organism to highly periodic behaviour (Goldberger et al. 2000) and it is error tolerant, allowing organisms to cope with stress and unpredictable environments (Goldberger et al. 1990). The analysis of behaviour hence critically requires approaches explicitly dealing with this complexity. This issue is particularly relevant in welfare assessment as most behavioural measures are not sensitive enough to detect subtle changes associated with mild or acute stress (Rutherford et al. 2004).

The field of behavioural ecology has recently begun to use novel analytical tools such as fractal analysis (Asher et al. 2009). Specifically, fractal

L. Seuront (✉)
Centre National de la Recherche Scientifique,
UMR 8187 LOG, Wimereux, France
e-mail: laurent.seuront@cnrs.fr

H.-J. Ceccaldi et al. (eds.), *Marine Productivity: Perturbations and Resilience of Socio-ecosystems*,
DOI 10.1007/978-3-319-13878-7_14, © Springer International Publishing Switzerland 2015

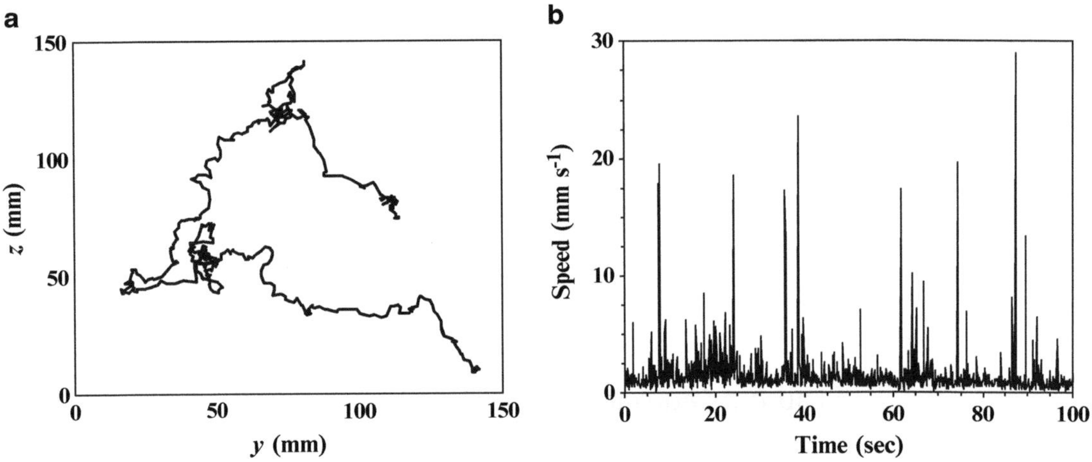

Fig. 1 Illustration of the intrinsic complexity perceptible in the spatial pattern (**a**) and temporal structure (**b**) of zooplankton swimming behaviour. (**a**) Two-dimensional projection of the three-dimensional trajectory of an adult male *Eurytemora affinis*. (**b**) Time series of the instantaneous speed of an adult *Temora longicornis* female. Both behaviours were recorded at 25 frames s^{-1} in a cubic (15 × 15 × 15 cm) glass chamber from *E. affinis* and *T. longicornis* individuals swimming freely in filtered estuarine and coastal waters, respectively

analysis has been introduced in the study of human physiology to distinguish between systems operating in normal vs. pathological states (Ivanov et al. 1999; Mishima et al. 1999). Both the temporal and structural complexity of a range of biological systems hence decrease under stressful conditions. For instance, the time series of beat intervals in healthy subjects have more complex fluctuations than patients with severe cardiac disease (Ivanov et al. 1999). Similarly, the geometry of the lung terminal airspace branching architecture is more complex in normal subjects than in patients with chronic obstructive pulmonary disease (Mishima et al. 1999). More specifically, stressed (e.g. diseased and parasited) animals typically reduce the complexity of their behavioural display (Alados et al. 1996). Fractal analysis has hence been extensively used as a non-invasive assessment of the general health of wild and captive animals (Rutherford et al. 2004; Alados et al. 1996), including copepods (Seuront 2011).

The quantitative assessment of changes in copepod swimming behaviour is critical as swimming and feeding are intertwined in most copepod species, hence any disruption of copepod swimming is predicted to have detrimental consequences to their biology and ecology (Seuront 2012), which in turn may affect ecosystem structure and function and geochemical fluxes. Behavioural changes have the potential to be used as indicators of ecosystem health. This issue is particularly relevant for sublethal toxicant concentration as behavioural changes provide sensitive non-invasive sublethal endpoint with short-response time for toxicity bioassays, which are more sensitive than mortality responses (Garaventa et al. 2010).

Over the last two decades, fractal analysis has increasingly been used to describe and provide further understanding to zooplankton swimming behaviour. This may be related to the fact that fractal analysis has the desirable properties to be independent of measurement scale and to be very sensitive to even subtle behavioural changes that may be undetectable to other behavioural variables (Rutherford et al. 2004; Coughlin et al. 1992). As early claimed (Coughlin et al. 1992), this creates 'the need for fractal analysis' in zooplankton behavioural ecology in general and in zooplankton ecotoxicology in particular.

In this context, I first briefly rehearse the very basic principles of fractal theory before describing a few fractally derived 'behavioural stress indexes' that can be directly applied to various aspects of zooplankton behavioural complexity and used to infer the stress experienced by these organisms under a range of conditions. I subsequently introduce the more elaborated and still seldom used, though more general, concept of multifractal that may conveniently be used as an objective and quantitative tool to thoroughly identify models of movement behaviour, such as Brownian motion, fractional Brownian motion, ballistic motion, Lévy flight/walk and multifractal random walk. I stress that fractal and multifractal analyses can detect differences in behavioural complexity, where traditional measures cannot. As such, I finally discuss their relevance as a practical tool to infer the nature of both natural and anthropogenic forcing.

2 From Fractal Theory to Stress Assessment: Behavioural Stress Indexes

2.1 A Few Words on Fractals

A fractal is 'a rough or fragmented geometric shape that can be split into parts, each of which is (at least approximately) a reduced-size copy of the whole' (13). This property is called *scale invariance* and means that the observed structure remains unchanged under magnification or contraction. This scale invariance can be observed in two distinct, though conceptually similar, forms referred to as *self-similarity* and *self-affinity*. Self-similarity has traditionally been illustrated using theoretical fractal objects (Mandelbrot 1982). A more realistic construction of a fractal object, a fractal tree, is shown in Fig. 2. In contrast, self-affinity characterises an object that may be written as a union of rescaled copies of itself, where the rescaling is anisotropic, that is, dependent on the direction. A typical example of self-affinity is given by the temporal patterns of the successive speed of copepods (Fig. 1b); it looks rough, like their trajectory (Fig. 1a), but with the two axes corresponding to physical quantities that are fundamentally different.

A fundamental consequence of scale invariance is, as originally described for the length of the coast of Britain (Mandelbrot 1982), that the length of, for example, copepod trajectories (Fig. 1a) does not converge towards a fixed value, but keeps increasing, *theoretically* without any upper limit, but see Rutherford et al. (2004) for a detailed discussion on the topic. As a consequence, in contrast to Euclidean lines, they cannot be differentiated or integrated, hence cannot be described by an integer dimension. The complexity of scale invariant patterns and processes can, however, be described by a dimension D, the so-called fractal dimension. In contrast to Euclidean dimensions, a fractal dimension is fractional. For instance, the Euclidean dimensions, d, of a line, a circle and a cube are, respectively, 1, 2 and 3. The trajectory of a copepod has, in turn, a fractal dimension, D, bounded between $D=1$ when the copepod swims along a completely linear path and $D=2$ when the movements

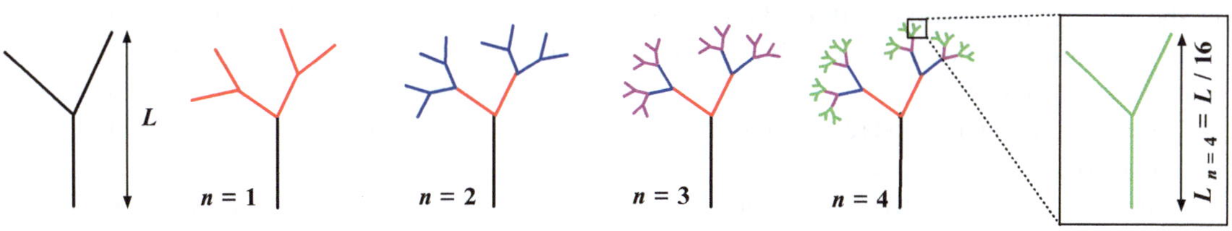

Fig. 2 Illustration of the first four successive steps of the iterative process leading to a self-similar fractal tree. At each step n of the process, each terminal branch of the tree is replaced by a rescaled version of the original tree. Here the scale ratio between two successive steps is 2, i.e. at a step n, each branch is replaced by a tree, which is a copy of the original tree reduced 2^n times. Hence, when $n=4$, the resulting ramifications are 2^4 times smaller than the original tree

are so complex that the trajectory fills the whole available space. The fractal dimensions reported in the literature for zooplankton trajectories (essentially cladocerans and copepods) typically fall in the range 1.0–1.8, indicating a range of behavioural strategies that may be related to the nature of the physical, chemical and biological cues present in the water (Rutherford et al. 2004; Seuront 2012, 2013; Garaventa et al. 2010; Shimizu et al. 2002).

Three types of behavioural data can be used in fractal analysis: (1) two- or three-dimensional movement pathways, (2) the temporal patterns of successive displacements (or equivalently speed) and (3) observation of the presence or absence of a behavioural state scored on a binary scale, i.e. whether an animal is active or inactive, or fluctuations between two behavioural states. In the next section, I provide a set of fractal techniques to analyse these behavioural data and briefly review how they were used to assess the stress experienced by zooplanktonic organisms.

2.2 Fractals as a Stress Assessment Tool in Zooplankton Behavioural Ecology

2.2.1 The Fractal Nature of Copepod Spatial Patterns

I describe hereafter three conceptually similar methods—the box-counting, the dividers and the mass dimension methods—that can be easily implemented to quantify the geometric complexity of copepod trajectories (Rutherford et al. 2004). Note that whilst these methods are discussed in the general framework of three-dimensional trajectories, they can be equivalently implemented in two dimensions.

The box-counting method relies on the δ cover of a trajectory, i.e. the number of boxes of length δ required to cover the trajectory. Practically, this procedure consists in superimposing a regular grid of boxes of length δ on the trajectory and counting the number of boxes that intersect the trajectory. This procedure is repeated using different values of δ. The volume occupied by a trajectory is then estimated using a series of boxes spanning a range of volumes down to some small fraction of the entire volume. The number of occupied boxes increases with decreasing box size, leading to the following power-law relationship:

$$N(\delta) \propto \delta^{-D_b} \tag{1}$$

where δ is the box size, $N(\delta)$ is the number of boxes intersecting the trajectory and D_b is the box fractal dimension; D_b is estimated from the slope of the linear trend of the log-log plot of $N(\delta)$ vs. δ.

The divider dimension D_d (also referred to as the compass dimension) is estimated by measuring the length of a trajectory at various scales δ. The procedure is analogous to moving a set of dividers (like a drawing compass) of fixed length δ along the trajectory. The estimated length of a trajectory $L(\delta)$ increases with decreasing δ as $L(\delta) \propto \delta^{1-D_d}$. As the estimated length $L(\delta)$ is also the product of $N(\delta)$ (the number of compass dividers required to cover the trajectory) and δ (i.e. $L(\delta) = N(\delta)\delta$), this can equivalently be written as

$$N(\delta) \propto \delta^{-D_d} \tag{2}$$

The divider dimension D_d is then estimated from the slope of the linear trend of the log-log plot of $N(\delta)$ vs. δ.

The mass dimension method counts the number of pixels occupied by a trajectory in sampling cubes ($\delta \times \delta \times \delta$). The mass $m(\delta)$ of occupied pixels is subsequently defined as $m(\delta) = N_O(\delta)/N_T(\delta)$, where $N_O(\delta)$ and $N_T(\delta)$ are, respectively, the number of occupied pixels and the total number of pixels within an observation window of size δ. These computations are repeated for various values of δ, and the mass dimension D_m is defined as

$$m(\delta) \propto \delta^{D_m} \tag{3}$$

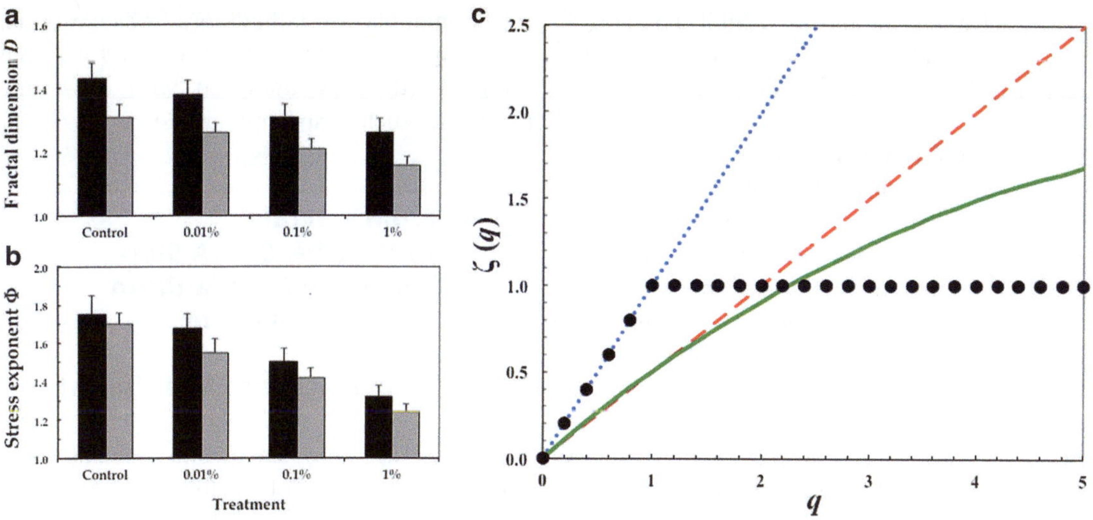

Fig. 3 The fractal dimension D (**a**) and stress index ϕ (**b**) estimated from the swimming behaviour of *Eurytemora affinis* adult males (*black*) and non-ovigerous females (*grey*) in control uncontaminated estuarine water and in estuarine water contaminated with the water-soluble fraction of diesel oil at 0.01, 0.1 and 1 % (Modified from Seuront 2010a, b). (**c**) The multifractal function $\zeta(q)$ allows to identify a range of movement behaviour (see text for details) such as ballistic motion (*dotted blue line*), Brownian motion (*red dashed line*), optimal Lévy flight (*black dots*) and multifractal random walk (*continuous green curve*) (Modified from Seuront and Stanley 2014)

where the fractal dimension D_m is estimated from the slope of the linear trend of the log-log plot of $m(\delta)$ vs. δ.[1]

The results of studies based on a geometric assessment of zooplankton behavioural complexity under various conditions of water contaminations—i.e. short-term exposure to copper, organophosphorus and carbamate (Shimizu et al. 2002), the water-soluble fraction of diesel oil (Seuront 2010a, b, 2012), nonylphenol, cadmium and a mixture of polycyclic aromatic hydrocarbons (Michalec et al. 2013a, b)—lead a variety of a priori conflicting conclusions, including no change (Michalec et al. 2013a, b), a decrease (Seuront 2010a, b, 2012; Fig. 3a) and an increase (Seuront 2010a, b) in the geometric complexity of swimming behaviour.

2.2.2 The Fractal Nature of Copepod Temporal Patterns

One of the most extensively used techniques to detect temporal self-affine patterns is power spectral analysis. Formally, a power spectrum is defined as the square of the amplitude of the Fourier transform of a time series of a descriptor; it is hence an expression of the variance of the descriptor at different temporal scales. In practice, the power spectral density $E(f)$ is given by $E(f) \propto f^{-\beta}$, where f is the frequency (s^{-1}; $f = 1/t$, where t is time). The spectral exponent β is estimated as the slope of a log-log plot of $E(f)$ vs. f. Specifically, the value of the exponent β provides an efficient way to classify the type of motion behaviour exhibited by zooplankton organisms.[2]

Spectral analysis has still barely been used in zooplankton behavioural ecology (Uttieri et al. 2008; Dur et al. 2010) but nevertheless suggests

[1] Note that it is readily seen from Eqs. (1) and (2) that $D_b = D_d$, whilst more convoluted developments show that $D_b = D_m$; hence $D_b = D_d = D_m$; see (Seuront 2010a) for details. Statistically inferring the absence of significant differences between fractal dimensions returned by different methods of analysis hence constitutes an additional guarantee of the trustworthiness of the fractal dimension estimates.

[2] Brownian motion (i.e. normal diffusion) is characterised by $\beta = 2$. Anti-persistent and persistent fractional Brownian motions are characterised by $\beta < 2$ and $\beta > 2$, respectively. Specifically, a motion is persistent in the sense that an organism moving in some direction at time t will tend to move in the same direction at the next time step.

that zooplankton organisms exhibit a range of behaviour including fractional Gaussian motion, fractional Gaussian noise and pure random noise.[3]

2.2.3 The Fractal Nature of Behavioural States

Self-affine techniques based on the analysis of frequency distributions of behavioural states were used to infer the response of zooplankton to a range of stressors. They include considerations of the scaling properties of the probability distribution functions (PDFs) of either the time t_x spent in a specific behavioural state x (i.e. $p(t_x) \propto t_x^{-c}$) (Schmitt et al. 2006) or the velocity v_x used to define different behavioural states x (i.e. $p(v_x) \propto v_x^{-c}$) (Michalec et al. 2010). A few studies investigated the scaling properties of the cumulative probability distribution functions (CDFs) of move duration greater than a determined duration t ($P(t \leq T) \propto t^{-\phi_1}$) (Seuront and Leterme 2007) and move lengths L greater than a determined length l ($N(l \leq L) \propto l^{-\phi_2}$) (Seuront 2010a, b, 2011). These studies consistently found a decrease in the exponents ϕ_1 and ϕ_2, hence in behavioural complexity, for a range of copepod species exposed to sublethal concentrations of naphthalene (Seuront and Leterme 2007) and the water-soluble fraction of diesel oil (Seuront 2010a, b; Fig. 3b). In contrast, the exponent c was shown to increase in *Eurytemora affinis* following a short-term exposure to sublethal concentrations of nonylphenols (Michalec et al. 2013a, b).

Note that the behaviour of an organism alternating between two behavioural states can also be assessed through the construction of a binary sequence $z_t(i)$ for each behavioural activity taken from continuous observations. When a specific activity is observed, $z_t(i)=1$, and $z_t(i)=0$ otherwise. The resulting time series of binary sequences can further be integrated as $w_i(t)=\sum_{i=1}^{N} z_t(i)$, where N is the number of behavioural observations. The temporal pattern of the integrated variable $w_i(t)$ can then be analysed with self-affine techniques such as spectral analysis.

[3] For instance, the velocity components of *Clausocalanus furcatus* were both characterised by $\beta \approx 0$ (Uttieri et al. 2008), indicative of a random process without internal serial correlation. In contrast, β ranged from 0.30 to 0.75 in *Temora longicornis* (Moison et al. 2012) and 1.4 to 1.5 in *Pseudodiaptomus annandalei* (Dur et al. 2010).

3 From Fractals to Multifractals: A Step Further in Zooplankton Stress Assessment

3.1 From Fractals to Multifractals

A measure (i.e. a physical quantity such as mass, energy, a number of individuals or more specifically the distance displaced by a copepod; Fig. 1b) has to be distinguished from its geometric support, which might or might not have a fractal geometry (Rutherford et al. 2004). Then, if a measure has different fractal dimensions on different parts of the support, the measure is a multifractal. Multifractals are hence a generalisation of fractal geometry initially introduced to describe the relationship between a given quantity and the scale at which it is measured. Whilst fractal geometry describes the complexity of a given pattern with the help of only one parameter (the fractal dimension), multifractals characterise its detailed variability by an eventually infinite number of sets, each with its own fractal dimensions.

An intuitive interpretation of multifractals is based on the spatial structure of modern cities (Rutherford et al. 2004). Consider a city viewed strictly from above, it can be considered as a succession of built (buildings) and unbuilt (streets and parks) areas. The only available information is hence the distribution of the built and the unbuilt areas. This is the geometric support of the city. Now, change the angle of vision by taking a position not directly above the city, but from the side. The city initially made of built and unbuilt areas is now a set of buildings with different heights. This is the measure we are now interested in. It is now possible to estimate the distribution of a wide range of building heights. Each height will (eventually) be characterised by a fractal dimension, hence the concept of multifractals.

3.2 Multifractals as a Diagnostic Tool to Assess a Family of Swimming Behaviours

The strongly non-Gaussian fluctuations perceptible in zooplankton successive displacements that range from very likely slow steps to rare and extremely rapid displacements (Fig. 1b) are inherently incompatible with classical self-affine approaches based, e.g., on the scaling behaviour of the power spectral density described above that are fundamentally limited to second-order moments. A more general approach is based on the analysis of qth order long-range correlations in displacements. Specifically, the norm $||\Delta X_\tau||$ of the three-dimensional displacements of a zooplanktonic organism is defined as

$$\Delta X_\tau = \sqrt{(x_{t+\tau} - x_t)^2 + (y_{t+\tau} - y_t)^2 + (z_{t+\tau} - z_t)^2},$$

where τ is the temporal increment and (x_t, y_t, z_t) and $(x_{t+\tau}, y_{t+\tau}, z_{t+\tau})$ are respectively the positions of the organism at time t and $t+\tau$. $||\Delta X_\tau||$ is a nonstationary process with stationary increments; its statistics do not depend on time, t, but on the temporal increment τ (Rutherford et al. 2004; Seuront and Stanley 2014). The moments of order q ($q>0$) of the norm of three-dimensional displacements $||\Delta X_\tau||$ depend on the temporal increment τ as

$$\Delta X_\tau^{\,q} \propto \tau^{\zeta(q)} \tag{4}$$

The exponents $\zeta(q)$ are estimated as the slope of the linear trend of $< \Delta X_\tau^{\,q} >$ vs. τ in log-log plots. The function $\zeta(q)$ characterises the statistics of the random walk $||\Delta X\tau||$ of the organism regardless of the scale and intensity (Rutherford et al. 2004; Seuront and Stanley 2014). Low and high orders of moment, q, characterise, respectively, smaller and more frequent displacements and larger and less frequent displacements.[4]

The shape of the function $\zeta(q)$ can be used as a direct, objective and quantitative diagnostic tool to unambiguously identify the type of motion exhibited by zooplankton organisms and ultimately any swimming organisms (Fig. 3c). Briefly, for Brownian motion, $\zeta(q)=q/2$, and fractional Brownian motion is defined as $\zeta(q)=qH$, where $H=\zeta(1)$ and the limits $\zeta(q)=0$ and $\zeta(q)=q$ corresponding, respectively, to confinement and localisation, and ballistic motion. Anomalous diffusion occurs when $H\neq 1/2$. Specifically, super-diffusion occurs when $H>1/2$ and sub-diffusion when $H<1/2$. For finite-length Lévy flights, the function $\zeta(q)$ is bilinear with $\zeta(q)=q/(\mu-1)$ for $q<\mu-1$ and $\zeta(q)=1$ for $q\geq\mu-1$; the exponent μ ($1<\mu\leq 3$) characterises the power-law tail of the probability distribution of the move-step length l as $P(l)\approx l^{-\mu}$, where $1<\mu\leq 3$. For $\mu\geq 3$, the mean and the variance of the move-step lengths are both finite; hence, as a consequence of the central-limit theorem, their distribution is Gaussian. For $1<\mu<3$, the scaling is super-diffusive; the value $\mu=2$ corresponds to a Lévy flight (i.e. the swimming behaviour is tailored to minimise the distance travelled whilst locating prey). Finally, a function $\zeta(q)$ that is nonlinear and convex is indicative of a multifractal random walk (Rutherford et al. 2004; Seuront and Stanley 2014).

The only study that used multifractals to assess the behavioural response of zooplankton to water contamination led towards an increase in *Pseudodiaptomus annandalei* behavioural complexity under conditions of stress induced by the presence of a diatom toxin (Michalec et al. 2013a, b). Specifically, *P. annandalei* swimming behaviour is very close to a (monofractal) ballistic motion in control water and progressively diverges towards an increasingly multifractal behaviour with increasing toxin concentrations.

4 Conclusions

The behavioural approach discussed in this contribution to assess zooplankton stress from the geometric and stochastic properties of their motion behaviour has the potential to become an efficient tool in zooplankton ecotoxicology as a sensitive, non-invasive and robust behavioural sublethal endpoint with short-response times

[4] Note the one-to-one correspondence between the function $\zeta(q)$ and the spectral exponent β for $q=2$, i.e. $\beta=1+\zeta(2)$ (Seuront 2010a).

for toxicity bioassays, in particular as it is very sensitive to subtle behavioural changes that may be undetectable to other behavioural variables (Rutherford et al. 2004; Coughlin et al. 1992).

Zooplankton behavioural complexity typically decreases under stress (Seuront 2010a, b, 2012; Seuront and Leterme 2007; Michalec et al. 2013a, b). Increases in complexity have, however, also been observed under certain stress conditions (Shimizu et al. 2002; Michalec et al. 2013a, b),[5] although such results seem to occur in response to acute or stimulatory challenges, quite apart from the chronic or inhibitory stressors that are associated to reduction in complexity (Alados et al. 1996; Seuront 2010a, b, 2012). As shown for several fractal and multifractal measures of environmental complexity (Seuront 2010a), regardless of the direction, it is ultimately the relative differences between the fractal and multifractal exponents observed for a given species under stressful and non-stressful conditions that may be more informative on the related behavioural changes.

Note that the approach described in this contribution is not limited to behavioural ecotoxicology but can be generalised to assess relative changes in the behavioural complexity of marine invertebrates in a wide range of ecologically relevant situation related to, e.g., the quality and the quantity or food and the presence of mates or predators (Seuront 2010a, b; Schmitt et al. 2006). It is finally stressed that the application of fractals to zooplankton behavioural ecology in general (Rutherford et al. 2004; Seuront 2011) and to zooplankton ecotoxicology in particular (Seuront 2010a, b, 2012; Shimizu et al. 2002; Michalec et al. 2013a, b; Seuront and Leterme 2007) is, however, still in its infancy. Further work is needed to entangle the fractal complexity of behavioural properties and to generalise the use of fractal and multifractal approaches to stress assessment in marine invertebrates.

[5] It is worth noting that the increase in the complexity of *Daphnia magna* trajectories in contaminated waters must be treated with caution as some of the fractal dimensions reported fall outside the theoretical range $1 \leq D \leq 2$, i.e. $D > 2$ (Shimizu et al. 2002).

References

Alados CL, Escos JM, Emlen JM (1996) Fractal structure of sequential behaviour patterns: an indicator of stress. Anim Behav 51:437–443

Asher L et al (2009) Recent advances in the analysis of behavioural organization and interpretation as indicators of animal welfare. J Roy Soc Interf 6:1103–1119

Coughlin DJ, Strickler JR, Sanderson B (1992) Swimming and search behaviour in clownfish, *Amphiprion perideraion*, larvae. Anim Behav 44:427–440

Dur G et al (2010) The different aspects in motion of the three reproductive stages of *Pseudodiaptomus annandalei* (Copepoda, Calanoida). J Plankton Res 32:423–440

Garaventa F et al (2010) Swimming speed alteration of *Artemia* sp. and *Brachionus plicatilis* as a sub-lethal behavioural end-point for ecotoxicological surveys. Ecotoxicology 19:512–519

Goldberger AL, Rigney DR, West BJ (1990) Chaos and fractal in human physiology. Sci Am 363:43–49

Goldberger AL et al (2000) Physiobank, physiotoolkit, and physionet: components of a new research resource for complex physiological signals. Circulation 101:215–220

Ivanov PC et al (1999) Multifractacilty in human heartbeat dynamics. Nature 399:461–465

Mandelbrot BB (1982) The fractal geometry of nature. Freeman, New York

Michalec FG et al (2010) Differences in behavioral responses of *Eurytemora affinis* (Copepoda, Calanoida) reproductive stages to salinity variations. J Plankton Res 32:805–813

Michalec FG et al (2013a) Behavioral responses of the estuarine calanoid copepod *Eurytemora affinis* to sub-lethal concentrations of waterborne pollutants. Aquat Toxicol 138/139:129–138

Michalec FG et al (2013b) Changes in the swimming behavior of *Pseudodiaptomus annandalei* (Copepoda, Calanoida) adults exposed to the diatom toxin 2-trans, 4-trans decadienal. Harmful Algae 30:56–64

Mishima M et al (1999) Complexity of terminal airspace geometry assessed by lung computed tomography in normal subjects and patients with chronic obstructive pulmonary disease. Proc Natl Acad Sci U S A 96:8829–8834

Moison M, Schmitt FG, Souissi S (2012) Effect of temperature on *Temora longicornis* swimming behaviour: illustration of seasonal effects in a temperate ecosystem. Aquat Biol 16:149–162

Rutherford KMD et al (2004) Fractal analysis of animal behaviour as an indicator of animal welfare. Anim Welf 13:99–103

Schmitt FG et al (2006) Scaling of swimming sequences in copepod behavior: data analysis and simulation. Physica A 364:287–296

Seuront L (2010a) Fractals and multifractals in ecology and aquatic sciences. CRC Press, Boca Raton

Seuront L (2010b) Hydrocarbon contamination and the swimming behavior of the estuarine copepod *Eurytemora affinis*. In: Marine ecosystems. Intech, Open Access Publisher, Croatia, pp 229–266

Seuront L (2011) Behavioral fractality in marine copepods: endogenous rhythms vs. exogenous stressors. Physica A 309:250–256

Seuront L (2012) Hydrocarbon contamination decreases mating success in a marine planktonic copepod. PLoS ONE 6(10):e26283

Seuront L (2013) Chemical and hydromechanical components of mate-seeking behaviour in the calanoid copepod *Eurytemora affinis*. J Plankton Res 35:724–743

Seuront L, Leterme SC (2007) Increased zooplankton behavioural stress in response to short-term exposure to hydrocarbon contamination. Open Oceanogr J 1:1–7

Seuront L, Stanley EH (2014) Anomalous diffusion and multifractality enhance mating encounters in the ocean. Proc Natl Acad Sci U S A 111:2206–2211

Shimizu N, Ogino C, Kawanishi T, Hayashi Y (2002) Fractal analysis of *Daphnia* motion for acute toxicity bioassay. Environ Toxicol 17:441–448

Uttieri M, Paffenhöffer GA, Mazzocchi MG (2008) Prey capture in *Clausocalanus furcatus* (Copepoda: Calanoida). The role of swimming behaviour. Mar Biol 153:925–935

A Subject of the Chlorine Management at a Thermal Power Plant on the Northwest Pacific Ocean in Japan

Toshio Iibuchi, Seiji Kobayashi, Sigenori Nanjou, Kanako Satou, Takeya Hara, and Michiyasu Kiyono

Abstract

The concentration of chlorine residue was monitored near the outlet of a large thermal power plant in Japan. The plant is required by the local government to ensure that the concentration of chlorine residue remains below 0.05 mg/L at the outlet. Background levels of chlorine residue were relatively high, though lower than 0.05 mg/L. There was a significant positive correlation between the concentrations of phytoplankton and the background concentrations of chlorine residue. About 0.3 mmol/L of H_2O_2 was required to produce a color reaction equivalent to the background signal in the *N*-*N*-diethyl-*p*-phenylenediamine assay for chlorine residue. However, reported concentrations of H_2O_2 in the Sargasso Sea are only 0.0002 mmol/L. H_2O_2, which is produced by phytoplankton, oxidizes Br^- and I^- to produce HBrO and HIO. We hypothesize that these oxidants react with reagents in the *N*-*N*-diethyl-*p*-phenylenediamine assay and are naturally produced chlorine residuals.

T. Iibuchi (✉) • M. Kiyono
Marine Ecology Research Institute, Towa-Edogawabashi Bldg., 347 Yamabuki-cho, Shinjyuku-ku, Tokyo 162-0801, Japan
e-mail: iibuchi@kaiseiken.or.jp

S. Kobayashi
JAPAN NUS Co., Ltd., Nishi-Shinjuku Kimuraya Bldg., 7-5-25 Nishi-Shinjuku, Shinjuku-ku, Tokyo 160-0023, Japan

S. Nanjou • K. Satou
Soma Kyodo Power Co., Ltd., Shinchi Power Station, 1-1 Imagami, Komagamine, Shinchi-Town, Soma County, Fukushima 979-2611, Japan

T. Hara
Marine Ecology Research Institute, Central Laboratory, Onjukumachi, Isumigun, Chiba 299-5105, Japan

1 Introduction

Power plants that use seawater for their cooling water usually suffer from biofouling. Chlorine is often used to lessen biofouling and the associated damage to the plant. When chlorine is added to seawater, it oxidizes bromide ions, the product being hypobromous acid (HBrO). Hypobromous acid reacts with ammonia, and a variety of oxidants (mainly bromamines) are subsequently generated. The sum of concentrations of these substances is referred to as the total oxidant residue (or just "chlorine residue") in seawater (Fig. 1). In many cases, Japanese power plants are required by local governments to keep the

H.-J. Ceccaldi et al. (eds.), *Marine Productivity: Perturbations and Resilience of Socio-ecosystems*,
DOI 10.1007/978-3-319-13878-7_15,

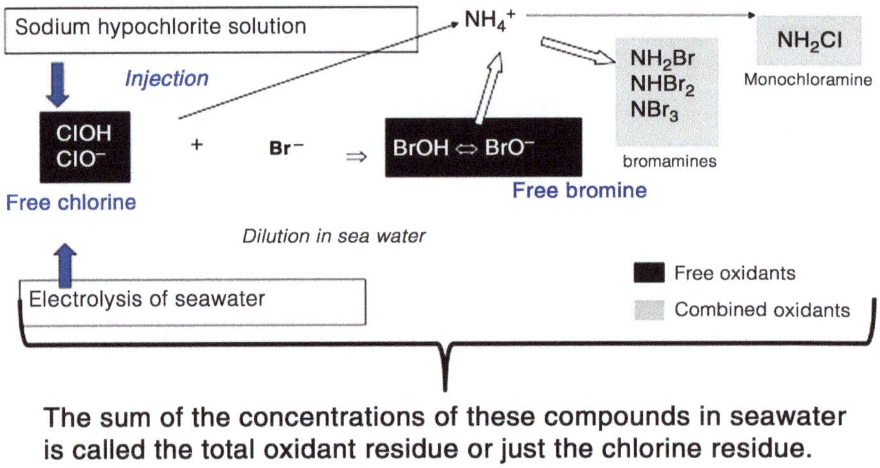

Fig. 1 In seawater, the total oxidant residue (chlorine residue) consists mainly of free bromine and bromamines (Redrawn from Khalanski 2002)

concentration of the chlorine residue under the limit of quantitation (LOQ) at the outlet. According to the Japanese Industrial Standards (JIS), the LOQs of the chlorine residue are 0.05 and 0.01 mg/L as Cl for the *N-N*-diethyl-*p*-phenylenediamine (DPD) and orthotolidine methods of analysis, respectively (Japanese Standards Association 2002).

2 Methods

2.1 Sampling of Seawater and Phytoplankton

Seawater samples from a depth of 0.5 m at stations 1–8 and from a depth of 6 m at stations 1–6 were obtained with a reciprocal pump in the vicinity of a thermal power plant on the coastline of Japan facing the Western North Pacific Ocean (Fig. 2). The dates of sampling were 30 September, 24 October, 2 December, and 20 December 2011 at stations 1–8 and 27 January, 21 February, and 15 March 2012 at stations 1–6. For about 1 year, including this period, the injection of chlorine into the intake of the power plant was stopped.

Phytoplankton samples were obtained from a depth of 0.5 m at stations 1, 2, and 4 at the same time as the seawater sampling. The phytoplankton in 10 L of seawater were filtered through a 20-μm mesh-size net on board the boat, concentrated into 200 mL seawater, and preserved with 30 mL of 33 % formaldehyde solution. Samples were stored in the dark under cool conditions. Professionals at the Plantbio Co., Ltd., Odawara, Japan, classified and counted the phytoplankton cells.

2.2 Measurement of Chlorine Residue

We used both the DPD and orthotolidine colorimetric methods to measure the chlorine residue in water samples on board the boat immediately after obtaining the samples. The detailed analytical methodology of the DPD method is described in Standard Methods (Rice et al. 2012), and that of the orthotolidine method is described in JIS Handbook Environmental Technology (Japanese Standards Association 2002).

2.3 Temperature Effects on Measurements of Chlorine Residuals

Sensitivity tests for the two colorimetric methods of chlorine quantitation were conducted on 28 June, 4 August, and 17 November 2012 and 2 February 2013. We used each colorimetric

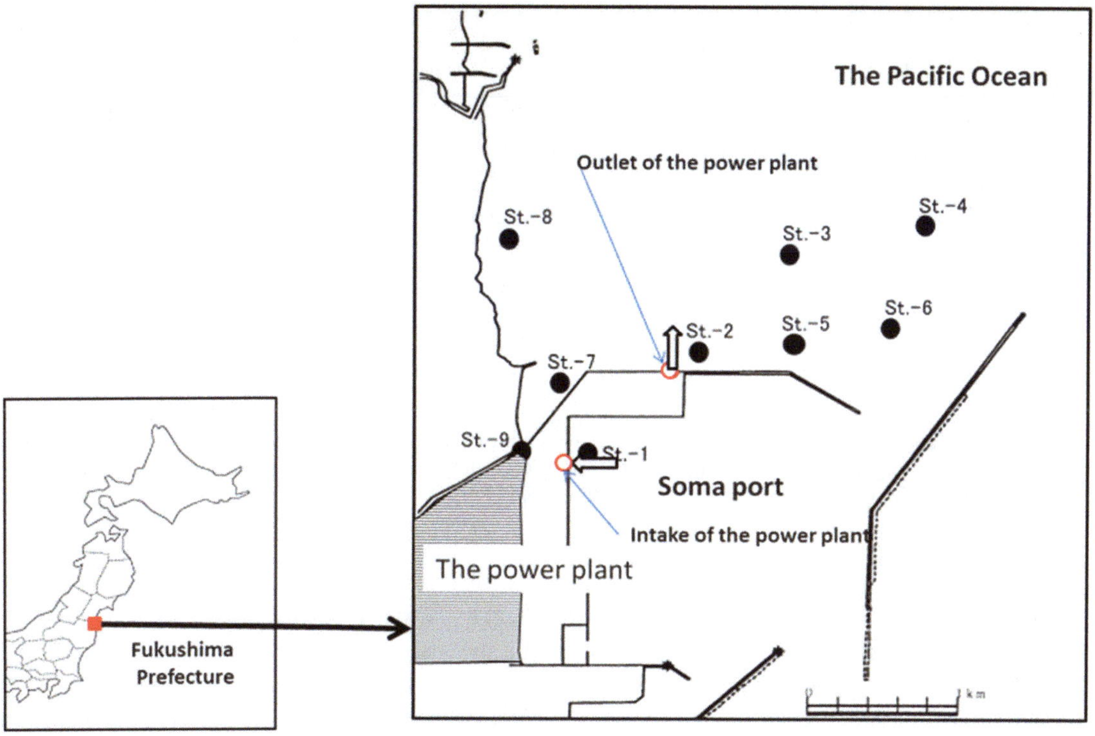

Fig. 2 Map of the sampling locations in northern Fukushima, Japan

method to measure the chlorine residue in the same sample solution at different temperatures. The concentration of the chlorine residue in each sample was predetermined using an iodometric titration, which is more accurate than either colorimetric method. The sample solutions were prepared by adding sodium hypochlorite to seawater and then leaving the solutions to stand for more than 60 min to allow the concentration of chlorine residue to stabilize. The solutions were held at a temperature of 10, 15, 20, 25, or 30 °C during the measurements. The temperature of 30 °C was used only in the summer. The analytical methodology of the iodometric titration is described in Standard Methods (Rice et al. 2012).

2.4 Interference by H_2O_2 in the Measurement of Chlorine Residue

To study the reactions between H_2O_2 and the reagents in the DPD and orthotolidine assays, we measured the color changes that occurred in a NaCl solution and in filtered seawater in which H_2O_2 was dissolved. The NaCl solution was prepared by dissolving 35 g of NaCl in 1 L of distilled water. In the case of the DPD method, 0.03 mL of 35 % H_2O_2 (11 mmol) was added to 1 L of NaCl solution and to 1 L of seawater. In the case of the orthotolidine method, 0.3 mL of 35 % H_2O_2 (105 mmol) was similarly added to 1 L of NaCl solution and to 1 L of seawater.

3 Results and Discussion

3.1 Field Survey of Chlorine Residue

The results of the field survey of chlorine residue based on the DPD method are shown in Fig. 3. The DPD reagent develops a weak red color even at the limit of quantitation. We ranked the extent of the redness as +, ++, +++, and ++++, corresponding to solutions that were made by diluting

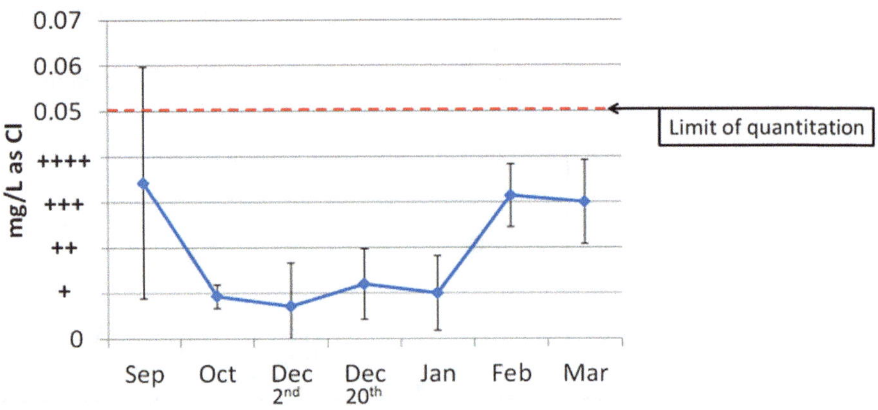

Fig. 3 Concentrations of chlorine residue measured by the DPD method during the period when chlorination was stopped. Concentrations below the limit of quantitation were characterized by the symbols +, ++, +++, and ++++, according to the extent of *red color* development in the DPD assay

the standard color solution (0.05 mg/L as Cl) to 0.01, 0.02, 0.03, and 0.04 mg/L as Cl, respectively. The results show a significant amount of background chlorine residue, even though the injection of chlorine at the power plant was stopped during the period of sampling. In the results obtained with the orthotolidine method, the background concentrations were lower (data not shown).

3.2 Temperature Effects on Measurements of Chlorine Residue DPD Method

The chlorine residue concentrations determined by the iodometric titration method (abscissa) are compared with those obtained by the DPD method (ordinate) in Fig. 4. An analysis of covariance indicated that there was no statistically significant temperature effect on the slopes of the linear relationships between the results obtained by the two methods. The elevations of the regression lines, however, were significantly different; the y-intercepts in spring and summer were higher than in autumn and winter (Fig. 5; $p<0.01$).

3.2.1 Orthotolidine Method

In the case of the orthotolidine method, the slopes of the lines were small (about 0.2–0.7) and differed among temperatures and between seasons (Fig. 6; $p<0.01$). The implication is that the chlorine residue estimated with the orthotolidine method was affected by the temperature of the water sample. The orthotolidine method tends to give a lower value than the iodometric titration and DPD methods.

3.3 Concentrations of Phytoplankton

The phytoplankton concentration data are shown in Fig. 7 (left). The concentrations at each site were similar and decreased from autumn to winter.

3.4 Interference by H_2O_2 in the Measurement of Chlorine Residue

The results of the assays for chlorine residue with H_2O_2 dissolved in NaCl solution and filtered seawater are shown in Fig. 8. In the DPD assay, with 11 mg/L of H_2O_2 dissolved in the water, the arti-

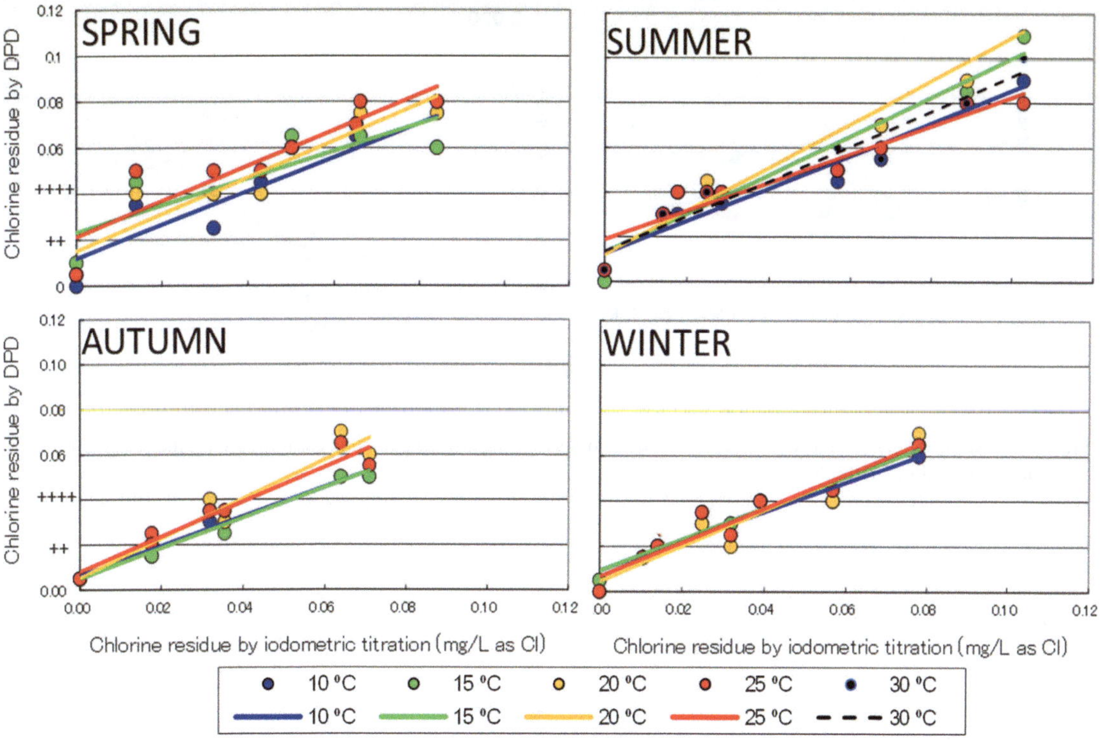

Fig. 4 Comparison of the chlorine residue estimated with the iodometric titration method (abscissa) and with the DPD method (ordinate)

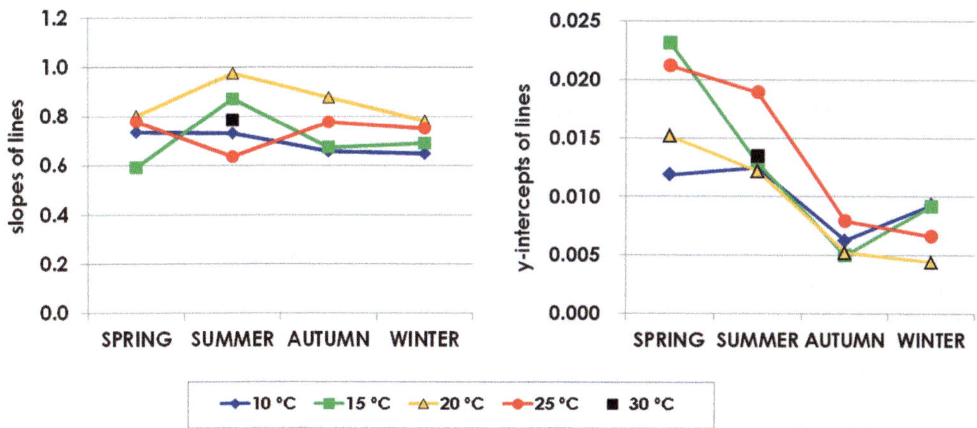

Fig. 5 The slopes (*left*) and the y-intercepts (*right*) of the lines in Fig. 4

factual chlorine residue concentrations were 0.09 mg/L as Cl and ++ (0.02 mg/L as Cl equivalent) in the NaCl solution and filtered seawater, respectively. In the orthotolidine assay, with 105 mg/L of H_2O_2 dissolved in the water, the artifactual chlorine residue concentrations were only

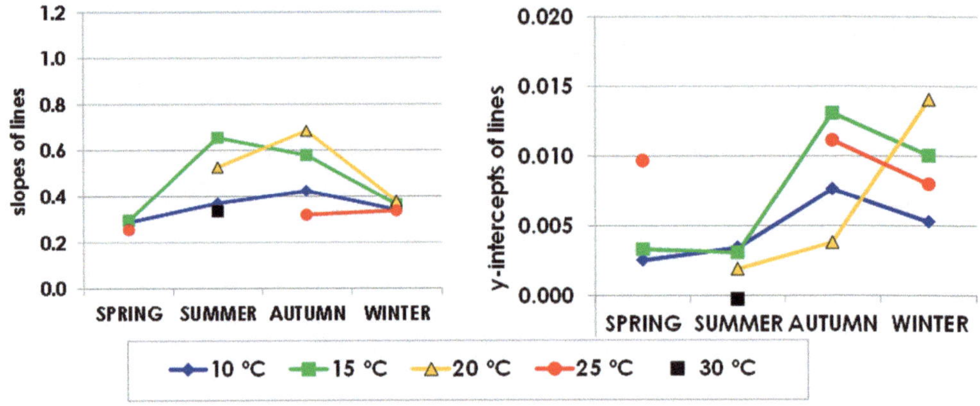

Fig. 6 The slopes (*left*) and the y-intercepts (*right*) of lines relating chlorine residue estimated with the iodometric titration and orthotolidine methods

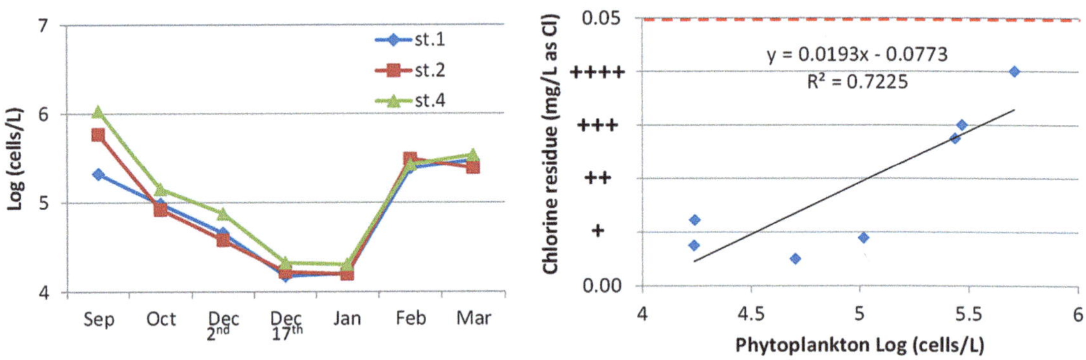

Fig. 7 Concentrations of phytoplankton (*left*) and the relationship between the concentrations of phytoplankton and chlorine residue (Data from Fig. 3) estimated by the DPD method (*right*)

Fig. 8 Interference by H_2O_2 in the measurement of chlorine residue. The H_2O_2 concentration in the water was 11 mg/L in the DPD assay (*left*) and 105 mg/L in the orthotolidine assay (*right*)

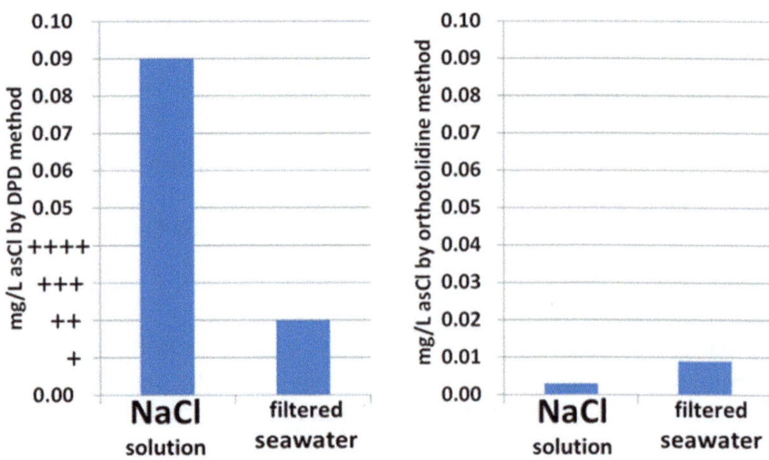

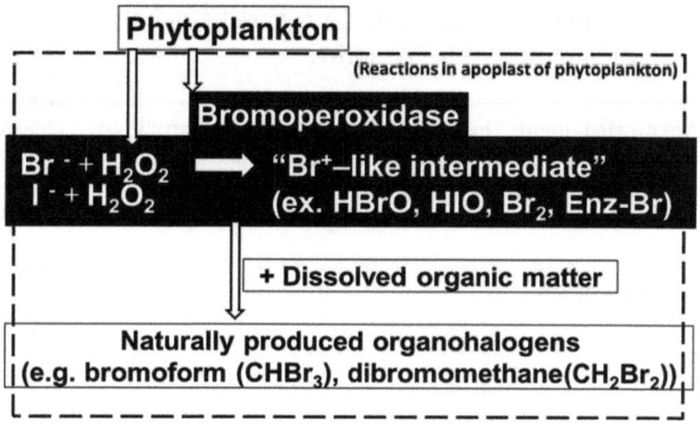

Fig. 9 A scheme for the production of chlorine residue (HBrO) and organo halogens by phytoplankton (Redrawn and modified from Tschirret-Guth and Butler 1994)

about 0.003 mg/L as Cl and 0.009 mg/L as Cl equivalent in the NaCl solution and filtered seawater, respectively.

4 Discussion

Comparison of the DPD and iodometric titration (Figs. 4 and 5) results indicates the existence of an unknown material that causes the background signal to be larger with the DPD method than with the iodometric titration method in spring and summer. On the basis of our experience, we hypothesized that the unknown material responsible for the chlorine residue blank was a by-product of phytoplankton photosynthesis. We believe that the positive correlation between the concentrations of phytoplankton and chlorine residue estimated with the DPD method (Fig. 7, right) supports our hypothesis. There is also a possibility that some abiotic mechanism is responsible for the blank; for example, strong solar radiation may produce oxidants in the atmosphere or in the sea (Plane et al. 1997). At first, we assumed that H_2O_2 was the naturally produced unknown material that was interacting with the DPD reagents because H_2O_2 has been reported to be one of the substances that interfere with the measurement of chlorine residue by the DPD method (Japanese Standards Association 2013).

According to the literature, H_2O_2 is produced in phytoplankton cells (Palenic et al. 1987; Twiner and Trick 2000). But to produce a red color in the DPD assay equivalent to a chlorine residue concentration of ++ (0.02 mg/L as Cl equivalent), 11 mg/L (0.3 mmol/L) of H_2O_2 was required (Fig. 8), which is about 1,500 times the reported concentration in the Sargasso Sea [at most 0.0002 mmol/L; Miller and Kester (1994)]. Recently, several species of diatoms have been reported to catalyze the oxidation of Br^- and I^- by H_2O_2, the products being HBrO and HIO (Hill and Manley 2009). Such reactions are catalyzed by bromoperoxidase and iodoperoxidase on the surface of the cell (in the apoplast; Fig. 9; Tschirret-Guth and Butler 1994; Lin and Manley 2012). These mechanisms probably account for the low concentration of H_2O_2 in the sea, because oxidation of Br^- and I^- (and the simultaneous reduction of H_2O_2) may occur continuously in the apoplast of the cell, where the concentration of H_2O_2 is as high as it is in higher plants and macroalgal species (Küpper et al. 2002). As mentioned above, the regulations relating to chlorination by power plants in Japan are very strict. If the results of this study are found to be generally applicable in the future, the strict regulation of chlorination at Japanese power plants should be reconsidered. The background concentration must be added to the threshold value (LOQ, i.e., 0.05 mg/L as Cl for the DPD method). We believe that identification of the substances responsible for the background concentration should be a priority.

Acknowledgment This research program was carried out under a contract with the Soma Kyodo Power Company, Ltd., Japan.

References

Hill VL, Manley SL (2009) Release of reactive bromine and iodine from diatoms and its possible role in halogen transfer in polar and tropical oceans. Limnol Oceanogr 54(3):812–822

Japanese Standards Association (2002) (eds) JIS Handbook Environmental Technology

Japanese Standards Association (2013) (eds) JIS Handbook of Environmental Technology II – water chemistry analysis (in Japanese)

Khalanski M (2002) Organic products generated by the chlorination of cooling water at marine power stations. Journées d'Etudes du Cebedeau, Tribune de l'Eau No. 619-620-621, France, 24–39

Küpper FC, Muller DG, Peters AF, Kloareg B, Potin P (2002) Oligoalginate recognition and oxidative burst play a key role in natural and induced resistance of sporophytes of Laminariales. J Chem Ecol 28(10):2057–2081

Lin CY, Manley SL (2012) Bromoform production from seawater treated with bromoperoxidase. Limnol Oceanogr 57(6):1857–1866

Miller WL, Kester DR (1994) Peroxide variations in the Sargasso Sea. Mar Chem 48:17–29

Palenic B, Zafiriou OC, Morel FMM (1987) Hydrogen peroxide production by a marine phytoplankter. Limnol Oceanogr 32(6):1365–1369

Plane JM, Blough NV, Ehrhardt MG, Water K, Zepp RG, Zika RG (1997) Photochemistry in the sea surface micro layer. In: Liss P, Duce R (eds) The sea surface and global change. Cambridge University Press, New York, pp 71–92, 536 pp

Rice EW, Baird RB, Eaton AD, Clesceri LS (eds) (2012) Standard methods for examination of water and wastewater 2012. American Public Health Association, Washington, DC, 1496 pp

Tschirret-Guth RA, Butler A (1994) Evidence for organic substrate binding to vanadium bromoperoxidase. J Am Chem Soc 116:411–412

Twiner MJ, Trick CG (2000) Possible physiological mechanisms for production of hydrogen peroxide by the ichthyotoxic flagellate *Heterosigma akashiwo*. J Plankton Res 22(10):1961–1975

Effects of Thermal Stressors on Growth-Related Gene Expressions in Cultured Fish

Toshiki Nakano, Masumi Kameda, Toshiyasu Yamaguchi, Minoru Sato, Luis O.B. Afonso, George K. Iwama, and Robert H. Devlin

Abstract

Growth in fish is regulated by the growth hormone (GH)-growth hormone receptor (GHR)-insulin-like growth factor-1 (IGF-1) axis. However, the effect of severe acute stressors on the GH-IGF-1 axis in fish is not well understood. The present study determined the changes in mRNA expression of growth-related genes *gh*, *ghr*, and *igf* and the redox state in coho salmon (*Oncorhynchus kisutch*), in response to severe acute stress. Severe stress consisted of exposure to heat shock (adequate rearing temperature +11 °C for 2 h). The plasma expression patterns of redox state-related biomarkers, such as glutathione, lipid peroxides, and superoxide dismutase, in response to heat shock suggest that heat shock might induce oxidative stress in fish. After exposure to heat shock, *ghr* mRNA levels in the pituitary glands and liver increased, whereas levels decreased 48 h post-stress. Hepatic *igf1* mRNA expression levels gradually decreased in response to the stressor. On the other hand, the pituitary *gh* mRNA expression did not change in response to the stressor. These findings showed that a heat shock-induced oxidative stress could affect the redox state and the expression of several growth-related genes in coho salmon. The results of this study also suggest that the expression of several growth-related genes in fish may be affected differently by the types and strength of stress.

T. Nakano (✉) • M. Kameda
T. Yamaguchi • M. Sato
Marine Biochemistry Laboratory, Graduate School of Agricultural Science, Tohoku University, Aoba-ku, Sendai 981-8555, Japan
e-mail: nakanot@bios.tohoku.ac.jp

L.O.B. Afonso
School of Life and Environmental Sciences, Faculty of Science and Technology, Deakin University, Warrnambool, VIC 3280, Australia

G.K. Iwama
Okinawa Institute of Science and Technology, Onna-Son, Kunigami, Okinawa 904-0495, Japan

R.H. Devlin
West Vancouver Laboratory, Center for Aquaculture and Environment Research (CAER- DFO/UBC), Fisheries and Oceans Canada, West Vancouver, BC V7V 1N6, Canada

H.-J. Ceccaldi et al. (eds.), *Marine Productivity: Perturbations and Resilience of Socio-ecosystems*, DOI 10.1007/978-3-319-13878-7_16,

1 Introduction

It is well known that fishing and aquaculture are very important to world food production. World market demand for high quality products has stimulated much of the growth in aquaculture, especially for salmonid, shrimp, and shellfish species (Nakano 2007). Cultured fish are exposed to biotic and abiotic stressors, such as toxicants and acute changes in temperature, which can increase the chances of these fish succumbing to infectious disease (Pickering 1993; Nakano and Takeuchi 1997; Nakano et al. 1999a; Iwama et al. 2006; Nakano 2007, 2011; Pankhurst 2011; Ellis et al. 2012; Prunet et al. 2012). Furthermore, perturbations due to global climate change, typhoon, tsunami, and artificial factors such as environmental pollutants, radioactive contaminants derived from nuclear power plant accident, in aquatic biological systems have recently become a serious problem (Pörtner 2002; Lesser 2006; Valavanidis et al. 2006; Hofmann and Todgham 2010; Urushihara 2013; Hara 2014).

In response to a particular stressor, a series of biochemical and physiological changes occur at both the cellular and organismal levels. These stress responses in fish can affect their general health, disease resistance, growth, and reproduction (Barton and Iwama 1991; Pickering 1993; Pickering and Pottinger 1995; Pankhurst and Kraak 1997; Barton 1997; Nakano 2011; Prunet et al. 2012).

The growth of fish is regulated to a large extent by liver-derived insulin-like growth factor (IGF)-1 in response to pituitary-secreted growth hormone (GH) binding to GH receptor (GHR) in the liver. The GH-IGF-1 axis has a critical role in regulating both fish growth and development (Kopchick and Andry 2000; Moriyama et al. 2000; Björnsson et al. 2002; Reineck et al. 2005; Klein and Sheridan 2008; Deane and Woo 2009; Reineck 2010).

Fish growth is genetically regulated and is also influenced by cellular, endocrinological, and environmental factors. The responses of endocrine tissue are affected by the integration of external stimuli with internal signals according to the physiological state (Peter 1979; Barton and Iwama 1991; Pickering 1993; Pickering and Pottinger 1995; Duan 1998; Moriyama et al. 2000; Mommsen and Moon 2001; Iwama et al. 2006; Kameda et al. 2008; Deane and Woo 2009; Reineck 2010; Nakano 2011; Prunet et al. 2012).

The physiological states of ectothermal organisms, such as fish, depend on the environmental temperature. Studies on thermal stress in fish have primarily focused on cellular molecular chaperones, heat shock proteins (HSPs), expression, and characterization (Iwama et al. 1998; Feder and Hofmann 1999; Basu et al. 2001, 2002; Pörtner 2002; Nakano 2011; Iwama et al. 2006). Little is known about the effects of severe acute stressors, such as heat shock, on the expression levels of genes that are related to growth in fish (Pörtner 2002; Lushchak and Bagnyukova 2006a; Kameda et al. 2008; Deane and Woo 2009; Reineck 2010; Nakano 2011; Beckman 2011; Nakano et al. 2013, 2014). Therefore, it is important to determine the effects of thermal stress on fish fitness and tolerance in order to improve their production and health in both natural and cultural conditions.

In this study, we examined changes in mRNA expression levels of the *gh*, *ghr*, and *igf1* genes in response to a severe acute stress derived from heat shock in coho salmon (*Oncorhynchus kisutch*). Coho salmon is known to be one of the most valued species used in aquaculture. We discuss the relationships between the thermal stress responses, expressions of growth-related genes, and the oxidative stress in fish in the context of our findings.

2 Materials and Methods

2.1 Fish, Rearing Conditions, Stress Performance, and Sampling

Coho salmon were purchased from a local hatchery, Sakai Hatchery Co., in Zao town,

Miyagi, Japan. After acclimatization for 2 weeks at the aquarium facility of Tohoku University, fish (approx. body weight, 144 g) were reared in 60-L flow-through glass tanks at 8 °C (light/dark = 12 h/12 h). The fish were fed by hand to apparent satiation twice a day with commercial feed (Nosan Co., Japan). Food was withheld for over 48 h before each sampling period. Fish were exposed to heat shock (+11 °C for 2 h) and sampled at 2.5, 17.5, and 48 h post-stress. Blood was collected from the caudal vessels under MS222 (m-aminobenzoic acid ethyl ester methanesulfonate) anesthesia. The plasma was separated by centrifugation and frozen at −80 °C. Fish were gutted; the tissues were quickly removed and frozen at −80 °C in RNA later (Ambion, Life Technologies, Austin, TX) until analysis.

Our experiments were conducted in accordance with the principles and procedures approved by the Animal Care Committee at Tohoku University (Sendai, Japan).

2.2 Plasma Cortisol and Glucose Levels

Plasma cortisol levels were measured using an enzyme-linked immunosorbent assay kit (Oxford Biomedical Research, UK) (Basu et al. 2001). Plasma glucose was measured using an enzymatic assay method with a Glucose CII-Test Wako kit (Wako Pure Chemical Industries, Ltd., Japan).

2.3 Plasma Lipid Peroxides, Glutathione, and Superoxide Dismutase Levels

Lipid peroxides (LPO) were determined as thiobarbituric acid-reactive substances (TBARS) by a HPLC-fluorescence method (Wong et al. 1987; Morliere et al. 1991). Glutathione (GSH) levels in plasma were determined by a glutathione reductase-recycling method with a Total Glutathione Quantification Kit (Dojindo Laboratories, Japan). This kit can measure the total amount of reduced GSH and oxidized form of GSH. The superoxide dismutase (SOD) activity was assayed by the formazan-WST method (Total SOD Assay Kit, Dojindo Laboratories, Japan).

2.4 RNA Extraction and cDNA Synthesis

Tissues were suspended in TRIzol Reagent (Invitrogen, Life Technologies, CA) and immediately homogenized using a polypropylene pestle. The resulting RNA pellet was dissolved in RNase-free water (UltraPure, Gibco, Life Technologies, NY), quantified by spectrophotometry (V-630-Bio, JASCO, Japan), and then diluted to 500 ng/μL for use in reverse transcription reactions. RNA samples were stored at −80 °C. Complementary DNA (cDNA) was synthesized using a ReverTra Ace qPCR RT Kit (Toyobo, Japan) with a mixture of random hexamers and oligo-dT primers or with a gene-specific primer for salmon GH (gh-reverse primer) and 250 ng of RNA (Nakano et al. 2013).

2.5 Real-Time qPCR for *gh, ghr, igf1,* and *arp* mRNA Levels

The mRNA expression levels of *gh*, *ghr*, and *igf1* in tissues were determined by a real-time quantitative PCR (qPCR) with an ABI Prism 7300 Sequence Detection System (Applied Biosystems, Life Technologies, Foster City, CA) using acidic ribosomal phosphoprotein P0 gene (*arp*) as an internal standard (Pierce et al. 2004; Nakano et al. 2013). mRNA values for *gh, ghr,* and *igf1* were normalized to those for *arp*. Accordingly, each sample amplification value for each gene was expressed as a relative gene expression ratio (relative mRNA level).

2.6 Statistical Analysis

All samples were run in duplicate and results were expressed as means ± SEM. All data were subjected to one-way analysis of variance

(ANOVA). Multiple comparisons between groups were made by the Tukey-Kramer method.

3 Results

3.1 Plasma Cortisol and Glucose Levels

Plasma cortisol levels increased at 2.5 h post-heat stress compared with those in control fish, but returned to basal levels at 17.5 h post-stress (Fig. 1a).

Plasma glucose levels increased at 2.5 and 17.5 h post-stress compared with those in control fish. However, at 48 h post-stress, plasma glucose levels in stressed fish decreased and were not significantly different from those in control fish (Fig. 1b).

3.2 Plasma LPO, GSH, and SOD Levels

Plasma LPO and GSH levels are shown in Fig. 2. As shown in Fig. 2a, the plasma LPO levels in stressed fish gradually increased after heat shock treatment and increased significantly compared with those in control fish at 17.5 and 48 h post-stress.

Plasma GSH levels decreased at 2.5 h post-heat stress, but returned to basal levels at 17.5 h post-stress (Fig. 2b). At 48 h post-stress, plasma glutathione levels in stressed fish increased significantly as compared to those in control fish.

Plasma SOD activity in stressed fish increased significantly compared with that in control fish at 17.5 h post-stress, but returned to basal levels at 48 h post-stress (data not shown).

3.3 *gh*, *ghr*, and *igf1* mRNA Levels

The mRNA expression levels of *gh*, *ghr*, and *igf1* in the pituitary glands and livers of stressed and control fish were compared (Figs. 3 and 4).

In the pituitary glands, *gh* mRNA levels were not significantly different between the control and stressed fish (data not shown). *ghr* mRNA expression levels in the pituitary glands of stressed fish gradually increased after heat stress treatment, with the highest level at 17.5 h post-stress. At 48 h post-stress, *ghr* mRNA expression levels returned to their basal levels (Fig. 3).

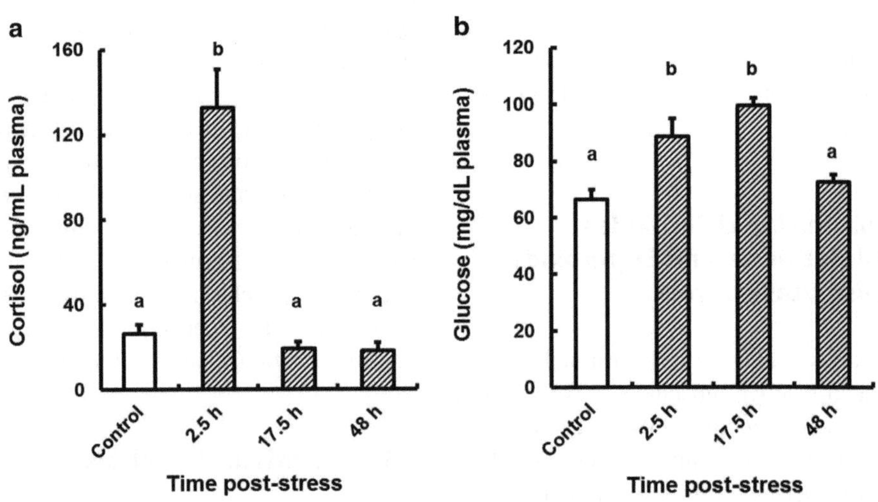

Fig. 1 (**a** and **b**) Effect of thermal stress on cortisol (**a**) and glucose (**b**) levels in plasma from coho salmon *O. kisutch*. Data represent means ± SEM ($n=8$). Statistical relationships between groups are indicated by *letters* where significant differences were detected ($p<0.05$)

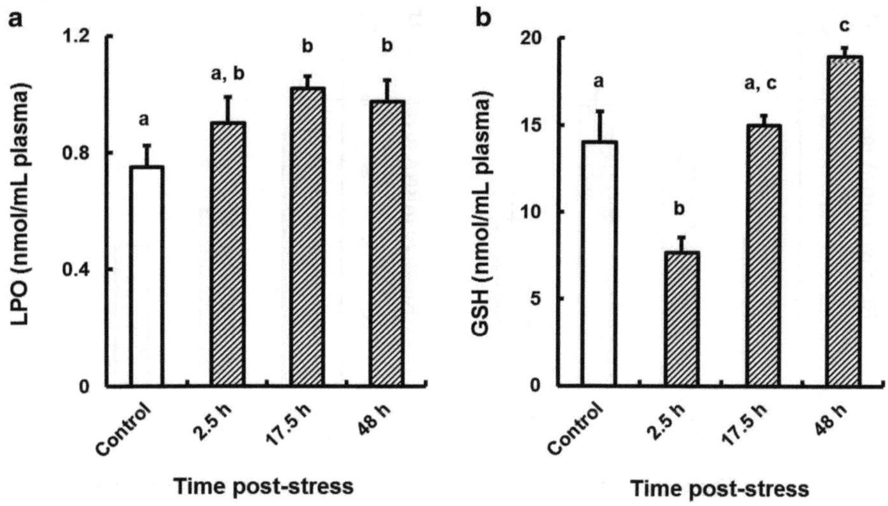

Fig. 2 (**a** and **b**) Effect of thermal stress on LPO (**a**) and GSH (**b**) levels in plasma from coho salmon *O. kisutch*. Data represent means ± SEM ($n=5$). Statistical relationships between groups are indicated by *letters* where significant differences were detected ($p<0.05$)

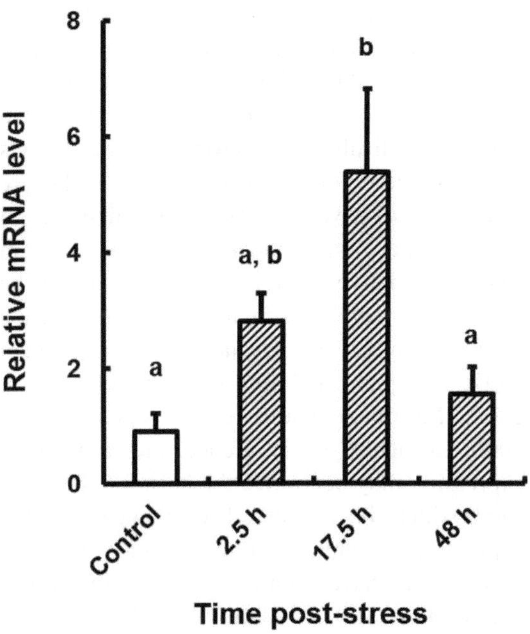

Fig. 3 Effect of thermal stress on expression level of *ghr* mRNA in the pituitary from coho salmon *O. kisutch*. The expressions of target gene were normalized by *arp* expressions. Data represent means ± SEM ($n=4$). Statistical relationships between groups are indicated by *letters* where significant differences were detected ($p<0.05$)

As shown in Fig. 4a, *ghr* mRNA levels in the livers of stressed fish increased. In contrast, the *igf1* mRNA levels in stressed fish livers decreased gradually after heat stress and decreased significantly as compared with those in the control fish at 48 h post-stress (Fig. 4b).

4 Discussion

The results of this study demonstrated that a severe acute thermal stressor could affect the redox state and the expression of growth-related genes in coho salmon.

The temperature can induce numerous changes in the biological functions of organisms. Increased environmental temperature results in increased oxygen consumption and stimulates various metabolic processes on the basis of known thermodynamic principles (Pörtner 2002; Lesser 2006; Lushchak and Bagnyukova 2006a, b; Lushchak 2011).

LPO, expressed as TBARS, in plasma are considered to be metabolites derived from various damaged tissues (Parihar and Dubey 1995; Nakano and Takeuchi 1997; Nakano et al. 1999a, b; Rau et al. 2004; Lushchak et al. 2005a, b; Olsen et al. 2005; Heise et al. 2006; Bagnyukova et al. 2007). In this study, the plasma TBARS levels of fish exposed to heat shock increased. TBARS levels in fish tissues

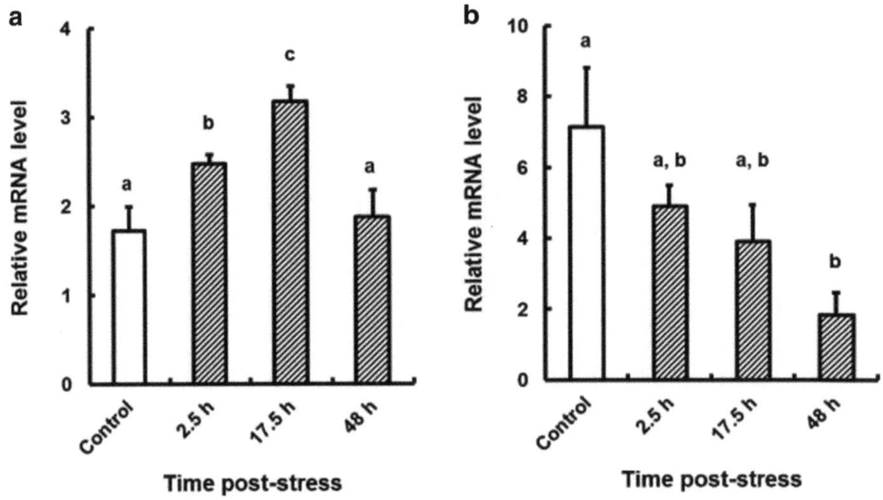

Fig. 4 (**a** and **b**) Effect of thermal stress on expression levels of *ghr* (**a**) and *igf1* (**b**) mRNA in the liver from coho salmon *O. kisutch*. The expressions of target gene were normalized by *arp* expressions. Data represent means ± SEM ($n=4$). Statistical relationships between groups are indicated by *letters* where significant differences were detected ($p<0.05$)

can change under several stressful conditions, such as heat exposure, handling stress, hyperoxia, oxidized oils administration, and heavy metal intake (Parihar and Dubey 1995; Nakano et al. 1999a; Ali et al. 2004; Rau et al. 2004; Lushchak et al. 2005a, b; Martínez-Álvarez et al. 2005; Olsen et al. 2005; Heise et al. 2006; Lesser 2006; Valavanidis et al. 2006; Bagnyukova et al. 2007).

The major nonprotein cellular thiol, reduced GSH, is a tripeptide (Glu-Cys-Gly) with reducing and nucleophilic properties that is one of the major regulators of the intracellular redox state (Niki 1988; Nakano and Takeuchi 1997; Arrigo 1999; Sies 1999; Lesser 2006; Valavanidis et al. 2006). GSH can act as a chain breaker of free radical reaction and is the substrate for glutathione peroxidase, an enzyme that scavenges reactive oxygen species (ROS) and LPO generated within cells. The plasma GSH levels observed in this study were similar to those in the livers of fish that were administered with an oxidant, such as t-butyl hydroperoxide, after heat exposure (Ploch et al. 1999; Ali et al. 2004; Lushchak and Bagnyukova 2006a, b; Heise et al. 2006; Valavanidis et al. 2006; Bagnyukova et al. 2007). At the initial post-heat stress stage, GSH may be consumed to eliminate ROS generated in blood.

Plasma SOD activity in heat-shocked fish showed a transient increase at 17.5 h post-stress in this study. Antioxidative enzymes, such as SOD, glutathione peroxidase, and catalase, can scavenge radicals and contribute to the body's antioxidative defenses. In particular, SODs are considered to have key roles in the first line of the enzymatic antioxidative defense system against oxidative injuries, as they catalyze the removal of oxygen radical, superoxide (Asada 1988; Oyanagui 1989; Nakano and Takeuchi 1997; Taniguchi and Endo 2000; Zelko et al. 2002; Martínez-Álvarez et al. 2005; Lesser 2006; Lushchak 2011). Hence, increased SOD expression may neutralize the harmful effects of superoxides in tissues. The changes in the expression of antioxidative enzymes, such as SOD, in fish have been observed with regard to stress (Poly 1997; Pörtner 2002; Martínez-Álvarez et al. 2005; Valavanidis et al. 2006; Craig et al. 2007; Lushchak 2011).

Heat exposure and enhanced oxygen consumption are considered to promote ROS generation in the tissue. The resulting ROS attack almost all cell components (Asada 1988; Nakano and Takeuchi 1997; Beckman and Ames 1998; Droge 2002; Lesser 2006; Valavanidis et al. 2006; Lushchak 2011). ROS production in cell, especially in mitochondria where spin-off of superoxides during mitochondria transfer of electrons to oxygen occurs, was increased in exercised mammalian muscle, heat-stressed bivalve gills, heat-shocked chicken muscles, and cultured cells as compared with non-stressed control (Ji 1995; Pörtner 2002; Heise et al. 2003; Martínez-Álvarez et al. 2005; Mujahid et al. 2005; Lesser 2006; Valavanidis et al. 2006; Shin et al. 2008). Inductions of hepatic HSP and GSH in stressed coho salmon have been observed in this study (Nakano et al. 2014). Thus, the present results regarding the plasma levels of cortisol and glucose and the expression patterns of redox state-related biomarkers, such as LPO, GSH, SOD, and HSP, in response to heat shock suggest that heat shock induces oxidative stress in fish. The resulting oxidative stress may enhance oxidation in the body and result in damage to tissues (Nakano and Takeuchi 1997; Nakano et al. 1999a; Martínez-Álvarez et al. 2005; Lesser 2006; Valavanidis et al. 2006; Nakano 2007, 2011; Lushchak 2011). Under oxidative stress conditions, the levels of antioxidative agents, such as GSH, SOD, and HSP, may increase due to their de novo synthesis.

The pituitary is the major organ of the GH-IGF-1 axis in fish (Kobayashi et al. 2002; Takei and Loretz 2006). In the present study, *ghr* gene expression in the pituitary increased in response to heat stress. In contrast, pituitary *ghr* gene expression significantly decreased in response to moderate acute physiological stress induced by handling (Nakano et al. 2013). This may involve interactions between glucocorticoids, such as cortisol and factors of the GH-IGF-1 axis, and be related to the differences observed between heat shock-induced oxidative stress and mild physiological stress derived from handling. The effects of stress on growth-related gene expression in fish may be affected differently by the strength of stress.

Hepatic *igf1* expression gradually decreased after heat shock treatment, whereas *ghr* expression levels increased after heat shock in this study. Thus, *ghr* and *igf1* genes responded differently to heat stress. Furthermore, pituitary *gh* mRNA expression did not change in response to oxidative stress. These observations suggest that igf1 gene expression may not be directly regulated by circulating GH levels alone.

A decrease in plasma IGF-1 level at 2–24 h post-stress has often been observed in many fish species (Beckman 2011). Our results for the changes in hepatic *igf1* gene expression levels in stressed fish are in agreement with those reported for fish that were administered exogenous cortisol and under confinement stress (Kajimura et al. 2003; Dyer et al. 2004; Saera-Vila et al. 2009). However, our results for hepatic *ghr* and *igf1* gene expression patterns in fish under oxidative stress appear to be different from those in coho salmon under mild physiological stress caused by handling (Nakano et al. 2013). These observations suggest that the hormonal regulation of hepatic both *ghr* and *igf1* gene expression can vary depending on the types of hormones and stress.

5 Conclusions and Perspectives

Oxidative stress could affect the expression of growth-related genes accompanying the changing in circulating glucocorticoids, such as cortisol, and alterations in signal transduction. Intercellular signaling is known to be affected by ROS (Schoeniger et al. 1994; Ji 1995; Franco et al. 1999; Arrigo 1999; Allen and Tresini 2000; Droge 2002; Lesser 2006; Valavanidis et al. 2006; Shin et al. 2008; Nakano 2011; Lushchak 2011). An antioxidative supplement that could dramatically reduce oxidative stress-induced damage in fish has been observed (Nakano et al. 1999a, b, 2004; Bell et al. 2000; Martínez-Álvarez et al. 2005; Nakano 2007, 2011). Accordingly, further studies are required on the possible beneficial effects of antioxidative nutraceutical supplements on growth-related factors in oxidative stressed fish.

Marine ecosystems could provide various products and services, including vital food resources for us (Holmlund and Hammer 1999; Worm et al. 2006). The ability of the ocean is thought to maintain water quality and regulate perturbations and other essential ecosystem services (Worm et al. 2006). Marine ecosystems can be influenced by exploitation, pollution, biodiversity loss, and habitat destruction or indirectly through global climate change and related perturbations (Worm et al. 2005, 2006; Berque and Matsuda 2013). In the northeastern (Tohoku) Pacific coastal area, Sanriku Coast, where fishing and farming are known to be essential to the industries, the Great East Japan Earthquake caused the perturbations in 2011 (Urushihara 2013; Hara 2014). The impact of the earthquake and massive tsunami on the Sanriku area and the subsequent processes of transition over time are yet to be determined. Facilitation of reconstruction of the coastal environment and fisheries at the Sanriku Coast has been required. Therefore, especially in a disaster-stricken area such as the Sanriku Coast, it is thought to be important to research on perturbations, recovery, and resilience processes in marine ecosystems and the management of socio-ecological system in fishing and aquaculture to take sustainable delivery of environmental benefits linked to human well-being (Hadjimichael et al. 2013). Furthermore, a study on the effect of stress in marine organisms is emerging as a worldwide common theme in relation to the perturbations and resilience on marine ecosystems (Pörtner 2002; Lesser 2006; Valavanidis et al. 2006; Hofmann and Todgham 2010; Nakano 2011; Nakano et al. 2013, 2014). Consequently, the results of this study on stress response in fish can provide information that is useful for improving fish fitness and production of fishing and aquaculture. To determine the relationships between oxidative stress, growth-related factors, antioxidant defenses, and the growth of fish, additional investigations are currently underway.

Acknowledgments The authors are grateful to Dr. S. Minami at Shirako Co. Ltd., Japan, and Miss. A. Yamauchi at Tohoku Univ. for the support of laboratory work. The authors also wish to thank Drs. N. Ito, M. Osada, and K. Takahashi at Tohoku Univ. for the technical assistance for qPCR analysis and Dr. I. Gleadall at Tohoku Univ. for the editorial assistance. TN would like to express a special thanks to Dr. E.M. Donaldson, Scientist Emeritus, at CAER, DFO/UBC, Fisheries and Oceans Canada, for the help. This work was supported in part by the Grants-in-Aid for scientific research (KAKENHI, #23580277) from Japan Society for the Promotion of Science (JSPS) to TN.

References

Ali M, Parvez S, Pandey S, Atif F, Kaur M, Rehman H, Raisuddin S (2004) Fly ash leachate induces oxidative stress in freshwater fish *Channa punctata* (Bloch). Environ Int 30:933–938

Allen RG, Tresini M (2000) Oxidative stress and gene regulation. Free Radic Biol Med 28:463–499

Arrigo A-P (1999) Gene expression and the thiol redox state. Free Radic Biol Med 27:936–944

Asada K (1988) Production, scavenging and action of active oxygen. Tanpakushitsu, Kakusan, Koso (Proteins Nucleic Acids Enzym) 33:7–12

Bagnyukova TV, Danyliv SI, Zin'ko OS, Lushchak VI (2007) Heat shock induces oxidative stress in rotan *Perccottus glenii* tissues. J Ther Biol 32:255–260

Barton BA (1997) Stress in finfish: past, present and future – a historical perspective. In: Iwama GK, Pickering AD, Sumpter JP, Schreck CB (eds) Fish stress and health in aquaculture. Cambridge University Press, Cambridge, pp 1–33

Barton BA, Iwama GK (1991) Physiological change in fish from stress in aquaculture with emphasis on the response and effects of corticosteroids. Annu Rev Fish Dis 1:3–26

Basu N, Nakano T, Grau EG, Iwama GK (2001) The effects of cortisol on heat shock protein 70 levels in two fish species. Gen Comp Endocrinol 124:97–105

Basu N, Todgham AE, Ackerman PA, Bibeau MR, Nakano K, Shulte PM, Iwama GK (2002) Heat shock protein genes and their functional significance in fish. Gene 295:173–183

Beckman BR (2011) Perspectives on concordant and discordant relations between insulin-like growth factor 1 (IGF1) and growth in fishes. Gen Comp Endocrinol 170:233–252

Beckman KB, Ames BN (1998) The free radical theory of aging matures. Physiol Rev 78:547–581

Bell JG, McEvoy J, Tocher DR, Sargent JR (2000) Depletion of α-tocopherol and astaxanthin in Atlantic salmon (*Salmo salar*) affects antioxidative defense and fatty acid metabolism. J Nutr 130:1800–1808

Berque J, Matsuda O (2013) Coastal biodiversity management in Japanese Satoumi. Mar Policy 39:191–200

Björnsson BT, Johansson V, Benedet S, Einarsdottir IE, Hildahl J, Agustsson T, Jönsson E (2002) Growth hormone endocrinology of salmonids: regulatory mechanisms and mode of action. Fish Physiol Biochem 27:227–242

Craig PM, Wood CM, McClelland GB (2007) Oxidative stress response and gene expression with acute copper exposure in zebrafish (*Danio rerio*). Am J Physiol Regul Integr Comp Physiol 293:R1882–R1892

Deane EE, Woo NYS (2009) Modulation of fish growth hormone levels by salinity, temperature, pollutants and aquaculture related stress: a review. Rev Fish Biol Fish 19:97–120

Droge W (2002) Free radicals in the physiological control of cell function. Physiol Rev 82:47–95

Duan C (1998) Nutritional and developmental regulation of insulin-like growth factors in fish. J Nutr 128:306S–314S

Dyer AR, Barlow CG, Bransden MP, Carter CG, Glencross BD, Richardson N, Thomas PM, Williams KC, Carragher JF (2004) Correlation of plasma IGF-I concentrations and growth rate in aquacultured finfish: a tool for assessing the potential of new diets. Aquaculture 236:583–592

Ellis T, Yildiz HY, Lopez-Olmeda J, Spedicato MT, Tort L, Øverli Ø, Martins CIM (2012) Cortisol and finfish welfare. Fish Physiol Biochem 38:163–188

Feder ME, Hofmann GE (1999) Heat-shock proteins, molecular chaperones, and the stress response. Annu Rev Physiol 61:243–282

Franco AA, Odom RS, Rando TA (1999) Regulation of antioxidant enzyme gene expression in response to oxidative stress and during differentiation of mouse skeletal muscle. Free Radic Biol Med 27:1122–1132

Hadjimichael M, Delaney A, Kaiser MJ, Edwards-Jones G (2013) How resilient are Europe's inshore fishing communities to change? Differences between the north and the south. AMBIO 42:1037–1046

Hara M (2014) Tohoku ecosystem-associated marine science project and marine ecosystems post disaster. Ocean News Lett 328:2–3

Heise K, Puntarulo S, Portner HO, Abele D (2003) Production of reactive oxygen species by isolated mitochondria of the Antarctic bivalve *Laternula elliptica* (King and Broderip) under heat stress. Comp Biochem Physiol 134C:79–90

Heise K, Puntarulo S, Nikinmaa M, Abele D, Portner H-O (2006) Oxidative stress during stressful heat exposure and recovery in the North Sea eelpout *Zoarces viviparus* L. J Exp Biol 209:353–363

Hofmann GE, Todgham AE (2010) Living in the now: physiological mechanisms to tolerate a rapidly changing environment. Annu Rev Physiol 72:127–145

Holmlund CM, Hammer M (1999) Ecosystem services generated by fish populations. Ecol Econo 29:253–268

Iwama GK, Thomas PT, Forsyth RB, Vijayan MM (1998) Heat shock protein expression in fish. Rev Fish Biol Fish 8:35–56

Iwama GK, Afonso LOB, Vijayan MM (2006) Stress in fishes. In: Evans DH, Claiborne JB (eds) The physiology of fishes, 3rd edn. CRC Press, Boca Raton, FL, pp 319–342

Ji LL (1995) Oxidative stress during exercise: Implication of antioxidant nutrients. Free Radic Biol Med 18:1079–1086

Kajimura S, Hirano T, Visitacion N, Moriyama S, Aida K, Grau EG (2003) Dual mode of cortisol action on GH/IGF-I/IGF binding proteins in the tilapia, *Oreochromis mossambicus*. J Endocrinol 178:91–99

Kameda M, Nakano T, Yamaguchi T, Sato M, Afonso LOB, Iwama GK, Devlin RH (2008) Effects of heat shock on growth hormone receptor expression in coho salmon. In: Proceedings of the 5th world fish congress, Yokohama, pp 3f–16, (CD-ROM), October 20–25

Klein SE, Sheridan MA (2008) Somatostatin signaling and the regulation of growth and metabolism in fish. Mol Cell Endocrinol 286:148–154

Kobayashi M, Kaneko T, Aida K (2002) Endocrinology. In: Aida K (ed) The fundamental of fish physiology. Kouseisha Kouseikaku, Tokyo, pp 128–154

Kopchick JJ, Andry JM (2000) Growth hormone (GH), GH receptor, and signal transduction. Mol Genet Metab 71:293–314

Lesser MP (2006) Oxidative stress in marine environments: biochemistry and physiological ecology. Annu Rev Physiol 68:253–278

Lushchak VI (2011) Environmentally induced oxidative stress in aquatic animals. Aquat Toxicol 101:13–30

Lushchak VI, Bagnyukova TV (2006a) Temperature increase results in oxidative stress in goldfish tissues. 1. Indices of oxidative stress. Comp Biochem Physiol 143C:30–35

Lushchak VI, Bagnyukova TV (2006b) Temperature increase results in oxidative stress in goldfish tissues. 2. Antioxidant and associated enzymes. Comp Biochem Physiol 143C:36–41

Lushchak VI, Bagnyukova TV, Husak VV, Luzhna LI, Lushchak OV, Storey KB (2005a) Hyperoxia results in transient oxidative stress and an adaptive response by antioxidant enzymes in goldfish tissues. Int J Biochem Cell Biol 37:1670–1680

Lushchak VI, Bagnyukova TV, Lushchak OV, Storey JM, Storey KB (2005b) Hypoxia and recovery perturb free radical processes and antioxidant potential in common carp (*Cyprinus carpio*) tissues. Int J Biochem Cell Biol 37:1319–1330

Martínez-Álvarez RM, Morales AE, Sanz A (2005) Antioxidant defenses in fish: biotic and abiotic factors. Rev Fish Biol Fish 15:75–88

Mommsen TP, Moon TW (2001) Hormonal regulation of muscle growth. In: Johnston IA (ed) Fish physiology vol. 18, muscle development and growth. Academic Press, San Diego, pp 251–308

Moriyama S, Ayson FG, Kawauchi H (2000) Growth regulation by insulin-like growth factor-I in fish. Biosci Biotechnol Biochem 64:1553–1562

Morliere P, Moyasan A, Santus R, Huppe G, Maziere JC, Dubertret L (1991) UVA-induced lipid peroxidation in cultured human fibroblasts. Biochim Biophys Acta 1084:261–268

Mujahid A, Yoshiki Y, Akiba Y, Toyomizu M (2005) Superoxide radical production in chicken skeletal muscle induced by acute heat stress. Poult Sci 84:307–314

Nakano T (2007) Microorganisms. In: Nakagawa H, Sato M, Gatlin DM III (eds) Dietary supplements for the health and quality of cultured fish. CAB International, Oxfordshire, pp 86–108

Nakano T (2011) Stress in fish. Yoshoku (Aquacult Mag) 48:64–67

Nakano T, Takeuchi M (1997) Relationship between fish and reactive oxygen species. Yoshoku (Aquacult Mag) 34:69–73

Nakano T, Kanmuri T, Sato M, Takeuchi M (1999a) Effect of astaxanthin rich red yeast (*Phaffia rhodozyma*) on oxidative stress in rainbow trout. Biochim Biophys Acta 1426:119–125

Nakano T, Miura Y, Wazawa M, Sato M, Takeuchi M (1999b) Red yeast *Phaffia rhodozyma* reduces susceptibility of liver homogenate to lipid peroxidation in rainbow trout. Fish Sci 65:961–962

Nakano T, Wazawa M, Yamaguchi T, Sato M, Iwama GK (2004) Positive biological actions of astaxanthin in rainbow trout. Mar Biotechnol 6:S100–S105

Nakano T, Afonso LOB, Beckman BR, Iwama GK, Devlin RH (2013) Acute physiological stress down-regulates mRNA expressions of growth-related genes in coho salmon. PLoS ONE 8:e71421

Nakano T, Kameda M, Shoji Y, Hayashi S, Yamaguchi T, Sato M (2014) Effect of severe environmental thermal stress on redox state in salmon. Redox Biol (Off J Soc Free Radic Biol Med and Soc Free Radic Res-Eur) 2:772–776

Niki E (1988) Ascorbic acid, glutathione. In: Nakano M, Asada K, Oyanagui Y (eds) Reactive oxygen. Kyoritsu Shuppan, Tokyo, pp 321–326

Olsen RE, Sundell K, Mayhew TM, Myklebust R, Ringø E (2005) Acute stress alters intestinal function of rainbow trout, *Oncorhynchus mykiss* (Walbaum). Aquaculture 250:480–495

Oyanagui Y (1989) Defensive system, reactive oxygen species and diseases. Kagaku Dojin, Tokyo, pp 21–38

Pankhurst NW (2011) The endocrinology of stress in fish: an environmental perspective. Gen Comp Endocrinol 170:265–275

Pankhurst NW, Kraak G (1997) Effects of stress on reproduction and growth of fish. In: Iwama GK, Pickering AD, Sumpter JP, Schreck CB (eds) Fish stress and health in aquaculture. Cambridge University Press, Cambridge, pp 73–93

Parihar MS, Dubey AK (1995) Lipid peroxidation and ascorbic acid status in respiratory organs of male and female freshwater catfish *Heteropneustes fossilis* exposed to temperature increase. Comp Biochem Physiol 112C:309–313

Peter RE (1979) The brain and feeding behavior. In: Hoar WS, Randall DJ, Brett JR (eds) Fish physiology vol. 8, bioenergetics and growth. Academic Press, New York, pp 121–159

Pickering AD (1993) Growth and stress in fish production. Aquaculture 111:51–63

Pickering AD, Pottinger TG (1995) Biochemical effects of stress. In: Hochachka PW, Mommsen TP (eds) Biochemistry and molecular biology of fishes vol. 5, environmental and ecological biochemistry. Elsevier, Amsterdam, pp 349–379

Pierce AL, Dickey JT, Larsen DA, Fukuda H, Swanson P, Dickhoff WW (2004) A quantitative real-time RT-PCR assay for salmon IGF-I mRNA, and its application in the study of GH regulation of IGF-I gene expression in primary culture of salmon hepatocytes. Gen Comp Endocrinol 135:401–411

Ploch SA, Lee Y-P, MacLean E, Di Guilo RT (1999) Oxidative stress in liver of brown bullhead and channel catfish following exposure to tert-butyl hydroperoxide. Aquat Toxicol 46:231–240

Poly WJ (1997) Nongenetic variation, genetic-environmental interactions and altered gene expression. II. Disease, parasite and pollution effects. Comp Biochem Physiol 117B:61–74

Pörtner HO (2002) Climate variations and the physiological basis of temperature dependent biogeography: systemic to molecular hierarchy of thermal tolerance in animals. Comp Biochem Physiol 132A:739–761

Prunet P, Overli O, Douxfils J, Bernardini G, Kestemont P, Baron D (2012) Fish welfare and genomics. Fish Physiol Biochem 38:43–60

Rau MA, Whitaker J, Freedman JH, Di Giulio RT (2004) Differential susceptibility of fish and rat liver cells to oxidative stress and cytotoxicity upon exposure to prooxidants. Comp Biochem Physiol 137C:335–342

Reineck M (2010) Influences of the environment on the endocrine and paracrine fish growth hormone-insulin-like growth factor-I system. J Fish Biol 76:1233–1254

Reineck M, Bjornsson BT, Dickhoff WW, McCormick SD, Navarro I, Power DM, Gutierrez J (2005) Growth hormone and insulin-like growth factors in fish: where we are and where to go. Gen Comp Endocrinol 142:20–24

Saera-Vila A, Calduch-Giner JA, Prunet P, Perez-Sanchez J (2009) Dynamics of liver GH/IGF axis and selected stress markers in juvenile gilthead sea bream (*Sparus aurata*) exposed to acute confinement. Differential stress response of growth hormone receptors. Comp Biochem Physiol 154A:197–203

Schoeniger LO, Andreoni KA, Ott GR, Risby TH, Bulkley GB, Udelsman R, Burdick JF (1994) Induction of heat-shock gene expression in post ischemic pig liver depends on superoxide generation. Gastroenterology 106:177–184

Shin MH, Moon YJ, Seo J-E, Lee Y, Kim KH, Chung JH (2008) Reactive oxygen species produced by NADPH oxidase, xanthine oxidase, and mitochondrial electron transport system mediate heat shock-induced MMP-1 and MMP-9 expression. Free Radic Biol Med 44:635–645

Sies H (1999) Glutathione and its role in cellular functions. Free Radic Biol Med 27:916–921

Takei Y, Loretz CA (2006) Endocrinology. In: Evans DHC, Laiborne JB (eds) The physiology of fishes, 3rd edn. CRC Press, Boca Raton, pp 271–318

Taniguchi N, Endo T (2000) Cross talk of SOD, NO, and glutathione metabolism. In: Taniguchi N, Yodoi J (eds) Biochemistry in oxidative stress and redox. Kyoritsu Shuppan, Tokyo, pp 1–11

Urushihara J (2013) Development of new sustainable fishing industries in disaster-stricken area, where huge tsunami occurred. Sci Window 4:20–23

Valavanidis A, Vlahogianni T, Dassenakis M, Scoullos M (2006) Molecular biomarkers of oxidative stress in aquatic organisms in relation to toxic environmental pollutants. Ecotoxicol Environ Saf 64:178–189

Wong SHY, Knight JA, Hopfer SM, Zaharia O, Leech JCN, Sunderman JW (1987) Lipoperoxides in plasma as measured by liquid-chromatographic separation of malondialdehyde-thiobarbituric acid adduct. Clin Chem 33:214–220

Worm B, Sandow M, Oschlies A, Lotze HK, Myers RA (2005) Global patterns of predator diversity in the open oceans. Science 309:1365–1369

Worm B, Barbier EB, Beaumont N, Duffy JE, Folke C, Halpern BS, Jackson JBC, Lotze HK, Micheli F, Palumbi SR, Sala E, Selkoe KA, Stachowicz JJ, Watson R (2006) Impacts of biodiversity loss on ocean ecosystem services. Science 314:787–790

Zelko IN, Mariani TJ, Folz RJ (2002) Superoxide dismutase multigene family: a comparison of the CuZn-SOD (SOD1), Mn-SOD (SOD2), and EC-SOD (SOD3) gene structures, evolution, and expression. Free Radic Biol Med 33:337–349

Part IV

Communications: Ecosystem Based Approach

The Channel Programme: A Large-Scale Integrated Approach

Paul Marchal, André Carpentier and Eric Foucher

Abstract

The Channel programme is a collation of interlinked multidisciplinary projects supporting ecosystem-based management in the English Channel by integrating works carried out in the past thirty years including novel approaches to the spatial modelling of ecosystems. The Channel programme has required a coordination of efforts and expertise by both scientists and stakeholders to address the challenges of the Marine Strategy Framework Directive and of its national implementations. The Channel programme has also consisted of harmonising information, regulations and habits in terms of ecosystem management in the English Channel.

1 Introduction

The English Channel is an economically and ecologically strategic area of North-West Europe. It represents a significant economic zone where a number of human activities take place and compete for space or resources: tourism and leisure, international ports and freight and exploitation of living and nonliving resources.

The English Channel maritime domain concentrates around 20 % of the worldwide maritime traffic, making it one of the most intensively used sea routes in the world, with 14 million passengers and 2.5 million crafts in 2008. The English Channel is rich in marine aggregates (sand and gravel) which are increasingly needed by British and French building industries. The English Channel is also rich in terms of biodiversity and commercial fishery resources. A number of commercial marine species can be found here, as well as many spawning, nursery and feeding areas, plus migration routes for fish, birds and marine mammals. The English Channel is the place where parts of the Atlantic and North Sea water

P. Marchal (✉)
Ifremer, Unité Halieutique de Manche mer du Nord,
150 quai Gambetta, BP 699,
F-62321 Boulogne-sur-mer Cedex, France
e-mail: Paul.Marchal@ifremer.fr

A. Carpentier
Ifremer, Unité Lagons, Ecosystèmes et Aquaculture Durable en Nouvelle-Calédonie, 101 Promenade Roger Laroque, BP 2059, F-98846 Nouméa Cedex, Nouvelle-Calédonie, France

E. Foucher
Ifremer, Unité Halieutique de Manche mer du Nord,
Avenue du Général de Gaulle,
14520 Port-en-Bessin, France

H.-J. Ceccaldi et al. (eds.), *Marine Productivity: Perturbations and Resilience of Socio-ecosystems*,
DOI 10.1007/978-3-319-13878-7_17,

masses intermingle. The physical configuration (e.g. bathymetry, coastline) of the English Channel contributes to specific hydrological structures (fronts, gyres, etc.) that control advection and the dispersal of marine organisms or pollutants.

The exploitation of natural resources (living resources such as exploited marine species but also mineral resources such as marine aggregates) generates substantial added value at national and European levels. As a result, there has been an increasing societal and public demand from the governance (European Union, national governments and regional councils), industries (e.g. fishing and marine aggregate sectors of activity) and nongovernmental organisations to provide a sound and integrated scientific support to ecosystem-based management in the English Channel. Of particular interest to the various stakeholders is the exploration of the English Channel ecosystem responses to global change.

An ecosystem-based approach to marine resources management requires the synthesis of existing scientific knowledge and its integration to new research. Such an approach is expected to improve the scientific basis to advice-giving and management. Key processes such as the interactions between species, between management strategies and exploited species, between environmental damage and the rest of the ecosystem and interactions between conflicting human activities cannot be understood by investigating each ecosystem component in isolation, but rather by exploring all aspects of the marine ecosystem, including the various sectors of activity, in an integrated fashion.

This objective requires a coordination of efforts and expertise by both scientists and stakeholders through international and multidisciplinary projects to address the challenges of the Marine Strategy Framework Directive and of its national implementations. Such a trans-régional approach also consists of harmonising information, regulations and habits in terms of ecosystem management in the English Channel.

The English Channel ecosystem has been investigated by several research teams. Some European projects have fostered a multidisciplinary approach targeted at marine living resources by integrating works carried out in the past 30 years including novel approaches to the spatial modelling of ecosystems. In this context, several scientific studies have engaged in the English Channel and the Southern North Sea to investigate:

- The assessment of key fish species
- The characterisation of their habitats
- The development of predictive modelling tools to evaluate the impact of human activities on the ecosystem
- The contribution of the ecological niche in structuring commercial fish communities
- The distribution and characteristics of spawning areas
- The biogeographic identification of sole, red mullet and sea bass
- The dynamics and strategies of fishing fleets
- The dynamics and consequences of *Phaeocystis* blooms
- The modelling of rivers inputs from the Seine, the Somme and the Escaut
- The trophic capacities and resilience of the ecosystem following planned or unplanned human perturbations

The objective of the ecosystem approach is therefore to understand more globally, at the scale of the Eastern Channel basin, the characteristics and the functioning of habitats and of the living resources they are home to. It will then become possible to apply an ecosystem approach to the management of marine resources in a broad sense (living, but also mineral and energetic resources) and to mitigate the pressure exerted by the different human uses and their adaptation to the ecosystem production capacities.

In this context, a multidisciplinary scientific initiative called the Channel programme was initiated in 2007, with the aim of increasing the knowledge on, and of better managing, this marine cross-border area. We present in this study the main objectives and provide insights into research projects that have contributed to these objectives since then.

2 Objectives of the Channel Programme

The main objectives of the Channel programme are the following:

- To collectively develop research capacity to achieve a multidisciplinary understanding of the functioning of marine and coastal ecosystems subject to strong human pressure
- To promote the sustainable management of the different parts of the English Channel by developing decision-taking tools with the aim of adapting human use intensity to the withstanding capacity of ecosystems
- To provide indicators to decision-makers with the aim of (1) characterising the status of ecosystems, (2) measuring the pressure exerted by diverse human activities on these ecosystems and (3) evaluating the efficiency of management measures

To achieve these objectives, the Channel programme has been structured into four main research axes:

- Describing ecological factors structuring the distribution of fishery resources: habitats (abiotic and biotic factors), biotopes, assemblages, life history of main species and environment
- Understanding the functioning of the food web: description, interactions between primary (including toxic phytoplanktonic species) and secondary production and fishes, functioning and sensitivity to perturbations
- Quantifying the impact of human pressure on habitat and food web: fisheries, marine aggregates extraction, offshore wind farms and pollutions
- Defining new management strategies for sustainable resource exploitation respectful of marine environment and biodiversity

3 Research and Governance Projects

The launching of the Channel programme in 2007 has fostered the development of several EU-funded projects with a central research or governance component. These complementary and interlinked projects (Table 1) have involved several hundreds of research scientists and several dozens of research organisms (e.g. governmental institutes, universities) in the EU and even abroad (e.g. collaboration between IFREMER[1] in France and DFO[2] in Canada).

We present below the background and objectives of a selection of these projects.

3.1 EMDI+ and CAMIS Projects

EMDI+ and CAMIS are two successive EU INTERREG-funded governance projects that have been coordinated by the Region Haute-Normandie in France.

The EMDI+ (Espace Manche Development Initiative) project – INTERREG IIIB North-West Europe – strongly contributed to the development of the political cooperation between France and the UK across the English Channel ('Channel Arc Manche' cooperation) between 2004 and 2008. The main objectives of the EMDI project were to work out a strategic vision, to test new cooperation avenues and to develop greater knowledge of the Channel area and the issues at stake, in particular through the creation of a cross-Channel atlas. This project brought some convincing arguments for the recognition of the Channel Arc Manche as a coherent cooperation area in Europe. It has also highlighted the opportunity for the Channel area to position itself as a demonstration maritime basin of the integrated maritime policy championed by the European Commission. More details on this project may be found on http://www.emdi.certic.unicaen.fr/en/news.html.

The more recent CAMIS (Channel Arc Manche Integrated Strategy) project (https://camis.arcmanche.eu/home/) has confirmed this position, with the objectives of (1) developing a framework for maritime governance, (2) providing expertise for Channel Arc Manche and

[1] Institut Français de Recherche pour l'Exploration de la Mer.

[2] Department of Fisheries and Oceans.

Table 1 Contribution of the 11 projects building in the Channel programme to the understanding of ecosystem elements and of the human pressures impacting them (1–7, 9, 11: EU-funded CHARM1, CHARM2, CHARM3, CRESH, CAMIS, VECTORS, DYMAPHY, MEMO, SOCIOEC; 8, 10: French National Research Agency (ANR) funded COMANCHE, EMIBIOS)

		1	2	3	4	5	6	7	8	9	10	11
Biology	Plankton	X	X	X						X	X	
	Invertebrates		X	X	X		X		X		X	
	Fish	X	X	X	X	X	X		X	X	X	X
Environment	Ecology	X	X	X	X	X	X	X	X	X	X	
	Physics	X	X	X	X	X	X	X	X	X	X	
Human activities	Fisheries	X	X	X	X	X	X		X		X	
	Aggregate extractions						X					
	Maritime transport					X	X	X				
	Renewable energies					X						
Water quality	Invasive species							X	X	X		
	Pollution					X						
Socio-economy				X		X			X	X		X
Governance	Législation	X	X	X		X						
	Spatial planning			X	X	X	X			X		
	Alternative scenarios					X	X			X	X	X
	Sustainable development						X			X	X	X
Technology / Tools	Selectivity					X	X			X		
	Plankton							X				

decision-makers in the English Channel area and (3) stimulating cooperation that addresses strategic issues at stake. To meet these objectives, the project integrated the contribution of other projects which developed scientific knowledge, in particular the CHARM project (see below), or sector-based strategies for the English Channel area, e.g. ports and transport and maritime safety.

3.2 CRESH Project

The CRESH project (Cephalopod Recruitment from English Channel Habitats) addresses the Priority 4 of the INTERREG IVA programme which concerns the sustainable environmental development of the common space.[3] Cephalopods are short-living species; their abundance depends on the annual recruitment (juveniles entering the exploited stage), and they are directly concerned by this topic. The project gathers eight French-British partners from universities, research institutes and fishermen's organisations. This brief introduction underlines complementary skills gathered to better understand these animals and the part that they play in the English Channel ecosystem.

[3] http://www.unicaen.fr/ufr/ibfa/cresh/?lang=en

3.3 CHARM1, CHARM2 and CHARM3 Projects

CHARM1 and CHARM2 have been two successive French-British cooperation projects, co-funded under the UE INTERREG programme, through the Haute-Normandie Region and the UK Government Office for the South-East. Each of these projects developed an atlas as a 'toolbox' to support decision-making and spatial planning in relation to the sustainable management of the English Channel marine resources. The first phase (2003–2005) covered the Dover Strait and the second one (2006–2008) covered the whole Eastern English Channel. These projects have also been a good opportunity to gather and harmonise French and UK ecological and environmental data. The Channel Habitat Atlas for marine Resource Management produced during CHARM2 can be downloaded on http://archimer.ifremer.fr/doc/2009/rapport-7377.pdf

The CHARM3 project is the logical follow-up to CHARM1 (2003–2005, 1.1 M€, 5 partners) and CHARM2 (2006–2008, 2.1 M€, 6 partners). Firmly grounded on results obtained in the previous phases, the overarching objective of CHARM3 was to conduct multidisciplinary studies through an ecosystem-based approach, where human beings are considered as an integral part of the ecosystem. Compared to CHARM1 and CHARM2, the CHARM3 has been extended to the entire English Channel and the southern part of the North Sea, and new research areas (economics, climate change) have been investigated. CHARM3 combined a wide range of expertise fields, including marine sciences (oceanography, marine biology, fishery science), economics, legislation, geography (e.g. cartography, spatial analysis, geographical information systems), statistical and conceptual modelling (habitat and trophic network models), marine conservation planning and information technologies (e.g. web technologies, programming). Many of these skills are interdisciplinary by nature, an essential aspect for any integrated approach. Seventeen French and British partners have been involved in the project and pursued three main groups of actions: (1) collection and standardisation of information, (2) integration of information and (3) development of tools to disseminate information.

3.4 VECTORS Project

The ongoing EU FP7-funded VECTORS project (Vectors of Change in Oceans and Seas Marine Life, Impact on Economic Sectors) (2011–2015) acquired considerable knowledge on the effects of past, present and future vectors of change on the marine environment. Pressures could be direct (e.g. maritime freight as invasive species vector, climate change and ocean acidification, intensive living and mineral resources exploitation) or indirect (e.g. energy-producing devices, demographic pressure on the coasts, tourism, recreational activities). The VECTORS project is particularly investigating how vectors change and their interactions result in changes in three specific processes: blooms of non-endemic and nonindigenous species, spreading of endemic species and changes of commercial species distribution and productivity. More details on the VECTORS project, including scientific production, may be found on http://www.marine-vectors.eu/

3.5 COMANCHE Project

King scallop (*Pecten maximus*) is an important species in the English Channel, notably as it constitutes the first species in landings (15–20,000 t per year) for the French fishing fleet (700 fishing vessels) based in this maritime area. Due to its economic importance, several scientific studies have been conducted on this species, but mainly at local scales. Moreover, despite the basic knowledge acquired through these studies, several points of the scallop dynamics remain poorly understood. Additionally, episodic perturbations (e.g. introduction of competitor, development of toxic algal blooms) have affected the king scallop populations in unpredictable ways, making it even more difficult to apprehend the dynamics of the populations and their interactions with the rest of the ecosystem. In this context, the ongoing COMANCHE (COquilles Saint-Jacques de la MANCHE) project (2012–2015), co-funded by the French National Research Agency, aims at improving our knowledge on the relationship between king scallop and its ecosystem, in an integrated fashion, and over the whole English Channel. The characteristics and the structure of the exploited populations of king scallop have been investigated, with the objective of determining the position of this species within the English Channel ecosystem and of quantifying the impacts associated to its exploitation. These objectives are being pursued through a multidisciplinary approach, involving physicists, biologists and economists, but also regional stakeholders (e.g. fisheries regional committees). More details on the COMANCHE project may be found on http://wwz.ifremer.fr/defimanche/Projets/En-cours/COMANCHE.

4 Partnership

The projects conducted within the Channel programme involved a wide and multidisciplinary partnership, which is presented in the table below.

5 Conclusion

The Channel programme feeds in long-term sustainable management of the marine resources in the English Channel through the development of

	Project partners	Financial partners
CHARM 3	*Leader*: Ifremer, Boulogne-sur-Mer	INTERREGIVA
	1. Ifremer – Brest, Dinard, Port-en-Bessin	
	2. Université des Sciences et Technologies de Lille Lab. D'Océanologie et Géosciences Wimereux	
	3. Université de Caen Basse-Normandie Lab. Biologie et Biotechnologies Marine – Caen	
	4. Agrocampus Rennes Pôle halieutique – Rennes	
	5. Université de Haute Bretagne (Rennes 2) Lab. Rennes – Espaces et Sociétés – Rennes	
	6. SAS Alkante – Rennes	
	7. Université de Bretagne Occidentale UMR AMURE – Centre de Droit et d'Economie de la Mer – Brest	
	8. University of Greenwich – Medway	
	9. University of Kent DICE – Canterbury	
	10. CEFAS – Lowestoft	
	11. Plymouth Marine Lab – Plymouth	
	12. SAHFOS CPR team – Plymouth	
	13. Marine Biological Association – Plymouth	
	14. University of Plymouth Marine Institute – Plymouth	
	15. Marinelife Ltd – Bridport	
	16. University of Exeter Centre for Ecology & Conservation –Cornwall campus	
	17. University of Portsmouth CEMARE – Portsmouth	
CRESH	*Leader*: Université de Caen Basse – Normandie	
	1. The Marine Biological Association of the UK	
	2. Devon Sea Fisheries Committee	
	3. Royal Holloway University of London	
	4. Centre for Environment, Fisheries & Aquaculture Science	
	5. Ifremer Boulogne/mer & Port-en-Bessin – Centre Manche Mer du Nord	
	6. Comité Régional des Pêches Maritimes de Basse – Normandie	
	7. University of Plymouth	

(continued)

CAMIS-EMDI +	*Leader*: Région Haute – Normandie
	1. Régions Bretagne
	2. Région Nord – Pas-de-Calais
	3. Région Picardie
	4. Région Basse – Normandie
	5. West Sussex County Council
	6. Cornwall County Council
	7. Kent County Council
	8. Southampton City County Council
	9. Devon County Council
	10. Conseil Général Côtes d'Armor
	11. Hampshire
	12. Université de Caen
	13. Marine South East
	14. SEEDA

ecosystem and conservation planning models. Model scenarios and outputs contribute to regional environmental planning and support the management of the living resources and assessments of potential impact of human pressures. The outputs of the many projects building in the Channel programme enhance our understanding of the English Channel marine ecosystem and, in doing so, improve management performances. The multidisciplinary character of the Channel programme is then particularly relevant in the context of the MSFD which, through a holistic approach of the seas and oceans, seeks to achieve a good ecological status for the marine environment by 2021.

The Channel programme has been successful bringing together scientists and stakeholders. It has demonstrated, through biannual conferences, during which scientists presented and discussed their results with managers, the value-added of grouping research teams of different backgrounds and fields of expertise (e.g. ecologists, sociologists, economists). It has further evidenced to decision-makers that managing a common maritime domain requires an ecosystem approach integrating all the human activities impacting the marine environment.

More details on the Channel programme may be found on the URL: http://wwz.ifremer.fr/defimanche.

From Data to End-to-End Models: 15 Years of Research to Describe the Dynamics of Exploited Marine Ecosystems in the Eastern Channel

Paul Marchal, Loïc Gasche, Raphaël Girardin, Olivier Le Pape, Martin Huret, Stéphanie Mahévas, Morgane Travers-Trolet, and Sandrine Vaz

Abstract

Considerable research has been conducted in the past 15 years around the Eastern English Channel ecosystem. Data collected since the 1970s on the biotic and abiotic compartments have been collated and mapped out in the mid-2000. This spatially explicit information formed a sound basis to improve knowledge on, and model, the functioning and dynamics of key ecosystem compartments, with a focus on flatfish species and fisheries and their interactions with other sectors of activity (aggregate extractions, maritime traffic). The more recent ongoing works are dedicated to the integration of those processes into several complementary end-to-end ecosystem models.

1 Introduction

The English Channel concentrates a rich ecosystem along with intense and diverse human activities. At the same time, the Channel is a relatively well-circumscribed geographical area, "a world within Europe", which makes it an appropriate site for investigating the complex interactions between intensive and diverse human activities and this ecosystem.

The English Channel is first a productive ecosystem coupling high and diversified pelagic and benthic productivity. It is home to many commercial species, encompassing their spawning and nursery areas and migratory routes.

P. Marchal (✉) • R. Girardin • M. Travers-Trolet
Ifremer, Unité Halieutique de Manche mer du Nord, 150 quai Gambetta, BP 699, F-62321 Boulogne-sur-mer Cedex, France
e-mail: Paul.Marchal@ifremer.fr

L. Gasche • S. Mahévas
Ifremer, Unité Ecologie et Modèles pour l'Halieutique, rue de l'Ile d'Yeu, BP 21105, F-44311 Nantes Cedex 03, France

O. Le Pape
AgrocampusOuest, UMR985ESE Ecologie et santé des écosystèmes, F-35042 Rennes, France

M. Huret
Ifremer, Unité Sciences et Technologies Halieutiques, ZI de la Pointe du Diable, CS 10070, F-29280 Plouzané, France

S. Vaz
Ifremer, Unité Halieutique de Méditerranée, avenue Jean Monnet, CS 30171, F-34203 Sète, France

H.-J. Ceccaldi et al. (eds.), *Marine Productivity: Perturbations and Resilience of Socio-ecosystems*, DOI 10.1007/978-3-319-13878-7_18,

The English Channel has also, for a long time, supported the activities of many users (e.g. fishing, maritime transport, aggregate extraction, discharges, wind parks, aquaculture, tourism), in a context of climate change. It is also considered one of the most intensively used sea areas in the world. Of these human pressures, fishing activity, maritime transport and aggregate extractions are probably on the top of the French and UK Governments' agenda for that area. Thus, there were in 2007 about 1,500 French fishing vessels registered in English Channel harbours, employing 4,300 fishers and generating a gross revenue of 1.4 million €. Maritime transport is also a major economic activity in the English Channel. Coming from around the planet and leaving in the direction of Asia, Africa and America, nearly 500 ships of over 300 t enter and leave the English Channel every day, making it 1 craft every 3 min. Perpendicular to this traffic, 90–120 daily rotations are operated by ferries between the continent and the British Isles, transporting 17 million passengers per annum. Marine aggregate extraction sites have for many decades been exploited along the UK coasts of the English Channel and more recently along the French coasts. In 2007, 5.5 Mt of marine aggregates were extracted from several tens of km^2 in UK southern coastal waters and 1 Mt from less than 10 km^2 along French coasts. Recently, this activity moved further offshore to areas trawled and dredged by French fishermen. Several 100 km^2 are presently prospected by French companies both in the Eastern and Central English Channel. All these activities have, in isolation or in combination, long been recognised to be major vectors of change for the ecosystem structure and functioning and also for related economic maritime sectors.

These multiple and diverse interactions between human activities and the pressure they exert on the marine ecosystem necessitate to manage the Eastern English Channel in an integrated and cross-sectorial fashion, consistent with the EU Marine Strategy Framework Directive (MSFD) and marine spatial planning (MSP).

This study presents, in a summarised fashion, the evolution of the research conducted in the past 15 years around the Eastern English Channel ecosystem, starting from data collation all the way through end-to-end ecosystem modelling, with a focus on flatfish species and fisheries and their interactions with other sectors of activity (aggregate extractions, maritime traffic).

2 From Data Collection to Mapping Information Layers and Spatial and Ecological Analyses

A considerable amount of information has been collected in the Eastern English Channel (EEC) over the period 2004–2010, mainly during three successive EU Interreg-funded projects: CHARM1 (2003–2005) covering the Dover Strait, CHARM2 (2006–2008) the whole EEC and CHARM3 (2009–2012) both the eastern and the western parts of the English Channel. An atlas including a variety of information layers related to the EEC physical environment (e.g. temperature, salinity, bed shear stress), fish and benthos habitats, trophic network and fisheries was in particular produced in the course of CHARM2 (Carpentier et al. 2009). The maps produced in the atlas were based on a collation of research surveys and commercial fisheries information. The raw information was interpolated and processed through a variety of statistical methods, including kriging, GLMs (generalised linear models), GAMs (generalised additive models) and quantile regressions, used for habitat suitability modelling and mapping. In addition, information on the life traits and diet of a large panel of commercial fish species was processed (geomorphometrics, stomach contents and stable isotope analyses) to characterise the EEC trophic network in a quantitative fashion. All these information processed during a 10-year period form a comprehensive and sound basis to calibrate models covering part and/or the totality of ecosystem components.

3 From Spatial Analyses to Modelling Ecosystem Compartments

3.1 The Common Sole (*Solea solea*) Population and Fishery, from Ecology Along the Life Cycle to Population Dynamics

Common sole is an abundant species in the Eastern Channel and sustains important fisheries (4,000–5,000 t per year). The common sole has a complex life cycle (Rochette et al. 2013); after the eggs have hatched, the larvae spend several weeks drifting in open water. Survivors go on to metamorphosis into benthic fish. Juveniles of common sole spend the first 2 years of their life in coastal nurseries before migrating to deeper areas, where they reproduce. To investigate the drivers of common sole abundance in the Eastern Channel, an integrated approach was developed, coupling different models describing the different life stages, to estimate the different sources of mortality throughout the whole life cycle.

3.1.1 Young Stages (Larval Stages and Nursery Habitat Dependence)

An individual-based model (IBM) coupled to a hydrodynamic model was used to simulate common sole larval supply from spawning areas to coastal and estuarine nursery grounds at the population scale on a 3-decade time series. Hydroclimate is the main driver of abundance patterns during early stages of the life cycle. As mortality (~1/1,000 survival) is particularly high at these stages (eggs and larvae), hydroclimate drives the year class strength of juvenile abundance, without viewable relation with the spawning biomass (Rochette et al. 2012).

3.1.2 Integrated Life Cycle Model Accounting for Various Pressures

Essential fish habitat suitability (EFHS) models and geographic information system were combined to describe nursery habitats for sole, using parameters known to influence juvenile flatfish spatial distribution (i.e. bathymetry, sediment, estuarine influence and wave exposure). Juveniles strongly depend on shallow soft-bottom sheltered coastal and estuarine nursery grounds and host a large proportion of total juvenile common sole, suggesting that these restricted habitats should be considered as essential habitats for sole (Rochette et al. 2010).

A hierarchical Bayesian framework was developed for modelling the life cycle of marine exploited fish with a spatial perspective. The approach combined within an integrated framework: (1) outputs of the model for larval drift and survival that provided yearly estimates of the dispersion and mortality of eggs and larvae, from spawning grounds to settlement in coastal nurseries (Rochette et al. 2012); (2) a habitat suitability model based on juvenile trawl surveys coupled with a geographic information system, to estimate juvenile densities and surface areas of suitable juvenile habitat in each nursery sector (Rochette et al. 2010) and (3) a statistical catch-at-age model for the estimation of the numbers at age and the fishing mortality on subadults and adults. Successive modelling approaches allowed to demonstrate that juvenile mortality on nursery grounds is high (~1/100 survival) and strongly limits the population size (Rochette et al. 2013), fishing pressure is a main source of mortality at subadult and adult stages (Rochette et al. 2013) and spatial segregation at the successive life stages (i.e. eggs/larvae, juveniles, adults) along the life cycle limits the connectivity between different subparts of the population. Perspectives include further development of the modelling framework on the common sole and applications to other fish species to disentangle the effects of multiple interacting stress factors (e.g. estuarine and coastal nursery habitat degradation, fishing pressure) on population renewal and to develop risk analysis in the context of marine spatial planning for sustainable management of fish resources.

3.2 Exploited Ecosystems

3.2.1 Benthic Effects of Aggregate Extractions

The effects of sediment dredging on benthic groups consumed by sole and plaice in the Eastern English Channel have been assessed. Benthic species were grouped according to their diet or mobility, and analyses were performed at the group level. First, a BACI (before-after control impact) analysis was conducted to determine whether dredging impacted benthos abundance. This approach allowed determining relationships between dredging intensity and decreases in benthic abundance and quantifying the recovery rate of the studied benthic groups. At the group level, it was shown that impacts of dredging are maximal from 6 to 12 months after dredging, these impacts being either positive (opportunistic groups) or negative. Concerning mobility, impacts of dredging were maximal on burrowers. On the contrary, mobile species seemed to be less impacted. Carnivorous species were amongst the most impacted trophic groups. Detritivores seemed to be positively impacted by dredging, at least for a few months after dredging. It was not possible to link increases or decreases in abundance after dredging to a recovery time, increases in abundance being linked to strong year effects.

3.2.2 Spatial Interactions Involving Fishing Fleets and Other Sectors of Activity

Discrete choice models building in a random utility function (RUMs) were developed to determine how fishing effort is allocated spatially and temporally by a selection of French, English and Dutch fleets fishing in the Eastern English Channel. Results showed that fishers tended generally to adhere to past annual fishing practices and to the areas where they experienced high revenues and also that French dredgers strongly interacted spatially with English vessels. Furthermore, results indicated that maritime traffic, aggregate activity and restricted areas negatively impact the choice of many of the fleets under investigation. Other spatially explicit statistical analyses have been conducted to evaluate, separately, the impacts of aggregate extraction and maritime traffic. The effects of both aggregate extraction intensity and the proximity to dredging sites on the distribution of fishing effort were investigated, for a broad selection of French and English demersal fleets operating in the Eastern Channel (Marchal et al. 2014). The most striking result is that neither dredging intensity nor the proximity to the extraction site had a major deterring effect on fishing activities. To the contrary, the fishing effort of dredgers and potters could be larger on aggregate sites than in their close neighbourhood, whilst the fishing effort of netters has increased substantially in the impacted area. The attraction of fishing fleets is likely due to a local and temporary concentration of target species. However, knowledge on the vulnerability and life-history characteristics of these species to aggregate extractions suggests that overextending the licensed areas would be detrimental to them and to their related fisheries in the longer term. Maritime traffic seems to be a perturbation for the fishing activities. However, in the case of the red mullet fishery, vessels do not avoid traffic lanes when they expect high fish densities. They then may take the risk of fishing inside the traffic lanes or in areas of high marine traffic densities. An effort has then been made to validate the outcomes of fleet dynamics models by administering a survey to French fishers operating in the Eastern English Channel. Although French fishers generally did not feel constrained in the amount of space they had available for fishing, some mentioned shipping lanes, aggregate extractions and competition with other fishers as constraining factors.

4 From Modelling Ecosystem Compartments to Holistic Ecosystem Modelling

Three modelling approach patterns have been pursued concurrently to model the impact of human activities on the Eastern Channel at the scale of the ecosystem, using the models ISIS-Fish, OSMOSE and ATLANTIS. An ISIS-Fish model was developed, including two flatfish populations (sole and plaice) and three benthic groups (Gasche et al. 2013). Fish populations interact with the fishing activity in the modelled area. Several fleets are modelled with an impact on both fish and benthos, and aggregate extractions are also built in with an impact on benthos only. No positive effects of MPAs on fish populations could be evidenced, either at the scale of the Eastern English Channel or at smaller scales gathering several bays on the French and English coasts. On the contrary, effects of MPAs are very important on benthic taxa. OSMOSE and ATLANTIS present a further degree of complexity compared to ISIS-Fish, by building in trophic interactions occurring between different ecosystem components. Both models are currently being calibrated in relation to the Eastern Channel ecosystem. The OSMOSE model is being applied to evaluate the effects of MPAs on fish populations. The ATLANTIS model will be applied to evaluate the effects of various area-based restrictions (including MPAs but also spatial interactions with maritime traffic) on all ecosystem compartments including fleet dynamics and economics.

5 Conclusions and Perspectives

Both the information on and the understanding of the Eastern English Channel marine ecosystem have been advanced substantially in the past 15 years through systematic mapping and subsequent modelling of key ecosystem compartments and related human pressures. These information and process knowledge are currently being integrated into comprehensive end-to-end ecosystem models. The next stages will be dedicated to the evaluation of ecosystem-based management strategies, building on the different modelling approaches once fully developed and calibrated, and last to the transfer of simulation outcomes to help stakeholders and support decision making in the Eastern English Channel.

References

Carpentier A, Martin CS, Vaz S (eds) (2009) Channel habitat Atlas for marine resource management, final report/Atlas des habitats des resources marines de la Manche orientale, rapport final (CHARM phase II). INTERREG 3a Programme, IFREMER, Boulogne s/ mer, France, 626 pp. & CD-rom

Gasche L, Mahévas S, Marchal P (2013) Supporting fisheries management by means of complex models: can we point out isles of robustness in a sea of uncertainty? PLoS ONE 8(10):e77566. doi:10.1371/journal.pone.0077566

Marchal P, Desprez M, Tidd A, Vermard Y (2014) How do fishing fleets interact with aggregate extractions in a congested sea? Estuar Coast Shelf Sci 149:168–177. doi:10.1016/j.ecss.2014.08.005

Rochette S, Rivot E, Morin J, Mackinson S, Riou P, Le Pape O (2010) Effect of nursery habitat destruction on flatfish population renewal. Application to common sole (*Solea solea*, L.) in the Eastern Channel (Western Europe). J Sea Res 64:34–44

Rochette S, Huret M, Rivot E, Le Pape O (2012) Coupling hydrodynamic and individual-based models to simulate long-term larval supply to coastal nursery areas. Fish Oceanogr 21:229–242

Rochette S, Le Pape O, Vigneau J, Rivot E (2013) A hierarchical Bayesian model to merge larval drift and juvenile habitat models with exploited fish population dynamics. Ecol Appl 23:1659–1676

Toward a Dynamical Approach for Systematic Conservation Planning of Eastern English Channel Fisheries

Yves Reecht, Loïc Gasche, Sigrid Lehuta, Sandrine Vaz, Robert J. Smith, Stéphanie Mahévas, and Paul Marchal

Abstract

In the past decade, systematic conservation planning tools have been increasingly and successfully used to set spatial conservation plans that meet quantitative protection targets while minimizing enforcement and socioeconomic costs. However, when applied to fisheries, systematic conservation planning fails to account for (1) changes in fleet dynamics induced by new conservation constraints and their associated feedbacks on conservation costs or (2) their influence on fish population dynamics and distributions, which may in turn alter the achievement of conservation targets. Such a static approach may therefore lead to short- or medium-term misestimates in forecasted costs and target achievements. In order to circumvent such limitations of systematic conservation planning, we present a first attempt to couple a conservation planning tool (Marxan with Zones) with a mixed fisheries dynamics simulation model (ISIS-Fish), applied to the Eastern English Channel fisheries. Broad principles and perspectives are discussed and anticipated future challenges of such an approach are presented.

1 Introduction

The Eastern English Channel is an important ecological area that is experiencing growing human pressure, mainly exerted by the bordering countries, France and the United Kingdom, and is subjected to a wide range of uses such as fishing, sediment extraction, and transport (Desprez 2000; Martin et al. 2010; Rochet et al. 2008). Spatial regulations, including marine protected areas (MPAs), are now increasingly introduced to legislate these perturbations (Leathwick et al. 2008; Sala 2002; Spalding et al. 2008;

Y. Reecht (✉) • P. Marchal
Ifremer, Unité Halieutique de Manche mer du Nord, 150 quai Gambetta, BP 699, F-62321 Boulogne-sur-mer Cedex, France
e-mail: Yves.Reecht@ifremer.fr

L. Gasche • S. Mahévas
Ifremer, Unité Ecologie et Modèles pour l'Halieutique, rue de l'Ile d'Yeu, BP 21105, F-44311 Nantes Cedex 03, France

S. Lehuta
Ifremer, unité Ecologie et Modèles pour l'halieutique, rue de l'ile d'Yeu, F-44311 Nantes 03, France

S. Vaz
Ifremer, Unité Halieutique de Méditerranée, avenue Jean Monnet, CS 30171, F-34203 Sète, France

R.J. Smith
Durrell Institute of Conservation and Ecology, University of Kent, Canterbury CT2 7NR, UK

H.-J. Ceccaldi et al. (eds.), *Marine Productivity: Perturbations and Resilience of Socio-ecosystems*,
DOI 10.1007/978-3-319-13878-7_19,

Wood et al. 2008). MPAs, to be successfully implemented, need to combine conservation objectives and socioeconomic features, such as fisheries (Gaines et al. 2010). Within this context, France and the United Kingdom are under obligations to create a consistent marine protected area (MPA) network that complies with several conventions, especially the Convention on Biological Diversity (review in Metcalfe et al. 2013) and the Bird and Habitat European Directives, while ensuring a viable future for the wide range of uses within this area.

In the past decade, systematic conservation planning tools have been increasingly and successfully used to develop spatial conservation plans – involving MPAs – which meet quantitative targets (e.g., a given protected percentage of each species distribution or habitat area) while minimizing enforcement and socioeconomic costs (Ban et al. 2011; Delavenne et al. 2012; Leslie 2005; Syakur et al. 2012). This approach thus provides a framework that is deemed suitable to design consistent MPA networks that are cost effective and minimize social costs, hence increasing their likelihood of effective implementation (Smith et al. 2009). However, systematic conservation planning applied to fisheries accounts neither for (1) changes in fleet dynamics induced by new conservation constraints and their associated feedbacks on conservation costs nor (2) their influence on fish population dynamics and distributions, which may in turn alter the achievement of conservation targets. Such a static approach may therefore lead to short- or medium-term misestimates in forecasted costs and target achievements.

Mixed fisheries simulation models are increasingly used to predict changes in fleet and fish population dynamics under various fisheries management scenarios (e.g., Batsleer et al. 2013; Lehuta et al. 2010), but lack, in most of the cases, the methodology to translate the results into advices to support spatial conservation measures.

In this context, coupling systematic conservation planning tools with mixed fisheries models (or other types of simulation models, in accordance with the type of issue tackled) seems a promising approach to test scenarios and build advice for the management of highly dynamic and complex systems such as coastal areas under intense human use.

2 Proposed Approach

2.1 Overview of Selected Tools

2.1.1 Systematic Conservation Planning

Several tools exist that are dedicated to systematic conservation planning (e.g., Marxan, Zonation) and can help to design MPAs. Most of them however rely on a binary and often unrealistic full protection strategy, therefore missing the complexity of management strategies which can be deployed through multiple types of MPAs. A recent tool, Marxan with Zones (MwZ) (Watts et al. 2009), allows this limitation to be overcome in an optimal way by extending the Marxan methodology (based on the minimum set principle aiming to achieve a given quantitative representation level of species and habitats at minimal cost) to multiple zone types. It provides the possibility of considering multiple, possibly concurrent, resource uses which are managed in different ways, while taking into account a variety of costs. It has been shown to be able to provide management scenarios that – compared to a standard Marxan analysis – ensure more equitable impacts among different uses while lowering the overall economic and social impact and still meeting conservation targets, thus increasing the likelihood of effective implementation (Klein et al. 2009). However, unlike Marxan, feedback on MwZ effectiveness remains scarce.

Marxan with Zones uses a simulated annealing algorithm to work as an optimization tool which meets complex constraints such as combinations of overall percent and/or absolute values of each feature (often species or habitats) to be protected in each type of "zone" (with varying protection levels corresponding to which human use is maintained). The objective function it minimizes has the form:

$$\sum_{PUs}\text{Cost} + \underbrace{\text{BLM} \times \sum_{Z}\sum_{PUs}\text{Boundary Cost}}_{\text{Connectivity costs}} + \sum_{Ft}\text{FPF} \times \text{FeaturePenalty}, \quad (1)$$

where Cost represents the sum of various costs associated with the selection of a "planning unit" (PU, the smaller spatial unit). These costs can be of any kind found to be relevant to each case study, for instance, surface area, enforcement, or socioeconomic costs (Smith et al. 2009). Additionally to these inherent costs, connectivity costs – of which the boundary length modifier (BLM) controls the overall contribution to the objective function value – allow control of the level of aggregation/fragmentation of conservation zones or increase the co-selection of connected PUs (Klein et al. 2009). The "connectivity" between conservation zones is also controlled in this term by zone boundary costs, which are formally multipliers of PU boundary costs between each combination of adjacent zones.[1] The last term of Eq. (1) refers to penalties for failing to achieve targets, summed over features (Ft; or species), and is controlled through feature penalty factors (FPF); the higher the FPF, the more likely the fulfillment of the target (Watts et al. 2009).

MwZ has been selected for this study owing to its ability to both reproduce complex management scenarios and include use-specific costs in a flexible way, which more accurately reflect mixed fisheries properties than other existing tools.

2.1.2 Mixed Fisheries Simulation Model

ISIS-Fish has been chosen because it is a modeling tool suitable for investigating the consequences of alternative policies on the dynamics of fish resources and fisheries (Mahévas and Pelletier 2004). This spatially explicit model allows quantitative policy screening for fisheries with mixed species harvests (Mahévas and Pelletier 2004; Pelletier et al. 2009). It may be used to investigate the effects of combined management scenarios including a variety of policies: total allowable catch (TAC), licenses, gear restrictions, and effort controls but also alternative ones such as the introduction of marine protected areas (Lehuta et al. 2010; Kraus et al. 2008; Drouineau et al. 2006) or individual quotas (Marchal et al. 2011). Fisher's response to management may be accounted for by means of decision rules based on population and exploitation parameters or explicit dynamics model with endogenous (e.g., fixed fish prices and variable costs that can be explicitly modeled) or exogenous variables (not affected by the model). This fisheries model is based on three submodels: (1) a fishing activity dynamics model, (2) a fish population dynamics model, and (3) a management dynamics model.

Each submodel is spatially and seasonally explicit, with a monthly time step to account for seasonal dynamics. The three submodels interact only if they overlap in space and time. The modeled area is represented by a grid, the resolution of which, in latitude and longitude, is chosen with respect to the dynamics being described and the available knowledge of the studied fishery. Within this region, zones (i.e., sets of grid cells) are defined independently and delimit the spatial scope for each population, each fishing activity, and each management measure. Finally, bioeconomic outputs can be simulated and their properties (including uncertainties) statistically analyzed to produce indicators of the relevance of management strategies (Lehuta et al. 2013).

2.2 Model Scopes and Implementations

Here we present the first highlights of an ongoing study which aims to couple a systematic conservation planning software package with a mixed fisheries model to evaluate the relevancy for fisheries management of the MPA network being implemented in the Eastern English Channel (Fig. 1) and, where relevant, provide advice for management strategies.

[1] Note that a PU can pertain to only one conservation zone at a time.

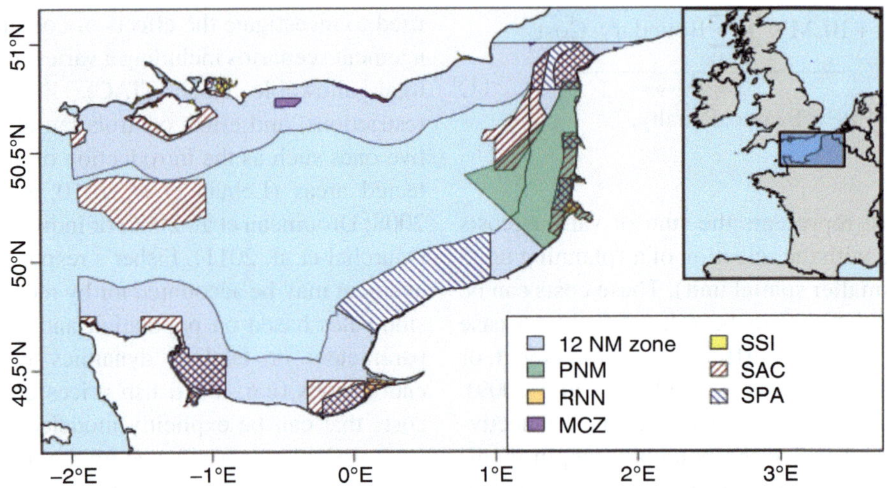

Fig. 1 MPAs (actual or planned) in the ICES VIId zone. *PNM* natural marine park (France); *RNN* national natural reserve (Fr.); *APPB* prefectural biotope protection (Fr.); *MCZ* marine conservation zone (UK); *SSSI* site of special scientific interest (UK); *OSPAR* OSPAR convention zone; *RAMSAR* RAMSAR convention zone; *SAC* special area of conservation (Natura 2000, habitat convention); *SPA* special protected area (Natura 2000, bird convention)

This study therefore focuses on ecosystem and socioeconomic features which are linked to fishing activities, restricted to professional fishing owing to data availability.

2.2.1 Marxan with Zones

Features to protect are:

- Abundance distributions (mean over 1990–2012) of two of the main targeted species (those accounted in the fleet dynamics model), common sole (*Solea solea*, Fig. 2a) and plaice (*Pleuronectes platessa*, Fig. 2b) assessed from the Channel Ground Fish Surveys (CGFS) data.
- Eighteen benthic habitats, on which towed gears can have a negative impact. Data used are those from Delavenne et al. (2012).
- Thirteen pelagic habitats, which contain communities that can be affected by most pelagic gears. The typology used is the one defined by Delavenne et al. (2013).

Among feature types, only exploited species distributions are planned to be dynamically linked to the simulation model, habitats being handled as a static part of the system.

As for costs, hours fished by type of gear and zone, estimated from data collected by vessel monitoring system (Lee et al. 2010), are used as proxies of value losses when a protection unit is selected for a type of zone which bans or limits some uses (Fig. 2c, d). As a first approach, costs will not be processed in a dynamic way because the selection of a PU would eliminate its cost on the next step (no or less fishing).

The spatial grid was defined so that each MPA was divided into as many subareas of unique administrative status and that remaining available areas were separated according to the same grid as the ISIS-Fish model (1/32nd of ICES statistical rectangle, Fig. 2) and further divided according to the 12 nautical miles zone.

A scenario was tested with only two kinds of protection zones – no ground-towed gears and no-take (all activities prohibited) – and where the 12 nautical miles zone was considered already contributing to conservation (limited access to vessels >24 m). Already planned MPAs were constrained to always apply one of the two protection levels. Zone contributions to conservation (Table 1) and cost multipliers (Table 2, formally corresponding to a proportion of effort reduction within protected zones) were set arbitrarily but should ideally be derived from quantitative study of the impact of each use on different features. Here, all benthic habitats were treated homogeneously with the removal of ground-towed gear

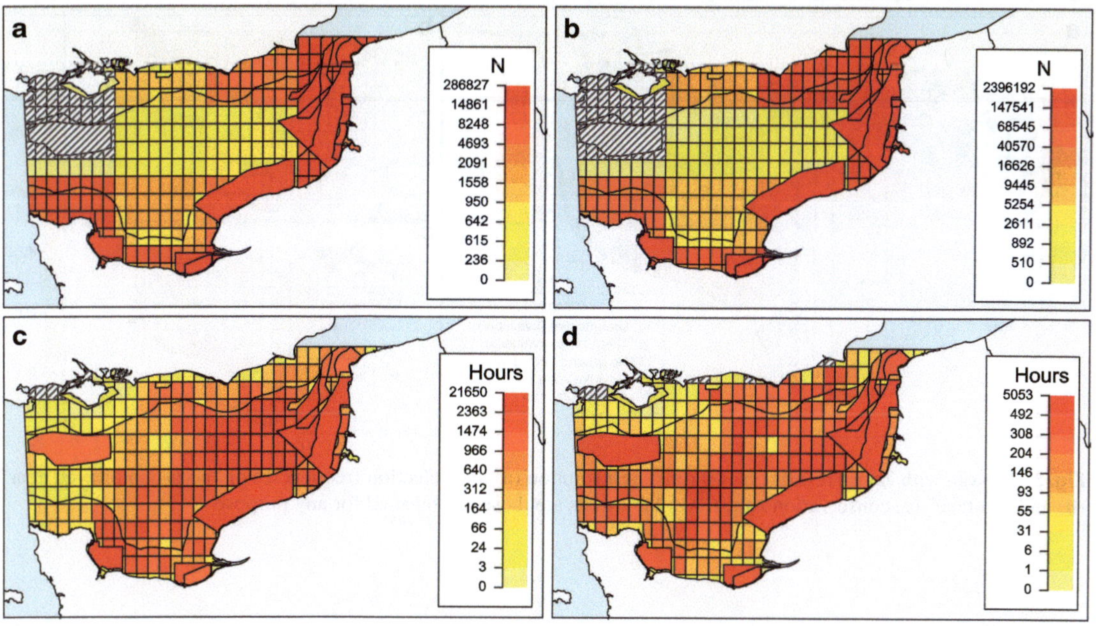

Fig. 2 Mean abundances (*N*, decile scale) by PU of *Solea solea* (**a**) and *Pleuronectes platessa* (**b**) estimated from the CGFS data and aggregated efforts by PU (hours fished, decile scale) from French (2008) and English (2007) VMS data for ground-towed gears (**c**) and all other types of gears (**d**) PUs with lacking data are hatched

Table 1 Conservation zone contributions to the protection of various types of features

	Zones			
Features	Unprotected	12 nm zone	No ground towed	No-take
Benthic habitats	0.0	0.1	1.0	1.0
Pelagic habitats	0.0	0.0	0.0	1.0
Target species	0.0	0.1	0.5	1.0

These are proportions of feature potentially protected under different types of restriction of fishing access

Table 2 Cost multipliers by zone and gear type

	Zones			
Gears	Unprotected	12 nm zone	No ground towed	No-take
Towed	0.0	0.1	1.0	1.0
Others	0.0	0.0	0.0	1.0

These are the proportions of effort reduction applied to each gear type under different types of restriction of fishing access

fully protecting them (no impact of other gears), while this measure is supposed to have no impact on pelagic habitats and yield a 50 % protection to target species (which can be also caught with other gears such as tremels). The exclusion of large vessels (mostly trawlers) from the 12 nautical miles zone is not expected to supply more than a 10 % protection to features impacted by ground-towed gears, since there are few such large vessels operating in the study area.

The results of this drastic scenario (few choices for management strategies, all of which prohibited ground-towed gears) are reported in Fig. 3.

2.2.2 Fisheries Dynamics Modelling

The ISIS-Fish model used is an improved version of the one developed by Gasche et al. (2013), characterizing population and exploitation dynamics of sole and plaice, with:

- A finer spatial resolution of 0.125° (latitude and longitude) that allows for a more realistic depiction of biological and exploitation processes

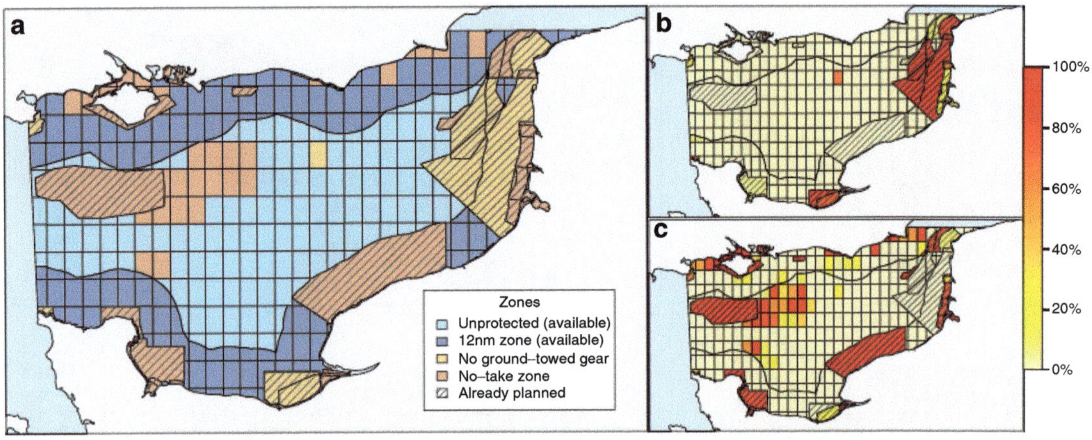

Fig. 3 Marxan with Zones results on 100 runs: best solution (**a**) and selection frequencies of "no ground-towed gear" (**b**) and "no-take" (**c**) conservation zones. Available cells are those not selected for any purpose

Fig. 4 ISIS-Fish grid and management zones defined in the model. These are an attempt to broadly reproduce MwZ outputs (Fig. 3)

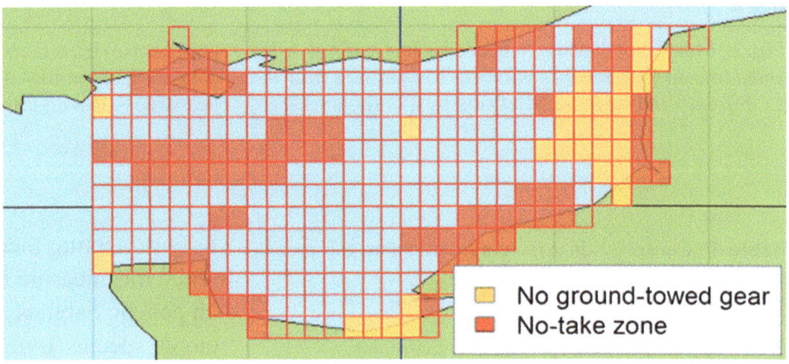

- Better account of population distributions across life stages – with several feeding grounds (three), nurseries (six), and reproduction (three) zones for each species (i.e., 12 zones for both plaice and sole) – according to Rochette et al. (2012) and Carpentier et al. (2009)
- Fishing activities updated according to Lehuta et al. (2015)

The model was tested over 12 years under two different scenarios: (1) one with only total allowable catches (TAC) as management measure (forced by 2008–2011 actual TACs, then dynamically set by a harvest control rule that aims at reaching F_{MSY} in five years following ICES advices) and (2) another with additional spatial conservation measures (Fig. 4) consistent with MwZ outputs (Fig. 3). No effort reallocation across métiers (defined by one target species on one zone with one gear) is assumed.

Results show very little differences between scenarios with TACs and TACs + MPAs for abundances (Fig. 5 left) and landings (Fig. 5 right) of both plaice and sole during the transition period toward management at F_{MSY} (years three to seven). However, differences in landings become more substantial afterward (when the TAC is less constraining), although no clear difference appears in trends.

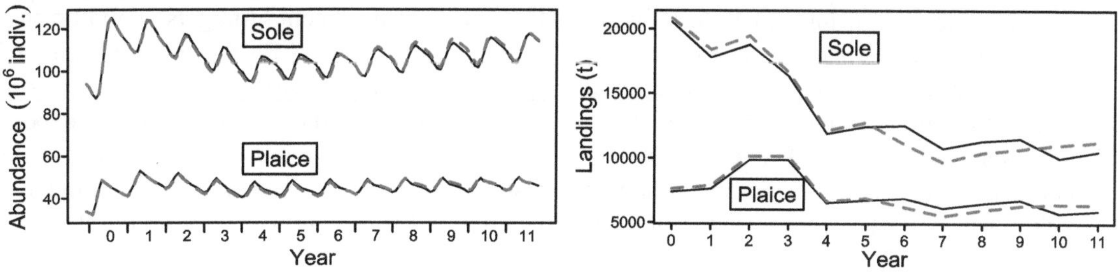

Fig. 5 ISIS-Fish model outputs of monthly abundances (*left*) and yearly landings (*right*) compared between TAC only (*black solid lines*) and TAC+MPAs (*gray dashed lines*) scenarios

2.3 Explicit Model Coupling in Practice

Marxan with Zones is a command-line software which works with text input files, all listed in a main input file and with fairly simple and documented structures (Watts et al. 2008). It is therefore easily controlled through any platform which allows running commands and is able to handle data and text files (e.g., we have easily controlled all the MwZ analysis sequence, from data formatting to output representation, with R; http://cran.r-project.org). Principal Marxan input files which would be subject to dynamical updates are:

- The planning unit file and more particularly the costs given to each planning unit (PU)
- The features (here exploited species) versus PU file which gives amount of each feature within each PU

Files such as those containing boundary lengths/connectivities between protection units or "zone boundary costs" pertain to the original design and are unlikely to be modified by iterative runs (except for sensitivity analysis).

As for ISIS-Fish, control from a third-party tool seems more difficult since most parameters are stored in embedded databases, with internal referencing of objects, as spatial units or populations, for instance. Therefore, even though ISIS-Fish simulations themselves can be run from command-line calls, the management zones of the model cannot be directly controlled through text files. However, ISIS-Fish is an open-source modeling platform, with an active development team, hence highly extendable.

For instance, concerning the translation from MwZ outputs to ISIS-Fish management strategies, the ISIS extensive scripting (Java script) abilities will be used to:

1. Load formatted MwZ outputs (preferably preprocessed by R scripts for easy handling)
2. Define as many management zones as different effort reductions by gear, calculated from MwZ outputs (pre-simulation script)
3. Apply for each cell an effort reduction by métier from its overlap with management zones for the gear used at each time step

The second point raises the issue of transferring costs, features, and optimized spatial management measures between two possibly different spatial scales. Indeed, there is no requirement for the spatial grid in MwZ to be regular, as it is the case for the one in ISIS-Fish. In fact, it is even convenient to keep existing – intricately shaped – MPAs as separated PUs (e.g., Fig. 3) for MwZ analyses. Therefore, even under simple homogeneity assumption regarding amounts within grid cells, the transfer of data from a grid to another requires extensive calculations, among which assessment of cell surface overlaps between the two model spatial grid layers (which only needs to be calculated once) and pro rata reallocation rates from one grid to the other. This is easily done with R scripts which can notably calculate an accurate effort reduction in ISIS spatial unit (cell) by gear, from a given MwZ solution, since gear limitations are given for each kind of conservation zone.

As for the automation of the coupling, two options emerge:

1. In order to iteratively run both of the tools and operate the data format conversion between them, a first step will be to use an external third-party tool. As suggested above, R is a good candidate because of its extended data-handling abilities and its capacity to interact with other software packages through so-called system calls.
2. It is nevertheless considered to further develop a module within ISIS-Fish to control MwZ directly from within simulation iterations, hence getting quick cost assessments and zoning optimizations to enhance the dynamic properties of the whole system. The existing ability of ISIS to connect to R could in particular be used to run scripts controlling the whole MwZ loop, from data preparation to processing of output to provide management zones that can be handled by ISIS. This way, costs and feature amounts could be updated from ISIS-Fish to MwZ, and as a feedback, zoning in ISIS-Fish could be updated according to MwZ outputs.

2.4 Types of Scenarios to Be Tested

Coupling these two tools would open the opportunity to test a wide range of scenarios regarding the dynamics of spatial conservation plans. The following propositions are far from exhaustive but focus on types of analyses already planned in the context of this study.

A first and fairly obvious type of analysis would consist of testing what the dynamics of the main fleets and fish populations would be under different conservation scenarios and how they would influence target achievement (no feedback, only the evaluation of species abundance proportions within protected areas after the simulated period) and then testing for the robustness of the MPA network through the stability in species representation. This would only require a one-way coupling from MwZ to ISIS-Fish that does not really require explicit and automated translation of a common management zone output to management scenarios within ISIS-Fish. Achievement of targets would be easily assessed from ISIS-Fish outputs in terms of species abundance distributions. More realistic economic costs, although completely independent at this stage from those used in MwZ, could also be assessed by comparison of simulated landings at the beginning and the end of the simulation period.

From the previous analysis, it may be possible to test whether any proposed MPA network with a particular set of conservation measures is suitable to ensure medium- to long-term viability of fleets and of the fish populations they harvest. At present, in the Eastern English Channel, enforcement measures are still to be defined in most proposed MPAs, and it may be very relevant to use MwZ on its own to provide near optimal management scenarios within the already planned MPA network. The addition of the ISIS-Fish simulations would enable the evaluation of the management strategy under which the fishing fleets will remain viable. For that purpose, the methodology proposed by Lehuta et al. (2013), based on bioeconomic indicators and their uncertainty to evaluate management strategies, could be used.

Other types of analyses would require a more intricate and fully dynamic coupling of the two tools than the preceding ones. It is, for instance, planned to test the effect of the chronological sequence of enforcement and how it could be optimized. Indeed, it would involve running ISIS-Fish over a given interval of time (e.g., one year), then testing which would be the best enforcement addition to the network using MwZ and running ISIS-Fish again from where it stopped, etc. Such an approach would have to be tested over various time lags.

All these types of analysis would benefit by also testing whether different near-optimal solutions, which differ noticeably in terms of selected sets of PU-zone pairs but not in terms of cost, would lead to different dynamics and viabilities of fleets and harvested populations. Such analyses would have to account for confidence in data. From a more general point of view, a sensitivity analysis on optimization and model parameters would be necessary.

3 Perspectives and Future Challenges

Challenges raised by this coupling approach pertain to (1) finding a relevant design and level of detail regarding processes embodied in each model, (2) getting a proper parameterization of both models, and (3) keeping an overall consistency, instead of technical issues regarding the coupling itself.

First of all, the methodology promoted here will be extended to a more representative set of the Eastern English Channel fisheries. For that purpose, the MwZ number of features will be extended to include seven target species, in order to be coupled with the ISIS-Fish model developed by Lehuta et al. (2015) and which includes five additional target species: cod *Gadus morhua*, whiting *Merlangius merlangus*, European sea bass *Dicentrarchus labrax*, squids *Loligo* spp., and scallop *Pecten maximus*. Another possible improvement of the ISIS-Fish model can be to add benthic and/or pelagic habitats to provide a dynamic description of habitat state impacted by human activities (including sediment extraction) like in Gasche et al. (in preparation).

Additional features such as bird or seal colonies, highly sensitive habitats, or endangered species known locations may also be added to increase the spatial constraint on the MwZ solutions and to better represent the local biodiversity and not only the exploited species.

The definition of management scenarios that are likely to be implemented is also of high importance. This raises the need for the identification of more realistic management measure regarding either the administrative status of zones (e.g., is it desirable to ban ground-towed gears from large zones designated under the bird directive only?) or the local characteristics of an MPA itself (e.g., in the case of particularly discrete natural bivalve beds where dredging cannot be banned without deep economic impacts). In that context, a consultation of managers is underway to move toward more realistic constraints on MwZ inputs.

Contributions of different zone types to the preservation of various features will also have to be characterized in a more explicit way, based on observed and quantified gear impacts, proposed effort reduction (by gear within the zone), and overall effort partition among gears.

This approach could also be improved by moving fishing activities from costs to features that also have to be protected (e.g., with a 90 % target on value landed by each type of activity that is no more than 10 % loss). This would ensure a better equitability of the conservation effort among métiers by applying a high penalty to scenarios inducing a larger loss than the one set by target for at least one type of activity.

These combined improvements should enhance our capacity to provide valuable diagnostics and advice regarding the effectiveness of MPA networks for the management of mixed fisheries.

Acknowledgments We wish to thank the French MPA Agency for providing up-to-date MPA limits. The study was funded by the Interreg IVa PANACHE (Protected Area Network Across the Channel Ecosystem) European project.

References

Ban NC, Adams VM, Almany GR, Ban S, Cinner JE, McCook LJ, Mills M, Pressey RL, White A (2011) Designing, implementing and managing marine protected areas: emerging trends and opportunities for coral reef nations. Coral Reefs Future Dir 408:21–31. doi:10.1016/j.jembe.2011.07.023

Batsleer J, Poos J, Marchal P, Vermard Y, Rijnsdorp A (2013) Mixed fisheries management: protecting the weakest link. Mar Ecol Prog Ser 479:177–190. doi:10.3354/meps10203

Carpentier A, Martin CS, Vaz S (eds) (2009) Channel habitat atlas for marine resource management, final report (CHARM phase II). INTERREG 3a Programme, IFREMER, Boulogne-sur-mer, France, 626 pp

Delavenne J, Metcalfe K, Smith RJ, Vaz S, Martin CS, Dupuis L, Coppin F, Carpentier A (2012) Systematic conservation planning in the eastern English Channel: comparing the Marxan and Zonation decision-support tools. ICES J Mar Sci J Conserv 69:75–83. doi:10.1093/icesjms/fsr180

Delavenne J, Marchal P, Vaz S (2013) Defining a pelagic typology of the eastern English Channel. Cont Shelf Res 52:87–96. doi:10.1016/j.csr.2012.10.016

Desprez M (2000) Physical and biological impact of marine aggregate extraction along the French coast of the Eastern English Channel: short- and long-term post-dredging restoration. ICES J Mar Sci 57:1428–1438. doi:10.1006/jmsc.2000.0926

Drouineau H, Mahévas S, Pelletier D, Beliaeff B (2006) Assessing the impact of different management options using ISIS-Fish: the French *Merluccius merluccius – Nephrops norvegicus* mixed fishery of the Bay of Biscay. Aquat Living Resour 19:15–29. doi:10.1051/alr:2006002

Gaines SD, Lester SE, Grorud-Colvert K, Costello C, Pollnac R (2010) Marine reserves special feature: evolving science of marine reserves: new developments and emerging research frontiers. Proc Natl Acad Sci 107:18251–18255. doi:10.1073/pnas.1002098107

Gasche L, Mahévas S, Marchal P (2013) Supporting fisheries management by means of complex models: can we point out isles of robustness in a sea of uncertainty? PLoS ONE 8:e77566. doi:10.1371/journal.pone.0077566

Gasche L, Lehuta S, Mahévas S, Marchal P, Savina-Rolland M, Baulier L (in preparation) Can spatial management measures help robustly managing human activities in the eastern English Channel?

Klein CJ, Steinback C, Watts M, Scholz AJ, Possingham HP (2009) Spatial marine zoning for fisheries and conservation. Front Ecol Environ 8:349–353. doi:10.1890/090047

Kraus G, Pelletier D, Dubreuil J, Mollmann C, Hinrichsen H-H, Bastardie F, Vermard Y, Mahevas S (2008) A model-based evaluation of marine protected areas: the example of eastern Baltic cod (*Gadus morhua callarias* L.). ICES J Mar Sci 66:109–121. doi:10.1093/icesjms/fsn166

Leathwick J, Moilanen A, Francis M, Elith J, Taylor P, Julian K, Hastie T, Duffy C (2008) Novel methods for the design and evaluation of marine protected areas in offshore waters: designing offshore MPAs. Conserv Lett1:91–102.doi:10.1111/j.1755-263X.2008.00012.x

Lee J, South AB, Jennings S (2010) Developing reliable, repeatable, and accessible methods to provide high-resolution estimates of fishing-effort distributions from vessel monitoring system (VMS) data. ICES J Mar Sci J Conserv 67:1260–1271. doi:10.1093/icesjms/fsq010

Lehuta S, Mahévas S, Petitgas P, Pelletier D (2010) Combining sensitivity and uncertainty analysis to evaluate the impact of management measures with ISIS–Fish: marine protected areas for the Bay of Biscay anchovy (*Engraulis encrasicolus*) fishery. ICES J Mar Sci J Conserv 67:1063–1075

Lehuta S, Mahévas S, Le Floc'h P, Petitgas P, Rose K (2013) A simulation-based approach to assess sensitivity and robustness of fisheries management indicators for the pelagic fishery in the Bay of Biscay. Can J Fish Aquat Sci 70:1741–1756. doi:10.1139/cjfas-2013-0066

Lehuta S, Vermard Y, Marchal P (2015) A spatial model of the mixed demersal fisheries in the eastern channel marine productivity: perturbations and resilience of socio-ecosystems. In: Ceccaldi H-J et al (eds) Proceedings of the 15th French-Japan oceanography symposium (in press)

Leslie HM (2005) A synthesis of marine conservation planning approaches. Conserv Biol 19:1701–1713. doi:10.1111/j.1523-1739.2005.00268.x

Mahévas S, Pelletier D (2004) ISIS-Fish, a generic and spatially explicit simulation tool for evaluating the impact of management measures on fisheries dynamics. Ecol Model 171:65–84. doi:10.1016/j.ecolmodel.2003.04.001

Marchal P, Little LR, Thebaud O (2011) Quota allocation in mixed fisheries: a bioeconomic modelling approach applied to the Channel flatfish fisheries. ICES J Mar Sci 68:1580–1591. doi:10.1093/icesjms/fsr096

Martin CS, Carpentier A, Vaz S, Coppin F, Curet L, Dauvin J-C, Delavenne J, Dewarumez J-M, Dupuis L, Engelhard G, Ernande B, Foveau A, Garcia C, Gardel L, Harrop S, Just R, Koubbi P, Lauria V, Meaden GJ, Morin J, Ota Y, Rostiaux E, Smith R, Spilmont N, Vérin Y, Villanueva C, Warembourg C (2010) The Channel habitat atlas for marine resource management (CHARM): an aid for planning and decision-making in an area under strong anthropogenic pressure. Aquat Living Resour 22:499–508. doi:10.1051/alr/2009051

Metcalfe K, Roberts T, Smith RJ, Harrop SR (2013) Marine conservation science and governance in North–West Europe: conservation planning and international law and policy. Mar Policy 39:289–295. doi:10.1016/j.marpol.2012.12.002

Pelletier D, Mahevas S, Drouineau H, Vermard Y, Thebaud O, Guyader O, Poussin B (2009) Evaluation of the bioeconomic sustainability of multi-species multi-fleet fisheries under a wide range of policy options using ISIS-Fish. Ecol Model 220:1013–1033. doi:10.1016/j.ecolmodel.2009.01.007

Rochet M-J, Prigent M, Bertrand JA, Carpentier A, Coppin F, Delpech J-P, Fontenelle G, Foucher E, Mahe K, Rostiaux E, Trenkel VM (2008) Ecosystem trends: evidence for agreement between fishers' perceptions and scientific information. ICES J Mar Sci 65:1057–1068. doi:10.1093/icesjms/fsn062

Rochette S, Huret M, Rivot E, Le Pape O (2012) Coupling hydrodynamic and individual-based models to simulate long-term larval supply to coastal nursery areas: modelling larval supply to coastal nurseries. Fish Oceanogr 21:229–242. doi:10.1111/j.1365-2419.2012.00621.x

Sala E (2002) A general model for designing networks of marine reserves. Science 298:1991–1993. doi:10.1126/science.1075284

Smith RJ, Eastwood PD, Ota Y, Rogers SI (2009) Developing best practice for using Marxan to locate marine protected areas in European waters. ICES J Mar Sci J Conserv 66:188–194. doi:10.1093/icesjms/fsn198

Spalding MD, Fish L, Wood LJ (2008) Toward representative protection of the world's coasts and oceans-progress, gaps, and opportunities: toward representative marine protection. Conserv Lett 1:217–226. doi:10.1111/j.1755-263X.2008.00030.x

Syakur A, Wibowo JT, Firmansyah F, Azam I, Linkie M (2012) Ensuring local stakeholder support for marine conservation: establishing a locally-managed marine area network in Aceh. Oryx 46:516–524. doi:10.1017/S0030605312000166

Watts ME, Klein CJ, Stewart RS, Ball IR, Possingham HP (2008) Marxan with Zones (V1.0.1): conservation zoning using spatially explicit annealing, a manual, http://www.uq.edu.au/marxan/documentation. Accessed 20 Jan 2015

Watts ME, Ball IR, Stewart RS, Klein CJ, Wilson K, Steinback C, Lourival R, Kircher L, Possingham HP (2009) Marxan with Zones: software for optimal conservation based land- and sea-use zoning. Spec Issue Simul Model Asia-Pac Reg SI ASIMMOD 24:1513–1521. doi:10.1016/j.envsoft.2009.06.005

Wood LJ, Fish L, Laughren J, Pauly D (2008) Assessing progress towards global marine protection targets: shortfalls in information and action. Oryx. doi:10.1017/S003060530800046X

A Spatial Model of the Mixed Demersal Fisheries in the Eastern Channel

Sigrid Lehuta, Youen Vermard, and Paul Marchal

Abstract

In the Eastern Channel, a diversity of fleets targets a large variety of species with various gears. The complexity of the fishery makes it difficult to develop and predict the effects of the new management measures promoted by the new Common Fisheries Policy, in particular harvest control rules and landing obligation. The evaluation of the impact of these new measures on fishing activities and resources requires a model which can capture the main features of the fishery. Here, we conduct an analysis of the Eastern Channel demersal fishery to collect information and identify the important aspects that need to be included in the model. The analysis shows the importance of six main species restricted to the Eastern Channel in the French demersal fleet's revenues: scallops, sole, plaice, red mullet, squid, and cuttlefish. Landings are very mixed which does not allow the identification of a main target species at the fishing trip level. We also noticed a strong spatial structuration of effort in space, mainly due to the characteristics of the boats and the distribution of fish. Effort is also structured in time because of the seasonality of species and of management constraints. The spatially and seasonally explicit model ISIS-Fish is consequently chosen to integrate this information and to allow simulating the evolution of the fishery under a new management regime. The spatial resolution and the definition of the fleets are selected in accordance with the results of the descriptive analysis. Métiers are described and catchability parameters are assessed. Spatially explicit population dynamics models are developed for each of the six species. Finally, a fishing behavior model is proposed to allow predictive simulations to account for changes in species distribution, management, and economic conditions.

S. Lehuta (✉)
Ifremer, unité Ecologie et Modèles pour l'halieutique, rue de l'ile d'Yeu, F-44311 Nantes 03, France
e-mail: Sigrid.Lehuta@ifremer.fr

Y. Vermard • P. Marchal
Ifremer, Unité Halieutique de Manche mer du Nord, 150 quai Gambetta, BP 699, F-62321 Boulogne-sur-mer Cedex, France

H.-J. Ceccaldi et al. (eds.), *Marine Productivity: Perturbations and Resilience of Socio-ecosystems*, DOI 10.1007/978-3-319-13878-7_20,

1 Introduction

The Common Fisheries Policy has been reformed in January 2014, resulting in new management measures. These new measures aim at overcoming the shortcomings of the previous management regime especially regarding the issues related to MSY (maximum sustainable yield), mixed fisheries, spatial dynamics, discarding behavior, and data-limited species. Creative tools are central to the management strategy evaluation (MSE) process. These include new spatially explicit harvest control rules and bio-economic models, which can account for fish population dynamics and fishing reaction to a variety of management scenarios. In this study, innovative harvest control rules (HCR) have been proposed to adapt catch quotas (TAC) to the value taken by a range of indicators traditionally used in EU fisheries management plans but also alternative status/pressure indicators that do not require analytical stock assessments (Butterworth et al. 2010; Apostolaki and Hillary 2009).

In the Eastern English Channel (EEC) (ICES Division VIId), a diversity of fleets targets a large variety of species with various gears. The spatial and temporal dynamics of these fleets depend both on fish population dynamics and constraints applied on fishermen (management, economics, etc.). The development of new HCRs is complicated by the mixed nature of the fisheries with risks of increasing discards and effort transfers from one species to another. To allow the evaluation of new HCRs on fish communities and fleets accounting for these risks, we developed a spatially explicit model, using the ISIS-Fish software. The framework embodies a spatialized operational model which simulates the dynamics of the mixed fisheries in the Eastern Channel including the spatial dynamics of the main species targeted (sole, plaice, scallops, red mullet, etc.) and the dynamics of the fleets driven by a fishing behavior model. We present the spatiotemporal analyses of the fishery dynamics that resulted in the modeling choices and parameterization of ISIS-Fish.

2 Analysis of the Fishery

In order to assess the impact of new management measures on the EEC fisheries, the spatio-temporal structure and dynamics of both fleets and fish populations have to be understood and described.

2.1 Structure of the Fleets and the Fishing Activity

We based our analysis of fishing activity on the segmentation created by the French Fishery Information System (SIH), which groups French vessels based on the main, or two main, gears used during the year (hereafter denominated IFR-fleet). The two most valuable species landed by French fleets in the EEC are common sole (*Solea solea*) and scallops (*Pecten maximus*). The majority of sole landings comes from netters and, to a more limited extent, from bottom or mixed trawlers. Scallops are mainly landed by dredgers. We focused on these four IFR-fleets, consisting of a total of 448 boats in average over 2008–2010, and pooled the others into an inexplicit fleet "OTHER." The rest of the value landed by these four fleets is mainly made on cephalopods, sea bass, whiting, red mullet, cod, and plaice (Fig. 1).

The cartography of fishing effort by fleet segment revealed that effort allocation depends largely on home region (north of France and Normandy) and vessel length class (<10 m, 10–12 m, 12–18 m, 18–24 m, 24–40 m, and >40 m). IFR-fleets were consequently further segmented according to length class and harbor (Fig. 2).

Some fleet segments contain less than three vessels which prevent access to economic data at the fleet scale. Possible grouping of fleet segments of approximate same size and displaying the same activity were investigated through a hierarchical ascendant classification on fishing time per métier (gear x statistical rectangle) within each fleet (IFR-fleets and harbor). The small segments that could not be grouped with others were not explicitly described and pooled with the "OTHER" fleet bringing the number of

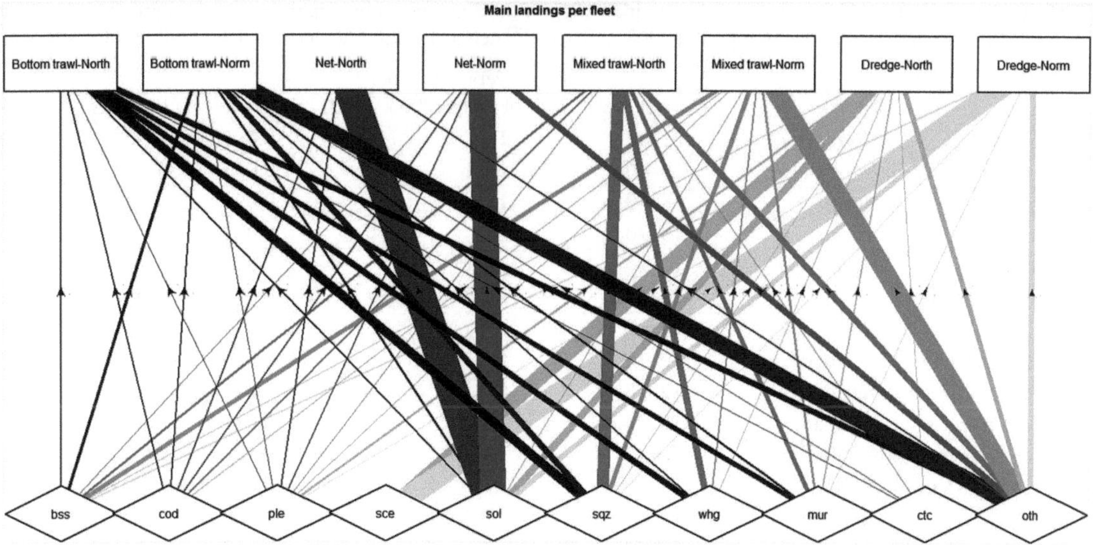

Fig. 1 Diagram flow of species importance in the revenues of the French fleets operating in the Channel. Arrow width is proportional to the proportion of the species in the revenues of the fleet. *bss* sea bass, *ple* plaice, *sce* scallops, *sol* sole, *sqz* squid, *whg* whiting, *mur* red mullet, *ctc* cuttlefish, *oth* other species, *norm* Normandy

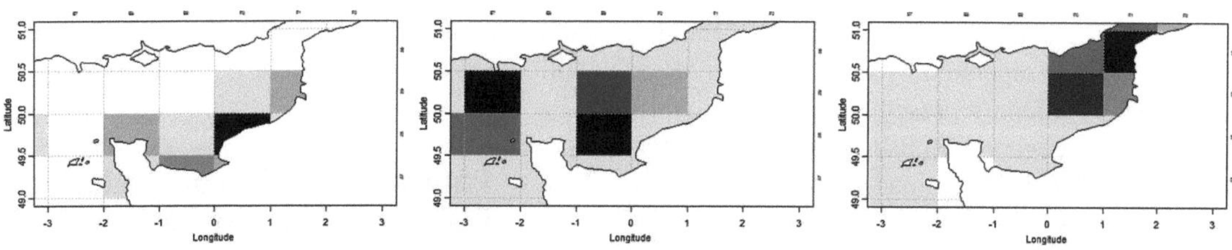

Fig. 2 Annual map of effort distribution for three different fleets when they operate métiers using bottom trawls (OTB). The figure shows how effort allocation in space depends on home harbor (*left* and *middle*, Normandy; *right*, north) and vessel length (*left*: <12 m; *middle* and *right*: 18–40 m) in relation with their range of operation

fleets considered to 17, hereafter referred to as "strategies" (Table 1).

2.2 Definition of Métiers and Strategies

The strategy of a fleet, that is, its activity along the year, is described by the set of métiers it practiced. Métiers are defined as the combination of a fishing gear, a fishing area, and when applicable a set of target species. The analysis of catch compositions did not allow the identification of target species a posteriori. It was therefore assumed that the choice of gear and area of practice reflects the choice of target species. However, depending on species distribution and availability, the relative amount of species targeted is not necessarily reflected in the catch composition. Within VIId, effort per gear per fleet is computed to identify the main gears used, and the others are pooled together in a métier OTH-VIId. For the purpose of the study, all the effort spent outside area VIId is considered a unique métier (OTH-OTH).

2.3 Identification of Fish Populations of Interest

Sole is the main species in value landed by French fleets in the Channel. It is managed by a TAC and a minimum landing size (MLS) of 24 cm. Plaice

Table 1 Strategies described in the model, with indication of the main gear, home harbor, vessel length class, average number of boats participating over the period 2008–2010, estimated technical efficiency, and availability of VMS data

IFR-fleet	Harbor	Vessel size	Average number of boats (2008–2010)	Technical efficiency	VMS
Bottom trawlers	Normandy	<12 m	16	0.081	No
Bottom trawlers	Normandy	18–40 m	17	0.013	Yes
Bottom trawlers	North	18–40 m	15	0.024	Yes
Mixed trawlers	Normandy	18–40 m	7	0.02	Yes
Mixed trawlers	North	18–24 m	11	0.021	Yes
Mixed trawlers	North	24–40 m	8	0.024	Yes
Dredgers–trawlers	Normandy	<10 m	9	0.054	No
Dredgers–trawlers	Normandy	10–12 m	50	0.056	No
Dredgers–trawlers	Normandy	12–18 m	103	0.041	Yes
Dredgers–trawlers	Normandy	18–24 m	5	0.04	Yes
Dredgers–trawlers	North	10–12	14	0.109	Yes
Dredgers–trawlers	North	12–18 m	9	0.034	Yes
Netters	Normandy	<10 m	22	0.009	No
Netters	Normandy	10–12 m	12	0.012	No
Netters	North	<10 m	7	0.009	No
Netters	North	10–12 m	51	0.04	No
Netters	North	12–18 m	12	0.027	Yes

is one of the main bycatches of sole, and its catches also are regulated by a TAC and an MLS of 27 cm. Scallops represent a highly valuable species for dredgers who fish it from November to April in five specific areas and then switch to bottom trawls. The fishery is extremely controlled through individual weekly maximum landings, MLS (110 mm), and a fishing closure during the spawning season. Red mullet is a relatively short-lived species, which is currently not subject to management. Red mullet has, however, become an important species for trawlers due to its high value, and new management measures are likely to be enforced in the next years. Cephalopods represent a large part of trawlers revenues. Squids and cuttlefish are fished sequentially; the main fishing season is October/November in the whole area and then June in coastal areas. Because cephalopods and red mullet are not subject to management, French trawlers have increasingly targeted these species when increased TAC constraints have severely capped the catches of cod and whiting. Among the other important species landed by French fleet, cod, whiting, and sea bass are widely distributed, and most of their life cycle and fisheries take place outside VIId. They were hence pooled together with the species of lesser economic importance, into species group "OTHER," which was not explicitly taken into account in the analysis.

3 Parameterization of the ISIS-Fish Model for the Eastern English Channel

ISIS-Fish is a modeling tool suitable for investigating the consequences of alternative policies on the dynamics of resources and fisheries (Pelletier et al 2009). This spatially explicit model is based on three sub-models: (1) a fishing activity dynamics model, (2) a population dynamics model, and (3) a management dynamics model. The three sub-models interact only if they overlap in space and time. In ISIS-Fish, fishing mortality is calculated from the distribution of fishing time over fish population areas, time, métiers, and fleets. The objective of the parameterization is thus to describe as accurately as possible the structuration of effort in space and time and over métiers and to predict its evolution under various environmental and management scenarios. Simulation times for the model increase with the complexity of the activity, thus making trade-offs necessary between accuracy and level of description.

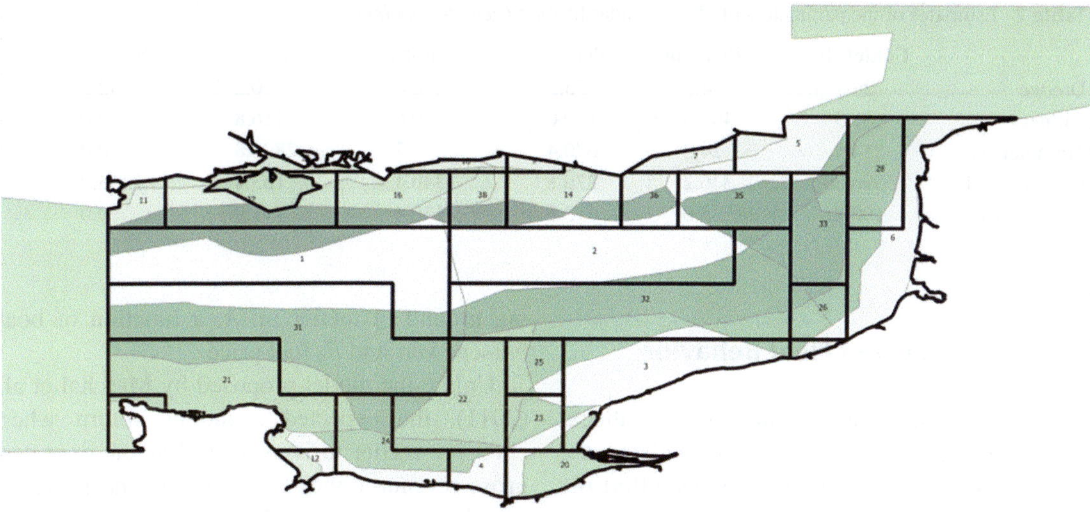

Fig. 3 Main habitats in the Eastern Channel as identified by Girardin (unpublished data) and corresponding ISIS-polygons (*black boxes*)

3.1 Spatial Resolution and Zones

To allow the description of spatial dynamics for a variety of species, population zones in ISIS-Fish are based on the habitat structure identified by Girardin (unpublished data) which consisted of 35 polygons. The polygons have been adjusted to match the regular spatial grid used in ISIS-Fish, the resolution of which is 0.25×0.25 degrees cells, resulting in 30 ISIS-polygons (Fig. 3).

Inside ICES Division VIId, two strategies were adopted to define "métier" and zones. For fleets monitored by VMS (vessels larger than 12 m), one "métier" per gear and ISIS-polygon is created to match closely species distribution. For the fleets of smaller vessels, logbooks helped identifying the main ICES-rectangles of practice for each "métier," and one "métier" per main rectangle is created (e.g., OTB-27E9, Fig. 2). ICES-rectangles with low effort for a given gear are pooled together in a unique métier (e.g., OTB-VIId).

3.2 Standardization of Effort

ISIS-Fish decomposes catchability into multiplicative effects related to fish accessibility and selectivity, gear efficiency, ability to specifically target a species, and technical efficiency. Technical efficiency is linked to vessel characteristics, mainly vessel length; it is therefore assumed unique within a strategy. We further assume that spatial differences in catch composition observed between métiers practiced with the same gear result from the heterogeneity in species distribution rather than from differences in fishing practices between areas. Selectivity curves are extracted from literature (Madsen et al. 1999) or derived from observer data, for each gear. Accessibility is assumed age dependent and calibrated on annual catch at age. Generalized linear models are used to assess these effects using logbook data transformed from catch in weight into catch in numbers. The model estimates a species-dependent effect of gear and an effect of the strategy as a proxy for technical efficiency. We use catch in numbers because ISIS-Fish applies the Baranov equation to abundance rather than biomass. Given the high frequency of occurrence of zeros in the dataset, individual trips were aggregated at the monthly scale, and a negative binomial distribution is used (log link). Results are presented in Tables 1 and 2:

$$C_{\text{month,year}}(\text{sp, gear, strategy}) \sim \text{gear} : \text{species} + \text{strategy} + \text{offset}(\text{Effort})$$

Table 2 Estimates of the parameters of effort standardization (*gear to species*)

	Cuttlefish	Red mullet	Plaice	Scallops	Sole	Squid
Dredge	29.7	4.3	55.5	15230.6	60.2	32.5
Gillnet	61.5	403.3	110.0	0.0	716.8	2.0
Trammel net	254.9	24.3	670.4	1.2	2898.8	0.0
Bottom trawl	590.8	346.2	279.8	340.4	133.4	1365.9
Beam trawl	190.8	8.0	613.3	1130.4	991.0	0.0

3.3 Fishing and Fleet Behavior

The strategy of a fleet is the pattern of effort allocation over métiers along the year. Initially, the model is set by computing the average effort per fleet, gear, and month based on logbooks for the period 2008–2010 and then further allocated spatially (either by ISIS-polygons or ICES-rectangles). The strategy is supposed to change in response to species availability, economic, and management constraints. Therefore, a gravity model is used to predict the changes in effort allocation over métiers within a strategy. Following Marchal et al. (2011), the proportion P of effort allocated to a métier is a linear combination of proxies for opportunism and traditional behavior and computed as follows:

$$P_{m,t} = \frac{\alpha\ \text{Opportunism}_{m,t}}{\sum_k \text{Opportunism}_{k,t}} + \frac{(1-\alpha)\text{Tradition}_{m,t}}{\sum_k \text{Tradition}_{k,t}}$$

$$\text{Opportunism}_{m,t} = \frac{\left[\left(\sum_s L_{s,m,t-1} * P_{s,t-1}\right) - C_{m,t-1}\right]}{\left(\sum_s L_{s,m,t-1} * E_{m,t-1}\right)}$$

$$C_{m,t} = \text{dist}_m * A * Pf_t$$

$$\text{Tradition}_m = \text{Effort}_{m,t-12}$$

with $L_{s,m,t-1}$, landings of species s by métier m in kg the previous month; $P_{s,t}$, price of species s at time t; $\text{Effort}_{m,t-12}$, fishing effort allocated to métier m the previous year; $C_{m,t-1}$, fuel costs for métier m; dist_m, distance from harbor to the fishing ground of métier m; A, a function of boat horsepower; and P_{ft} fuel price.

Unlike the model proposed by Marchal et al. (2011), the expected economic return when choosing métier m is expressed as the expected gross revenue divided by yield in the previous month minus a proxy for fuel price associated with the métier. The model thus assumes that fishermen try to limit landed quantities if they do not provide a significant rise in revenues. It is supposed to reflect their behavior in the case of the implementation of landing obligation. Fuel costs associated to traveling to fishing grounds are also taken into account to represent the incentive of fishing close to port in a context of increasing fuel price. Finally, it was chosen to use the previous month as a proxy for expected future conditions, instead of the previous year in Marchal et al. (2011) to represent the knowledge fishermen have of fish availability and current economic conditions.

3.4 Modeling Population Dynamics

The dynamics of sole, plaice, red mullet, scallops (4 populations), squid, and cuttlefish are described using age-structured models to match information available in surveys (scallops, red mullet, cephalopods) (Carpentier et al. 2009) and that derived from ICES stock assessments (sole, plaice) (ICES 2012). At each time step, fish can grow according to any chosen growth equation, recruit, migrate, and die from other causes than fishing or be caught (Table 3). The distribution of species in

Table 3 Main parameters of the dynamics of fish populations

	Age group number	Growth	Weight	Natural mortality	Spawning period	Juveniles	Migration pattern	Origin of spatial distribution	Price elasticity parameter
Sole	11 (age 1–11)	VBGF: *Linf*=37.7; *K*=0.3; *T0*=−0.84	*a*=4.9e-07; *b*=3.79	0.1 ICES (2012)	February–June	Juveniles quit nurseries at age 3	Assumed annual redistribution in February for spawning	BTS survey, July	−0.24 (2.8e-3) Log–log
Plaice	7 (age 1–7)	VBGF: *Linf*=71.65; *K*=0.23; *T0*=−0.83	*a*=3.46e-5; *b*=2.45	0.1 ICES (2012)	December–March	Juveniles quit nurseries at age 3	Dec.: to spawning area; Apr.: to coastal areas	CGFS October; Coull et al. (1998)	−2.6e-6 (8.8e-8) linear
Red mullet	4 (age 1–4)	VBGF: *Linf*=51.35; *K*=0.186; *t0*=−1.21	*a*=3.28e-6; *b*=3.24	–	May–July	Juveniles quit coastal areas at age 1		CGFS October;	−4.1e06 (8.4e-8) semi-log
Scallops (all populations)		VBGF Antoine (1979)	*a*=2.74e-7; *b*=2.9	0.1	July	Maturity at age 2	na	COMOR survey	
Scallops Baie de Seine (coast)	5 (age 2–6)	*Linf*=139.23; *K*=0.569; *T0*=0.527							Month effect only
Scallops Baie de Seine (out)	5 (age 2–6)	*Linf*=139; *K*=0.475; *T0*=0.494							−4.1e-8 (2.7e-8) linear
Scallops Beachy Head	4 (age 3–6)	*Linf*=129; *K*=0.484; *T0*=0.648							−3.1e-6 (1.0e-6) linear

(continued)

Table 3 (continued)

	Age group number	Growth	Weight	Natural mortality	Spawning period	Juveniles	Migration pattern	Origin of spatial distribution	Price elasticity parameter
Scallops Vergoyer	4 (age 3–6)	*Linf*=122.45 *K*=0.495 T0=0.746							−0.099 (0.009) log–log
Scallops Dieppe	4 (age 3–6)	*Linf*=122.45 *K*=0.495 *T0*=0.746							−4.3e-7(1.7e-8) linear
Squid	1	na	Mean=0.077 Ifremer (2014)	Adults die after breeding	End of autumn			BTS and CGFS surveys	−8.6e-6 (1e-7) linear
Cuttlefish	1 (age 1)	*LDinf*=30.5 *k*=1.25 *t0*=n/a	Mean=0.26 Ifremer (2014)	Adults die after breeding	Spring		Winter migration offshore, inaccessible to fishing	BTS and CGFS surveys	Month effect only

VBGF Von Bertalanffy Growth Function. Except otherwise specified, biological parameters are from Carpentier et al. (2009)

ISIS-polygons is derived from scientific surveys (UK Bottom Trawl Survey for sole, Cefas; France Channel Ground Fish Survey for plaice and red mullet, IFREMER; France COMOR (Coquille Manche Orientale) survey for scallops, IFREMER). It is assumed that fish distribution is homogeneous within ISIS-polygons and that local depletion occurs between migration episodes.

A price equation is estimated for each population based on sales slip, assuming a month effect and a negative relationship with landings at the monthly scale. The shape of the relationship (linear or log–log) is chosen according to the value of the coefficient of determination of the model.

4 Conclusions and Perspectives

The modeling exercise allows the integration of information from different sources (logbooks, VMS, and survey data, literature) on fish populations and fleets in the Eastern English Channel at a fine spatial and temporal scale. The model as set describes spatial and temporal heterogeneities in population and effort distribution and allows the prediction of their likely evolution under the influence of fishing, economic, and management pressures. Management scenarios, alternative to current management, will be investigated, including management plans building in harvest control rules, and discard limitations. Management objectives focus on sole with its common bycatch species plaice and on red mullet as an example of a data-limited species. Particular attention will be paid to the impacts of single-species management measures on the dynamics of all species.

References

Antoine L (1979) La croissance de la coquille Saint-Jacques Pecten maximus (L) et ses variations en mer Celtique et en Manche. PhD thesis, Université de Bretagne Occidentale. http://archimer.ifremer.fr/doc/00005/11673/

Apostolaki P, Hillary R (2009) Harvest control rules in the context of fishery-independent management of fish stocks. Aquat Living Resour 22:217–224

Butterworth DS, Johnston SJ, Brandão A (2010) Pretesting the likely efficacy of suggested management approaches to data-poor fisheries. Mar Coast Fish Dynam Manage Ecosyst Sci 2(1):131 145. doi:10.1577/C08-038.1

Carpentier A, Martin CS, Vaz S (eds) (2009) Channel habitat Atlas for marine resource management, final report/Atlas des habitats des ressources marines de la Manche orientale, rapport final (CHARM phase II). INTERREG 3a Programme, IFREMER, Boulogne-sur-mer, France, 626 pp

Coull KA, Johnstone R, Rogers SI (1998) Fisheries sensitivity maps in British waters. Published and distributed by UKOOA Ltd

ICES (2012) Report of the working group on the assessment of demersal stocks in the North sea and Skagerrak (WGNSSK) ICES CM 2012/ACOM:13

Ifremer (2014) Indices de populations et de communautés issus des campagnes de surveillance halieutique de l'Ifremer. http://www.ifremer.fr/SIH-indices-campagnes. 25 avril

Madsen N, Holst R, Wileman D, Moth-Poulsen T (1999) Size selectivity of sole gill nets fished in the North Sea. Fish Res 44(1):59–73

Marchal P, Little RL, Thébaud O (2011) Quota allocation in mixed fisheries: a bioeconomic modeling approach applied to the Channel flatfish fisheries ICES. J Mar Sci 68(7):1580–1591. doi:10.1093/icesjms/fsr096

Pelletier D, Mahevas S, Drouineau H, Vermard Y, Thebaud O, Guyader O, Poussin B (2009) Evaluation of the bioeconomic sustainability of multispecies multi-fleet fisheries under a wide range of policy options using ISIS-fish. Ecol Modell 220:1013–1033. doi:10.1016/j.ecolmodel.2009.01.007

The Estuary-Bay of Seine Continuum: The Need for an Ecosystem-Based Management

Jean-Claude Dauvin

Abstract

Estuaries and coastal zones reflect conflicting processes due to their positions at the interface between terrestrial and marine environments. The estuary of the Seine is an example of such a continuum and is affected by numerous human activities. Although degradation started in the middle of nineteenth century, the Seine estuary remains an important area for biodiversity as well for the food web involving fish and birds. The estuary area has been affected by intense historic anthropogenic pressures and, more recently, by new human activities, including harbour extension, aggregate extraction and the development of offshore wind farms. Nowadays, harbour projects continue to highlight the existing ecological compartments and are still studied independently. French and European regulations, including reforms of the harbour status, should ensure a better integration of natural heritage protection into the integrated coastal zone management process. However, there is a paradox between the declaration of the French state, which encourages a global management plan for the Seine Estuary, and sectoral territorial approaches adopted by the managers. This paper presents an analysis of the current contrasted situation in the administration of territories, which favours sectoral management rather than a unique ecosystem-based management process.

1 Introduction

The Seine estuary represents a territory associated with dense activities, but which remains an area of exceptional natural heritage, corresponding to a space of numerous conflicts crystallising many different interests: industrial and harbour activities, fishing, aggregate extraction, tourism and agriculture. Multiple actors are concerned: the French state, the regions of Lower and Upper

J.-C. Dauvin (✉)
Normandie Université, UNICAEN, Laboratoire Morphodynamique Continentale et Côtière, UMR CNRS 6143 M2C, 24 rue des Tilleuls, F-14000 Caen, France
e-mail: jean-claude.dauvin@unicaen.fr

H.-J. Ceccaldi et al. (eds.), *Marine Productivity: Perturbations and Resilience of Socio-ecosystems*,
DOI 10.1007/978-3-319-13878-7_21, © Springer International Publishing Switzerland 2015

Normandy, the departments of Calvados, Eure and Seine-Maritime, the 'Conservatoire des Espaces Littoraux et des Rivages Lacustres', 'Grand Port Maritime du Havre' and 'Grand Port Maritime de Rouen', municipalities and associations of local authorities or urban areas such as Le Havre and Rouen (to develop their touristic and urban capacity), as well as associations of ecologists, farmers, manufacturers, hunters, fishermen, etc. The various users are in a situation of competition, each trying to promote/defend particular regulations and appropriate the estuarine territory (Lozachmeur and Dauvin 2008). First of all, the actors do not share a common perception of the estuarine territory, and the accumulation of regional, national and European laws, as well as directives and territory management approaches, does not favour the promotion of an overall vision of the estuary. As a result, estuarine organisation is currently understood in terms of sector-based management. The present study aims to analyse the contrast between this accumulation of sectoral management approaches and a unique ecosystem-based management process.

2 Main Characteristics of the Seine Estuary

2.1 General Features

The Seine estuary was formed at the end of the Tertiary, on a rocky substratum of Mesozoic age; it extends from the Poses weir, 160 km upstream of Le Havre, at the upstream limit of tidal penetration into the estuary, to the eastern part of the Bay of Seine (Fig. 1). The estuary can be divided into three main sections: the fluvial (or upstream) estuary, the middle estuary and the marine (or downstream) estuary. The freshwater discharge of the river is relatively small (400 $m^3.s^{-1}$), with flood stage > 2,200 $m^3.s^{-1}$ (autumn/winter) and low flow stage < 100 $m^3.s^{-1}$ (at the end of the summer). The Seine estuary is situated on the coast of the eastern basin of the English Channel, and its hydrodynamics is strongly influenced by the tide. The tidal range can reach 7 m during spring tides in the downstream part of the estuary, where a maximum turbidity zone acts as a

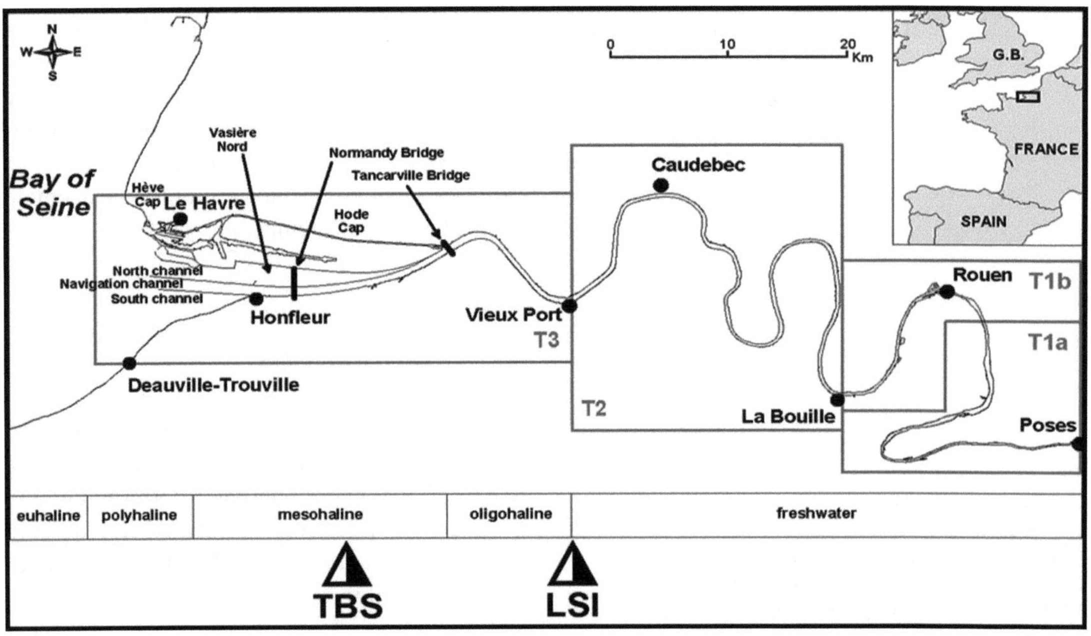

Fig. 1 Location of the Seine estuary and main sites mentioned in the text. *TBS* transverse border of the sea, *LSI* limit of saltwater intrusion. *T1*, *T2* and *T2* waterbodies defined for the implementation of the Water Framework Directive

physicochemical regulator for numerous natural chemical species as well as pollutants, notably metals. But, due to the tide, there is no anoxic zone in the downstream estuary (Guézennec et al. 1999).

The Seine valley is of major economic importance for France, notably due to the presence of two maritime harbours, Rouen and Le Havre. The Seine estuary lies at the discharge point of a catchment's area covering 78,000 km^2. This area is inhabited by 16 million people (>25 % of the French population, with Paris the French capital and two major cities, Rouen and Le Havre) and accounts for 50 % of the river traffic in France, 40 % of the country's economic activity and 30 % of the agricultural activities.

Anthropogenic influences in the Seine estuary began during the early mid nineteenth century and continue up to the present day (Guézennec et al. 1999; Dauvin 2006, 2009; Fisson 2014). The estuary's ecosystems became more fragile as a result of cumulative human activities, mainly due to harbour and industrial development (dredging and disposal of sediment dredged from the harbours, embankment works and construction of dykes and creation of industrial development areas). These activities have led to the extreme compartmentalisation of the units, a reduction of the intertidal zone (loss of more than 100 km^2 between 1850 and the present day), the downstream migration of the maximal turbidity zone and the increase of muddy deposits in the subtidal areas of the Bay of Seine (Guézennec et al. 1999). At the same time, the physicochemical conditions of the estuarine environment have continually deteriorated over more than a century. By the 1980s, the estuary had become highly contaminated: levels of metal pollutants (e.g. Cd, Hg), hydrocarbons (PAHs) and polychlorinated biphenyls (PCBs) were among the highest recorded worldwide. Moreover, insufficient or inadequate water treatment facilities created an oxygen deficit downstream of Paris and Rouen (Abarnou et al. 2000; Fisson 2014). In the marine estuary, abnormal biological functioning led to the collapse of the fisheries sector, particularly in the case of the brown shrimp Crangon (declining from 1,000 t in 1970s to < 100 t since the beginning of the 1990s), while, in the fluvial estuary, professional fishing ceased entirely in the 1970s due to the almost total disappearance of migratory fish species (Dauvin 2002). Nevertheless, we can nowadays observe a significant decrease in metal contamination, an improvement in oxygenation and a reduction of the phosphorus and ammonia fluxes (Fisson 2014). Despite this, some chemical contaminants persist such as PAHs, PCBs and pesticides, while new questions are arising on emerging contaminants such as phthalates and medicinal residues (Fisson 2014).

2.2 Multiple Perceptions of the Seine Estuary

Because estuaries have no true specific legal status, there are multiple perceptions of the continuum between the estuary and the Bay of Seine. Hence, the Seine estuary is characterised in many different ways (Desroy et al. 2004; Dauvin and Ruellet 2009): for geographers and physicists, the estuary extends from Poses to the sea, while biologists distinguish several zones according to the salinity (ranging from the freshwater zone to oligo-, meso-, poly- and finally euhaline zones); three transitional water bodies are defined by the Seine-Normandy Water Agency for the implementation of the Water Framework Directive; the Territorial Directive Plan for the French government takes into account a wide area including a large part of Lower Normandy and the Seine valley. This area covers a very extensive territory including the Paris region and a large part of historical Normandy from Dieppe in the east to Cherbourg in the west. In the communes bordering the Seine estuary, the association of elected officials only considers the lower part of the estuary. On the other hand, the Upper Normandy Regional Council (Rouen), like the geographers and the French government, considers that the estuary extends up to Poses. Estuary managers are interested in the part of the estuary that borders their locality, while, finally, legislators define the estuary as the waters between the transverse border of the sea (TBS) [separation of the maritime public domain (MPD) and the fluvial public domain (FPD)] and the limit of saltwater intrusion (LSI) [boundary between the maritime and freshwater fisheries] (Desroy et al. 2004).

2.3 Accumulation of Protective Measures

Nonetheless, despite the diverse negative environmental impacts, the Seine estuary is still a highly favourable habitat for the juveniles of numerous fish species, and the ornithological richness is one of the main positive characteristics of its natural heritage (Dauvin 2002). The richness of this natural heritage can be judged from the overabundance of regulatory measures and inventories that have been developed over the years. But this 'overprotection' is more apparent than real. In fact, out of a total of nine statuses accorded for protection and knowledge of the environment, only three appear essential: the national natural reserves, the regional nature park '*Boucles de la Seine Normande*', and the territorial acquisitions of the coastline and lakeshore conservancy (Conservatoire des Espaces Littoraux et des Rivages Lacustres, CELRL). These statuses correspond to more or less extended zones with their own management and governance (Alard 2002). Due to this organisation, there are six main scientific committees with different limits of competence: the scientific committee of the 'Groupement d'Intérêt Public Seine Aval' (GIP-SA), the scientific committee of the National Natural Reserve of the Seine Estuary, the scientific committee of the regional natural park, the scientific committee of the natural heritage of the Lower Normandy region and Upper Normandy region and the scientific committee of the Seine estuary (Lozachmeur and Dauvin 2008). No institutional links exist between these committees.

3 The Port 2000 Project and the Emergence of a Global Plan of the Seine Estuary Management

3.1 Port 2000

On 5 December 1998, the French Government decided to take into consideration the Le Havre harbour expansion project called 'Port 2000'. The decree stated that, since the development of Le Havre harbour would create jobs, the project was important for the French economy as a whole and for the economy of the lower Seine valley in particular. Nonetheless, the French state recognised the importance of the balance between the economic development objectives connected with Port 2000 and the environmental objectives connected with the protection of the natural heritage (Hamm et al. 2001). The Port 2000 extension (2000–2005) and its compensatory actions not only contributed to morpho-sedimentary modifications of the estuary but also changed the perception of the actors as regards the vulnerability of the Seine estuary and the need to promote a global plan for its management (Hamm et al. 2001; Revue Travaux 2006).

The construction of Port 2000 was marked by the importance of the public debate (the first of its type in France), which can be considered as giving an opportunity to introduce a global environmental discussion on the whole estuary from the Poses weir to the sea. This project also led to the setting up of the council of the Estuary, whose aim was to develop a basis for an improved vision of the sustainable development of this estuarine territory (Lozachmeur and Dauvin 2008). The monitoring committee provided a venue for this debate.

The Seine estuary nature reserve was initially created on 31 December 1997 covering a territory of 3,800 ha. However, at the request of the European Commission, this was extended by the Decree No 2004-1187 of 9 November 2004 to an overall surface area of 8,528 ha.

The main milestones of the Port 2000 project were, September 1995, start of the project; end of 1997, public hearing; winter 1999, creation of an expert committee; July 2001, start of works on Port 2000; February 2002, start of rehabilitation works; July 2005, end of the dyke works; and April 2006, first container ship arrives in Port 2000.

3.2 The Years 2000: An Optimistic Decade for an Ecosystem-Based Management

At the beginning of the 2000s, there was an effective governance of the Seine estuary (Lozachmeur and Dauvin 2008). The Estuary Council (about 10 members), chaired by the prefect of the Upper Normandy region, has the decisional power to implement a global management plan for the estuary. The monitoring committee (over 100 members, including associations of ecologists) provides an arena for debate. The Estuary Council has a scientific and technical committee to provide scientific and technical expertise. The operational part of the Seine-Aval programme was reinforced in 2003 by the creation of the GIP-SA, which coordinates the Seine-Aval programme, transfers the scientific results to decision makers and policymakers, develops research applications, helps in the use of models and communicates the scientific results and their applications to a wide public (Dauvin 2009). The GIP-SA began in 1995 and was first renewed in 2007 and then again in 2012 until 2020. (http://seine-aval.crihan.fr).

In 2004, the Estuary Council asked the Seine-Normandy Water Agency to coordinate a forecast study on the perspectives for restoring the environmental functions of the Seine estuary by 2025, taking into account all the Seine Estuary actors. The analysis concerned the socioeconomic and environmental systems of the Seine estuary by considering two main points of convergence (Dauvin 2006):

- A shared awareness of the geographic dimension of the estuary, based on the understanding that the upstream/downstream and north/south relationships impose forms of solidarity that should be discussed and acted upon.
- A recognition of the interdependence of the local stakeholders, given the influence of external factors that play a role in the long-term status of estuarine functions. These factors depend not only on the stakeholders in the Seine catchments area (notably regarding the quality of waters entering the system) but also the national, European and international context (Dauvin 2006).

Four scenarios correspond to four key strategic questions: (1) what would be the consequence for environmental functions if present trends continue, (2) can environmental functions be restored to their 1980 level and under what conditions, (3) what conditions will lead to a catastrophic scenario, and what would be its consequences and (4) what are the limits of local resolve and local means? The following scenarios were analysed by (Dauvin 2006):

- The dynamic trend scenario, taking into account both the socioeconomic and political context as well as the behaviour and interactions of the local stakeholders.
- The voluntarist restoration scenario, requiring the successful completion of a major ecological engineering project.
- The economic crisis scenario, in which a major economic crisis hits the region affecting the activities of the Rouen Port Authority, clearly demonstrating the negative consequences on environmental and socioeconomic functions, and (4) the local initiative scenario, which involves working together to create a global collective project for the estuary and bringing together all the resources necessary for its successful execution. This is the only scenario that has any hope of restoring environmental functions.

In the second stage of the process, the study needed to be continued to determine the monetary cost of scenarios 2 and 4. But this exercise remained confidential and lacked any real financial propositions from the stakeholders to act in favour of a global management of the estuary.

The GIP-SA and the city of Le Havre responded to the 2005 call for projects related to integrated coastal zone management launched by the French state. The objectives were to develop a knowledge-management and decision-making tool that would federate the different data providers and develop a global management plan, focusing on estuarine habitat restoration (Lozachmeur and Dauvin 2008). An analysis of the mode of governance was carried out to develop a tool for data centralisation and decision-making support.[1]

The Grenelle forum on the Seine estuary (2008–2009), organised by the association of elected officials from the communes on both sides of the downstream part of the estuary, gave rise to a wide debate among all the actors of the territory. This forum was a local manifestation of the national Grenelle approach concerning French policy on the environment, which allowed us to identify the specificities of the territory and the issues at stake. The Grenelle forum on the Seine estuary organised more than 15 workshops to share knowledge and discuss the future of the estuary and gave additional food for thought on the future challenges and governance of the territory.[2]

So, by the end of the year 2000, there was an accumulation of projects and studies that took stock of the high level of scientific knowledge on the Seine estuary and its degradation, as well as the multiplication of regulations on a complex transitional territory between landscape and seascape, involving mainly the natural heritage, users, stakeholders, policymakers, etc. All of the actors agreed with the need to establish a shared global management of the estuary.

4 The 2010s: A Time of Disappointment

After a decade of optimistic and integrative shared approaches towards an ecosystem-based management, the end of the years 2000 and the early 2010 saw a period of disillusionment and self-interest on the part of managers of natural heritage zones. This was probably due to external factors such as the international economic crisis of 2008 and the reduction of harbour activities but also the emergence of new French regulations, new sectoral management plans and a new vision of the French state for the Seine valley territory, as well as the appearance of new actors in the eastern part of the Bay of Seine in relation to the disposal of dredged sediment, the implementation of offshore wind farms and new aggregate extraction areas.

The Law No 2008-660 of 4 July 2008 concerning harbour reform gave increased responsibilities on environmental space management to the harbours at Le Havre and Rouen (each of which became a 'grand port maritime'). The article L. 101-3.-I. stipulated that, within their territory limit, harbours should integrate sustainable development in the management and conservation of the marine public domain and the natural areas of which they are owners or which are allocated to harbours.

Following the Law of 4 July 2008, the scientific committee of the Seine estuary was created in the place of the scientific and technical committee; its role is to have an overview on all questions concerning the natural heritage and management of the estuary. Both the Le Havre and Rouen harbour authorities manage their own territories, especially their natural heritage. Both harbours (Le Havre and Rouen) are responsible for drafting their own natural heritage management plan and setting up separate scientific and technical committees. Every 5 years, both harbours submit a strategic project (for 2009–2013, and then 2014–2019) with an environmental section, presented in two separate documents, called the 'Plan de Gestion des Espaces naturels, PGEN' for the Rouen harbour and 'Schéma de Développement du Port et de la nature, SDPN' for the Le Havre harbour. Figure 2 illustrates the complexity of drawing up the SDPN for Le Havre Harbour in compliance with all the French and European regulations. Nevertheless, the three main harbours of the Seine valley (Paris, Rouen and Le Havre) have created a single economic entity known as HAROPA to ensure a better international competition with other major northern European harbours. But, so far, HAROPA has not developed an integrated vision of a single

[1] http://www.aurh.fr/; http://seine-aval.crihan.fr

[2] http://www.aurh.fr/animation/le-grenelle-de-l-estuaire/

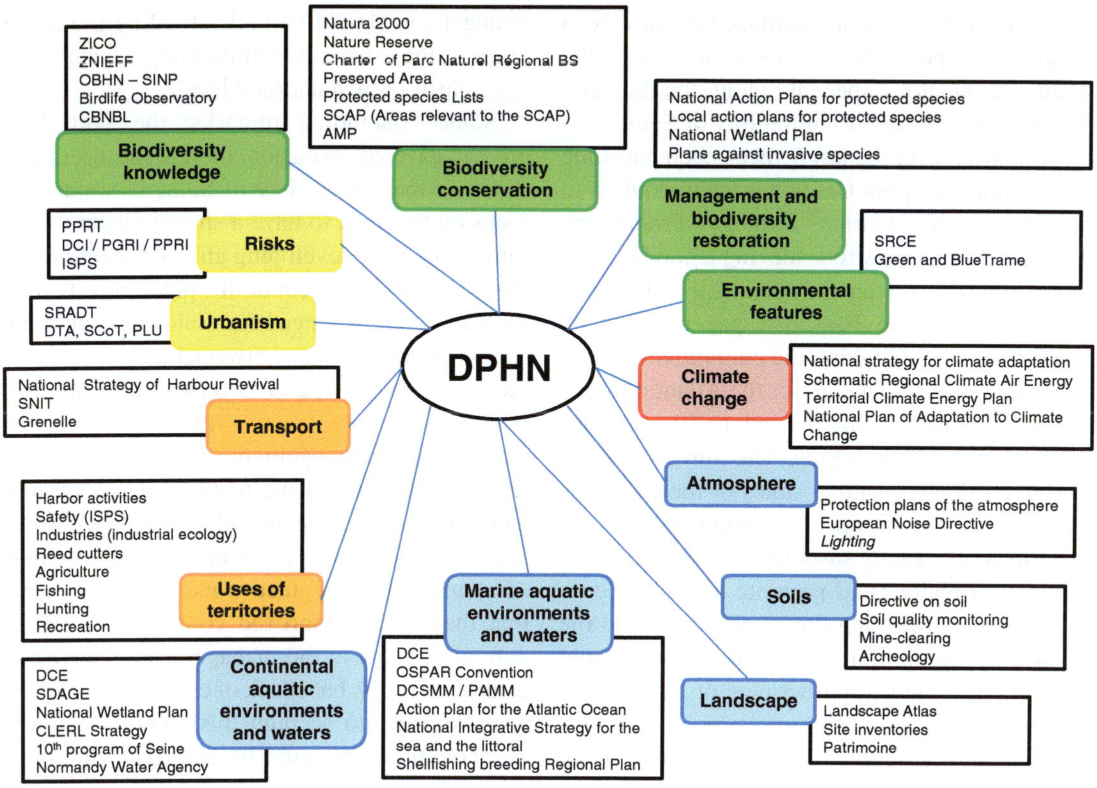

Fig. 2 Development perspective for the harbour and natural heritage (DPHN) of Le Havre harbour: links between French and European regulations and strategies (from Le Havre harbour, personal communication 2014)

management plan for the three harbour territories (Le Havre, Rouen, Paris).

The publication of Decree No 0095 of 23 April 2013 makes provision for the creation of an inter-ministerial delegation for the development of the Seine valley, placed under the authority of the Prime Minister. The piloting of the governance of the Seine valley (a large territory extending from Cherbourg in the west to Dieppe in the east and including the Paris region in the upstream part of the Seine catchments) is placed clearly under the French state and the three Regions (Ile-de-France, Upper and Lower Normandy) with an executive committee. The prefect of Upper Normandy was designated as the coordinator for a duration of 5 years. This committee drew up a strategic plan for the management and the economic development of the Seine valley, without any particular attribution of its environmental management. Hence, this recent decree defines a new perimeter for the Seine valley, as well as a further new actor in the context of a plethora of participants in the management of the Seine estuary. There is no information given about the links with the other existing administrative, scientific or technical structures such as the scientific committee of the Seine estuary.

Recently, new development projects have involved the search for an alternative area for the disposal of sediment dredged from Rouen harbour outside the Seine estuary in the eastern part of the Bay of Seine (Marmin et al. 2014), the implementation of offshore wind farms off Courseulles-sur-Mer in the central part of the Bay of Seine and a new prospecting area for aggregate extraction to the north of the Le Havre harbour navigation channel. Each of these activities has produced independent impact studies, and separate scientific and technical committees and working groups have been created. Supplementary working groups on natural heritage have also been created to support the management plan of the Natura 2000 sites, especially along the Augeron shore in the eastern part of the Bay of Seine.

Moreover, over the last 5 years, each maritime state of the European Union has been required to implement the Marine Strategy Framework Directive with a view to achieving a good quality status for the marine environment. This new European directive had also generated many working groups and scientific committees. At the same time, the action plan for the Atlantic Ocean aims to revitalise the marine and maritime economies in the Atlantic region. The aim is to show how the Atlantic member states of the EU, their regions and the European Commission can all contribute to creating sustainable development in the coastal regions and promote the 'blue economy', while protecting the environment and the ecological equilibrium between economy and protection of the natural heritage of the Atlantic Ocean.

Hope for the future! Article 67 of the French Law on biodiversity presented in the Council of Ministers on 26 March 2014 authorises the government to take provisional measures that will allow the carrying out of experimentation on the management of protected natural areas where there is an accumulation of protective measures, as in the case of the Seine estuary. There is a lack of clarity as well as a failure in the pooling resources and synergy in the management of these protected territories, while diverse protected natural spaces may overlap with each other on the same territory. An experiment is planned that aims to simplify the management of these protected natural spaces, in cases where they overlap on the same territory. Running for a duration of 4 years, the experimentation would allow the testing of three levels of simplification in the management of these overlapping protected natural spaces: (1) drafting of a unique summary document giving the main guidelines, (2) federation of the different advisory bodies for the spaces concerned to form a single scientific and technical consultative authority, and (3) designation of a single administrator for all the spaces concerned. There is no doubt that this new legislative arrangement should be understood by the actors of the Seine estuary as representing a benchmark example at French national level.

Some concluding remarks: the Port 2000 project gave an occasion for all the users and actors of the estuary-bay of Seine continuum to focus on the need to have a shared vision for the future, aimed at developing an integrated tool for the global management of this precious estuarine territory: there is a need for dialogue, communication, mutual training between partners and promotion of a new type of governance. From 2010 onwards, a new era began with the persistence of sector-based management approaches and a dramatic increase in the number of workgroups and scientific and technical committees which take local sectors into account in the management of a multitude of protective measures. It is clear that there is an improved consideration of the natural heritage in the management of the Seine estuary territory, but a lack of expertise at a global scale taking into account this complex estuary-bay continuum affected by anthropogenic pressures. At present, it is evident that numerous difficulties remain with the governance of the estuary due to the emergence of new laws, decrees and directives at European and French national levels. We can note the absence of a global estuary management plan in spite of repeated commitments by the French state (as recently expressed in a letter from the Upper Normandy prefect to the scientific committee of the Seine estuary on 5 May 2014).

However, for the future, it will be necessary to envisage an innovative capacity for the governance and to have a real leader involved in the development of the Seine estuary and bay. We also need to establish an ecosystem-based management for this complex territory considered on a global scale and not sector by sector, while taking into account the cumulative degradation of the Seine estuary over more than century by multiple human activities.

Acknowledgements Thanks are due to all our Seine-Aval colleagues, to A. Baffreau and S. Marmin for her help with the figures and M.S.N. Carpenter for correcting the English grammar, syntax and style.

References

Abarnou A, Burgeot T, Chevreuil M, Leboulenger F, Loizeau V, Madoulet- Jaouen A, Minier C (2000) Les contaminants organiques: quels risques pour le monde vivant ? Fascicule Seine-Aval 13:1–35

Alard D (2002) Zones humides de la basse vallée de la Seine. Programme Scientifique Seine Aval, édition Ifremer 15:1–36

Dauvin JC (coord.) (2002) Patrimoine biologique et chaînes trophiques. Programme Scientifique Seine Aval, édition Ifremer 7:1–48

Dauvin JC (2006) Estuaires Nord-Atlantiques: problèmes et perspectives. Bulletin spécial Seine-Aval, 80 pp

Dauvin JC (2009) Le programme Seine-Aval 3. Contexte, bilan et enjeux. Fascicule Seine-Aval 2(1):1–56

Dauvin JC, Ruellet T (2009) The estuarine quality paradox: is it possible to define an ecological quality status for specific modified and naturally stressed estuarine ecosystems? Mar Pollut Bull 59:38–47

Desroy N, Baron M, Payen C, Dauvin JC (2004) L'estuaire de la Seine: zone de conflit entre développement économique et préservation de l'environnement. Oceanis 30:71–93

Fisson C (2014) L'estuaire de la Seine: état de santé et evolution. Fascicule Seine-Aval 3(1):1–4

Guézennec L, Romana LA, Goujon R, Meyer R (1999) Seine-Aval: un estuaire et ses problèmes. Fascicule Seine-Aval 1:1–29

Hamm L, Romana LA, Lerat F (2001) Maintien des fonctionnalités biologiques de la vasière nord de l'estuaire de la Seine. Actes de Colloques Ifremer 29:157–167

Lozachmeur O, Dauvin JC (2008) Réflexions sur la restauration et la gouvernance de l'estuaire de la Seine dans une perspective de gestion intégrée des zones côtières. Rapport GIZC au GIP Seine-Aval, 61 pp

Marmin S, Dauvin JC, Lesueur P (2014) Collaborative approach for harbour dredged sediment management in the Bay of Seine (France). Ocean and Coastal Management 102:328–339

Revue Travaux (2006) n°828, Le Havre, Port 2000, Editions Science et Industrie

Use of Bio-fluorescent Characteristics for Ecosystem Monitoring on Hydrothermal Deposits

Masahiko Sasano, Yasuharu Nakajima, Joji Yamamoto, and Yasuo Furushima

Abstract
A new technology is necessary for effective monitoring of deep-sea hydrothermal ecosystems towards the environmental impact assessment of marine mining projects. For this purpose, we develop a new deep-sea observation technique using bio-fluorescent characteristics. Ultraviolet (UV)-LED illumination and violet laser illumination are used for fluorescent video recording of a variety of deep-sea organisms. This study reports the bio-fluorescent patterns and colours observed for a variety of deep-sea organisms and their usefulness for monitoring.

1 Introduction

Anthropogenic pressures on the marine ecosystem are, to date, mostly limited to coastal zones. Landfill, urban effluent, overfishing, oil spills, etc. have most effect on coastal areas. Global climate change is expected to affect all of the Earth's ecosystems, but the greatest threats are predicted for coastal ecosystems such as coral reefs and Arctic seashores (IPCC report 2007).

Since the deep submergence vehicle (DSV) Alvin discovered the first hydrothermal vent on the Galapagos Rift in 1977 (Weiss et al. 1977), hydrothermal vents have been reported in many deep-sea areas. The biological community around the hydrothermal vent is remarkable, because the ecosystem functions independent of solar energy and has a unique ecological structure based on chemosynthetic bacteria.

In recent years, anthropogenic impacts have extended to deep-sea areas. In the hydrothermal area, hydrothermal deposits are formed by sulphides and heavy-metal components settled out from hydrothermal plumes, which are expected to provide viable ore resources for gold, silver, copper, lead, zinc and other heavy metals.

There is an urgent need for environmental impact assessment of deep-sea mining projects and for technical improvements to enable the necessary monitoring of deep-sea ecosystems (Fig. 1).

M. Sasano (✉) • Y. Nakajima • J. Yamamoto
National Maritime Research Institute,
Mitaka, Tokyo, Japan
e-mail: sasano@nmri.go.jp; nakajima@nmri.go.jp; yamamoto-j@nmri.go.jp

Y. Furushima
Japan Agency for Marine-Earth Science and Technology, Yokosuka, Kanagawa, Japan
e-mail: furus@jamstec.go.jp

H.-J. Ceccaldi et al. (eds.), *Marine Productivity: Perturbations and Resilience of Socio-ecosystems*,
DOI 10.1007/978-3-319-13878-7_22, © Springer International Publishing Switzerland 2015

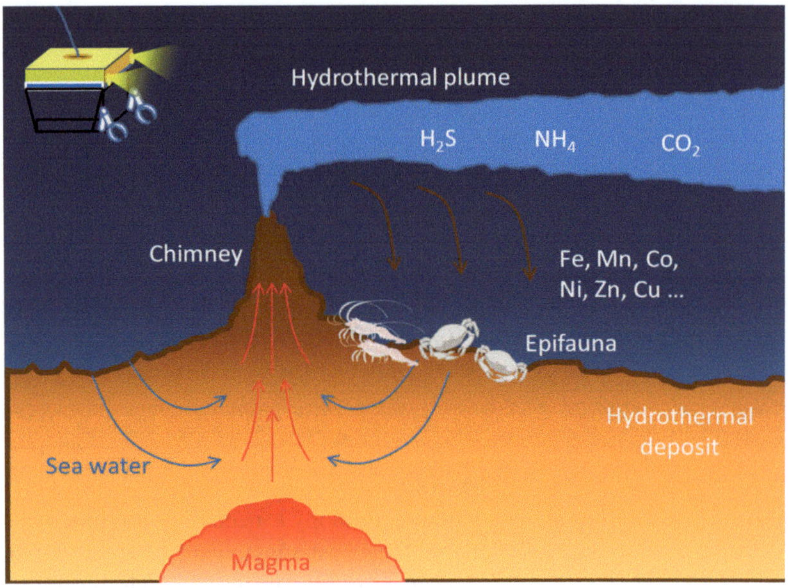

Fig. 1 Conceptual diagram of hydrothermal vent, deposit and deep-sea ecosystem

2 Mining of Hydrothermal Deposits

2.1 Japanese Seafloor Mining Project

In contrast to offshore oil fields, the resources from hydrothermal deposits are solid and heavier than seawater. In this case, the major technical challenge is not only drilling but also rifting. To date, a range of technical challenges has prevented the commercialization of deep-sea mining, but it has recently become regarded as viable, along with the development of deep-sea exploration and drilling techniques.

Within the Japanese exclusive economic zone (EEZ), identified hydrothermal deposits are distributed between depths of 500–2,000 m, which is a relatively shallower zone for hydrothermal deposits and therefore advantageous for mining. In Japan, research into the understanding of hydrothermal vents and hydrothermal deposits is mainly conducted by Japan Agency for Marine-Earth Science and Technology (JAMSTEC). They operate the "Artificial Hydrothermal Vents" project in the Iheya North field in Okinawa Trough (Takai et al. 2013). The technical development of marine mining is mainly conducted by Japan Oil, Gas and Metals National Corporation (JOGMEC). They operate test mining of the hydrothermal deposits at Izena Hole in Okinawa Trough in 2012. The Japanese government proposes to begin commercialised mining within the next 10 years or so (The final report of the first term in development program of hydrothermal deposits 2013).

2.2 Ecosystem Investigation, Model Building and Impact Assessment

Prior to the discovery of hydrothermal vents, the deep-sea floor was thought to have negligible bio-productivity. Naturally, there are no phytoplankton (the marine primary producer) in such dark areas. However, deep-sea hydrothermal vents host another type of marine primary producer (chemosynthetic bacteria) that utilises sulphides and other chemical components. Rich ecosystems develop around hydrothermal vents including shrimps, crabs, etc. based on chemosynthetic bacteria.

As a procedure for mining these areas, it is first necessary to conduct detailed ecological investigation from a position of natural science; second, to build ecosystem models to estimate the impact of a proposed mining project from a position of environmental conservation; and third, to obtain social consensus on the deep-sea mining project based on the environmental impact assessment. Figure 2

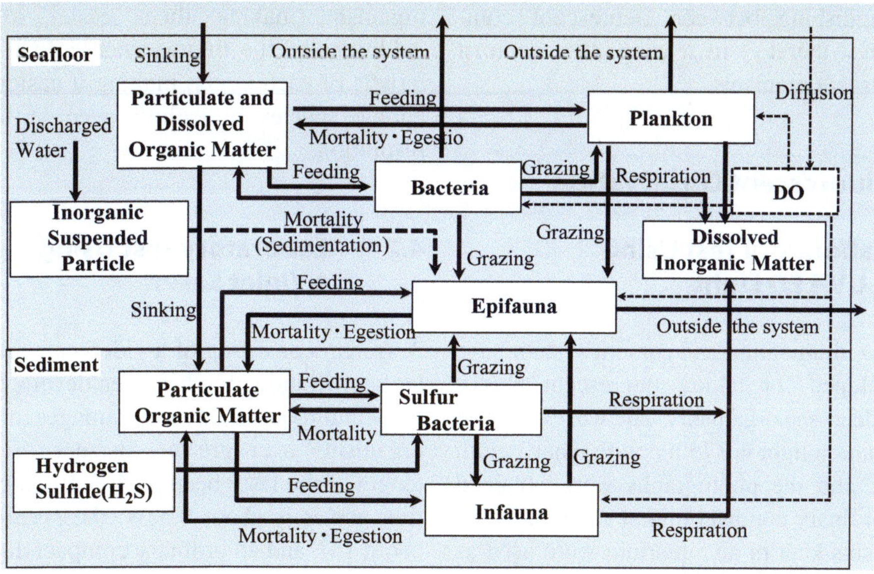

Fig. 2 Diagram of an ecosystem model in deep-sea hydrothermal field

shows a diagram of the ecosystem model in the deep-sea hydrothermal field.

These procedures follow those adopted for terrestrial mining, but there is presently much less scientific knowledge of deep-sea areas and detailed marine investigation is much more difficult than on land.

Meanwhile, the topic of the sustainable use of hydrothermal resources has been discussed with regard to black ore culturing (JAMSTEC Press Release 2012) and hydrothermal power generation (JAMSTEC Press Release 2013), but has rarely been discussed in terms of the sustainable use of deep-sea ecosystems for uses such as fisheries.

3 Bio-fluorescence of Marine Organisms

3.1 Biological Roles, Fluorescent Materials

Sunlight does not reach the deep-sea floor, and many deep-sea organisms use self-generated bioluminescence or detect that generated by other organisms. In general, deep-sea organisms use light to intimidate, dazzle, manoeuvre, etc.

The basic compounds required for luminescence are called luciferins; these are catalysed by luciferase, resulting in luminescence. Some luciferins show fluorescent properties in response to UV excitation. Additionally, it is known that some anaerobic bacteria and cnidarians show fluorescence under UV excitation, but the detailed purposes are unclear.

3.2 Use of Bio-fluorescence Monitoring

Besides the identification of fluorescent substance, clarification of fluorescent mechanisms and ecological understanding of deep-sea bio-fluorescence, bio-fluorescence is considered to have important applications for monitoring deep-sea ecosystem. In ordinal deep-sea investigation, video recording of the seafloor is conducted using white light. However, it is not essential to use pseudo-sunlight for deep-sea floor monitoring: deep-sea organisms, particularly macrobenthos, tend to have little pigmentation and it is therefore difficult to identify species according to their colours.

In this study, fluorescence is used instead of elastic scattering of light illumination for deep-sea floor monitoring near hydrothermal vents. It is expected to be an effective method for monitoring deep-sea ecosystems by identifying and distinguishing between fluorescent compounds and thereby increasing the contrast between target organisms.

4 Fluorescent Observation

4.1 Laboratory Test Using a UV-LED Light

A UV-LED illumination and camera system have been developed for taking fluorescent photographs of deep-sea organisms. The wavelength of the illumination light is 375 nm with a half-width of 10 nm, and the photographs were obtained using an ordinary compact digital camera. Deep-sea organisms kept in an aquarium were used as targets. The geometric arrangement of the system is shown in Fig. 3, and the distance between the target and camera was a few tens of centimetres. Fluorescence photographs are shown in Figs. 4 and 5 with ordinary photographs for comparison.

Figure 4 shows a pair of photographs of a hydrothermal vent shrimp (*Alvinocaris longirostris*). The red fluorescence from the inner parts is confirmable on the UV-illuminated photograph. Figure 5 shows a pair of photographs of a Yunohana crab (*Gandalfus yunohana)*; the green fluorescence from the bristle of the body surface is confirmable on the UV-illuminated photograph. In both figures, the fluorescence photographs show better image contrast of the target organism, making them easier to identify. Additionally, the fluorescence colour is characteristic of each target, making it easier to differentiate species by colour even at low image resolution.

4.2 Laboratory Test Using a Violet Laser

A system consisting of a video camera with violet laser illumination has been developed for taking continuous fluorescent images of deep-sea organisms over greater distances. The wavelength of the laser beam is 405 nm, the illumination power is about 0.5 W, the spread angle is about 15°, and an ordinary compact digital video camera was used for recording. Deep-sea organisms kept in an aquarium were used as targets. The geometric arrangement of the system is shown in Fig. 6, and the distance between the target and video camera was several metres. Figure 7 shows images from the fluorescence video captured for three types of deep-sea organisms.

The left panel of Fig. 7 shows a hydrothermal vent galatheid crab (*Shinkaia crosnieri*), confirming green fluorescence on its bristle. The centre panel of Fig. 7 shows *a Bathynomus giganteus*, with green fluorescence on its joints and edges of the shell. The right panel of Fig. 7 shows a cat shark (*Cephaloscyllium umbratile)*, with yellow-green fluorescence on its surface.

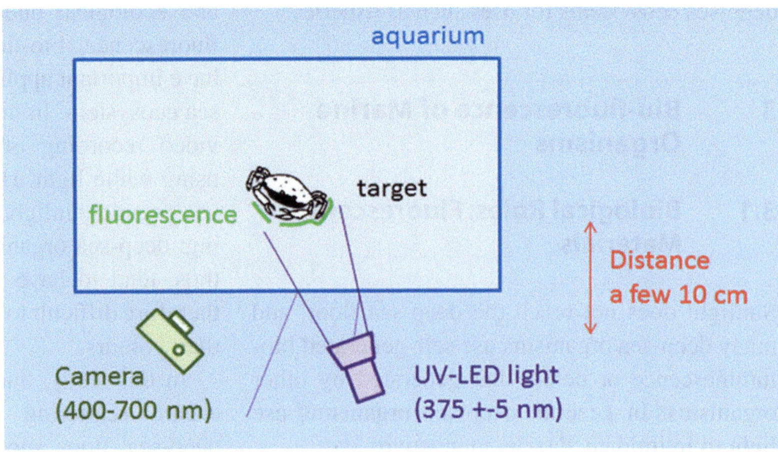

Fig. 3 Layout of UV-LED light, camera and target in laboratory test

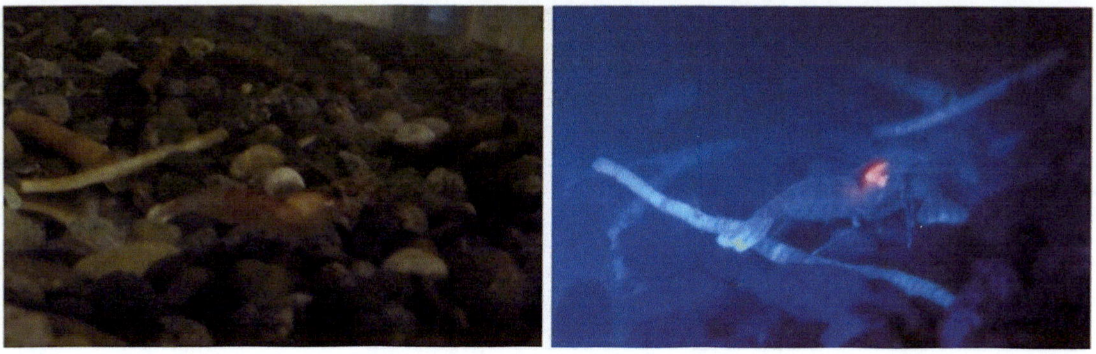

Fig. 4 Ordinary photograph (*left*) and fluorescence photograph (*right*) of a hydrothermal vent shrimp (*Alvinocaris longirostris*)

Fig. 5 Ordinary photograph (*left*) and fluorescence photograph (*right*) of a Yunohana crab (*Gandalfus yunohana*)

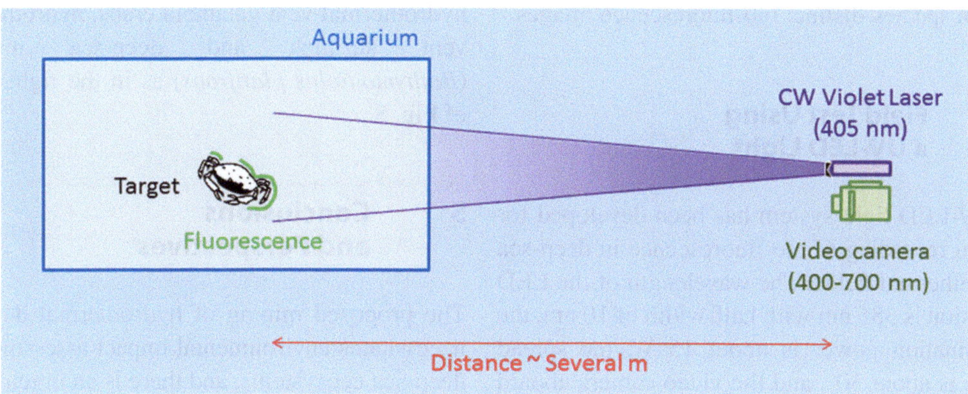

Fig. 6 Layout of violet laser, video camera and target in laboratory test

Fig. 7 Fluorescent video images of hydrothermal vent galatheid crab (*left*), giant isopod (*centre*) and cat shark (*right*)

Fig. 8 ROV 'Hyper Dolphin' (*left*), deep-sea UV-LED light (*centre*) and a UV-excited fluorescent image of hydrothermal vent galatheid crabs, hydrothermal vent shrimps and deep-sea mussels at Iheya North field, Okinawa Trough, Japan

The results demonstrate that light at a wavelength of 405 nm provides suitable excitation to obtain species-distinct bio-fluorescence images.

4.3 Field Test Using a UV-LED Light

A UV-LED light system has been developed for in situ recording of bio-fluorescence in deep-sea hydrothermal fields. The wavelength of the LED emission is 385 nm with half-width of 10 nm, the illumination power is about 15 W, the spread angle is about 30°, and the video camera aboard the ROV 'Hyper Dolphin' (JAMSTEC Web page 2013) was used for recording. Deep-sea observations were obtained in Iheya North field, Okinawa Trough, at a depth of 990 m. The difference in colour by species is less clear than in the aquarium due to light attenuation by seawater between targets and video camera. However, it is still possible to differentiate the fluorescence colours of hydrothermal vent galatheid crabs, hydrothermal vent shrimps and deep-sea mussels (*Bathymodiolus platifrons*) as in the right panel of Fig. 8.

5 Conclusions and Perspectives

The proposed mining of hydrothermal deposits necessitates environmental impact assessment of deep-sea ecosystems, and there is an urgent need for technological development to assist such detailed investigation. Current methods of monitoring hydrothermal ecosystems are operationally intensive. We therefore suggest a new monitoring method using the bio-fluorescent characteristics of deep-sea macrobenthos.

In the laboratory tests, fluorescent images of some deep-sea species were obtained using a UV-LED light with the wavelength of 375 nm and a violet laser with the wavelength of 405 nm. In the field trial, the fluorescent images of a biological community on hydrothermal vents were obtained using a UV-LED light mounted on the ROV, and three species were distinguished on the fluorescent image. These results implied that the bio-fluorescent characteristics could be applied to in situ observation for deep-sea organisms.

In the near future, Raman spectroscopy will be applied to deep-sea benthos, hydrothermal plumes and ore investigation as a new in situ observation method in addition to UV-excited fluorescent image investigation.

Acknowledgements This work was supported by JSPS KAKENHI Grant Number 12929746. The authors wish to thank the Enoshima Aquarium for the fluorescent observation tests. The authors are also grateful to Prof. B. Thornton at University of Tokyo and the 'Hyper-Dolphin' team at JAMSTEC for the deep-sea fluorescence observations.

References

IPCC: annual report 4, Working Group 2. http://www.ipcc.ch/publications_and_data/ar4/wg2/en/contents.html (2007)

JAMSTEC Press Release. http://www.jamstec.go.jp/j/about/press_release/20120323/ (2012). (in Japanese)

JAMSTEC Press Release: electricity generation and illumination via an environmental fuel cell in deep-sea hydrothermal vents. http://www.jamstec.go.jp/j/about/press_release/20130903/ (2013). (in Japanese)

JAMSTEC Web page: 3,000 m class remotely operated vehicle Hyper-Dolphin. http://www.jamstec.go.jp/e/about/equipment/ships/hyperdolphin.html (2013)

Takai K, Kawagucci S, Miyazaki J, Watsuji T, Ishibashi J, Yamamoto H, Nozaki T, Kashiwabara T, Shibuya T (2013) Post-drilling hydrothermal vent and associated biological activities seen through artificial hydrothermal vents in the Iheya North field. Okinawa Trough. In: International symposium on underwater technology 2013

The final report of the first term in development program of hydrothermal deposits. http://www.meti.go.jp/press/2013/07/20130705003/20130705003-2.pdf (2013). (in Japanese)

Weiss RF, Lonsdale P, Lupton JE, Bainbridge AE, Craig H (1977) Hydrothermal plumes in the Galapagos Rift. Nat Mag 267(5612):600–603

Indicators for Ecosystem-Based Management: Methods and Applications

Verena Trenkel, Anik Brind'Amour, Sigrid Lehuta, Pascal Lorance, Stéphanie Mahévas, and Marie-Joëlle Rochet

Abstract

Indicators are essential tools for policy making, public communication and the provision of scientific advice. In fisheries science, indicators have been increasingly used to advising on fish and shellfish stock management, especially since the precautionary approach to fisheries management was developed. They are now becoming a cornerstone of the wider ecosystem approach to the management of all human activities. In this section, we provide some recent examples of methods we developed for creating and selecting pressure indicators and ecological indicators derived from different types of information (scientific survey data, commercial fisheries data) for a range of ecosystems, covering pelagic, demersal and deep-water systems.

1 Introduction

The DPSIR framework (Fig. 1) is the backbone of many indicator approaches (FAO 1999). It is based on categorizing indicators into (D) driving forces creating pressures (P) that act on ecosystem states (S) to create impacts (I) which in turn lead to management responses (R). The choice of suitable indicators requires careful consideration of the context in which they are to be used, as well as how they can be interpreted together to inform management decisions.

Here, we present examples of recent methodological developments and case studies we have carried out in the area of indicator selection and validation for fisheries and wider marine ecosystem–based management.

2 Developing Food Web Indicators

There is an increasing interest in analysing the state of fish communities as sets of functional groups (FG). FGs are groups of species that play a similar role in the food web and whose dynamics can be considered consistent. FGs have proved useful for investigating the influence of fishing

V. Trenkel (✉) • A. Brind'Amour • P. Lorance
S. Mahévas • M.-J. Rochet
Ifremer, rue de l'Ile d'Yeu, BP 21105,
F-44311 Nantes Cedex 03, France
e-mail: verena.trenkel@ifremer.fr

S. Lehuta
Ifremer, unité Ecologie et Modèles pour l'halieutique,
rue de l'ile d'Yeu, F-44311 Nantes 03, France

H.-J. Ceccaldi et al. (eds.), *Marine Productivity: Perturbations and Resilience of Socio-ecosystems*,
DOI 10.1007/978-3-319-13878-7_23, © Springer International Publishing Switzerland 2015

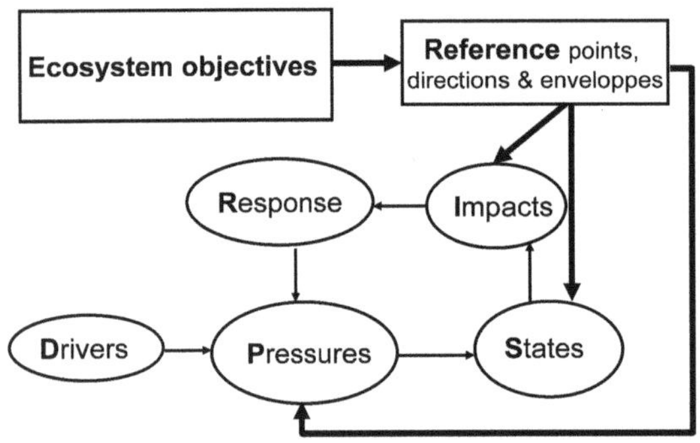

Fig. 1 DPSIR indicator framework for ecosystem-based management

pressure (Essington et al. 2006; Rochet et al. 2010, 2013b), climate change (Ainsworth et al. 2011; Travers-Trolet et al. 2014) and other environmental perturbations (Lokrantz et al. 2010).

Functional groups are often defined in an ad hoc manner, based on the literature or local diet data or data from other areas. A priori groups of species are frequently used in ecosystem models, e.g. taxonomic groups such as mollusc, Clupeidae, etc. Some models also include diet-based groups of species (large piscivores, Banaru et al. 2013). In recent years, there has been renewed interest in grouping species using local and species-specific morphological measurements (Bremner et al. 2003; Dumay et al. 2004); this ecomorphological approach is founded on the relationship between morphology and function (e.g. Reecht et al. 2013).

Recent studies identifying and mapping ecosystem diversity and anthropogenic pressures in Mediterranean Sea have underlined significant geographical patterns (Coll et al. 2010, 2012; Granger et al. 2015). This suggests that species composition of the Mediterranean marine communities varies among geographic areas.

In this example, we characterized the temporal changes of three contrasted Mediterranean areas using a posteriori functional groups of species defined by species-specific morphological traits. The three areas vary in terms of total surface, degree of enclosure, nutrient inputs and/or fishing efforts (e.g. Caddy et al. 1995; Coll et al. 2012). The Gulf of Lions is a highly productive area with a wide shelf in the western Mediterranean basin. Its productivity is notably due to high run-offs by the Rhône river and local hydrographic features. It supports numerous fisheries and ranges among the most fished areas of the Mediterranean Sea together with the Aegean Sea (Coll et al. 2012). The Aegean Sea is a semi-enclosed area with a north to south decreasing gradient of productivity, highly correlated with nutrient enrichment patterns. The Ionian Sea is more open and less productive than the other two. According to Coll et al. (2012), trawling and dredging disturbances are less important than in the two other areas.

The biological data (fish and invertebrate biomass) were collected during annual bottom trawl surveys conducted in May to July from 1994 to 2012 for the Gulf of Lions, 1995–2012 for the Ionian Sea and 1996–2008 for the Aegean Sea. Ten morphological traits (8 continuous and 2 categorical) were measured in the field and on photographs.

Cluster analysis of morphological traits identified ten FGs (Fig. 2). It first discriminated between benthic and pelagic species, and within each group, it distinguished species by their diet (e.g. strictly invertebrate and mixed piscivores and invertebrate feeders). For instance, FG5 grouped small pelagic species such as sardines and anchovies which feed mainly on zooplankton and inhabit the water column, while FG2 was characterized by flatfish which feed mostly on macroinvertebrates and live in or on soft substratum. This classification was partly correlated with trophic (5 %) and habitat guilds (19 %) as defined in the literature. These low percentages confirm that the traits are not solely related to one of the two functions and that similar species in terms of morphology may be grouped together although having some differences in their diet (Albouy et al. 2011).

Temporal changes of functional group biomass in each area were assessed by computing the difference in biomass between the beginning (3 first years) and the end (final 3 years) of the time series. The three Mediterranean areas displayed differences in their FG time trends, with the Ionian Sea contrasting highly with the two other areas. Almost all FGs from the Ionian Sea increased over time by up to 378 %. Concurrently, nearly all the pelagic FGs of the three areas

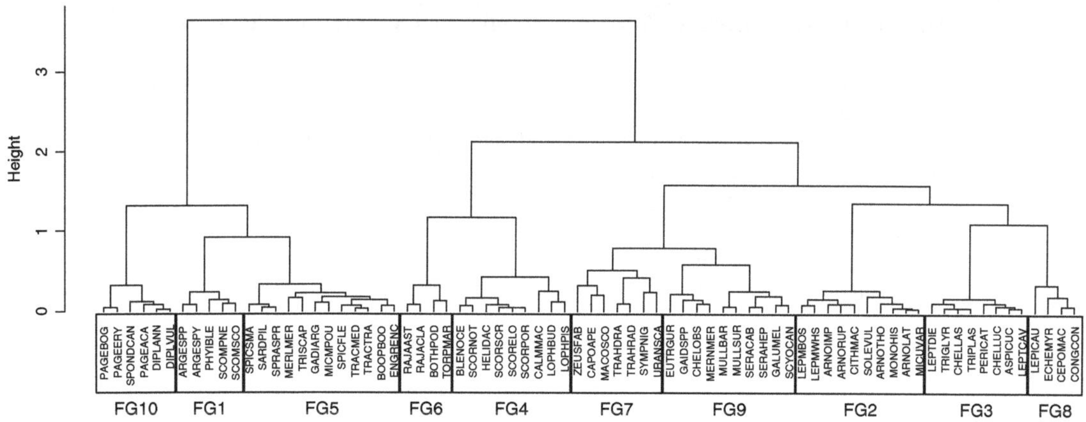

Fig. 2 Classification of 75 species into ten functional groups obtained by hierarchical cluster analysis of the average values of ten morphological traits

(i.e. FG12, FG5, FG1, FG10) showed a significant increase in time, with the cephalopods increasing by over 120 %. Simultaneously, benthic and particularly demersal FG decreased in the Gulf of Lions and Aegean Sea but showed opposite trends in the Ionian Sea.

It is generally accepted that moderate enrichment leads to an increase in ecosystem production (Caddy et al. 1995), while over-enrichment may reduce benthic and demersal production by oxygen depletion and other undesirable effects (Seitz et al. 2009). The three study areas show differences in production (Coll et al. 2010) and to a lesser extent fishing efforts (Coll et al. 2012) which might explain some of the observed patterns.

3 Selecting Pelagic Ecosystem Indicators

Pelagic ecosystems set distinct requirements for indicators, since the constituent species can exhibit substantial, environmentally influenced, fluctuations in abundance and wide-ranging mobility. Small pelagic fish communities consist of few species. In contrast to many demersal mixed-species fisheries, pelagic fishing generally targets single species, so direct fishing impacts affect single stocks, though indirect effects may cause food web perturbations (Rochet et al. 2013a, b).

In this example, we identified high level ecological, economic and social objectives for European exploited pelagic ecosystems in a stakeholder consultation process and proposed operational management objectives and suitable indicators based on the literature on pressures and impacts (Trenkel et al. in press) (see Fig. 3 for an example). We found that given the strong but species-specific links of pelagic species with the environment and the large geographic scale of their life cycles, pelagic indicators are needed at the level of stocks, independent of area, while community indicators may be set for certain (sub-)areas.

4 Evaluating Spatial Population Indicators

The distribution area of many populations is positively related to abundance such that, at large population size, more habitat space is occupied and abundance-occupancy relationships form well-known macroecological patterns (e.g. Frisk et al. 2011). Various theories exit to explain abundance-occupancy relationships including density-dependent habitat selection (ideal free distribution theory), local vital rates (birth and death) and range position (Gaston et al. 2000). In the absence of abundance information, indicators of spatial distribution have been proposed as a direct measure of population status (EC 2010).

Pelagic ecological objective: Maintain pelagic food supply for higher trophic levels
Scientific knowledge: Few top predators and piscivorous fishes have strong pelagic prey dependance
Operational objectives: Maintain herring and sprat in the Baltic Sea as prey for cod; Maintain total forage fish biomass in North Sea
State & impact indicators: Prey biomass in critical area or a proxy predator condition/growth/productivity
Reference level: x tonnes of prey biomass
Pressure management: TAC, effort control, escapement rules

Fig. 3 Example for linking higher-level ecological pelagic management objectives to operational objectives and indicators using scientific knowledge

In this example, we evaluated the reliability of commercial fishing data for deriving occupancy indicators that could serve as proxies for stock abundance using a simulation approach (Trenkel et al. 2013). For this, four questions were investigated: (1) Occupancy changes with stock biomass, but is this change strong enough to make occupancy a sensitive indicator of population biomass? (2) Fishing boats follow fish, but when does such activity alter the positive macroecological relationship between occupancy and abundance? (3) When does the activity of pursuing fish adversely affect occupancy estimates derived from catch and effort data? (4) How does uncertainty in fishing effort data affect occupancy estimates?

The spatial simulations mimicked the dynamics of four deep-water fish species. The results showed that biomass-occupancy relationships can be weak and fishers following fish can modify the spatial distribution of target species, even reversing the sign of the biomass-occupancy relationship in certain cases. They can also affect the reliability of occupancy indicators, which can also be impaired by error in effort data. Using commercial catch and effort data and abundance indices for deep-sea fish populations to the west of the British Isles, it was found that only for roundnose grenadier might occupancy provide insights into biomass changes. Thus, care should be taken when using occupancy for evaluating range changes in cases where fishing might have modified spatial distributions, when uncertain commercial data are used or when the abundance-occupancy relationship is too flat, i.e. when occupancy changes little with abundance.

5 Investigating the Link Between Fishing Pressure and Fishing Impact Indicators

There is increasing awareness that for developing an ecosystem approach to fisheries management it might be required to broaden our perspective on fisheries selectivity from the gear and haul level to the fishery and exploited community scale (Garcia et al. 2012). Indeed, recent modelling results suggest that selectively targeting restricted ranges of species or sizes may be more harmful to marine communities than a more "balanced" exploitation apportioning extraction across the food web. This example sought empirical evidence supporting these modelling results by means of a comparative study across communities exploited in different ways.

A comparative analysis of fishing pressure versus fishing impact metrics across a range of temperate, exploited shelf communities was undertaken to investigate empirically the link between fisheries selectivity and the biodiversity in exploited communities (Rochet et al. 2013a, b). In this analysis, individuals were "ecological units" – distinct communities with defined fishing patterns. The temporal units were periods of time with a consistent fishing pressure, with a time lag between fishing pressure and impact. The question studied was whether there was evidence of a link between metrics of fishing pressure, including selectivity metrics, and metrics of fishing impacts. Time series of fishing pressure and impact metrics (FPMs and FIMs) were calculated from 13 temperate shelf sea communities from the Western and Eastern North Atlantic and the Mediterranean. Catch statistics were used to calculate FPM. Bottom trawl survey data were used to estimate FIMs. Time series of FPMs were examined to identify ~10-year time periods with consistent levels, or at least consistent trends, in fishing pressure. FPMs were averaged over these periods; FIMs were averaged across the subsequent 10-year period, allowing for a 10-year lag between pressure and impact. The relationship between pressure and impact metrics was examined by a canonical correlation analysis with 27 data points (number of ecological units × number of time periods). Although stark contrast was found between FPMs across places and/or time periods, only a weak link (22 % of total variance) was found between fishing selectivity and the community biodiversity 10 years later: communities from which a more diverse catch was taken had higher biodiversity, while communities from which more predators were extracted had a higher total biomass. Although we examined fishing impacts with a reasonable time lag after fishing pressure metrics, there is still a suspicion that these results mostly reveal that fisheries extract from a community what is available – if there are more species in a community, then more species may have some value and be targeted and caught.

6 Evaluating Management Performance Indicators

In this example, we used a simulation approach to study the properties of fisheries indicators in view of selecting robust and relevant indicators for the evaluation of the performance of management measures. A simulation approach allows disentangling the effects of several factors and identifying the sensitivity to a particular pressure (Lehuta et al. 2013b). Our methodology has two steps: step 1, review and select metrics for evaluating the long-term performance of management measures using management strategy evaluation, and step 2, simulate management strategies using selected management performance metrics to assess their capacity of reaching objectives and their robustness to uncertainties. To carry out the second step, we used the ISIS-Fish model (Pelletier et al. 2009), a list of management strategies to be assessed with explicit management objectives and a list of uncertainties and a simulation design to run efficiently the operating model (Mahévas and Iooss 2013; Mahévas and Lehuta 2013), to compute the selected performance metrics and finally to assess their properties. Sensitivity analysis was used to derive sensitivity indices of model outputs to model inputs based on variance decomposition.

The approach was applied to the pelagic fishery in the Bay of Biscay. The population model included a spatially explicit description of anchovy dynamics (Lehuta et al. 2010) and global surplus production models for four other stocks also targeted by the fishery. The fishing behaviour model provided spatial predictions of fishing time allocation (Vermard et al. 2008). The overall model was validated for the period 2000–2004 (Lehuta et al. 2013a). We then simulated seven management measures as defined in the long-term management plan proposed for the anchovy fishery and considered six management performance metrics: (1) P1, the number of years when biomass drops below Blim; (2) P2, the number of years of fishery closure (should not excess one in 10 years); (3) P3, the number of years with catch of anchovy higher than 7,000 t which is the threshold of economical profitability expressed by the sector; (4) P4, the linear trend in SSB (should show no degradation to guaranty reproductive capacity); (5) P5, the interannual variability in landings (which should be as low as possible as a proxy of fishery stability, to guaranty continuous supply to the industry); and (6) P6, the interannual variability in the proportion of recruits in the population (as low as possible).

Management scenarios significantly influenced the fishery's dynamics as far as landings values and stability, prevention of fishery closure and stock collapse were concerned. These indicators thus clearly reflected the effect of management and can be used to monitor management performance and adapt regulations (Fig. 4).

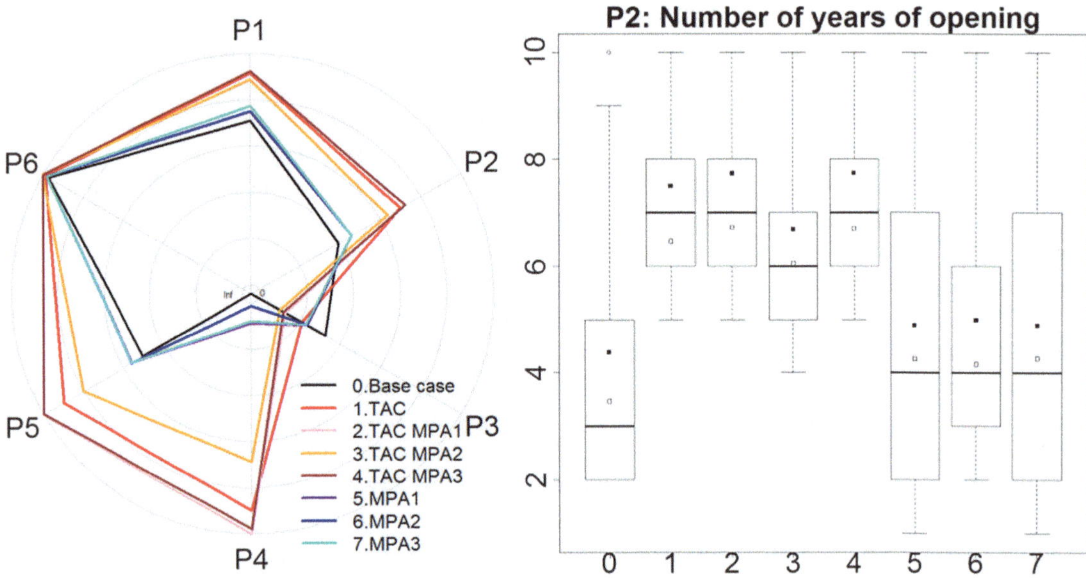

Fig. 4 *Left*, radar plot of average value of management performance metrics for eight management scenarios. Metrics were scaled by maximum values: P1, no. of years with biomass lower than Blim ∈ [0;10]; P2, no. of years of fishery closure ∈ [0;10]; P3, no. of years with anchovy catch >7,000 t ∈ [0,max=3.47]; P4, trend in biomass ∈ [min;max]; P5, variation in landings ∈ [0;max]; P6, 1/variability of age structure ∈ [0;max]. "max" stands for maximum. *Right*, simulated values of P2 metric by management scenario (x axis) (*boxes* represent the median and first and third quartile derived from uncertainty of parameters)

In contrast, population growth (P4) and structure (P6) were mainly influenced by endogenous processes (mainly variability in natural mortality and migration). The variations observed in the corresponding metrics thus reflected the influence of these processes rather than the effect of management. They are good indicators of the conditions experienced by the population, but no conclusion on management performance could be drawn from their values. High sensitivity indices related to uncertainties are indication of a lack of robustness of the metric (Lehuta et al. 2013b). They point out the sources of uncertainty that need to be reduced to allow the use of the metric as an indicator. More generally, the sensitivity of performance metrics to factors other than management measures should warn managers about the potential misinterpretation of "good values" as good management performance and stress the need for completing the vision with robust metrics of management impact.

References

Ainsworth CH, Samhouri JF, Busch DS et al (2011) Potential impacts of climate change on Northeast Pacific marine food webs and fisheries. ICES J Mar Sci 68:1217–1229

Albouy C, Guilhaumon F, Villéger S et al (2011) Predicting trophic guild and diet overlap from functional traits: statistics, opportunities and limitations for marine ecology. Mar Ecol Prog Ser 436:17–28

Banaru D, Mellon-Duval C, Roos D et al (2013) Trophic structure in the Gulf of Lions marine ecosystem (north-western Mediterranean Sea) and fishing impacts. J Mar Syst 111:45–68

Bremner J, Rogers SI, Frid CLJ (2003) Assessing functional diversity in marine benthic ecosystems: a comparison of approaches. Mar Ecol Prog Ser 254:11–25

Caddy JF, Refk R, Do-Chi T (1995) Productivity estimates for the Mediterranean: evidence of accelerating ecological change. Ocean Coast Manag 26:1–18

Coll M, Piroddi C, Steenbeek J et al (2010) The biodiversity of the Mediterranean Sea: estimates, patterns, and threats. PLoS ONE 5:e11842

Coll M, Piroddi C, Albouy C et al (2012) The Mediterranean Sea under siege: spatial overlap between marine biodiversity, cumulative threats and marine reserves. Glob Ecol Biogeogr 21:465–480

Dumay O, Tari PS, Tomasini JA, Mouillot D (2004) Functional groups of lagoon fish species in Languedoc Roussillon, southern France. J Fish Biol 64:970–983
EC (2010) Commission Decision of 1 September 2010 on criteria and methodological standards on good environmental status of marine waters. Off J Eur Union L 232:14–24
Essington TE, Beaudreau AH, Wiedenmann J (2006) Fishing through marine food webs. Proc Natl Acad Sci U S A 103:3171–3175
FAO (1999) Indicators for sustainable development of marine capture fisheries, vol 8, FAO technical guidelines for responsible fisheries. Food and Agriculture Organization of the United Nations, Rome, 68 pp
Frisk MG, Duplisea DE, Trenkel VM (2011) Exploring the abundance-occupancy relationships for the Georges Bank finfish and shellfish community from 1963–2006. Ecol Appl 21:227–240
Garcia SM, Kolding J, Rice J et al (2012) Reconsidering the consequences of selective fisheries. Science 335:1045–1047
Gaston KJ, Blackburn TM, Greenwood JJD et al (2000) Abundance-occupancy relationships. J Appl Anim Ecol 37:39–59
Granger V, Fromentin JM, Bez N et al (2015) Large spatio-temporal monitoring highlights shift in Mediterranean fish diversity hotspots. Progress in Oceanography 130:65–74
Lehuta S, Mahévas S, Petitgas P, Pelletier D (2010) Combining sensitivity and uncertainty analysis to evaluate the impact of management measures with isis–fish: marine protected areas for the Bay of Biscay anchovy (*Engraulis encrasicolus*) fishery. ICES J Mar Sci 67:1063–1075
Lehuta S, Petitgas P, Mahévas S, Huret M, Vermard Y, Uriarte A, Record NR (2013a) Selection and validation of a complex fishery model using an uncertainty hierarchy. Fish Res 143:57–66
Lehuta S, Mahévas S, Petitgas P, LeFloch P (2013b) A simulation-based approach to assess sensitivity and robustness of fisheries management indicators for the pelagic fishery in the Bay of Biscay. Can J Fish Aquat Sci 70:1741–1756
Lokrantz J, Nystrom M, Norstrom AV, Folke C, Cinner JE (2010) Impacts of artisanal fishing on key functional groups and the potential vulnerability of coral reefs. Environ Conserv 36:327–337
Mahévas S, Iooss B (2013) Grilles de sélection d'une méthode d'analyse de sensibilité globale. In: Faivre R, Iooss B, Makowski D, Mahévas S, Monod H (eds) Analyse de sensibilité et exploration de modèles – Applications aux modèles environnementaux. Quae, Versailles
Mahévas S, Lehuta S (2013) Illustration d'une analyse de sensibilité pour l'utilisation d'un modèle complexe en gestion des pêches. In: Faivre R, Iooss B, Makowski D, Mahévas S, Monod H (eds) Analyse de sensibilité et exploration de modèles – Applications aux modèles environnementaux. Quae, Versailles
Pelletier D, Mahévas S, Drouineau H et al (2009) Evaluation of the bioeconomic sustainability of multi-species multi-fleet fisheries under a wide range of policy options using isis-fish. Ecol Model 220:1013–1033
Reecht Y, Rochet M-J, Trenkel VM, Jenning S, Pinnegar JK (2013) Use of morphological characteristics to define functional groups of predatory fishes in the Celtic Sea. J Fish Biol 83:355–377
Rochet M-J, Trenkel VM, Carpentier A et al (2010) Do changes in environmental pressures impact marine communities? An empirical assessment. J Appl Ecol 47:741–750
Rochet MJ, Burny C, Spedicato MT et al (2013a) Does fisheries selectivity affect food web dynamics in a way that needs to be managed? ICES CM 2013/E:03. In: ICES annual science conference, Reykjavik, Iceland, 23–27 September 2013
Rochet M-J, Collie JS, Trenkel VM (2013b) How do fishing and environmental effects propagate among and within functional groups? Bull Mar Sci 89:285–315
Seitz RD, Dauer DM, Llanso RJ, Long WC (2009) Broad-scale effects of hypoxia on benthic community structure in Chesapeake Bay, USA. J Exp Mar Biol Ecol 381:S4–S12
Travers-Trolet M, Shin YJ, Shannon LJ, Moloney CL, Field JG (2014) Combined fishing and climate forcing in the southern Benguela upwelling ecosystem: an end-to-end modelling approach reveals dampened effects. PLoS ONE 9:e94286
Trenkel VM, Beecham JA, Blanchard JL, Edwards CTT, Lorance P (2013) Testing CPUE-derived spatial occupancy as an indicator for stock abundance: application to deep-sea stocks. Aquat Living Resour 26:319–332
Trenkel VM, Hintzen NT, Farnsworth KD et al (in press) Identifying marine pelagic ecosystem objectives and indicators for management. Marin Policy. doi:10.1016/j.marpol.2015.01.002
Vermard Y, Marchal P, Mahévas S, Thébaud O (2008) A dynamic model of the Bay of Biscay pelagic fleet simulating fishing trip choice: the response to the closure of the European anchovy (*Engraulis encrasicolus*) fishery in 2005. Can J Fish Aquat Sci 65:2444–2453

How to Size "Fair" Compensatory Mitigation Due to Coastal Area Destruction for Fisheries Resources: Visual Habitat Equivalency Analysis Software, a Scoring Method to Sizing Compensatory Mitigations Applied in Offshore Windmill Renewable Energy Project Case Study

Pioch Sylvain

Abstract
In France four offshore windmill projects are planned before 2017. For the three Northeast projects, fisheries fields are likely to be impacted, due to foot foundation. The environmental impact studies state that impacts to fishermen will be offset with subsidies from windmill taxes and with specific mitigation like artificial reef or stock enhancement for fisheries losses mainly for scallops (*Pecten sp.*). Losses should be offset, for scallops, by the release of juveniles and for fishes with effective artificial reefs (as new functional habitats) with management project developed by fishermen. We propose to scale the size of projects likely to mitigate the losses for scallop fisheries, in the aim to reach an ecological equivalency between the losses and the gains. To do so, we propose the use of "Visual Habitat Equivalency Analysis" (Visual_HEA) assessment method, a software proposed by the EU commission and the French Ministry of Environment in the case of ex post damage compensation in application of Environmental Liability Directive 2004/35/EC. This tool was developed by the Nova Southeastern University and the CEFE–University Montpellier 3, for its last versus and French translation, the concept is based on the assessment of ecosystemic services through indicator species called "proxy." A brief presentation of the tool will be proposed before to expose preliminary results with a case study for an ex ante impact:

P. Sylvain (✉)
UMR 5175, Centre d'Ecologie Fonctionnelle
et Evolutive, Université Montpellier 3,
Route de Mende, 34199 Montpellier, Cedex 5, France
e-mail: sylvain.pioch@univ-montp3.fr

H.-J. Ceccaldi et al. (eds.), *Marine Productivity: Perturbations and Resilience of Socio-ecosystems*,
DOI 10.1007/978-3-319-13878-7_24,

offshore windmill fictitious authorized impact. A criticism of the approach will be exposed to define what could be the arguments to a "fair" mitigation for fisheries resources, as basis for future negotiation with fishermen.

1 Introduction

How to offset biodiversity destruction due to infrastructures from growing human activities? The rapid and tremendous losses of biodiversity known since the last 15 years, especially in marine ecosystem, are the most problematic scientific and social challenge for the next generation (Worm et al. 2006). The contradictory conflict between the human development from the end of the twentieth century to the beginning of the twenty-first century called "concrete era" and the necessity to protect biodiversity by a world "green garden" model seems to be impossible in the long term, if our development model, based on growing consumption and market-based theory, is not adapted to our limited planet and its resources (MEA 2005). Nonetheless, for the short term, a new way of thinking the land planning with biodiversity conservation consideration, especially to offset human footprint, could be interesting to explore. Since the early 1990s, the "no net loss" concept is the US government's overall policy goal regarding wetlands and coastal ecosystems (Sibbing 2008). "No net loss" was recommended in 1987 at the National Wetlands Policy Forum to be adopted in 1989 by President George H. W. Bush's administration. The policy, "a compromise between development and conservation," was based on the needs to protect aquatic ecosystem by creating[1] and restoring replacement from the same ecosystems. The overall objective is to strive to maintain a global functional ecosystem balance by compensatory mitigation in the same quantity as a destroyed ecosystem: the losses equal the gains. An additional positive impact could even be produced like additional conservation actions (called ACA, see Fig. 1).

Ideally, this is intended to compensate for the same ecosystem as those which were lost; however, this is ecologically impossible (Moreno-Mateos et al. 2012). A best effort is to attempt to reach a social compromise and standard of performance with the restoration project (Levrel et al. 2012). In early 1990, Dunford (et al. 2004), on behalf of NOAA, developed a service/service method called "Habitat Equivalency Analysis (NOAA 1995)." The method consists of sizing habitat losses by scoring the level of ecosystemic services (ES) lost by an authorized (NEPA 1969; CWA 1972) or non-authorized (National Resource Damage Assessment protocol, under Oil Pollution Act from 1990) operation, including ES gained with a compensatory mitigation area. To help and facilitate repetitive and fastidious calculations, the National Coral Reef Institute created the HEA software in 2006 called Visual_HEA (Kohler and Dodge 2006). In Europe, this approach is based on "service to service" assessment and was introduced into national law in 2006 with the passage of Environmental Liability Law (Bas and Gaubert 2010). The idea is quite the same as the "no net loss" in the USA, by mitigating human impact with the offset of the same ecosystem plus the interim losses calculated with an additional discount rate (see Fig. 2 and Dunford et al. 2004).

The French government adopted the European Directive in 2008[2] (implementing decree in 2009)[3] and, based on the European program REMEDE results (REMEDE 2008), recommend HEA as a standard method for scoring loss and gain (Gaubert and Hubert 2013). Our objective, in addition to comply with the French Ministry of Environment (MEDDE), was to facilitate future end users of the service/service approach for Environmental Liability Law in France and Europe. In this aim, we reform Visual_HEA[4] to improve the software as well as adapt Visual_HEA to the European context. The 2.6 version of

[1] Ecological glossary from Society for Ecological Restoration International Science & Policy Working Group (2004).

[2] Loi n 2008-757 du 1er août 2008.

[3] Décret n°2009-468 du 23 avril 2009.

[4] The Visual_HEA V2.6 in English is a freeware; website at the following location: https://www.nova.edu/webforms/ocean/software_request.html?visualhea

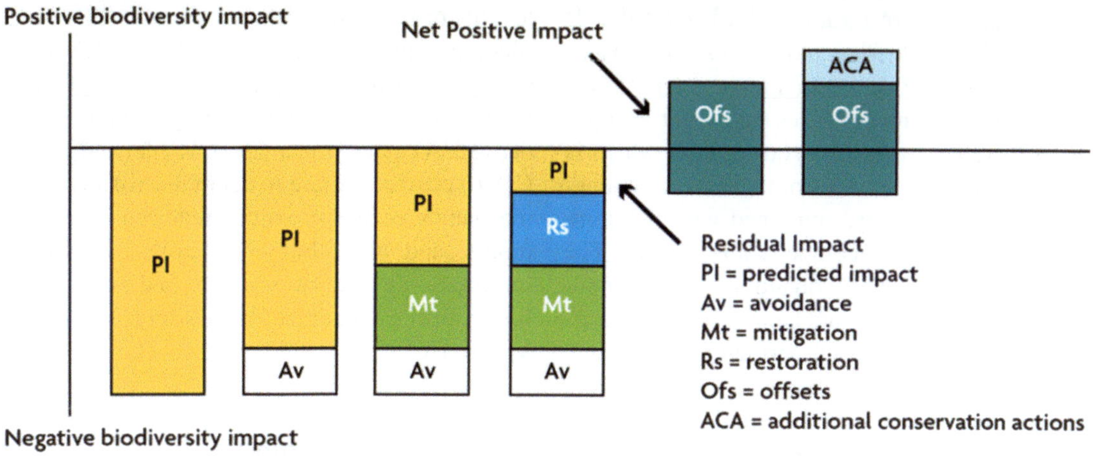

Fig. 1 The mitigation hierarchy adapted from Rio Tinto and Western Australia EPA (BBOP 2009)

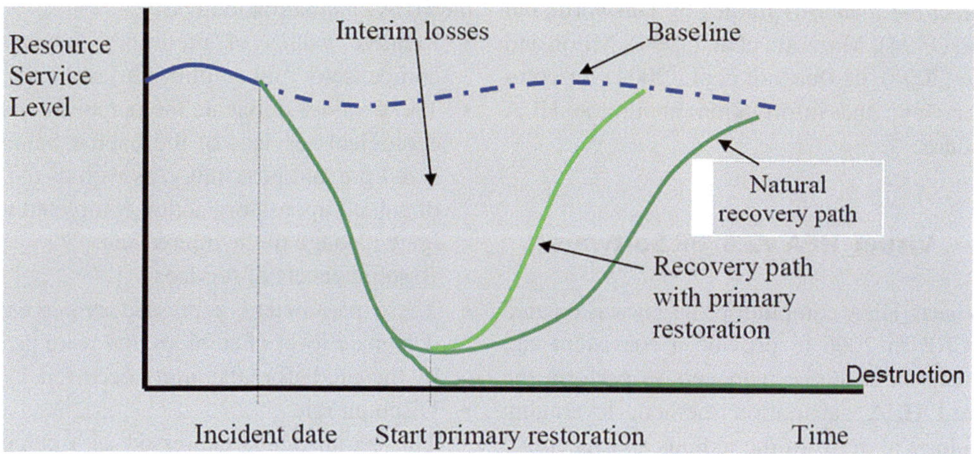

Fig. 2 "Service to service" approach and graphical representation of ecosystem service loss and recovery through natural restoration over time (in Milon and Dodge 2001)

Visual_HEA[5] adheres to the standard HEA calculation method with major changes. The use of this method under a software facilitates the idea to test several different projects, from the viewpoint of their impacts. This could be very helpful in the case of major infrastructure development, such as offshore windmill projects, occupying 100 square kilometers, in relatively stable and specific marine ecosystems (both benthic and pelagic), constituted by flat soft bottom due to foundation design and work requirements.

[5] The Visual_HEA V2.6_FR in French is a freeware; website at the following location: http://www.developpement-durable.gouv.fr/IMG/zip/saveLogiciel_visual_HEA.zip

2 Materials and Methods

After a brief presentation of the method, we will explain how the software Visual_HEA V2.6_FR runs.

2.1 The HEA Method

Habitat Equivalency Analysis (HEA) is a method used by resource managers to compute the quantity of compensatory restoration of a habitat that is required to provide for ecological loss of services of the resource due to an injury event (Pioch 2010). The HEA method uses a discounting algorithm to value a natural resource asset which is equal to all future services of that asset after degradation due to injury, combined with the computed value of any compensatory action. The HEA method provides a quantitative and temporal measure of the loss and gain of ecological services of a habitat for a set period of time (Roach and Wade 2006). The purpose of this paper is to provide an overview of the application of this method to an actual case study, using an enhanced and translated (to the French language) version of the Visual_HEA software, which is a useful tool to automate the HEA process. For conciseness, we direct the reader to studies by Unsworth and Bishop (1994), Mazzotta et al. (1994), Milon and Dodge (2001), or Dunford et al. (2004) for complete review and information about the HEA procedure.

2.2 Visual_HEA v2.6_FR Software

The Visual_HEA computer program was created by NCRI[6] in 2006 to provide a consistent and easy way for resource managers to perform the standard HEA calculation method, leveraging technology to perform the tedious and repetitive calculations required. The program was written with a robust programming language (Visual Basic) and targets the Windows operating system platform. Visual_HEA provides a rich graphical user interface which accepts user-defined parameters that are required to perform the HEA. These input parameters are based on assumptions of loss and gain in relation to pre- and post-injury of a resource combined with any compensatory action performed to mitigate the injured resource. Input parameters are graphically depicted in the user interface and calculations are automatically performed based on user inputs. Resultant values from the analysis are presented graphically, with the option to output the analysis results to an ascii text file or Portable Document Format (PDF) file. Since its introduction in 2006, Visual_HEA has been downloaded 5,000 times and is used globally to value habitat loss due to injury using the standard HEA calculation method. Kohler and Dodge (2006) provide a more in depth discussion on the mechanics of the program, required parameter inputs, and algorithm calculations. Additional information about the software and a download link can be obtained by visiting the NCRI Visual_HEA.[7]

2.3 Input Parameters

The following is a brief review of the input parameters required by Visual_HEA to perform the HEA calculation method:

- Relative values of pre-injury services and compensatory (at equilibrium) services
- These values delineate the perceived value of ecological services of the habitat being analyzed pre and post injury, as well as the value of any compensatory action completed to mitigate damage to the injured site.
- Baseline levels of services
- These parameters, expressed as percentages, designate level of services that were provided by the site before the injury occurred.
- Discount rate
- This parameter is expressed as a percentage rate per time unit. The discount rate functions under the presumption that future restored services are more highly valued initially and then discounted as time lapses over the duration of the analysis period. Conversely, the values of past services are increased over the analysis period subject to the discount rate. Future and past ecological service calculations function independently and can be computed for different temporal durations.
- Year of claim
- This parameter indicates the year the claim is made, which can be actual or arbitrary to provide a starting point for analysis.

[6] National Coral Reef Institute, Florida, USA

[7] Website at the following location: http://www.nova.edu/ocean/visual_hea/

- Service loss parameters from the injury
- This suite of parameters is composed of the actual size of the injured area and level and duration of habitat loss from the point in time of injury until recovery if applicable.
- Service gain parameters from the compensatory action (restoration)
- These parameters consist of the level and duration of services gained due to compensatory action for the period analyzed.

Given these parameters as inputs, the Visual_HEA software applies the standard HEA method and displays the results of analysis within a detailed viewer, showing different data and their representation (visually).

2.4 Improvements to Visual_HEA

The French version of Visual_HEA developed in collaboration with Nova Southeastern University (Matt Johnston and Dick Dodge) and the laboratory CEFE–University Montpellier 3 (Fanny Berger and Sylvain Pioch) retains the same overall parameter inputs, interface structure, and processing algorithms as version 2.5, with added enhancements that allow the software to adhere to established French government guidelines. While producing this new version, several software bugs were also identified and addressed. All enhancements and fixed bugs, made to the software, were also applied to the English language version of the software and both designated as version 2.6 (French and English versions, respectively).

3 Applications and Results for Experimental Case Study: Impact Mitigation for Fictitious Case of Offshore Windmill Project

To test this new version of Visual_HEA 2.6_FR, here exposed is an experimental case study for an offshore windmill project located in tempered sea area (ground dominated by scallops and sandy bottom habitat).

3.1 Sizing a Compensatory Mitigation: Fictitious Offshore Windmill Project

For this case study in application of HEA software, we propose to size the compensatory mitigation of an offshore windmill project. The main impacted socioeconomical sector will be the fisheries of scallops. The main impacted ecosystem will be the sandy bottom habitat, represented by a dominant species: the scallops (*Pecten sp.*).

3.1.1 Material

The project will be held in a temperate coastal water area. The renewable energy infrastructure is constituted by 100 windmill foundations, distributed within an area around 77 km^2. At the feet of each offshore windmill, a concrete block is installed on the sea ground as anchorage, with an acreage of 10×10 m for 100 m^2. The total surface destroyed is 100 (foundation) $\times$ 100 m^2 (each surfaces destroyed) for 10000 m^2, or 1 ha.

The destruction is total (100 % of ecosystem services and ecological functions lost), without any future access from the benthic fauna and flora for a soft bottom: the sand is replaced by an artificial hard bottom surface (the concrete of the foundation). After the previous impact, the level of services is considered to be 0 % (all the marine habitat and species are destroyed under the foundation).

The proxy species,[8] keystone ecological species, for this kind of sandy bottom habitat, is the scallop (*Pecten sp.*). It also mainly insures services of provisioning (fisheries) and regulating with water filtration and purification for a healthy ecosystem (fishing, flora and fauna life, etc.). This species needs good hydromorphologic and water quality and that is a strong economic asset. There are also several studies and data about this species, especially from halieutics studies.

The metric used to assess the chosen proxy is the number of scallops by m^2. The average level of scallops per square meter is around 0.67/m^2 in

[8] Proxy species: other species are dependent with them. They represent a good ecological indicator.

this area exploited by fishermen and managed for a sustainable production.

According to Fifas (2004) the first sexual maturity for this population of scallops occurs after 2 to –3 years, but we estimate that 100 % of services could be produced after 10 years (total life longevity). The project will, hypothetically, need 10 years to produce the full services and offset the 100 % of losses occurred in the offshore windmill project due to the foundation's impact.

The compensation project is an in-kind (equivalency for habitat/species and ecosystemic area) restoration project, located in the same ecological area (sandy bottom habitat dominated by scallops). It is an overfished area, where the scallops are disappeared or relict and the level of services produced is 0 %.

The fishermen are agreeing with the project and the fishery management proposed before the restoration project begins. The aim is that the compensatory sandy bottom area produces a level of 100 % of services after 10 years. The restoration project is an enhancement, with juvenile scallops released. The amount of juveniles and the periodicity of release is based on experimental sea-enhancement project in the same area to reach 0.67 scallops/m^2, with the mortality rate (predation, human pressure, disease, etc.). The release of juveniles in an overfished area (0 % of services produced) will restore 30 % of services in the first last year (regulating and supporting services will be restored), and 10 % from the 2nd until the 10th year, before reaching the full service rate for the cohort.[9]

The compensatory ratio for this project is 1:2 because the restoration by enhancement (releasing juvenile scallops) is ecologically risky, with a rate of success positive but strongly dependent of complex social aspect (agreement of fishery management), with few failures in recent experiences.

[9]Cohort: in statistics and demography, a cohort is a group of subjects who have shared a particular event together during a particular time span.

Table 1 Summary of the data used for the application of the HEA to fictitious marine renewable energy offshore windmill project in tempered coastal area

Proxy	*Number of scallops (Pecten sp.)/m²*
Year of reference for discounting	*2015*
Damaged surface area	*1 ha*
Annual discount rate	*4 %*
Level of services supplied before the impact	*100 %*
Level of services supplied after the impact	*0 %*
Regeneration pace of the sandy bottom in the compensatory mitigation area	*30 % after the 1st year and 10 % the 9 years following*
Lifetime of the compensatory measure	*25 years (from 2015 to 2045)*
Gains of service obtained from the compensatory measure	*100 % additional, within 10 years, linear function*
Compensatory ratio	*1:2*

3.1.2 Results (Table 1)

Using Visual_HEA v2.6, Fig. 3 shows the evolution of the level of services supplied under the foundation over time.

In the same, this Fig. 4 shows the evolution of the level of services supplied by the compensatory mitigation in-kind project over time, due to the ecological restoration (realized in the same kind of ecosystem).

The results obtained from the HEA with the use of Visual_HEA v2.6 are summarized in Table 2.

Intermediary losses that occurred in the offshore windmill project are estimated at 260,000 discounted services per acreage and per years (DSAYs), between the year of the damage and the year when the initial state has been recovered.

After 25 years, the gains obtained from the compensatory measure are estimated at 226,561,213 DSAYs, i.e., 22,656 DSAYs per restored hectare. Hence, at the end of the project lifetime, each restored hectare would have supplied a gain of service of 2,265, 6 % in relation to the initial level of services provided by the sandy bottom habitat before the authorized impact.

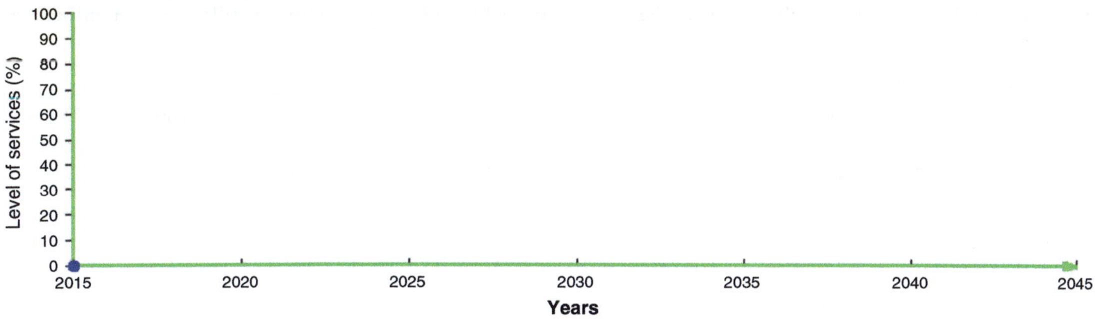

Fig. 3 Evolution of the level of services on the damaged area over time after the accident. Intermediary losses of services appear in *blue* (Snapshot of Visual_HEA v2.6)

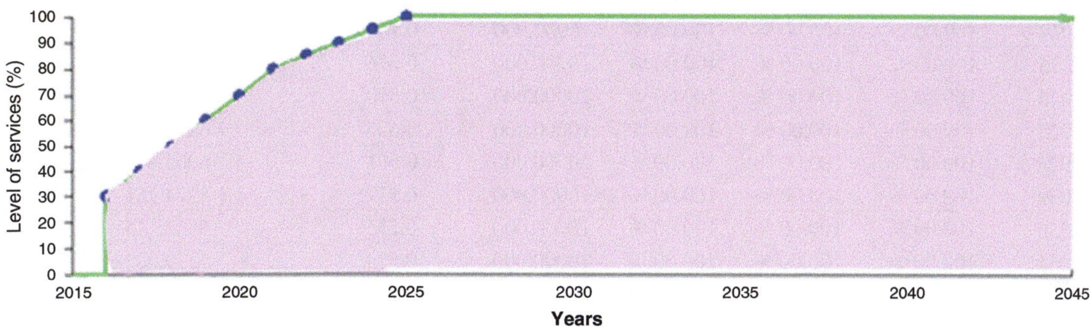

Fig. 4 Evolution of the level of services supplied by the compensatory mitigation in-kind project over time, due to the ecological restoration realized. Gains of services appear in *pink* (Snapshot of Visual_HEA v2.6)

So, in order to compensate in equivalency, with the intermediary losses of services, from the total destruction of 1 ha of sandy bottom (Fig. 4), the restoration project has to be carried out over 2,295 ha of sandy bottom area by releasing juvenile scallops, with a compensatory ratio of 1:2 due to the risk.

3.2 Discussion: Limitations of These Conceptual Approaches to Assess a Compensatory Mitigation by Equivalency

This tool (and all the over ones based on an equivalency in "nature") is based on concepts interesting to keep in mind. Actually, they enlighten the limits of the compensatory theoretical field and the "avoid, reduce, mitigate" process underlain by the "no net loss" goal, which had to overcome, to be applicable, 3 fundamental problems:

- Social aspect, *preference homogeneity of persons*: "no net loss" doesn't mean "nobody loss"; actually, to agree with the compensatory mitigation concept, we must assume that injured parties will agree a compensatory project in a different location, far away from their needs.
- Ecological aspect, *substitutability of resources and services*: a worldwide consensus in ecological restoration; assume that nobody can restore at 100 % an ecosystem lost (Moreno-Mateos et al. 2012). The ecological dynamic trajectories, especially the interaction between biological and biogeochemical or landscape effect, are so complex that each ecosystem is ecologically unique.

Table 2 Results of the HEA with the data of the previous impact for a fictitious authorized offshore windmill project (Snapshot of Visual_HEA v2.6)

Services losses at the Injury Area

Year	Intermediary services losses (%) Beginning	End	Average	Intermediary services losses average (SSA)	Discount factor	Discounted Intermediary services losses average (SSA)
2015	100.00 %	100.00 %	100.00 %	10000.000	1.000	10000.000
2016	100.00 %	100.00 %	100.00 %	10000.000	0.962	9615.385
2017	100.00 %	100.00 %	100.00 %	10000.000	0.925	9245.562
2018	100.00 %	100.00 %	100.00 %	10000.000	0.889	8889.964
2019	100.00 %	100.00 %	100.00 %	10000.000	0.855	8548.042
2020	100.00 %	100.00*	100.00 %	10000.000	0.822	8219.271
2021	100.00 %	100.00 %	100.00 %	10000.000	0.790	7903.145
2022	100.00 %	100.00 %	100.00 %	10000.000	0.760	7599.178
2023	100.00 %	100.00 %	100.00 %	10000.000	0.731	7306.902
2024	100.00 %	100.00 %	100.00 %	10000.000	0.703	7025.867
2025	100.00 %	100.00 %	100.00 %	10000.000	0.676	6755.642
2026	100.00 %	100.00 %	100.00 %	10000.000	0.650	6495.809
2027	100.00 %	100.00 %	100.00 %	10000.000	0.625	6245.970
2028	100.00 %	100.00 %	100.00 %	10000.000	0.601	6005.741
2029	100.00 %	100.00 %	100.00 %	10000.000	0.577	5774.751
2030	100.00 %	100.00 %	100.00 %	10000.000	0.555	5552.645
2031	100.00 %	100.00 %	100.00 %	10000.000	0.534	5339.082
2032	100.00 %	100.00 %	100.00 %	10000.000	0.513	5133.732
2033	100.00 %	100.00 %	100.00 %	10000.000	0.494	4936.281
2034	10000 %	100.00 %	100.00 %	10000.000	0.475	4746.424
2035	100.00 %	100.00 %	100.00 %	10000.000	0.456	4563.869
2036	100.00 %	100.00 %	100.00 %	10000.000	0.439	4388.336
2037	100.00 %	100.00 %	100.00 %	10000.000	0.422	4219.554
2038	100.00 %	100.00 %	100.00 %	10000.000	0.406	4057.263
2039	100.00 %	100.00 %	100.00 %	10000.000	0.390	3901.215
2040	100.00 %	100.00 %	100.00 %	10000.000	0.375	3751.168
2041	100.00 %	100.00 %	100.00 %	10000.000	0.361	3606.892
2042	100.00 %	100.00 %	100.00 %	10000.000	0.347	3468.166
2043	100.00 %	100.00 %	100.00 %	10000.000	0.333	3334.775
2044	100.00 %	100.00 %	100.00 %	10000.000	0.321	3206.514
2045	100.00 %	100.00 %	100.00 %	10000.000	0.308	3083.187
Beyond						77079.667
Total discounted service unit years lost average (SSAs):						**260000.000**

Services gains at the injury area

Year	Intermediary services gains (%) Beginning	End	Average	Intermediary services gains average (SSA)	Discount factor	Discounted intermediary services gained average (SSA)
2016	30.00 %	40.00 %	35.00 %	3500.000	0.962	3365.385
2017	40.00	50.00 %	45.00 %	4500.000	0.925	4160.503
2018	50.00 %	60.00 %	55.00 %	5500.000	0.889	4889.480

(continued)

Table 2 (continued)

Services gains at the injury area						
	Intermediary services gains (%)			Intermediary services gains		Discounted intermediary services gained average
Year	Beginning	End	Average	average (SSA)	Discount factor	(SSA)
2019	60.00 %	70.00 %	65.00 %	6500.000	0.855	5556.227
2020	70.00 %	80.00 %	75.00 %	7500.000	0.822	6164.453
2021	80.00 %	85.00 %	82.50 %	8250.000	0.790	6520.095
2022	85.00 %	90.00 %	87.50 %	8750.000	0.760	6649.281
2023	90.00 %	95.00 %	92.50 %	9250.000	0.731	6758.884
2024	95.00 %	100.00 %	97.50 %	9750.000	0.703	6850.221
2025	100.00 %	100.00 %	100.00 %	10000.000	0.676	6755.642
2026	100.00 %	100.00 %	100.00 %	10000.000	0.650	6495.809
2027	100.00 %	100.00 %	100.00 %	10000.000	0.625	6245.970
2028	100.00 %	100.00 %	100.00 %	10000.000	0.601	6005.741
2029	100.00 %	100.00 %	100.00 %	10000.000	0.577	5774.751
2030	100.00 %	100.00 %	100.00 %	10000.000	0.555	5552.645
2031	100.00 %	100.00 %	100.00 %	10000.000	0.534	5339.082
2032	100.00 %	100.00 %	100.00 %	10000.000	0.513	5133.732
2033	100.00 %	100.00 %	100.00 %	10000.000	0.494	4936.281
2034	100.00 %	100.00 %	100.00 %	10000.000	0.475	4746.424
2035	100.00 %	100.00 %	100.00 %	10000.000	0.456	4563.869
2036	100.00 %	100.00 %	100.00 %	10000.000	0.439	4388.336
2037	100.00 %	100.00 %	100.00 %	10000.000	0.422	4219.554
2038	100.00 %	100.00 %	100.00 %	10000.000	0.406	4057.263
2039	100.00 %	100.00 %	100.00 %	10000.000	0.390	3901.215
2040	100.00 %	100.00 %	100.00 %	10000.000	0.375	3751.168
2041	100.00 %	100.00 %	100.00 %	10000.000	0.361	3606.892
2042	100.00 %	100.00 %	100.00 %	10000.000	0.347	3468.166
2043	100.00 %	100.00 %	100.00 %	10000.000	0.333	3334.775
2044	100.00 %	100.00 %	100.00 %	10000.000	0.321	3206.514
2045	100.00 %	100.00 %	100.00 %	10000.000	0.308	3083.187
Beyond					77079.667	
Total discounted service unit years gained average (SSAs):						**226561.213**
Discounted services average gained per surface unity (SSAs):						**22.656**
Compensatory mitigation habitat size (sq. m): 2.000 × 260000/22.656						**22951.855**

- Economical aspect, *consistency and equal value of environmental goods over the time*: this last point assumes that people will consider the value of "*res natura*" without any fluctuation both economic and social over the time (the value of gas or oil, for instance, will probably change from now to few decades).

It's also important to clearly focus the engineering efforts on the very first phases of the "avoid, reduce, mitigate" design of a project. The stress must be maintained on these two first phases, because the best compensatory mitigation to offset residual impact is the one that we do not need to do. For offshore windmill, the position of the project, the technical construction choices, the planning, and of course the design of the foundations clearly need to be more effective to ensure their better integration in coastal ecosystem (see below Fig. 5, from Lacroix and Pioch 2011).

Fig. 5 Eco-designed offshore windmill foundation for fisheries enhancement and renewable energy production (in Lacroix and Pioch 2011)

4 Conclusion

Different methods have been developed to assess losses and gains related to biodiversity offsets. Despite the use of common factors (distinctiveness, condition, surface occupied by the species/habitat impacted, risk associated with the restoration technique used, time discount rate) to size ecological mitigation projects, very few software have emerged.

The use of consistent and robust methods has been widespread through this type of user-friendly devices combining accessibility and transparency. The main interest of HEA method is to try to provide a rational and reproducible process, based on simple and accessible data even for a "nonexpert" public. The use of an "integrator" or systemic proxy allows, with little financial means or techniques, to obtain (prior to the project) knowledge about the impact expected, and therefore what could be the compensatory mitigation surfaces and project to implement for a "fair" compensation: no more but also no less (Strange et al. 2002; Zafonte and Hampton 2007). The goal is obviously to choose the best alternative project, with less impact to the environment.

But HEA couldn't (1) help to choose the "best" compensatory mitigation option nor to (2) take account of the landscape effect (3) or anticipate the future use dynamics of the area.

Nevertheless, the score given to assess the gains or losses of ecosystem services can promote negotiations on surfaces to compensate in nature and not on issues related to financial losses, unrelated to the restoration of a natural environment or achieving the goal of "no net loss." This financial compensation may be done under accompanying measure proposals, but it cannot be used as part of compensatory mitigation by ecological equivalency, as it could be seen in several projects. Finally, passing through a method aims to assess the surface to restore in equivalency the losses of nature, with ecological engineering techniques, rather to put a price on nature or assess the monetary losses from ecosystemic services destruction, avoiding a sense of "entitlement to destroy" when the applicant simply pays to offset the economic impacts of its ecosystem destruction, for instance.

In offshore windmill projects, for fishermen, we assume that it is more valuable to (1) negotiate in-kind mitigation for sandy bottom with scallop enhancement and (2) even additional out of kind restoration with artificial reef (hard substratum) dedicated to lobsters or valuable fishes over-targeted by fisheries or conservation actions for their fishing grounds (BBOP 2009). We propose this hierarchy in compensatory mitigation sized with Visual_HEA, rather than monetary compensatory income which is difficult to explain for the next generation of fishermen as a "fair" offset for the destruction of their "working tool": the fishing ground.

Nature has no price, but it has a value!

Acknowledgments Our thanks to Hélène Gaubert and Doris Nicklaus from MEDDEM–CGDD (French Ministry of Environment). We also would like to thank the NOVA Oceanographic Center team members who helped share their network of relations. For testing and helpful remarks, we thank Adeline Bas and Anne-Charlotte Vaissière from IREMER/UBO – AMURE and Celine Jacob From UM3–CEFE.

The bases of this communication for the French–Japanese symposium in Marseille and its proceedings are also developed in a submitted article, in review for publication at the time of this communication: *Pioch, S., Johnston, M., Vaissière, A.C., Berger, F., Jacob, C., Dodge, R. (in review) "Scoring method to sizing ecological mitigation projects based on rapid ecological function assessment: Visual_HEA V2.6 free software."*

References

Bas A, Gaubert H (2010) La directive « responsabilité environnementale » et ses méthodes d'équivalence. Etudes et documents n°19, Service de l'Economie, de l'Evaluation et de l'Intégration du Développement Durable (SEEIDD) du Commissariat Général au Développement Durable (CGDD), MEDDAT, 92 p

BBOP (Business and Biodiversity Offsets Programme) (2009) Business, biodiversity offsets and BBOP: an overview. BBOP, Washington, DC, 45 p

CWA (Clean Water Act) (1972) 33 U.S.C. §1251 et seq.

Décret n°2009-468 du 23 avril 2009 relatif à la prévention et à la réparation de certains dommages causés à l'environnement, Journal Officiel de la République Française, 26 avril 2009, texte 1 sur 42 p

Directive 2004/35/CE L143: 56–75, European Commission

Dunford RW, Ginn TC, Desvousges WH (2004) The use of habitat equivalency analysis in natural resource damage assessments. Ecol Econ 48(1):49–70

Fifas S (2004) La Coquille Saint-Jacques en Bretagne. Direction des ressources vivantes, Ifremer:15

Gaubert H, Hubert S (2013) The Environmental Liability Law (ELL) and the equivalency methods. Methodological Guide, 213 p

Kohler KE, Dodge RE (2006) Visual_HEA: Habitat Equivalency Analysis software to calculate compensatory restoration following natural resource injury. In: Proceedings of 10th international coral reef symposium, Japan, pp 1611–1616

Lacroix D, Pioch S (2011) The multi-use in wind farm projects: more conflicts or a win-win opportunity? Aquat Living Resour 24:129–135

Levrel H, Pioch S, Spieler R (2012) Compensatory mitigation in marine ecosystems: which indicators for assessing the "no net loss" goal of ecosystem services and ecological functions? Mar Policy 36(6):1202–1210

Loi n°2008-757 du 1er août 2008 relative à la responsabilité environnementale et à diverses dispositions d'adaptation au droit communautaire dans le domaine de l'environnement. Journal officiel de la République Française, 2 août 2008, texte 2 sur 107

Mazzotta MJ, Opaluch JJ, Grigalunas TA (1994) Natural resource damage assessment: the role of restoration. Nat Resour J 34(Winter):153–178

MEA (2005) Ecosystems and human well-being: synthesis, Millennium Ecosystem Assessment. Island Press, Washington, DC, 155 p. MEA, 2005b

Milon JW, Dodge RE (2001) Applying habitat equivalency analysis for coral reef damage assessment and restoration. Bull Mar Sci 69(2):975–988

Monnery J, Hubert S, Gaubert H (2011) Application des méthodes d'équivalence à la pollution accidentelle du Gave d'Aspe. Etudes et documents n°47, Service de

l'Economie, de l'Evaluation et de l'Intégration du Développement Durable (SEEIDD) du Commissariat Général au Développement Durable (CGDD), MEDDAT, 187 p
Moreno-Mateos D, Power ME, Comin FA, Yockteng R (2012) Structural and functional loss in restored wetland ecosystems. Plos Biol 10(1), 8 p
NEPA (National Environmental Policy Act) (1969) Pub. L. No. 91-190, § 102, 83 Stat. 852 (1970) (codified as amended at 42 U.S.C. §§ 4321–4370a (1988))
NOAA (1995) Habitat equivalency analysis: an overview. NOAA damage assessment and restoration program, policy and technical paper series no. 95(1) (up to date in 2000 & 2006)
Pioch S (2010) Mesures compensatoires dans les écosystèmes marins en Floride. Exemple et expériences d'aménagements pour la production de services écosystémiques. Rapport EGIS-EAU, IFREMER, NSU, Agence des AMP, CDC Biodiversité, 131 p
REMEDE "Resource Equivalency Methods for assessing Environmental Damage in the European United" (2008) Toolkit for performing Habitat Equivalency Analysis to assess and scale environmental damage in the European Union, 254 p
Roach B, Wade WW (2006) Policy evaluation of natural resource injuries using habitat equivalency analysis. Ecol Econ 58(1):421–433
Sibbing JM (2008) Nowhere near no-net-loss. National Wildlife Federation. http://www. nwf. org/wildlife/pdfs/NowhereNearNoNetLoss.pdf. Reviewed: June
Society for Ecological Restoration International Science & Policy Working Group (2004) The SER international primer on ecological restoration. www.ser.org &Tucson: Society for Ecological Restoration International
Strange E, Galbraith H, Bickel S, Mills D, Beltman D, Lipton J (2002) Determining ecological equivalence in service-to-service scaling of salt marsh restoration. Environ Manag 29(1):290–300
Unsworth RE, Bishop RC (1994) Assessing natural resource damages using environmental annuities. Ecol Econ 11:35–41
Worm B, Barbier EB, Beaumont N, Duffy JE, Folke C, Halpern BS, Jackson JBC, Lotze HK, Micheli F, Palumbi SR, Sala E, Selkoe KA, Stachowicz JJ, Watson R (2006) Impacts of biodiversity loss on ocean ecosystem services. Science 314:787
Zafonte M, Hampton S (2007) Exploring welfare implications of resource equivalency analysis in natural resource damage assessments. Ecol Econ 61(1):134–145

Ecosystem-Based Versus Species-Based Approach for Assessment of the Human Impact on the Mediterranean Seagrass *Posidonia oceanica*

Charles F. Boudouresque, Sébastien Personnic, Patrick Astruch, Enric Ballesteros, Denise Bellan-Santini, Patrick Bonhomme, Delphine Botha, Eric Feunteun, Mireille Harmelin-Vivien, Gérard Pergent, Jérémy Pastor, Jean-Christophe Poggiale, Florent Renaud, Thierry Thibaut, and Sandrine Ruitton

Abstract

Biotic indices, which reflect the quality of the environment, are widely used in the marine realm. The *Posidonia oceanica* seagrass meadow is a benthic sublittoral habitat characteristic of the Mediterranean Sea. The Ecosystem-Based Quality Index (EBQI) is based upon the overall functioning of the *Posidonia oceanica* ecosystem, from primary producers and the litter to top predators. It is compliant with the Marine Strategy Framework Directive (MSFD) of the European Union (EU). It provides managers with a response that is very different from that of the indices only based upon the key species *P. oceanica* and a few affiliate species.

C.F. Boudouresque (✉) • S. Personnic • D. Botha
M. Harmelin-Vivien • J.-C. Poggiale • T. Thibaut
S. Ruitton
Aix-Marseille University and Toulon University (France), Mediterranean Institute of Oceanography (MIO), CNRS/IRD, UM 110, Pytheas Institute, Campus of Luminy, Marseilles, France
e-mail: charles.boudouresque@mio.osupytheas.fr

P. Astruch • P. Bonhomme
GIS Posidonie, Pytheas Institute, Aix-Marseille University, Campus of Luminy, Marseilles, France

E. Ballesteros
Centre d'Estudis Avançats de Blanes, CSIC, Blanes, Spain

D. Bellan-Santini • F. Renaud
Institut Méditerranéen de Biodiversité et d'Ecologie (IMBE), Pytheas Institute, Aix-Marseille University, Station Marine d'Endoume, Marseilles, France

E. Feunteun
Museum National d'Histoire Naturelle, Station Marine de Dinard, Dinard, France

G. Pergent
Equipe Ecosystèmes Littoraux, University of Corsica Pasquale Paoli, Corti, France

J. Pastor
Centre de Formation et de Recherche sur les Environnements Méditerranéens, Perpignan University, Perpignan, France

H.-J. Ceccaldi et al. (eds.), *Marine Productivity: Perturbations and Resilience of Socio-ecosystems*,
DOI 10.1007/978-3-319-13878-7_25, © Springer International Publishing Switzerland 2015

While the species-based indices are mainly designed for assessing the water quality of a waterbody (e.g. pollution), the EBQI provides a picture of the actual status of the ecosystem functioning, including not only pollution but also the whole range of human impacts, from anchoring to overfishing.

1 The Issue

Biotic indices are extensively used in the marine realm (1) to assess the quality of a waterbody; (2) to assess processes such as currents, sedimentation and climate under natural and anthropogenic forcing; and (3) to monitor the status of species or communities of interest, either emblematic species, indicators of ecosystem health or indicators of pollution (e.g. Orfanidis et al. 2001, 2003, 2011; Pergent-Martini et al. 2005). Sometimes, species used as biological indicators are key species or ecosystem engineers. This is the case of the seagrass *Posidonia oceanica* (Linnaeus) Delile (Boudouresque et al. 2006; Romero et al. 2007; Gobert et al. 2009; Lopez y Royo et al. 2010, 2011; Marbà et al. 2012).

However, does an indicator based on the health of a species, or a few species, give information about the health of the entire ecosystem to which it belongs? Here, we use a recently proposed ecosystem-based index of the *P. oceanica* seagrass meadow (EBQI) and compare it with species-based indices.

2 The Conceptual Model and the EBQI

The seagrass *Posidonia oceanica* is endemic to the Mediterranean Sea. It constitutes extensive meadows between the mean sea level and down to 30–40 m in depth (Molinier and Picard 1952; Boudouresque and Meinesz 1982; Boudouresque et al. 2009, 2012; Pergent et al. 2012).

We used an updated version (Fig. 1; Ruitton et al. 2013; Personnic et al. 2014) of the conceptual model of the functioning of the *P. oceanica* ecosystem proposed by Boudouresque et al. (2012). Interestingly, *P. oceanica* ecosystem itself is closely coupled with the pelagic coastal ecosystem (Plankton, POM, planktivorous teleosts) and more loosely coupled with terrestrial ecosystems through seabirds (Morat et al. 2011) and dead *P. oceanica* leaves (Cardona et al. 2007).

The rationale governing the EBQI (Ecosystem-Based Quality Index) is (1) trying to quantify and assess some compartments (boxes 1 through 13; Fig. 1) of the conceptual model by means of a set of parameters, (2) determining their relative weight and (3) by using a simple algorithm, calculating a rank for the ecosystem status within a given area, matching the five classes of the ecological status of the European Union Water Framework Directive (WFD) (Water Framework Directive 2000), from bad to high (Ruitton et al. 2013; Personnic et al. 2014).

The status of each functional compartment (box) was assessed by means of a semi-quantitative scale (4 through 0), from very good (4) to very bad (0). Calibration of the scale was based upon the available literature (e.g. UNEP-MAP-RAC/SPA 2011). Compartments were weighted according to their relative importance in the ecosystem functioning, from 5 (highest weighting) to 1 (lowest weighting). The grade for each compartment was given by its status (0 through 4), multiplied by its weighting (1 through 5) and was therefore graded from 0 to 4 and 0 to 20 (depending upon the weighting of the considered compartment). The grades of all compartments were added up, which gave the final grade of the ecosystem status (EBQI) at a given site. For practical purposes, the EBQI was converted to a scale from 0 to 10 (Table 1). The EBQI was applied to 17 localities of the northwestern Mediterranean, from the Balearic Islands to the French Riviera and Corsica (Table 1; Personnic et al. 2014).

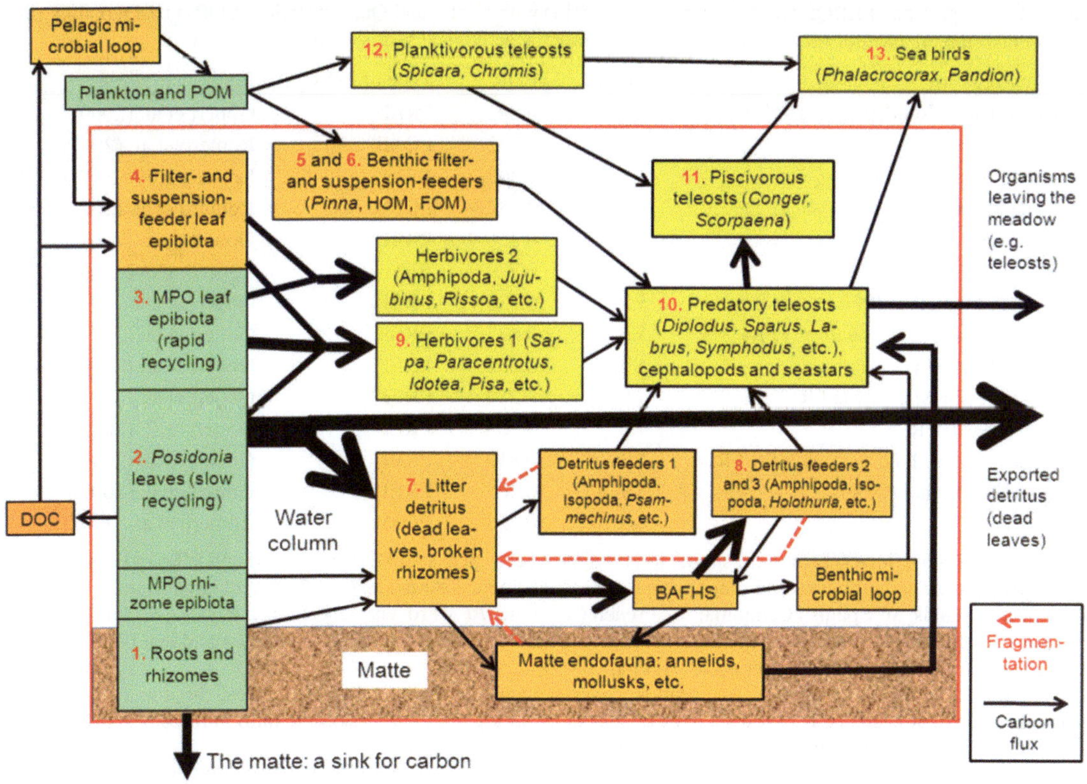

Fig. 1 A conceptual model of the functioning of the *Posidonia oceanica* seagrass ecosystem. Functional compartments (*boxes*): primary producers are in *green*; filter feeders, suspension feeders, litter, detritus feeders, dissolved organic carbon (DOC) and microbial loops are in *orange*; predators (including herbivores) are in *yellow*. *POM* particulate organic carbon, *BAFHS* bacteria, archaea, fungi and heterotrophic stramenopiles involved in the litter degradation. The width of the *arrows* roughly represents the volume of the carbon flow. The *P. oceanica* ecosystem properly speaking is included within the *red rectangle*. Figures (*1* through *13*, in *red*) correspond to the compartments (*boxes*) taken into account by the EBQI

Table 1 Status of conservation of the *P. oceanica* ecosystem (EBQI) in Balearic Islands (BI), Spanish Catalonia (SC), French Catalonia (FC), Provence and French Riviera (PFR) and Corsica (C)

	Compartment	1	2	3–4	5	6	7	8	9	10	11	12	SRDI	13	EBQI
	Weight	3	5	4	2	2	2	2	5	5	5	3	3	1	
LOCALITIES	Espardell (BI)	4	4	3	3	3	3	3	3	1	1	1	3	2	6.4
	Sitges (SC)	2	0	0	0	2	2	2	3.5	0	0	1	0	0	2.3
	Tossa de Mar (SC)	2	3	4	0	2	2	4	3	2	0	2	3	1	5.6
	Medes Islands (SC)	2	3.5	4	3	2	2	2	2.5	4	4	3	4	2	7.9
	Peyrefite Bay (FC)	2	3.5	2	4	2	2	2	2	3	1	1.5	4	0	5.8
	Niolon (PFR)	2	2.5	2	0	1.5	1	3	2	1	0	2	2	1	3.9
	Prado Bay (PFR)	2	2.5	2	0	2.5	2	3	2.5	3	1	1.5	3	2	5.3
	Plateau des chèvres (PFR)	2	2.5	4	0	1.5	2	3	2.5	2	1	0.5	2	2	5.0
	Saint-Cyr Bay (PFR)	1	3	2	1	2	2	2	2	2	2	2	2	0.5	4.9
	Gulf of Giens (PFR)	3	4	2	2	2	1	3	1.5	1	0	1	1	0.5	4.3
	Porquerolles North (PFR)	3	2	3	2	2	0	1	1.5	1	1	2	2	1	4.3
	Porquerolles South (PFR)	3	4	4	3	3	3	2	2	2	2	3	3	1	6.9
	Bagaud Pass (PFR)	4	3	2	4	3	4	4	3	3	2	3	4	1	7.6
	Port-Cros South (PFR)	4	4	4	4	3	4	3	3.5	4	4	3	4	1.5	9.3
	Villefranche Bay (PFR)	2	1.5	2	1	3	0	0	2	3	2	1.5	4	0	4.8
	Scàndula, Elbu Bay (C)	4	3	1	4	3	2	2	2	2	1	1.5	3	4	5.7
	Valincu Gulf (C)	4	3	2	2	2	2	3	2	2	2	2	2	1	5.4

For each compartment (see Fig. 1): the weighting (1 through 5) and the status mark (0 through 4) at the 17 studied localities. EBQI ranges from 0 to 10. *SRDI* Specific Relative Diversity index of fish. Ecological status classes: high (deep blue), good (light blue), moderate (green), poor (orange) and bad (red)

Table 2 Comparison of EBQI (from Personnic et al. 2014) with Ecological Quality Ratios (EQRs) based mainly upon *P. oceanica* (the organism itself) and aimed at establishing the ecological status of the seawater body

Locality (region)	EBQI	EQR	Type of EQR	Reference
Port-Cros Island, southern coast (PFR)	9.3	0.802	PREI[a]	Gobert et al. (2009)
Medes Islands (SC)	7.9	0.752	POMI[b]	Romero et al. (2007)
Scandola, Elbu Bay (C)	5.7	0.802	BiPo[c]	Lopez y Royo et al. (2010)
Tossa de Mar (SC)	5.6	0.682	POMI	Romero et al. (2007)
Valincu Gulf (C)	5.4	0.386	PREI	Gobert et al. (2009)
		0.729	BiPo	Lopez y Royo et al. (2010)
Prado Bay, Marseilles (PFR)	5.3	0.636	PREI	Gobert et al. (2009)
Plateau des chèvres, Marseilles (PFR)	5.0	0.477	PREI	Gobert et al. (2009)
Saint Cyr Bay (PFR)	4.9	0.682	PREI	Gobert et al. (2009)
Villefranche-sur-Mer Bay (PFR)	4.8	0.280	PREI	Gobert et al. (2009)
Gulf of Giens (PFR)	4.3	0.708	PREI	Gobert et al. (2009)
Porquerolles Island, northern coast (PFR)	4.3	0.819	PREI	Gobert et al. (2009)
Niolon (Côte Bleue) (PFR)	3.9	0.465	PREI	Gobert et al. (2009)
Sitges (SC)	2.3	0.238	POMI	Romero et al. (2007)

EBQI ranges from 0 (lowest ecological status) to 10 (highest ecological status). PREI, POMI and BiPo indices are based upon distinct but similar metrics and range from 0 (lowest ecological status) to 1 (highest ecological status)

BI Balearic Islands, *SC* Spanish Catalonia, *FC* French Catalonia, *PFR* Provence and French Riviera, *C* Corsica

[a]The metrics of PREI are shoot density, shoot leaf surface area, ratio between epibiota biomass and leaf biomass, depth of the lower limit and type of this limit (Gobert et al. 2009)

[b]The metrics of POMI are shoot density, meadow cover, percentage of plagiotropic rhizomes, shoot leaf surface area, percentage of foliar necrosis, P, N and sucrose content in rhizomes, ^{15}N and ^{34}S isotopic ratio in rhizomes, N content in epiphytes, Cu, Pb and Zn content in rhizomes (Romero et al. 2007)

[c]The metrics of BiPo are shoot density, shoot leaf surface area, lower depth limit and lower limit type (Lopez y Royo et al. 2010)

The five ecological status classes of the EBQI are (Personnic et al. 2014): bad (EBQI<3.5), poor (3.5≥EBQI<4.5), moderate (4.5≥EBQI<6), good (6.0≥EBQI<7.5) and high (EBQI≥7.5) (Table 1).

3 A Strongly Contrasting Response Compared with that of Species-Based Indices

A number of species-based biotic indices (Ecological Quality Ratios, EQRs), based upon *P. oceanica* alone, or upon *P. oceanica* and leaf epibiota, have been proposed, e.g. PREI, POMI and PiBo (Romero et al. 2007; Gobert et al. 2009; Lopez y Royo et al. 2010). There is no significant correlation between these indices and the EBQI (Personnic et al. 2014). It is worth noting that the former indices were not designed to assess the ecosystem health, but other parameters such as the water quality of a waterbody. The contrasting ranks of Porquerolles Island (northern coast) from EBQI (poor) to EQR (high, first ranking, Table 2), together with those of Valincu Gulf and the Gulf of Giens, may be due to impacts other than the water quality, e.g. artisanal and recreational overfishing and anchoring of pleasure boats.

The ecosystem response can therefore be quite different from the simple key species one. This is far from unexpected. Species-based indices only take into account the seagrass or sometimes the seagrass and its leaf epibionts. In nonpolluted areas, the seagrass can look healthy (e.g. coverage, shoot density, leaf length), while most of the functional compartments of the ecosystem (e.g. top predator fish) have collapsed (Fig. 2). This may, of course, look like a purely theoretical view, since the collapse of a compartment (e.g. top predator fish) should affect all other

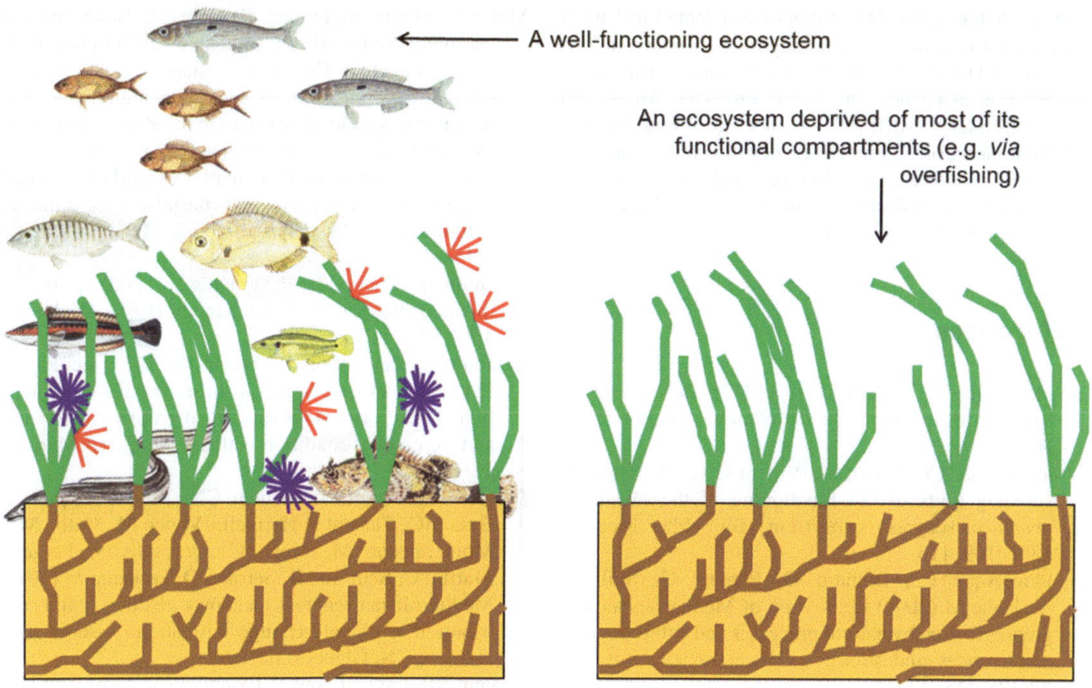

Fig. 2 *Left*. A supposed pristine *P. oceanica* ecosystem, with species belonging to all functional compartments: the seagrass in *brown* (rhizomes) and *green* (leaves) and leaf epibionts in *red* (primary producers); the sea urchin *Paracentrotus lividus* in *purple* (herbivores) and teleosts (predators, top predators and planktivores). *Right*. A *P. oceanica* meadow deprived of most of its functional compartments (e.g. via overfishing), which could be considered as healthy on the basis of seagrass descriptors, such as shoot density and meadow coverage

compartments, including the seagrass itself via cascade effect (Sala et al. 1998). However, this is not in fact such a theoretical approach, as there is no correlation between the EBQI and the species-based indices (EQRs).

4 Conclusion

The noncongruence of the EBQI, i.e. the quality of the ecosystem functioning, with the empirical idea one may have of the *P. oceanica* ecosystem at a given site, due to, e.g. the health of the seagrass and the clearness of the water (and even the beauty of the landscape and the seascape), confirms the usefulness of the EBQI index, based upon the whole ecosystem rather than solely upon the seagrass itself, sometimes together with some affiliate species. Other anthropogenic impacts, e.g. overfishing, are putatively more important than the above-mentioned parameters with regard to the structure and functioning of the ecosystem.

The EBQI that is proposed here for the *P. oceanica* ecosystem could constitute a model for similar indices designed for other marine ecosystems, such as coralligenous outcrops, underwater caves, soft bottoms and sublittoral reefs. Such indices would constitute an innovative category of biological indices, namely, ecosystem-based indices, both in the terrestrial and the marine realms. As in the marine environment, the health assessment of terrestrial habitats is usually based upon a few species or a single functional compartment (e.g. primary producers or trees) rather than on the whole ecosystem functioning, including wolves, bears and eagles, as far as Western Europe is concerned.

Acknowledgements The authors are indebted to Fabien Morat for data concerning the shag *Phalacrocorax aristotelis*, Boris Daniel (AAMP, Agence des aires marines protégées; French agency for Marine Protected Areas) and Pierre Boissery (AERMC, Agence de l'Eau Rhône-Méditerranée-Corse; French agency for water management in south-eastern France and Corsica) for contributions and valuable comments and to Michael Paul for improvement of the English text.

References

Boudouresque CF, Meinesz A (1982) Découverte de l'herbier de posidonie. Parc National de Port-Cros, Hyères

Boudouresque CF, Mayot N, Pergent G (2006) The outstanding traits of the functioning of the *Posidonia oceanica* seagrass ecosystem. Biol Mar Mediterr 13(4):109–113

Boudouresque CF, Bernard G, Pergent G, Shili A, Verlaque M (2009) Regression of Mediterranean seagrasses caused by natural processes and anthropogenic disturbances and stress: a critical review. Bot Mar 52:395–418

Boudouresque CF, Bernard G, Bonhomme P, Charbonnel E, Diviacco G, Meinesz A, Pergent G, Pergent-Martini C, Ruitton S, Tunesi L (2012) Protection and conservation of *Posidonia oceanica* meadows. RAMOGE and RAC/SPA, Tunis

Cardona L, Revelles M, Sales M, Aguilar A, Borrell A (2007) Meadows of the seagrass *Posidonia oceanica* are a significant source of organic matter for adjoining ecosystems. Mar Ecol Prog Ser 335:123–131

Gobert S, Sartoretto S, Rico-Raimondino V, Andral B, Chery A, Lejeune P, Boissery P (2009) Assessment of the ecological status of Mediterranean French coastal waters as required by the Water Framework Directive using the *Posidonia oceanica* Rapid Easy Index: PREI. Mar Pollut Bull 58:1727–1733

Lopez y Royo C, Casazza G, Pergent-Martini C, Pergent G (2010) A biotic index using the seagrass *Posidonia oceanica* (BiPo), to evaluate ecological status of coastal waters. Ecol Indic 10:380–389

Lopez y Royo C, Pergent G, Alcoverro T, Buia MC, Casazza G, Martínez-Crego B, Pérez M, Silvestre F, Romero J (2011) The seagrass *Posidonia oceanica* as indicator of coastal water quality: experimental intercalibration of classification systems. Ecol Indic 11:557–563

Marbà N, Krause-Jensen D, Alcoverro T, Birk S, Pedersen A (2012) Diversity of European seagrass indicators: patterns within and across regions. Hydrobiologia 704:265–278

Molinier R, Picard J (1952) Recherches sur les herbiers de Phanérogames marines du littoral méditerranéen français. Ann Inst Océanogr 27:157–234

Morat F, Mante A, Drunat E, Dabat J, Bonhomme P, Harmelin-Vivien M, Letourneur Y (2011) Diet of the Mediterranean European shag, *Phalacrocorax aristotelis desmarestii*, its ecological significance and interaction with local fisheries in the Riou Archipelago (Marseilles, France). Vie Milieu 61(2):77–86

Orfanidis S, Panayotidis P, Stamatis N (2001) Ecological evaluation of transitional and coastal waters: a marine benthic macrophytes-based model. Mediterr Mar Sci 2:45–65

Orfanidis S, Panayotidis P, Stamatis N (2003) An insight to the ecological evaluation index (EEI). Ecol Indic 3:27–33

Orfanidis S, Panayotidis P, Ugland KI (2011) Ecological Evaluation Index continuous formula (EEI-c) application: a step forward for functional groups, the formula and reference condition values. Mediterr Mar Sci 12:199–231

Pergent G, Bazairi H, Bianchi CN, Boudouresque CF, Buia MC, Clabaut P, Harmelin-Vivien M, Mateo MA, Montefalcone M, Morri C, Orfanidis S, Pergent-Martini C, Semroud R, Serrano O, Verlaque M (2012) Mediterranean seagrass meadows: resilience and contribution to climate change mitigation. A short summary. IUCN, Gland/Málaga

Pergent-Martini C, Leoni V, Pasqualini V, Ardizzone GD, Balestri E, Bedini R, Belluscio A, Belsher T, Borg J, Boudouresque CF, Boumaza S, Bouquegneau JM, Buia MC, Calvo S, Cebrian J, Charbonnel E, Cinelli F, Cossu A, Di Maida G, Dural B, Francour P, Gobert S, Lepoint G, Meinesz A, Molenaar H, Mansour HM, Panayotidis P, Peirano A, Pergent G, Piazzi L, Pirrotta M, Relini G, Romero J, Sanchez-Lizaso JL, Semroud R, Shembri P, Shili A, Tomasello A, Velimirov B (2005) Descriptors of *Posidonia oceanica* meadows: use and application. Ecol Indic 5:213–230

Personnic S, Boudouresque CF, Astruch P, Ballesteros E, Blouet S, Bellan-Santini D, Bonhomme P, Thibault-Botha D, Feunteun E, Harmelin-Vivien M, Pergent G, Pergent-Martini C, Pastor J, Poggiale JC, Renaud F, Thibaut T, Ruitton S (2014) An ecosystem-based approach to assess the status of a Mediterranean ecosystem, the *Posidonia oceanica* seagrass meadow. Plos One 9(6):1–17 (e98994)

Romero J, Martínez-Crego B, Alcoverro T, Pérez M (2007) A multivariate index based on the seagrass *Posidonia oceanica* (POMI) to assess ecological status of coastal waters under the water framework directive (WFD). Mar Pollut Bull 55:196–204

Ruitton S, Boudouresque CF, Personnic S, David R et al (2013) An ecosystem-based approach to evaluate the status of Mediterranean ecosystem habitats. Rapp Comm Int Mer Méditerr 40:577

Sala E, Boudouresque CF, Harmelin-Vivien M (1998) Fishing, trophic cascades and the structure of algal assemblages: evaluation of an old but untested paradigm. Oikos 82:425–439

UNEP-MAP-RAC/SPA (2011) Draft guidelines for the standardization of mapping and monitoring methods of marine magnoliophyta in the Mediterranean. Working document, prepared by C. Pergent-Martini for the tenth meeting of Focal Points for SPAs, Marseilles, France, 17–20 May 2011, UNEP(DEPI)/MED WG 359/9. RAC/SPA, Tunis

Water Framework Directive (2000) The European Union Water Framework Directive (WFD). Directive 2000/60/EC of the European Parliament and of the Council of 23rd October 2000 establishing a framework for Community action in the field of water policy. Off J Eur Union L 327/1:1–72

Measurement of Fish Habitat Use by Fish-Mounted Data Loggers for Integrated Coastal Management: An Example of Japanese Sea Bass (*Lateolabrax japonicus*) in Tokyo Bay

Hideaki Tanoue, Nobuyuki Miyazaki, Takashi Niizawa, Koichiro Mizushima, Michihiko Suzuki, Sandrine Ruitton, Sebastián Biton Porsmoguer, Natheer Alabsi, Sara Gonzalvo, Masahiko Mohri, Akira Hamano, and Teruhisa Komatsu

Abstract

Effective integrated coastal management requires an understanding of how natural organisms use coastal areas, to protect those species. In this study, we used fish-mounted data loggers to examine the habitat use of the Japanese sea bass (*Lateolabrax japonicus*), a representative fish species in Tokyo Bay. Footage from the micro-video cameras on the fish showed Japanese sea bass swimming with other conspecifics. The three loggers obtained a time series of water temperature, salinity, swimming depth, and three-axis acceleration for a total of 60 h. These records indicated that the river, the ship port, and the fishing port were all important habitats for the Japanese sea bass. Based on our results, we discuss how such data can be used for conservation policies for natural organisms in integrated coastal management.

H. Tanoue (✉)
Department of Fisheries Science and Technology, National Fisheries University, Shimonoseki, 2-7-1, Nagata-Honmachi 759-6595, Japan

Aix-Marseille University, Mediterranean Institute of Oceanography (MIO), Université du Sud Toulon-Var, CNRS/INSU, IRD, UM 110, Campus universitaire de Luminy, case 901, 13288 Marseille cedex 09, France
e-mail: h-tanoue@09.alumni.u-tokyo.ac.jp

N. Miyazaki • N. Alabsi • S. Gonzalvo • T. Komatsu
Atmosphere and Ocean Research Institute, The University of Tokyo, Kashiwa 277-8564, Japan

T. Komatsu
Atmosphere and Ocean Research Institute (AORI), University of Tokyo, President of S.F.J.O. (*) of Japan, 5-1-5, Kashiwanoha, 277-8564 Kashiwa-shi, Japan

T. Niizawa • K. Mizushima
Technical Headquarters, Sanyo Techno Marine Co., Tokyo 103-0012, Japan

M. Suzuki
Headquarters, Little Leonardo Co., Tokyo 113-0021, Japan

S. Ruitton • S. Biton Porsmoguer
Aix-Marseille University, Mediterranean Institute of Oceanography (MIO), Université du Sud Toulon-Var, CNRS/INSU, IRD, UM 110, Campus universitaire de Luminy, case 901, 13288 Marseille cedex 09, France

M. Mohri • A. Hamano
Department of Fisheries Science and Technology, National Fisheries University, Shinomoseki, 2-7-1, Nagata-Honmachi 759-6595, Japan

H.-J. Ceccaldi et al. (eds.), *Marine Productivity: Perturbations and Resilience of Socio-ecosystems*,
DOI 10.1007/978-3-319-13878-7_26,

1 Introduction

To sustainably utilize coastal areas, human coastal activities must be harmonized with the natural environment (Komatsu et al. 2012). Integrated coastal management perceives coastal areas as natural systems, involves stakeholders in a proactive role, provides integrated management plans, and promotes the coexistence of humans and the natural environment in coastal waters in an integrated and planned manner. Effective integrated coastal management requires an understanding of how natural organisms use coastal areas, to protect those species.

Tokyo Bay has historically been a productive fishery. Commercial fisheries peaked at approximately 140,000 tons in 1960, but the catch has been decreasing since then (Shimizu 1997). A long-term scientific study of fish biomass conducted in Tokyo Bay using bottom trawling showed an increase in the biomass of Japanese sea bass (*Lateolabrax japonicus*) over time (e.g., Kodama et al. 2010). However, Japanese sea bass spend their lives not only in the sea but also in brackish waters (Shoji 2002). Knowing precisely what areas they use would be useful so that these areas could be appropriately managed. Because fishing is prohibited in many brackish waters of Tokyo Bay, such as around ports and waterways where large vessels pass or anchor, habitat use of fish in these areas cannot be monitored by fishing.

Recently, data loggers and a micro-underwater video camera have become small enough to be attached to small fish to measure acceleration, water temperature, depth, and conductivity (e.g., Alabsi et al. 2011; Komatsu et al. 2011; Tanoue et al. 2013; Kudo et al. 2007). The acceleration sensor measures the intensity of swimming behavior and the number of active burst events by fish such as feeding (e.g., Tanoue et al. 2012). Small ultrasonic and VHF transmitters can be attached to the fish along with the data loggers to track fish movement and aid in the recovery of the data loggers after a time-scheduled release from the fish. In addition, salinity data obtained with a conductivity sensor are useful to detect habitat use near river mouths. These technological advances make it practical to use data loggers to observe fish behaviors.

In this study, we examined habitat use of Japanese sea bass, a representative fish species in Tokyo Bay, by using micro-data loggers with sensors for three-axis acceleration, depth, water temperature, and conductivity, in addition to micro-underwater video cameras and ultrasonic and VHF transmitters.

2 Materials and Methods

2.1 Study Site

The catch of Japanese sea bass in Chiba Prefecture (Fig. 1) is the greatest in Japan. Between 1958 and 1997, 85 % of the fish that landed from Tokyo Bay were in Chiba Prefecture (Shoji 2002). Banzu Tidal Flat is an area of 1,400 ha at the mouth of the Obitsu River, which discharges freshwater into the east of Tokyo Bay (Fig. 1). This area is near fishing and general ports and is a habitat of young Japanese sea bass. The river mouth of the Obitsu River and its neighboring areas are a fishing ground for adult Japanese sea bass. For these reasons, we conducted our survey in the waters near the river mouth of the Obitsu River (Fig. 1).

2.2 Survey of Habitat Use by Japanese Sea Bass by Using Data Loggers

Field surveys of habitat use by Japanese sea bass were conducted in October 2011. Several data collection devices were used in two combinations. The first system included a salinity data logger (DSL, Little Leonardo Co., Japan), a three-axis micro-acceleration data logger with depth and water temperature sensors (ORI380-D3GT, Little Leonardo Co.), an underwater video camera (SSP Co., Japan), an ultrasonic transmitter (V9, Vemco Co., Canada), and a releasing device (Little Leonardo Co.) with a float (NiGK Co., Japan) connected to a mount. The second system included the same salinity and three-axis micro-acceleration data loggers, the same releasing

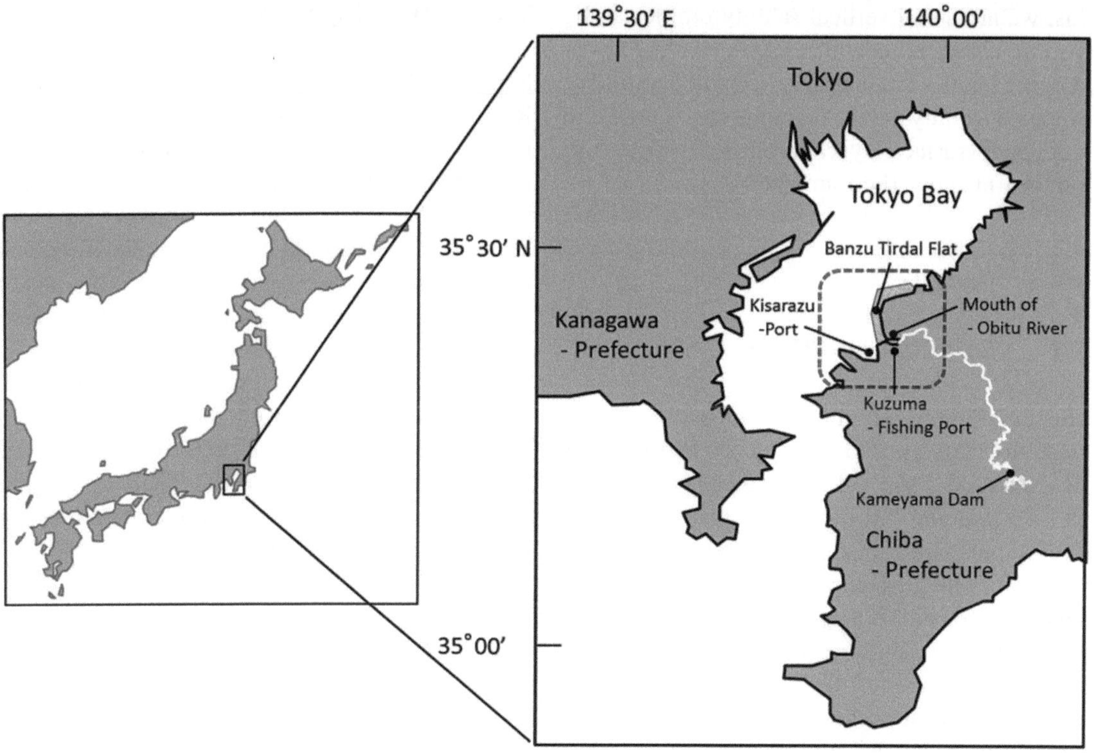

Fig. 1 Map of the studied area in Tokyo Bay

device, and a VHF transmitter (mm130, ATS Co., Australia) connected to a mount. The floats were used to adjust the weight of the systems to near neutral in seawater to avoid putting extra weight on the fish.

Four Japanese sea bass caught in Tokyo Bay between 25 and 27 October 2011 (ID1: total length (TL), 60.8 cm; ID2: TL, 52.3 cm; ID3: TL, 57.5 cm; ID4: TL, 56.4 cm) were kept in a cage at the mouth of the Obitsu River for 24 h. They were outfitted with the data collection equipment by sewing the mount to their backs with a surgical needle and biodegradable thread. The fish were then released about 500 m upstream from the river mouth (35°24′57″N, 139°54′26″E). The releasing device was set to cut a plastic cable that fixed the system to the mount at a scheduled time, which allowed the system to float to the surface.

The first system was retrieved by tracking signals from the ultrasonic transmitter with a receiver (VR100, Vemco Co., Canada) on a boat. The second system was retrieved about 14 h after release by surveying radio signals emitted from the system with Yagi antenna receivers (Ham Center Sapporo Co., Japan) on a boat and on the shore. The first system is referred to as the tracking survey and the second as the releasing survey.

Three-axis acceleration was recorded at 32 Hz, and swimming depth, water temperature, and salinity were recorded at 1 Hz. Data were downloaded from the loggers to a computer and analyzed using an Igor Pro (ver. 6.0 J, WaveMetrics, Lake Oswego, OR, USA) and Igor Filtering Design Laboratory (IFDL: ver. 4, WaveMetrics) software.

During the tracking survey, we visually observed the surrounding land and sea environments. In the releasing survey, we observed the surrounding environment at the time of retrieval. We also surveyed the salinity distribution in the Obitsu River with a salinity–temperature–depth device (ASTD102, JFE Advantech Co., Japan) during the tracking and releasing surveys. To do this, we measured vertical salinity profiles at the river mouth (6 m in depth), 500 m upstream (4 m in depth), and 1 km upstream (3 m in depth). We averaged salinity values from the river surface to 1 m depth (surface layer) and from the river bottom to 1 m above (bottom layer).

3 Results

3.1 Salinity Conditions

The salinities in the surface and bottom layers were respectively 19.1 psu and 30.8 psu at the river mouth, 6.5 psu and 28.7 psu 500 m upstream, and 1.9 psu and 5.0 psu 1 km upstream of the river mouth, showing a clear decrease in salinity upstream. The difference in salinity between the surface layer and bottom layer was greatest at the river mouth and lowest 1 km upstream of the river mouth due to freshwater and shallow bottom depths upstream. Seawater intruded into the bottom layer 500 m upstream of the river mouth, but was much lower 1 km upstream.

3.2 Tracking Survey

Sea bass ID1 was released in the lower Obitsu River (Fig. 2). It moved 200 m downstream of the release point and then moved back and forth between the release point and 200 m downstream (Fig. 2). We observed anglers targeting Japanese sea bass in the area. Swimming depths of the fish ranged from 0.3 to 2.0 m (Fig. 3). Salinity ranged from 10 to 20 psu. Snapshots from the video camera mounted on the fish showed ID1 swimming with other Japanese sea bass (Fig. 3). Another snapshot showed that ID1 looked up toward the water surface from the middle layer of the river where an angler was fishing from the riverbank (Fig. 3). The acceleration logger recorded four distinct bursts (Fig. 3).

3.3 Release Survey

Three Japanese sea bass (ID2, ID3, ID4) for the release survey were released near the Kuzuma fishing port in the lower Obitsu River (Fig. 4).

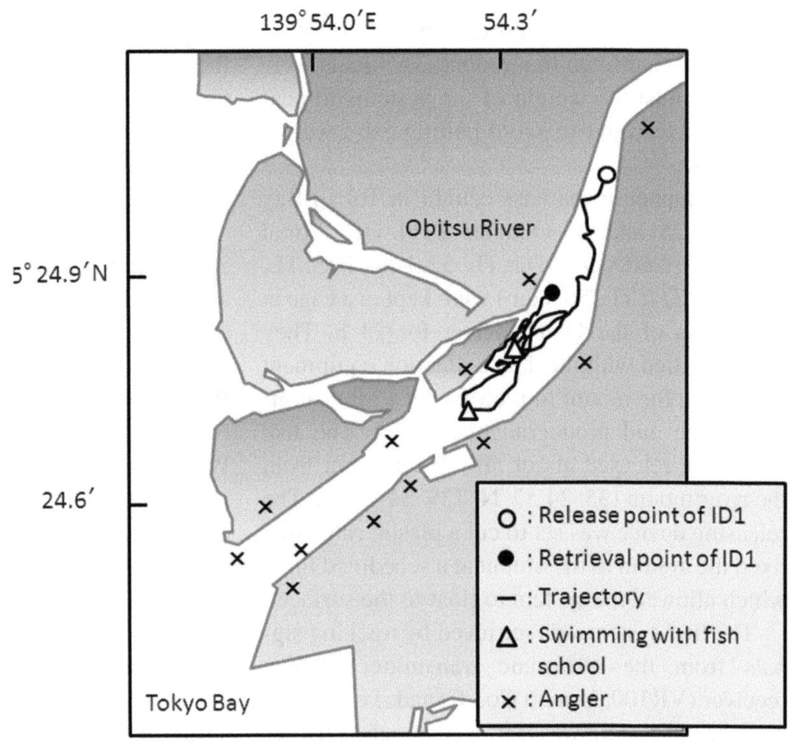

Fig. 2 Area of the tracking survey of ID1 around the river mouth of the Obitsu River (Tanoue 2014)

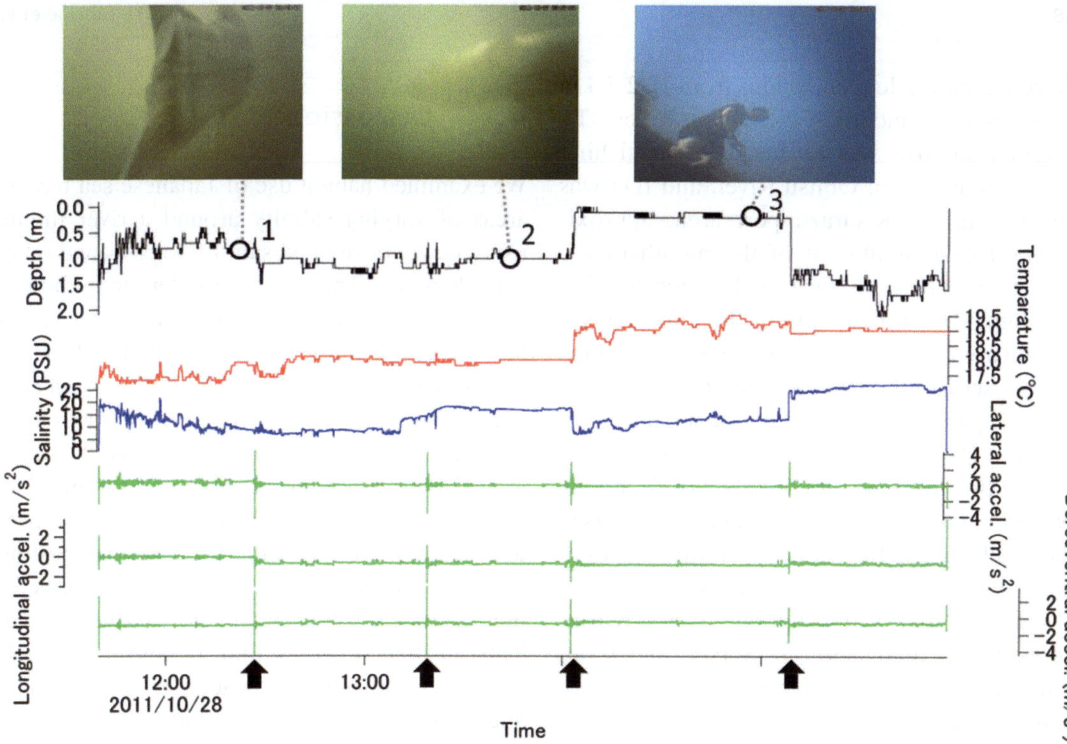

Fig. 3 Time series of swimming depths, water temperatures, salinities, and three-axis accelerations from loggers and snapshots from underwater video mounted on the Japanese sea bass (ID1). Snapshots 1 and 2 show other fish swimming together with the video-mounted individual. Snapshot 3 shows an angler fishing from the riverbank. *Black arrows* indicate bursts of three-axis acceleration (Tanoue 2014)

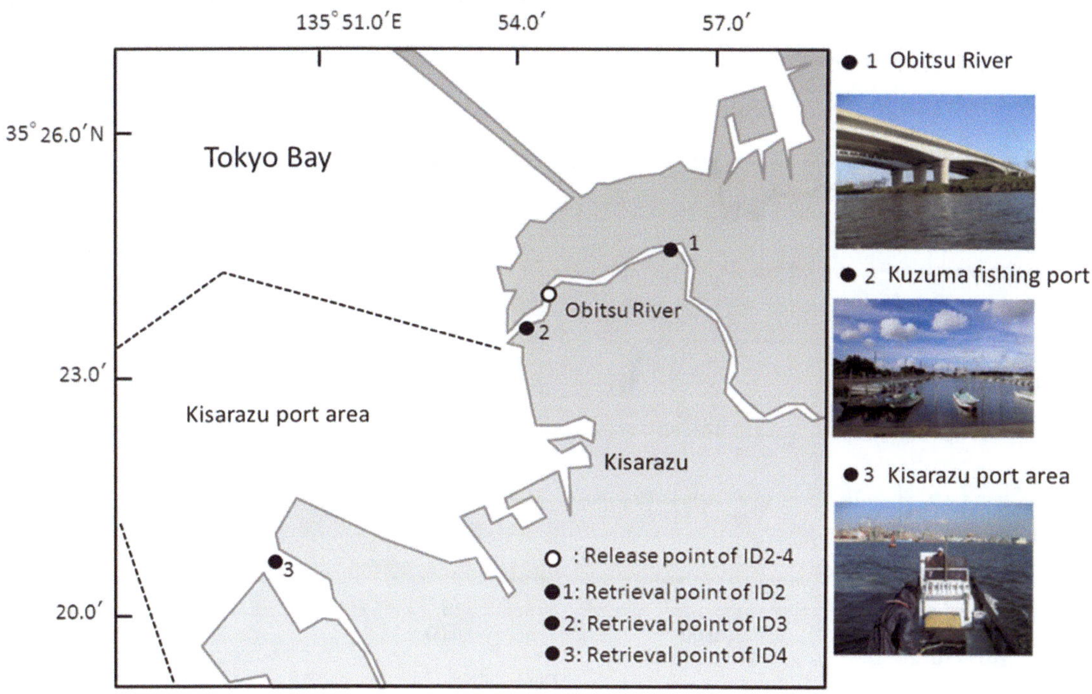

Fig. 4 Map showing survey area of Japanese sea bass west of Tokyo Bay, Japan. Individuals ID2, ID3, and ID4 were released at the *open circles*. *Closed circles* indicate the retrieval positions of data logger systems released from the fish (modified from Tanoue 2014)

We retrieved the logger system from ID2 3 km upstream of the mouth of the Obitsu River. The system from ID3 was retrieved in the fishing port at the mouth of Obitsu River, and ID4 was retrieved in the Kisarazu port area, approximately 10 km southwest of the mouth of the Obitsu River (Fig. 4). One angler reported that he caught ID3 at the mouth of the Obitsu River and released it there. The three loggers obtained a time series of water temperature, salinity, swimming depth, and three-axis acceleration for a total of 60 h (Figs. 5, 6, and 7). These records showed that all three individuals moved among sea and brackish waters. In particular, ID2 used sea, brackish, and freshwaters. All time series of three-axis acceleration showed burst events in all salinity and water temperature environments (Figs. 5, 6, and 7). Bursts sometimes moved fish between greatly different salinity or temperature environments. For example, in a burst by ID3 on 28 October at 22 h, salinity was ~10 psu before the burst and ~25 psu after. On 29 October at 3 h, water temperature was 17 °C before the burst and 18.5 °C after the burst at ~25 psu.

4 Discussion

We examined habitat use of Japanese sea bass in areas of varying salinity around a river mouth using newly developed salinity loggers and other data loggers. Three-axis acceleration showed swimming burst events by the fish. Japanese sea bass are carnivorous, feeding on small fish. Thus, bursts likely indicate predation events as sea bass ingested small baitfish by creating suction as reported in akame (*Lates japonicus*) (Tanoue et al. 2012). Water temperature and salinity data were recorded along with burst events, and conditions often changed after the burst, suggesting that many feeding events occurred at a haline or thermal fronts, where small baitfish concentrated. This further suggests that Japanese sea bass use fresh and brackish waters in river mouths and upstream as feeding grounds, in addition to seawater habitats.

Footage from the micro-video cameras on the fish showed Japanese sea bass swimming with other conspecifics. This is the first observation of schooling behavior in Japanese sea bass by

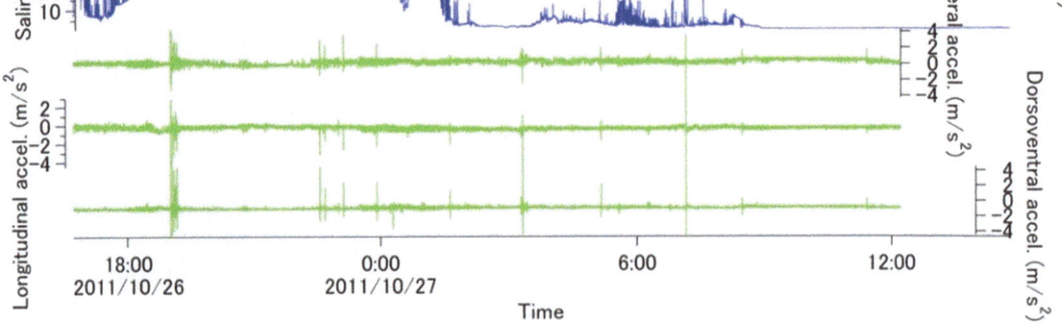

Fig. 5 Time series of swimming depth, water temperature, salinity, and three-axis acceleration from data loggers mounted on Japanese sea bass (ID2) retrieved approximately 3 km upstream of the mouth of the Obitsu River (Tanoue 2014)

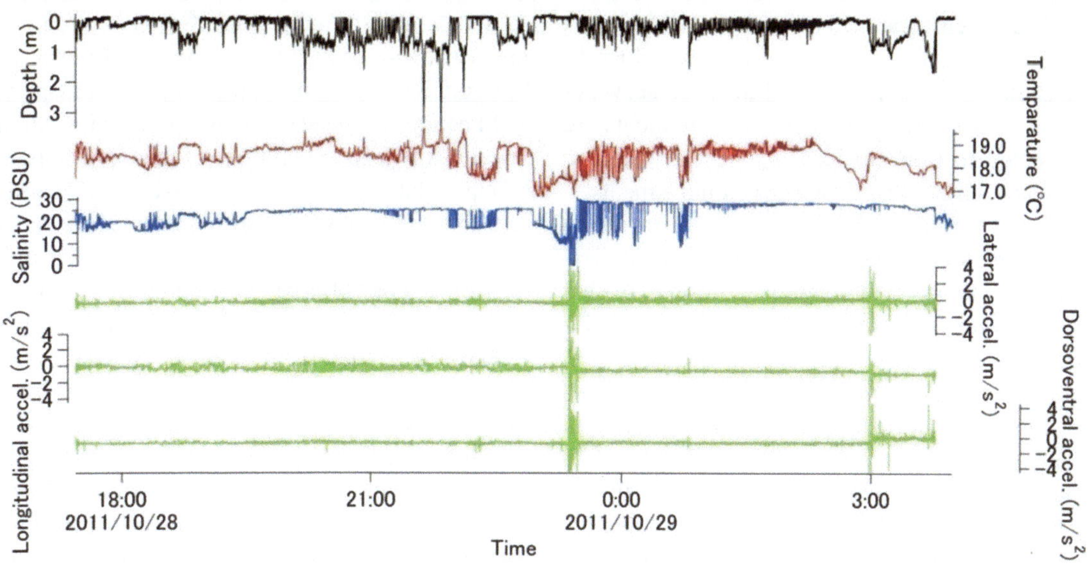

Fig. 6 Time series of swimming depth, water temperature, salinity, and three-axis acceleration from data loggers mounted on the Japanese sea bass (ID3) released and retrieved in the Kuzuma fishing port near the river mouth of the Obitsu River in Tokyo Bay (Tanoue 2014)

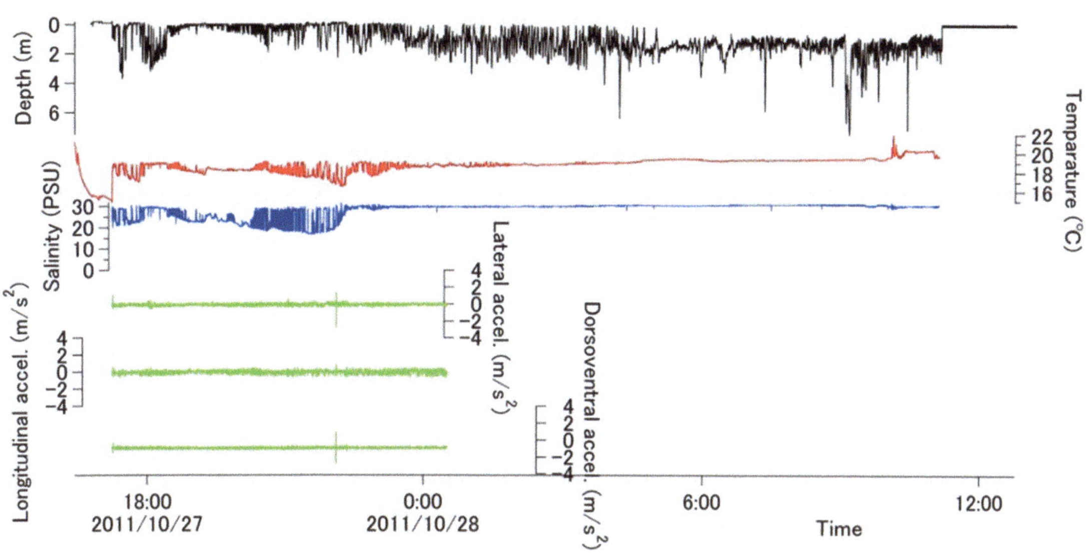

Fig. 7 Time series of swimming depth, water temperature, salinity, and three-axis acceleration from data loggers mounted on the Japanese sea bass (ID4) released and retrieved in the port area of Kisarazu in Tokyo Bay (Tanoue 2014)

fish-borne cameras in brackish waters and the sea (Fig. 3). The camera also filmed an angler fishing from the riverbank above the fish. The cameras allowed observation of the Japanese sea bass in the middle layer of the river, more clearly detecting the behavioral ecology of the fish than other methods.

Depth data of all individuals in the release surveys showed frequent vertical movements. Time series of salinities and water temperatures indicated many pulses when they stayed in the surface layer with low salinity water. The pulses of ID2 were remarkable. It moved 3 km upstream of the river mouth, where differences in salinity and water temperature between the surface and bottom layers are great due to less mixed freshwater in the surface layer and intruded seawater in the bottom layer. The vertical movements are suspected to indicate foraging activity because bursts occurred during the vertical movements.

These surveys on habitat use by Japanese sea bass in Tokyo Bay revealed that these fish use rivers, river mouths, and ports, in addition to fishing grounds. Resource management for this species has focused only on commercial fishing zones, which overlooks many other important habitats for the conservation of this and similar species. For a more effective resource management of this species, barriers to cooperation among different jurisdictions and administrations surrounding the habitats of Japanese sea bass must be overcome. In particular, the jurisdictions and administrations in coastal waters are complex due to many different activities, including fisheries, aquaculture, ports, waterways, influent rivers, and water quality. Each of these components has different management groups and objectives. Therefore, to realize effective management of the habitats of some fish, objective data are needed on habitat use by the fish to accommodate the diverse uses of coastal waters. Managers widely recognize that biodiversity is one of the most important elements for sustainable development, but they do not have the objective data needed to determine appropriate habitat protection. These observations that Japanese sea bass use rivers, ports, and the sea may provide a motive to examine cooperative management among managers of different habitats. For example, these fish frequently encounter leisure anglers in river mouths, but no regulations currently exist on fishing in these areas.

Fishermen and local governments play the primary role in the management of fish in the sea. Nevertheless, no management has been instituted for this species in rivers, although sport fishers generally respect catch-and-release rules. The Kameyama Dam is located in the upper reaches of the Obitsu River (Fig. 1) and has become a popular fishing spot for black bass (*Micropterus salmoides*), an invading species. Black bass were intentionally introduced by anglers and have caused damage of indigenous river ecosystems in Japan (Senou 2005). The Japanese Ministry of Environment prohibited the import, transport, and keeping of the species by the Invasive Alien Species Act in 2005 (Ministry of the Environment of Japan 2005) because the black bass is a carnivore that feeds on small fish and alters indigenous ecosystems (e.g., Brown et al. 2009) through top–down control like Nile perch (e.g., Barel et al. 1985). This fishing spot is located in the reservoir separated from the river by Kameyama Dam. However, the black bass may move from the reservoir to the lower Obitsu River when the dam is drained. If they move into the lower reach of the river, the black bass could compete with the Japanese sea bass for small baitfish. Therefore, we recommend that Chiba Prefecture institute strict supervision of or abolish fishing of black bass in this area to protect indigenous resources. To do this, organizing a discussion among stakeholders in the river catchment (e.g., dam and river managers, citizens, and anglers) is necessary to discuss to persuade anglers to abandon game fishing of black bass.

Ethics Statement

This study was carried out with permission from the Ethics Committee of the University of Tokyo, Japan (Permit No. A11-6, A12-9).

Acknowledgments We express our thanks to Mr. T. Mori and Mr. N. Miyata for their help with the fieldwork and Dr. H.-J. Ceccaldi, Dr. Y. Henocque, Dr. Y. Koike, Dr. G. Stora, Dr. M. H. Tusseau-Vuillemin for editing our manuscript. This study was supported by the Ocean Policy Research Foundation, Sanyo Techno Marine Corporation, and the Program of Bio-logging Science of the University of Tokyo (UTBLS).

References

Alabsi NM, Tanoue H, Komatsu T, Charef A, Mitani I, Kato M, Horii T, Aoki I, Miyazaki N (2011) Measurement of the swimming behavior of a deep-water fish, the splendid alfonsino (*Beryx splendens*), in captivity using micro data loggers. J Fish Aquat Sci 6:309–321

Barel CDN, Dorit R, Greenwood PH, Fryer G, Hughes N, Jackson PBN, Kawanabe H, Lowe-McConnell RH, Nagoshi M, Ribbink AJ, Trewavas E, Witte F, Yamaoka K (1985) Destruction of fisheries in Africa's lakes. Nature 315:19–20

Brown TG, Runciman B, Pollard S, Grant ADA, Bradford MJ (2009) Biological synopsis of smallmouth bass (*Micropterus dolomieu*). Can Manuscr Rep Fish Aquat Sci 2887:1–55

Kodama K, Oyama M, Lee J, Kume G, Yamaguchi A, Shibata Y, Shiraishi H, Morita M, Shimizu M, Horiguchi T (2010) Drastic and synchronous change in megabenthic community structure concurrent with environmental variation in a eutrophic coastal bay. Prog Oceanogr 87:157–167

Komatsu T, Tanoue H, Mohammad N, Watariguchi K, Osswald T, Hill D, Miyazaki N (2011) Relation between body tilt angle and tail beat acceleration of a small fish, *Parapristipoma trilineatum* (Threeline grunt), during mobile and immobile periods. Measured with a micro data logger. In: Ceccaldi H-J, Dekeyser I, Giraut M, Stora G (eds) Global change: mankind-marine environment interactions, proceedings of the 13th French-Japanese oceanography symposium. Springer, Berlin, pp 261–264

Komatsu T, Sagawa T, Sawayama S, Tanoue H, Mohri A, Sakanishi Y (2012) Mapping is a key for sustainable development of coastal waters: examples of seagrass beds and aquaculture facilities in Japan with use of ALOS images. In: Ghenai C (ed) Sustainable development. In Tech Publishing Co., Rijeka, pp 145–160

Kudo T, Tanaka H, Watanabe Y, Naito Y, Otomo T, Miyazaki N (2007) Use of fish-borne camera to study chum salmon homing behavior in response to coastal features. Aquat Biol 1:85–90

Ministry of Environment of Japan (2005) The Invasive Alien Species Act. (in Japanese). http://law.e-gov.go.jp/htmldata/H16/H16HO078.html

Senou H (2005) Conservation and use of biodiversity – stumbling blocks found in the case of black bass. Seibutukagaku 56:90–100. (in Japanese)

Shimizu M (1997) Fished species in Tokyo Bay. In: Numata M, Furota T (eds) Biology in Tokyo Bay (Tokyo Wan no Seibutsushi). Tsukiji Shyokan, Tokyo, pp 143–155. (in Japanese)

Shoji N (2002) Distribution and use of resources. In: Tanaka K, Kinoshita I (eds) Japanese sea bass and biodiversity: new developments of biology of aquatic resources, Fisheries science series no.131. Kouseisha-Kouseikaku, Tokyo, pp 9–20. (in Japanese)

Tanoue H (2014) Application of advanced technology to integrated coastal management – assessment of fish habitat use by bio-logging. Ocean Policy Stud 12:53–83 (in Japanese)

Tanoue H, Komatsu T, Tsujino T, Suzuki I, Watanabe M, Goto H, Miyazaki N (2012) Feeding events of Japanese lates *Lates japonicus* detected by a high-speed video camera and three-axis micro-acceleration data-logger. Fish Sci 78:533–538

Tanoue H, Komatsu T, Natheer A, Mitani I, Watanabe S, Watanabe Y, Hamano A, Miyazaki N (2013) Measurement of diurnal body tilt angle distributions of threeline grunt *Parapristipoma trilineatum* using micro-acceleration data loggers. J Mar Sci Eng 1(1):3–9

Coral Observation by the Boat-Based Fluorescence Imaging Lidar

Masahiko Sasano, Motonobu Imasato,
Hiroya Yamano, and Hiroyuki Oguma

Abstract

A boat-based coral observation system was developed using lidar (light detection and ranging) technique for large-area coral monitoring. The system comprises an ultraviolet (UV) pulsed laser and a gated image-intensified CCD (ICCD) camera to obtain fluorescent images of the seafloor. Coral observations were conducted using a glass-bottom boat at Taketomi Island, Okinawa, Japan, and the distributions of live corals along the boat tracks were obtained.

1 Introduction

Reef-building corals (hereafter corals) are at risk from global climate change and other threats. In particular, the Japanese marine area forms the northern limit of global coral distribution, and previous studies have confirmed that corals are rapidly decreasing (Okamoto et al. 2000) and moving northward (Yamano et al. 2011). Such changes emphasise the increasing importance of monitoring regional coral distribution. Most reef-building corals have fluorescent proteins that emit blue-green fluorescence on UV excitation (Hedley and Mumby 2002). Therefore, the spatial distributions of live coral can be obtained from observations made using fluorescence imaging lidar (Fig. 1).

2 Lidar Technique

2.1 Observation Method

Lidar is an active remote sensing system that emits pulsed laser to a target and receives scattered light from it. The distance r from the system to the target is expressed via the time difference Δt between emitting the laser pulse and receiving the scattered light,

$$r = \frac{c}{2n}\Delta t \qquad (1)$$

Here, c is the speed of light in vacuum and n is the absolute refractive index, which is 1.0003 in air and ranges from 1.33 to 1.34 in seawater.

M. Sasano (✉) • M. Imasato
National Maritime Research Institute,
Mitaka, Tokyo, Japan
e-mail: sasano@nmri.go.jp; imasato@nmri.go.jp

H. Yamano • H. Oguma
National Institute for Environmental Studies,
Tsukuba, Ibaraki, Japan
e-mail: hyamano@nies.go.jp; oguma@nies.go.jp

H.-J. Ceccaldi et al. (eds.), *Marine Productivity: Perturbations and Resilience of Socio-ecosystems*,
DOI 10.1007/978-3-319-13878-7_27,

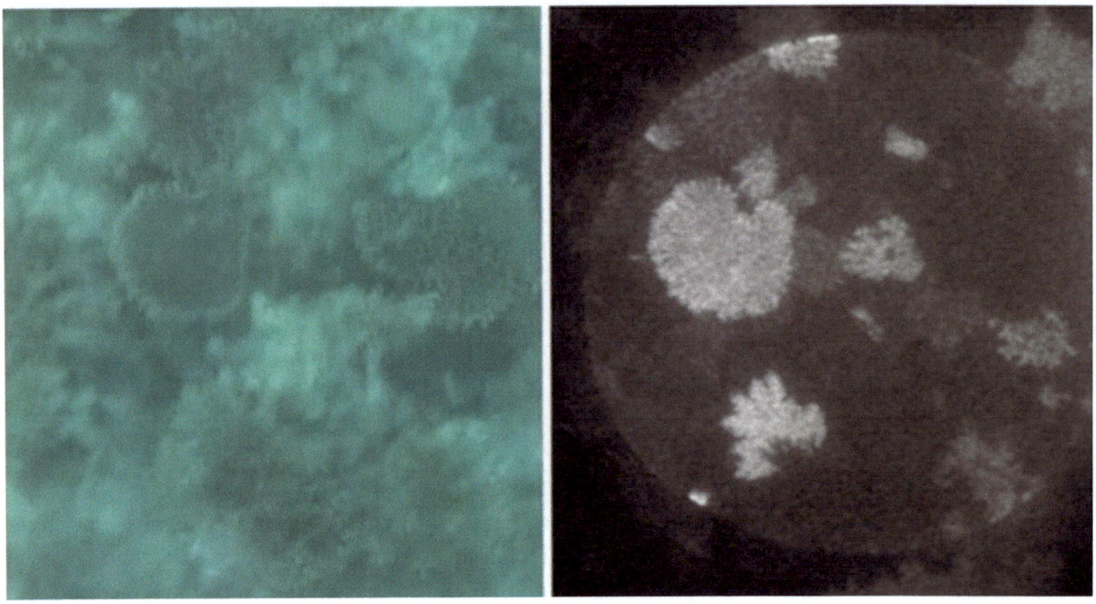

Fig. 1 Example of coral observation by video (*left*) and fluorescence imaging lidar (*right*)

Recently, lidar is applied to three-dimensional remote measurements of urban buildings and investigation of forestry biomass using airborne platform. The present study employs lidar for observing corals and is expected to help extend the range of this technology to various marine ecosystem investigations.

2.2 Bathymetry, Oil Spill Detection and Ocean Water Quality

One of the main applications of lidar for marine environmental investigation is airborne scanning system for shallow water bathymetry. The water depth can be measured by the difference of time arrivals of laser signals scattered from the sea surface and the seafloor. The observable depth can extend to 50 m in transparent seawater. The scanning lidar system can maintain a wide observation swath, thereby permitting rapid bathymetric measurements in shallow water. In Japan, airborne scanning lidar has been used to investigate underwater topological variations resulting from the earthquake and tsunami of 11 March 2011.

Airborne lidar has also been used to detect oil spills. The system is similar to bathymetry but uses a UV pulsed laser for illumination. Since oil has greater fluorescence than seawater, the oil spill can be detected by fluorescence measurement on the sea surface. This observation is particularly helpful for monitoring oil spills at night and for estimating oil spill thickness.

Additionally, it is possible to use this system to monitor seawater quality. The fluorescence of seawater varies with the concentration of coloured dissolved organic matter. The authors have observed the ratio of fluorescence to water-Raman signal at the sea surface from Sagami Bay to Tokyo Bay via helicopter-based fluorescence lidar (Sasano et al. 2008). The results (shown in Fig. 2) confirm that the seawater shows increasing fluorescence closer to populated areas.

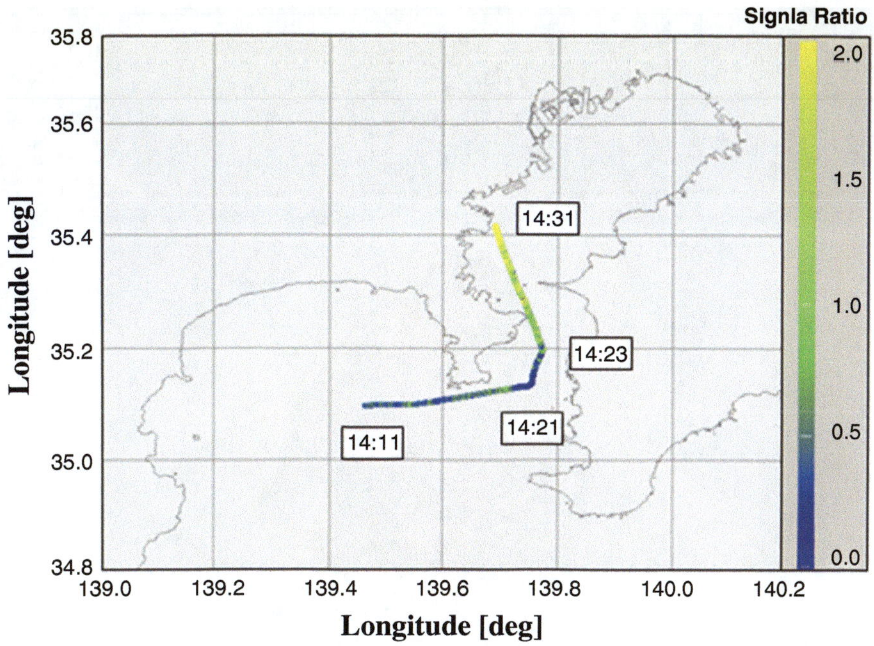

Fig. 2 Pseudo-colour map of seawater fluorescence characteristics *via* helicopter-based fluorescence lidar at Sagami Bay and Tokyo Bay (fluorescence to water-Raman signal ratio)

3 Coral Observation and Viability Check

Most reef-building corals have fluorescent proteins that emit blue-green fluorescence on UV excitation. Therefore, information on live corals can be obtained by a fluorescence imaging lidar survey of the seafloor. However, to obtain fluorescent images of corals, it is necessary to avoid the distortion effect by sea surface waves, and therefore an airborne system is problematic in this respect. Therefore, underwater lidar requires a platform that facilitates reliable imaging, such as a glass-bottom boat.

The very short exposure time (about 100 ns) used in fluorescence imaging lidar suppresses not only background noise caused by sunlight but also image blurring caused by the rapid ship motion. The spatial distribution of live corals along boat tracks can be obtained by setting DGPS position data and assessing the seafloor substrate by analysing data from each fluorescent image. Figure 3 shows the observed coral distributions at Taketomi Island, Okinawa, Japan.

4 Conclusion and Perspectives

A fluorescence imaging lidar system that utilises a glass-bottom boat was developed, and the system's potential to observe live coral distribution is demonstrated. In future, coral observation at various sites in the Japanese ocean area is planned to monitor coral reef ecosystems and assess the impacts of global climate change such as ocean warming and ocean acidification.

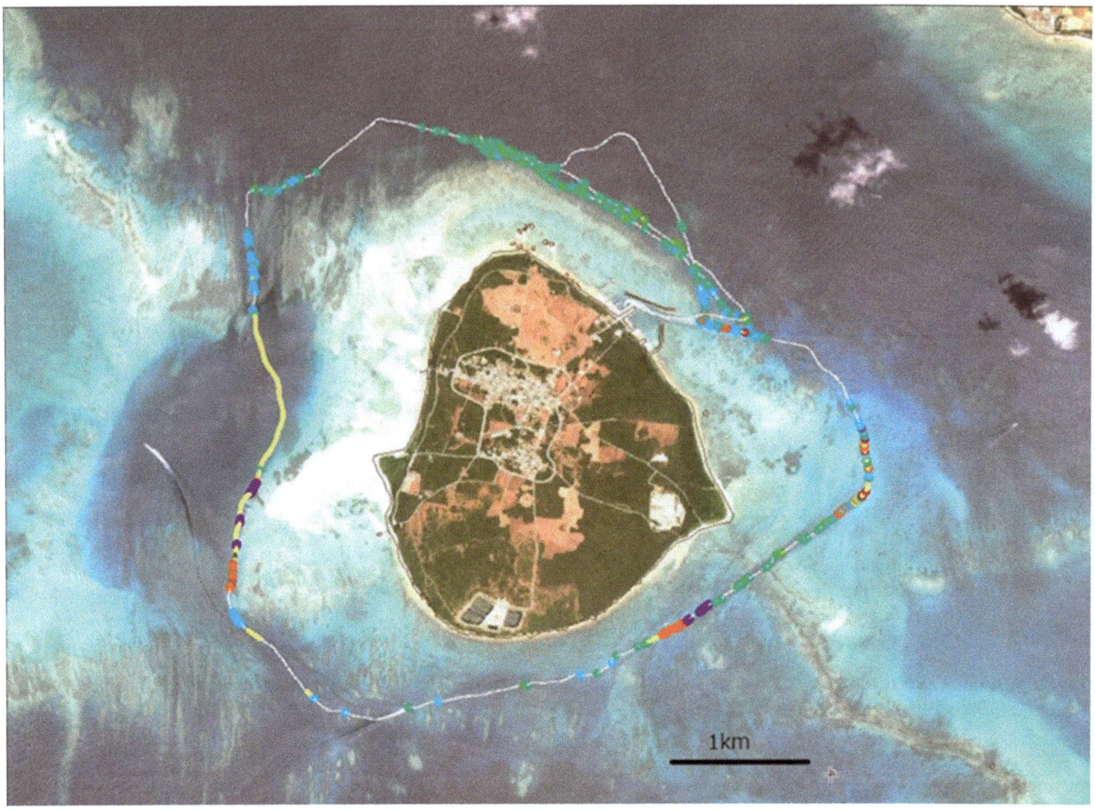

Fig. 3 Coral distribution observed by glass-bottom-boat-based fluorescence imaging lidar on 30 June 2011 (circuit) and 15 December 2011 (to and from north area) with a satellite image (WorldView-2, 24 August 2011) at Taketomi Island, Okinawa, Japan. •*Blue circle*: branch-type coral. •*Green circle*: table-type coral. •*Yellow circle*: sand. •*Red circle*: dead coral. •*Orange circle*: dead coral rubble. •*Purple circle*: seagrass, algae

Acknowledgements This study was funded by the Global Environment Research Account for National Institute, Japan. The authors express their appreciation to all those who participated in this study and thank the Ministry of Land, Infrastructure, Transport and Tourism of Japan (MLIT) and the Ministry of the Environment of Japan (MOE) for permitting publication of this paper.

References

Hedley JD, Mumby PJ (2002) Biological and remote sensing perspectives of pigmentation in coral reef organisms. Adv Mar Biol 43:277–317

Okamoto M, Sato T, Morita S (2000) Basic coral distribution data for long term monitoring at Sekisei Lagoon. In: OCEANS 2000 MTS/IEEE conference and exhibition, vol 2, pp 1383–1387

Sasano M, Hitomi K, Yamanouchi H (2008) The fluorescence lidar for monitoring of oil spill. J Vis Soc Jpn 28:9–14. (in Japanese)

Yamano H, Sugihara K, Nomura K (2011) Rapid poleward range expansion of tropical reef corals in response to rising sea surface temperatures. Geophys Res Lett 38:L04601

Part V

Communications: People, Institutions and the Sea

From Global to Local: A Comparative Ocean and Coastal Management Approach in Western Europe, France, and East Asia, Japan

Yves Hénocque

Abstract

Ocean and coast are considered like the very place of exacerbating global changes and their consequences including climate change, bio-invasion, wastes, pollutions, piracy, migrations, etc. Responses are applied at the local level but should be thought from a global perspective, requiring a shared governance underpinned by efficient coordination between state, interstate and supra-state actors as well as cooperation with non-instituted actors. The awareness of global issues and related role of coasts and seas is still recent and rather blur. It takes shape through mobilising the law, science and technology in order to feed new forms of governance in a fluid and much uncertain world. Such is the challenge of the recently instituted maritime strategies or ocean policy in both France and Japan, taking place in very different context but also showing convergences that could lead to fruitful collaboration between the two countries and the regions they belong to.

1 Introduction

France and Japan are two very different countries from the geographical, environmental, socio-economical and cultural points of view.

In Japan, the 'Bill for the Basic Ocean Law' took effect in July 2007. The contents of this new law 'define the basic principles of Japan on the ocean, clarify responsibilities of the national government, local governments, business operators, and citizens, specify the basic items concerning measures on the ocean, and stipulate the establishment of the Headquarters for

Y. Hénocque (✉)
Institut Français pour l'Exploitation de la Mer (Ifremer), Conseiller Principal Politique maritime et Gouvernance, General-Secretary of S.F.J.O. (*) of France, Correspondant Asie-Pacifique (JAMSTEC), 155 rue Jean-Jacques Rousseau, 92138 Issy-les-Moulineaux cedex, France
e-mail: yves.henocque@ifremer.fr

H.-J. Ceccaldi et al. (eds.), *Marine Productivity: Perturbations and Resilience of Socio-ecosystems*,
DOI 10.1007/978-3-319-13878-7_28, © Springer International Publishing Switzerland 2015

Comprehensive Ocean Policy for the purpose of promoting these measures in a comprehensive and systematic manner, aiming at realizing a new ocean-oriented nation' (Kuribayashi 2008). As the basis of the Basic Ocean Plan, 12 basic measures have already been agreed upon. They include the promotion of development and conservation of the marine environment within the EEZ and other areas, securing the safety and security of the oceans including maritime transport, promotion of ocean science and technology, ocean industries and their international competitiveness, integrated coastal management and enhancement of citizen's understanding of the oceans, and conservation of remote islands and international cooperation.

In France, the recent national move (2009) for the management of the coast and ocean has been largely influenced by the new European Integrated Maritime Policy and its Marine Strategy Framework Directive. It was initiated within the framework of the 'Grenelle Environment' and then 'Grenelle of the Sea' national consultations ending up with the publication of a Blue Book giving the outlines of the national maritime strategy related to knowledge enhancement, governance arrangement, maritime activities development, education and training and international commitments in metropolitan France and its overseas territories.

A first Basic Act (Loi Grenelle 1) enacted in 2009 was complemented in 2010 by a second more detailed one (Loi Grenelle 2) setting out the institutions and governance mechanism for implementation of the national maritime strategy. The last decree of application of the law, devoted to the preparation and enactment of a national maritime strategy and its implementation at marine ecoregions and subregions, was published in February 2012.

In these respective national contexts, both countries have numerous ongoing local ICM-related initiatives taking place in specific regional and local context. The question is therefore how to progressively build up from national to local and vice versa an adequate, viable and well-supported governance and management process that can help each nation to consolidate their national framework, hence promoting their ocean-state position in their respective maritime region and in the international arena.

2 Converging Concerns

2.1 From Global and National Perspectives

2.1.1 Setting the National Context from a Regional Perspective

Since 2007 and 2008, Japan has an Ocean Basic Law and a Basic Ocean Plan developing 6 principles and 12 measures. It is more recently that France, in the framework of the EU integrated maritime policy, took the political decision to prepare its own maritime strategy through a short consultation process immediately followed by the preparation and adoption of a Blue Book (Anonymous 2009a, b) on seas and oceans' national strategy.

Looking at both national maritime policies (though a formalised French maritime strategy has still to be drafted and enacted), respective goals and objectives look somewhat similar except the specifics of the French overseas territories, which are quite strategic since their EEZ represents around 97 % of the French entire EEZ (Hénocque 2010a, b) (Table 1).

Japan Political Shift: From a Reactive to a Proactive Stance – The Japanese archipelago constitutes part of the Asian continent offshore island chain creating a series of marginal seas along the Eurasian continent, including the Sea of Okhotsk, the Sea of Japan, the Yellow Sea, the East and South China Seas and the Philippine Sea. In such a position, there are three strategic lines of communication that are vital to Japan: the one to the Indian Ocean, the Pacific Ocean and the Eurasian landmass with China becoming a major maritime power in the region and soon in the world (Kotani 2012) (Fig. 1).

In spite of strong fishing and maritime industries, a long tradition of ocean research and development and the entrustment of the world's 6th largest EEZ under UNCLOS, many Japanese scientists and government and administrative representatives consider

Table 1 National maritime policies of France/Japan

France (Blue Book on Ocean Policy)	Japan (Basic Plan on Ocean Policy)
Invest into the future	**Enhance the knowledge of the sea**
1. Generate the people's passion for the sea	*1. Promotion of marine surveys*
2. Improved knowledge for improved management	*2. Promotion of marine science and technology R&D*
3. Maritime education and professional training	*3. Enhance citizens' understanding of the sea and foster human resources*
4. Protect the coastal and marine environment	
5. Develop coastal and ocean monitoring	
Develop a maritime sustainable economy	**Harmonise sea development and environmental protection**
1. Natural resources sustainable development	*1. Promote development and use of marine resources*
2. Sustainable fisheries and aquaculture development	*2. Preserve marine and coastal environment*
3. Innovating and competitive shipbuilding industry	Sound development of maritime industries
	1. Secure maritime transport
4. Rethink maritime transport	*2. Promote maritime industries and strengthen international competitiveness*
5. Develop port international dimension	
6. Strategy for leisure boats and aquatic sports development	
Promote maritime France in Europe and the world	**International partnership with regard to the sea**
1. France and international governance development	*1. Secure international coordination and promote international cooperation*
2. France and EU integrated maritime policy building up	
3. Maintain sovereignty and fulfil responsibilities	
4. Strengthen France's intervention capacity for defence and security	
Set up a renewed governance	**Comprehensive governance of the sea**
1. Coastal and ocean governance setting and policy instruments development	*1. Promote development in EEZ and continental shelves*
	2. Comprehensive management of the coastal zones
2. Foster the state operative capacity at sea	*3. Preserve the islands*
3. Foster operative efficiency at international level	**Secure safety and security at sea**
	1. Secure safety and security at sea
Promote the French overseas territories' maritime dimension	
1. Territorial governments, developers of the national policy	
2. Marine environment: an asset as well as responsibilities	
3. Marine resources as one of the key economic sectors	

that Japan's contributions to ocean governance 'were uninspiring until the enactment of the Basic Act on Ocean Policy in 2007' (Terashima 2010).

This important move cannot be disconnected from the concern of 'stabilizing the security environment of the seas in East Asia where China's naval build up and maritime expansion are affecting the regional sea power balance, while interstate disputes are being intensified over territorial sovereignty of islands, delimitation of jurisdictional waters, and sovereign rights over ocean resources'.

In the background is the vital US-Japan Alliance that, since its 50th anniversary (2010) and both countries' forces joint response to the Great East Japan Earthquake on March 11, 2011, is considered to be at a turning point in renewing a partnership in response to a strategic environment of increasing complexity.

France Tardy Recognition of Its Maritime Heritage – As it was recalled in the first presidential speech ever entirely devoted to ocean policy in 2009 (President Sarkozy at the time), France has often turned its back on the sea (though not

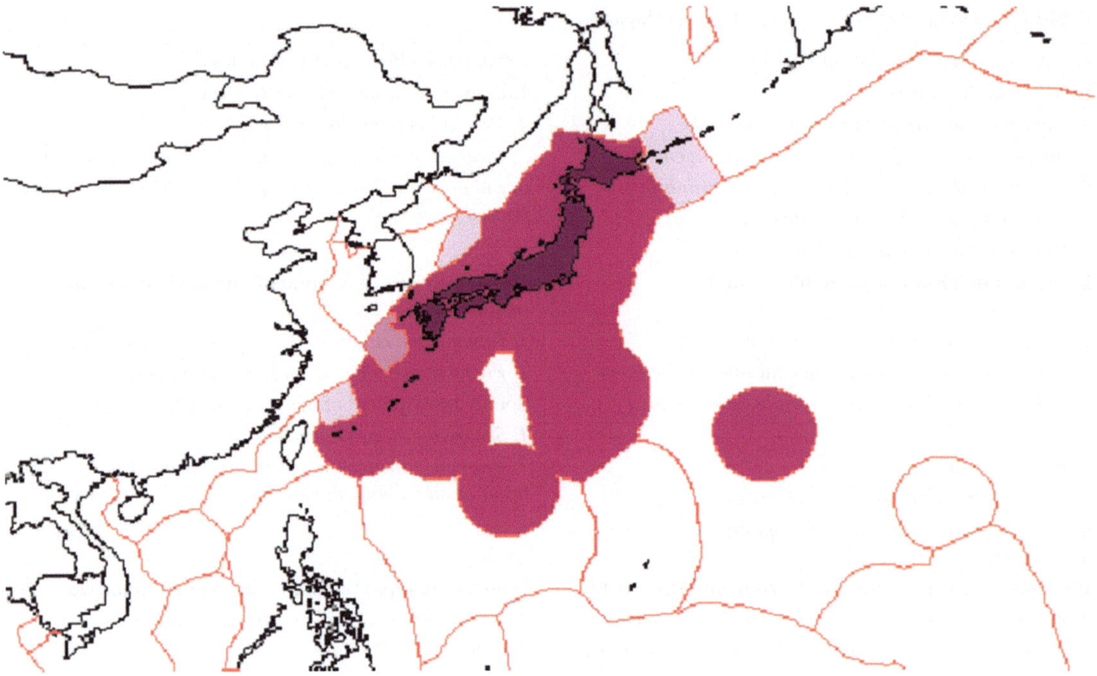

Fig. 1 Japan Exclusive Economic Zone (EEZ) boundaries

necessarily becoming totally 'sea blind' from the viewpoint of sea power[1]) and somewhat forgot its maritime heritage till it was decided in 2009 to launch a national consultation and work out the contours of a national ocean policy (Fig. 2).

France is an EU Member State having a coastline on the Atlantic Ocean/Channel Sea and the Mediterranean Sea. In addition, it has overseas territories inherited from the colonial period in the three oceans including the Antarctic area. As shown on the map (Fig. 2), those overseas territories' EEZ represent 97 % of France total maritime superficy. Nevertheless, while acting along the principles of sovereignty and subsidiarity, France is deeply involved in the EU integrated maritime policy process.

Although the European Union (EU) has acted like a land power for the past 50 years, it actually has the geographical potential to become a sea power as well (Rogers 2010).

In its European Security Strategy,[2] the EU clearly acknowledged that as a union of 25 states (27 nowadays) with over 500 million people producing a quarter of the world's gross domestic product (GDP), it is inevitably a global player. The same strategy states that the primary objectives of the EU are to craft a ring of well-governed countries around its borders and to strive for more 'effective multilateralism', a rule-based international system predicated on the rule of law. This 'ring of friends' is fundamentally forged through the European Neighbourhood Policy, the Eastern Partnership and the Union for the Mediterranean and applies not only to land areas but also to almost enclosed or semi-enclosed large water bodies which are the Baltic Sea towards the northeast, the Mediterranean Sea towards the south and the Black and Caspian Seas towards the east (Hénocque 2011). All of them are located at the boundaries of the European Union with therefore great significance from the geography and geopolitics point of view, particularly because they belong to the

[1] President of the French Republic, The French White Paper on Defence and National Security, 2008

[2] December 12, 2003

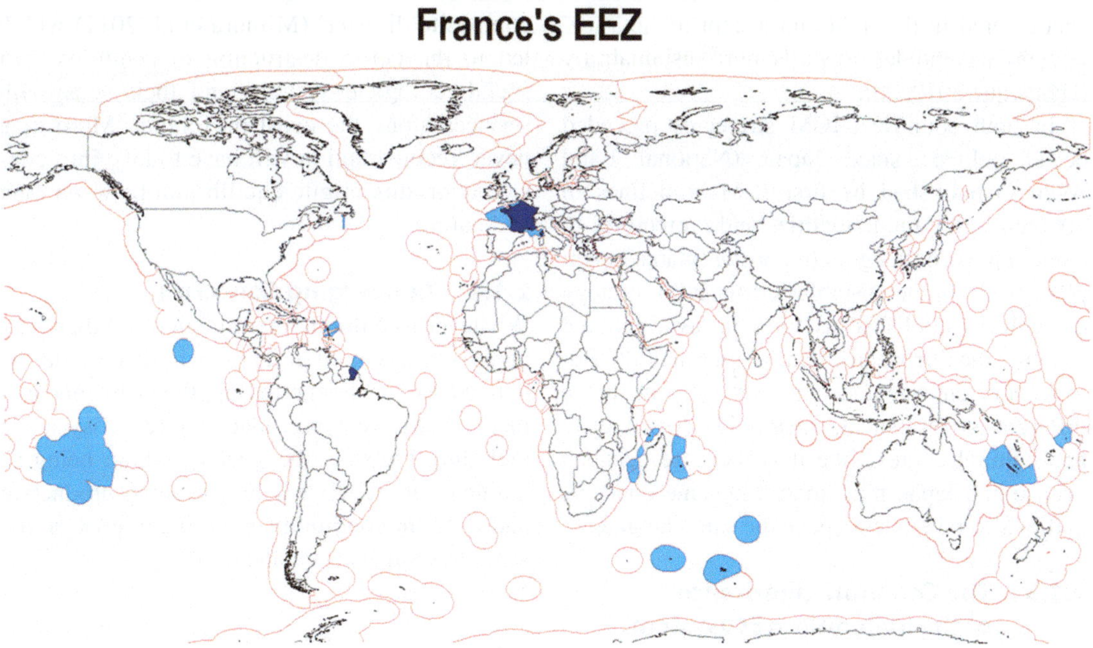

Fig. 2 France Exclusive Economic Zone boundaries (in *blue*)

maritime domain and hence represent gateway between the various continental and coastal powers of the 'Eurasian coastline' (Rogers 2009) composed of 7 of the 15 biggest trading partners of the EU (China, Japan, South Korea, India, Taiwan, Singapore and Saudi Arabia).

If the power balance in the Asia-Pacific is quickly changing, it has a direct impact on the European side: with the rise of China and the refocusing of American maritime assets into East Asia, Europeans will have to more and more assume the burden that was left to the USA after the Second World War as clearly demonstrated during the recent intervention related to Libya in the Mediterranean.

2.1.2 Why ICM Counts into National Ocean Policies?

When launching its strategy to implement the EU Integrated Coastal Management (ICM) Recommendation, the Commission indicated that coastal areas are particularly in need of an integrated territorial approach, but notwithstanding the continued need for ICM on shore, further emphasis should be placed on the implementation of ICM across the land-sea boundary and in a regional sea context (Anonymous 2007).

This consideration is particularly important, while ICM is nowadays a popular concept that is under implementation in many locations and riparian countries of regional seas all over the world. Moreover, in many regions like the Baltic Sea, ICM initiatives are increasingly getting transboundary thus, in this case, enhancing the joint regional cooperation between the Baltic Sea regional users (Ourcoast 2010). In the Mediterranean, within the Barcelona Convention system, it went as far as ratifying the first pan-regional ICM Protocol. These numerous ICM-related projects and initiatives represent a regional potential capital provided they are given the opportunity and the means to get networked together. Further, in the case of the French national maritime strategy, it has been made clear that the ICM principles and approach should be extended off to the sea to cover the entire EEZ in an articulated continuum from the watershed throughout the varied maritime areas defined under the UN Convention on the Law of the Sea. The best way to do it is to start dealing with the actual multiple management practices encountered in the field and accept to 'journey' with the stakeholders towards more sustainability (Hénocque 2010a, b).

In both countries, ICM strategies preceded ocean policies since Japan (National Land Agency) published its first ICM guidelines in 2000 engaging municipalities and prefectures to come up with integrated coastal management plan, while France established its ICM strategy (though not published as such but as a national report to the European Commission) in 2006. But to what extent scientists and experts from such a different geographical, historical, sociological and cultural context like it is the case between France and Japan may share the same concepts and to a certain extent speak the same language?

2.1.3 The Common Inheritance of the Millennium Ecosystem Assessment

Based on its assessments, the 2005 Millennium Ecosystem Assessment (MEA) founded the ecosystem-based approach in describing the society-nature interactions based on the concept of ecosystem services which is directed equally to the social and natural sciences.

It is this very framework that has been used in a national study of the interaction between humans and terrestrial-aquatic landscape ecosystems (satoyama) and marine-coastal seascape ecosystems (satoumi) in Japan (Anonymous 2010). While the outcomes of this study have been highly publicised during the Nagoya COP 10 (Convention on Biological Diversity) event in 2010, one of its main recommendations is to strengthen the satoyama-satoumi approach (ecosystem-based management approach) in managing biodiversity and ecosystem services, recognising the mosaic composition of ecosystem types and their inherent interlinkages while pointing out the need for networking 'unconnected and piecemeal' initiatives and setting up new forms of governance allowing the decentralised management of the 'commons', which is particularly true for the marine areas.

As regards the drivers of change, the 11 March 2011 Great East Japan Earthquake and Tsunami correspond to an abrupt and 'giant composite disaster' (Mimura et al. 2011) which led to the entire destruction of centuries-built social-ecological systems and their ecosystem services along the coast of Tohoku. More than never, reconstruction will have to take into consideration this fragile equilibrium between man and nature.

2.1.4 Converging Concern 1

Whatever its cultural interpretation, we then may agree on a common approach that could be defined as 'an integrated approach to management (ICM) that considers entire ecosystems, including humans, the goal of which being to maintain an ecosystem in a healthy, productive and resilient condition so that it can provide the services humans want and need'.

2.2 Landscape/Seascape: Perception and Cultural Values

2.2.1 Going at Sea with Satoumi

'Satoyama' in Japan refers to a mosaic of ecosystems including wetlands, grasslands, woodlands, farmlands, paddy fields and settlements. The interaction of humans with nature has led to the emergence of these ecosystems that provide significant habitats for a great variety of wild animals and plants. Thus, the interaction of humans with the satoyama social-ecological systems has played a vital role in biodiversity conservation, socio-economic progress and the emergence of traditional knowledge on different components of these ecosystems.

In France, a Burgundy landscape with its vineyards in the foreground, the result of centuries of human interaction with vineyards and their 'terroir', may be considered as a kind of satoyama cultural landscape. But, while in Japan satoyama is never far from the sea because of the country insularity, it may be quite far and seen as totally disconnected as it is the case for an inner region (Burgundy) in a continental country like France.

'Satoumi' in Japan refers to 'high productivity and biodiversity in the coastal sea area with human interaction'. In this case, history, hence traditional knowledge, is much shorter than in satoyama, but, particularly in Japan, there is already a good deal of practice regarding the use of ecological engineering technique that is increasingly addressing the ecosystem and not just one specific production. In Shiraishi-jima (Okayama Prefecture), marine farming is about 'enhancing the whole ecosystem and not just the fish taking into consideration the intrinsic ecological features of the area and trying to maintain the ecological niches'.

In France, where the perception of coastal traditional landscape remains quite conservative, a salt marsh landscape is nowadays considered as part of this traditional coastal landscape, forgetting that it was once the result of human strong interaction with nature, targeting only one product, salt, which has been since then nicely reborn as the 'fruit of ocean, sun and wind'.

In comparison, Satoumi in Japan is already 'at sea', probably because there was no choice in regard to the search for vital and scarce space for development on the coast, but also certainly because of the unique system of fishing rights in Japan and the whole compensation system which goes with it, and strong national and regional policies towards a 'learning by doing' kind of marine ecological engineering.

As a matter of fact, besides the rather recent involvement of environment-led NPOs, 'common fishing right deeply connects with satoumi' making 'the Japanese common fishing right institution a prototype of satoumi' (Hidaka 2011). Therefore, the discourse on satoumi cannot be disconnected from this unique and very original trait of fishing right in Japan where local communities and fishermen co-manage their coastal and marine resources through fisheries and/or aquaculture activities.

2.2.2 Converging Concern 2

In their respective context, both fishing communities in Japan and in France know very well what is and will be at stake tomorrow: the ecosystem health. But this is true mainly at local level while it is not yet apparent at the next bigger scale where market short-term considerations are still the rule.

2.3 Sectoral Policies and Protection

The maritime sector of France is marked by the predominance of coastal tourism. Recreational boating, shipbuilding, marine equipment, navy and seaports are also important players not only in the maritime domain but in the French economy as a whole.

In Japan, the situation is somewhat different since the fishing rights law gives the fishermen the pre-eminence in the coastal areas, while the defence of the coast and ports development are vital to the survival of coastal populations and commercial exchanges with the world. Nevertheless, the conservation and sustainable use of biodiversity have been gaining strength in policymaking in the last years.

2.3.1 Marine Protected Areas as an ICM Tool

The newly established Marine Protected Areas Agency in France (2006) has developed a national strategy for MPA deployment in metropolitan France and overseas territories. While they were initially envisioned as a means of managing coastal resources where they were not specifically managed, it quickly appeared that to take into account the continuum from the watershed to the EEZ boundaries, a wider approach of integrated coastal and ocean management was necessary. Under the French maritime policy and its coastal and sea management strategy, marine protected areas are thus treated as tools for achieving the overall strategy. Any ICM strategy should then take into account the planning, designation and management of MPAs as a means for ensuring biodiversity conservation and sustainable management of coastal resources as in the case of fisheries (no-take zones).

As reported in Nagoya COP 10 (October 2010), there are various types of MPAs along the coast of Japan, the majority of them focusing on the conservation and sustainable use of fishery resources and are managed by the Fishery Cooperative Agencies (FCA) themselves. Along the principle of 'use and build-up on what already exists', there is a great potential as a tool for ICM but which would necessitate their mapping to locate them and develop an overall vision, ecoregion by ecoregion, that would allow to develop, with the fishers and their FCA, networks of MPAs at greater scale.

2.3.2 Converging Concern 3

Marine protected areas, whatever they are, should be considered as tools for integrated coastal management. This is in agreement with the Aichi target 11 which stipulates that 'by 2020, at least 17 % of terrestrial and inland water areas and 10 % of coastal and marine areas, especially areas of particular importance for biodiversity and ecosystem services, are conserved through effectively and equitably managed, ecologically representative and well-connected systems of protected areas and other effective area-based conservation measures, and integrated into the wider landscape and seascape'.

2.4 Small Islands and Their Surrounding Ocean Areas

Due to their geographical characteristics, remote islands and their surrounding seas, especially when they are not inhabited, are often a 'treasure trove' (Kagami 2011) of valuable ecosystems and biodiversity and not just used solely for the purpose of claiming countries' EEZ.

In Japan, the establishment of the Basic Act on Ocean Policy does include a measure for the Conservation of the Remote Islands which is now entering into implementation.

At a much bigger scale, the 'Pacific Oceanscape', endorsed by the Pacific Forum leaders in August 2010 (Rayfuse 2011), is intended to provide a framework for cooperation in ocean conservation and management across a large part of the Pacific. This agreement urges the Pacific Island Countries and Territories (PICTs) to incorporate principles relating to sustainable ocean and coastal development into national policy and planning, along with the establishment of coordinated institutional mechanisms for transparent implementation of the Pacific Islands Regional Oceans Policy (PIROP). Importantly enough, the Pacific Oceanscape concept also highlights the need to incorporate strengths and traditions of coastal communities to attain 'sustainable island life' (Tsamenyi and Jit 2011).

This move in the South Pacific is an excellent example of a very large marine ecoregion approach including island states' national development.

It is undeniably of high interest to Japan, the reason why OPRF has made the subject one of its research subject, but also for France in the frame of its new maritime policy, not so much as regards metropolitan France but its overseas territories which are mostly small islands with different political status. The oceanscape, ecoregion and/or large marine ecosystem approaches should well underpin such a French political initiative as the already published 'Southern Indian Ocean Blue Book' to have a chance to be shared with the other island states from the region.

2.4.1 Converging Concern 4

In the case of both countries (Japan and France), islands are strategic in regard to their EEZ boundary and should be seen as the strategic core of a large marine area sustainable development.

3 Looking at Common Issues

3.1 The Crucial Involvement of Local Governments and Stakeholders

Out of the four convergences previously identified, there is a major and central issue that is common to both countries in regard to their ocean/maritime policy/strategy implementation: beyond local initiatives, in regard to the implementation of a national ICM strategy, what is the appropriate governance and planning system that may link the land and the sea allowing local authorities to get into the process and to develop a sense of ownership?

3.1.1 Current State in France

In France, there are no coastal partnerships at local level but a number of sectoral organisations themselves assisted and controlled by the state administration. To face this lack of communication, local authorities are increasingly creating their own institutional arrangement for allowing a cross-sectoral dialogue, but they are doing so in a rather disorderly manner, without much consultation between them. One of the best examples is the Sea Forum created by the Brittany Region under its Coastal Areas of Brittany Charter where the state is co-chairing the Forum. In other coastal regions, various kinds of governance arrangement are gradually building up as well.

At a larger scale, the new national maritime policy legal framework has set a new consultative body at national level, the Coast and Sea National Board which, under the chairmanship of the Prime Minister, gathers representatives from the local governments (elected officials), private sector (all maritime activities) and the civil society (NGOs). The regional ramification of this national body is ensured through the setting up of inter-regions Coast and Sea Maritime Councils chaired by the State (Prefect of Region/Maritime prefect) in charge of preparing the corresponding interregional strategic plan. The latter will be articulated with the region (sea and coast regional strategy) and then with the local implementation arrangements between municipalities and departments.

3.1.2 Current State in Japan

In Japan, there have been a series of almost simultaneous amendments of the River Act (1997), the Coast Act (1999), the Harbour Act (2000) and the Fishery Port Act (2001) towards the inclusion of environmental conservation. In 2000, these amendments were completed by the National Land Agency with the 'Guidelines for Integrated Coastal Management Plan' addressed to prefectures and municipalities.

The main principles put forward by these guidelines were as follows: (1) participation and cooperation, concerning stakeholder groups such as the central and local governments (prefectures and municipalities), private sector, non-profit organisations (NPO), fishermen and local communities; (2) wide overview, giving full consideration to entire bays, inland seas and river estuaries; (3) long-term view, setting a future vision of coastal areas following natural cycle analysis; and (4) continuous implementation, based on the results of regular monitoring and evaluation.

An Integrated Coastal Zone Management Commission was planned as well at the level of each coastal prefecture and/or municipality with a representation of all stakeholder groups.

It seems that since then, there have been only a few initiatives from local governments, most of them coming from municipalities and almost none from prefectures. There may be a number of reasons to explain such a situation (Ebara 2000):

- The articulation between land use and urban planning (municipal master plan) and the coastal zone management plans is uncertain. This has taken place mainly in the Seto Inland Sea, in regard to reclamations, impact assessment and marine environment recovery, while the development of non-coordinated sectoral policies remained the rule.
- The articulation between the 48 predefined coastal areas (which most of them correspond to one or two prefecture boundaries) and the bigger marine ecoregions that surround Japan is currently unknown as the respective role of local and central governments.
- The new coastal zone management plan has no statutory basis contrary to other administrative plans (e.g. municipal master plans, prefecture, coastal management plan mainly related to the coastline defence).
- Although some municipal governments have tried to incorporate innovative methods for promoting public participation in the planning process, the average level of participation remains limited, often reduced to the use of passive channels such as written survey.
- There are many overlapping administrative statutory plans which make the integration process quite difficult to achieve and to enforce.
- Users other than fishers have a few legal channel to sue engineering projects like reclamation since most of them are considered of 'public interest' (Kobutsu) and therefore entirely depend on the governor's or administration's decision.

In both countries, networking of local initiatives and governance arrangement to achieve it need to be part of the ICM development strategy.

3.2 Fisheries and Marine Spatial Planning

In Japan, besides the rather recent involvement of environment-led NPOs, 'common fishing right deeply connects with satoumi' making 'the Japanese common fishing right institution a prototype of satoumi' (Hidaka 2011). Therefore, the discourse on satoumi cannot be disconnected from this unique and very original trait of fishing right in Japan where local communities and fishermen co-manage their coastal and marine resources through fisheries and/or aquaculture activities.

As already stated by a Japanese law specialist in 1997 (Nakayama 1997) in the case of the Seto Inland Sea, 'with the issue of environmental protection becoming increasingly important to fisheries today, it is now necessary to aggressively position environmental protection laws within the legal system concerned with fisheries', and further, 'because the protection of marine resources is an integral part of overall environmental conservation, and because it is a particularly vital issue for fisheries, fishery operators must be cast as major players in the environmental conservation struggle'. In other words, fishermen must realise that they are one of the main components in the marine resource use system and that they therefore must actively participate in ecosystem-based fishery based on mutual agreements through existing or to-be-created governance forms and procedures (e.g. Shiretoko World Heritage).

The kind of co-management which characterises the Japanese fishing rights system is certainly one of the enabling conditions that may lead to ecosystem-based management, but local co-management through Fishery Cooperative Associations often lacks a global perspective and ecosystem-scale knowledge to set management and conservation goals (Pitcher and Lam 2010). Traditional ecological knowledge and the capacity to reconstruct past ecosystem states in designing effective restoration strategy (e.g. Hinase, Bizen City, in Okayama Prefecture) may also contribute to shift from mere fisheries to an ecosystem approach of fisheries.

Another strong specificity of the fishery system in Japan is the co-management (even without any legal framework) of marine protected areas or seasonal no-take zones. While demonstrating a strong capacity in managing the targeted resources, Fishery Cooperative Associations remain in a one single sector approach with a clear recovery goal through the maintenance of a certain spawning stock biomass to support future fishing efforts, while it is more difficult to define recovery goals in an ecosystem framework.

More generally speaking, place-based integrated coastal management using the Maritime Spatial Planning (MSP) tool may represent a promising approach to implement ecosystem-based management provided there is no unique path to conserve marine ecosystems and sustain livelihoods.

Like Japan and contrary to some other EU Member States, France did not enter into a systematic use of maritime spatial planning at sea. At each level of maritime and coastal governance, strategic planning will be determined by compatible strategic documents:

- At national level: a national strategy for sea and coastal zones serving as a reference frame for environmental protection, the development of marine resources and the concerted and integrated management of activities related to the sea and coastal zones.
- At infra-national levels: strategic documents dealing with marine basins, coastlines, archipelagos or island regions.
- At local level: action plans related to the sea and coastal zones will be specified in the documents associated with the various legal tools for planning and management, more particularly the *Coherent Territorial Plan (SCOT)* and its maritime component (Anonymous 2009a, b).

4 Conclusion

France and Japan are two very different countries from the environmental, socio-economical and cultural points of view. France is at the edge of the European continent, bordered by three oceans or semi-enclosed sea bodies, the Atlantic Ocean (west), the Channel/North Sea (north) and the Mediterranean Sea in the south, while Japan is an archipelago located offshore the Asian continent and thus surrounded by ocean and semi-enclosed sea bodies, mainly the Okhotsk Sea (north), the Pacific Ocean (east), the Japan Sea (west) and the South China Sea in the south.

Nevertheless, we have shown that there were many convergences of concern in the frame of the respective national ocean policies and their implementation, and whatever the policies at stake (sectoral or cross-cutting), implementation necessarily brings back to local consideration from the governance and management points of view.

In both countries, emerging principles and practices of participation and subsidiarity both support the idea that decision-making should be taken as near to the local level as possible, but a broader framework needs to be in place to ensure adequate perspective and coherence between different local initiatives and to provide appropriate technical and financial support to make them sustainable.

The French region or the so-called prefecture, in Japan, associated with strategic planning may hold the key to resolving the problem of territorial and sectoral integration in a manner which both reflects national and international policies and is adapted to local conditions. This is not necessarily the level at which detailed action planning and management should take place since they are essentially local tasks, but the region (France) or the prefecture (Japan) is potentially a critical enabling level in terms of ICM initiatives' coherence in between the local and national level.

References

Anonymous (2007) Communication from the Commission – report to the European Parliament and the Council: an evaluation of Integrated Coastal Zone Management (ICZM) in Europe/*Eur Lex COM/2007/0308 final */Doc: 52007DC0308

Anonymous (2009a) Secretariat General of the Sea (Prime Minister) Blue Book. A national strategy for the sea and oceans. http://www.sgmer.gouv.fr

Anonymous (2009b) General Secretary of the Sea. Blue Book. A national strategy for the seas and oceans, p 79. http://www.sgmer.gouv.fr

Anonymous (2010) Japan Satoyama Satoumi assessment. Satoyama-Satoumi ecosystems and human well-being: socio-ecological production landscapes of Japan – summary for decision-makers. United Nations University, Tokyo

Ebara A (2000) Coastal management system and ecological preservation in Japan. In: Proceedings of the international symposium on protection and management of coastal marine ecosystems, Bangkok, Thailand, 12–13 December 2000

Hénocque Y (2010a) The crafting of coastal and ocean governance. A journey across Europe France and East Asia Japan. OPRF, Tokyo, 35 pp

Hénocque Y (2010b) Toward the global governance of coast and ocean social-ecological systems. In: Techno-Ocean 2010 conference, Kobe, 14 October 2010. Keynote speech. OPRF, Tokyo

Hénocque Y (2011) Enhancing social capital for sustainable coastal development: is Satoumi the answer? In: EMECS 9 symposium, August 2011, Baltimore, USA. Submitted to Estuarine, Coastal and Shelf Science Journal

Hidaka T (2011) Rules and regulations supporting Satoumi as social system. Satoumi Session. In: 9th international conference on the Environmental Management of Enclosed Coastal Seas (EMECS 9)

Kagami Y (2011) Remote islands and the international regime of the protected areas. In: Proceedings of 2nd international seminar on islands and the oceans, OPRF, Tokyo, 29 November–1 December 2010

Kotani T (2012) Reluctant sea power: geopolitics in Asia and Japan's Maritime strategy. In Ross R et al. (ed) Twenty-first century seapower: cooperation and conflict at sea. Routledge, London

Kuribayashi T (2008) Thinking about the Basic Ocean Law; enactment of the Basic Ocean Law and its significance. Ship & Ocean Newsletter no 10, September 2008

Mimura N, Yasuhara K, Kawagoe S, Yokoki H, Kazama S (2011) Damage from the Great East Japan Earthquake and Tsunami – a quick report. Mitig Adapt Strateg Glob Chang 16:803–818

Nakayama M (1997) Legal systems and coastal management. In: Okaichi T, Yanagi T (eds) Sustainable development in the Seto Inland Sea, Japan – from the viewpoint of fisheries. Terra Scientific Publication Co., Tokyo, pp 189–214

Ourcoast (2010) ICZM focus in the Baltic Sea region. Newsletter 5, Ourcoast – ICZM in Europe

Pitcher TJ, Lam ME (2010) Fishful thinking: rhetoric, reality, and the sea before us. Ecol Soc 15(2):12

Rayfuse R (2011) The Pacific Oceanscape: a secure future for Pacific island nations based on ocean conservation and management. In: Proceedings of 2nd international seminar on islands and the oceans, OPRF, Tokyo, 29 November–1 December 2010

Rogers J (2009) From Suez to Shanghai: the European Union and Eurasian maritime security, Occasional paper 77. European Union Institute for Security Studies, Paris

Rogers J (2010) To rule the waves: why a maritime geostrategy is needed to sustain European Union. EGMONT Royal Institute for International Relations. Egmont's Grand Strategy Project. Security Policy Brief

Terashima H (2010) Management of islands and their surrounding oceans. In: Proceedings of 2nd international seminar on islands and oceans, OPRF, Tokyo, 261 pp

Tsamenyi M, Jit J (2011) Evaluation of the Pacific oceanscape to manage the Pacific islands and ocean environment. In: Proceedings of 2nd international seminar on islands and the oceans, OPRF, Tokyo, 29 November–1 December 2010

Capture Fisheries and Aquaculture in Japan and the World: Current Status and Future Concern

Takeshi Yamane

Abstract
Global capture fishery and aquaculture production reached 90.4 and 63.6 million tons, respectively in 2011. For the past 5 years, capture fisheries have been gradually decreasing, while, on the other hand, aquaculture shows an increasing trend. When viewed from human consumption, the demand for aquatic products has been increasing. However, per capita food fish supply remained from 17.4 kg in 2006 to 18.8 kg in 2011. Although there is currently a rising demand for aquatic products, an increase in global capture fisheries remains sluggish. In the present study, the global trend was compared and analyzed with the status of capture fishery and aquaculture production in Japan.

1 Introduction

Total fishery products of Japan reached $4{,}860\times 10^3$ tons in 2012 (Fisheries white paper 2013) with marine fishery products supplying $3{,}760\times 10^3$ tons and aquaculture $1{,}040\times 10^3$ tons. Actual consumption of fishery products in Japan is $8{,}890\times 10^3$ (equal amount of domestic, $4{,}860\times 10^3$ tons, and imported products, $4{,}590\times 10^3$ tons) in 2012. Total Japanese fishery product peaked in 1984 at $12{,}820\times 10^3$ tons, but from 1988 to 1995, it decreased rapidly to $7{,}480\times 10^3$ tons. Since then, the total Japanese fishery product has showed a decreasing trend, and in 2012, it decreased to $4{,}860\times 10^3$ tons eventually. This value is comparable to 38 % of the peak level in the 1984 fishery production. On the other hand, total aquaculture products were greater than $1{,}000\times 10^3$ tons in 2003 and reached a record volume of $1{,}250\times 10^3$ tons, but since then, it showed signs of leveling off reaching $1{,}030\times 10^3$ tons in 2012 due to decreasing domestic fishery production.

Overall global fishery production, as derived from the FAO capture database, continues to remain stable (inland and marine). However, marine fishery production is in a declining trend. On the other hand, inland fishery production is in a gradual increasing trend. Global aquaculture production has been increasing rapidly in the past 6 years, indicating a twofold increase from the 2006 aquaculture production. It is particularly worth noting that inland aquaculture production increased rapidly. On the other hand, marine

T. Yamane (✉)
Planning and Development, Lake Biwa Institute for the Future (Ltd), Obanagwa 1-17, Otsu City 520-0031, Japan
e-mail: yamanety@yahoo.co.jp

H.-J. Ceccaldi et al. (eds.), *Marine Productivity: Perturbations and Resilience of Socio-ecosystems*, DOI 10.1007/978-3-319-13878-7_29, © Springer International Publishing Switzerland 2015

aquaculture production remains at 1/3 of inland aquaculture. When focusing on the global aquaculture production, inland aquaculture production is way ahead of marine aquaculture production. The state of global aquaculture is profoundly different with that in Japan. Fishery products are globally utilized, basically, for human consumption as food and for nonfood uses such as feeding material, oil and feed for aquaculture, etc.

Today, both capture fishery and aquaculture productions in Japan indicate a gradual decreasing trend. The state of Japanese fishery production is also very different than the state of global fishery production. The global utilization of human consumption shows a rising trend, while on the other hand, nonfood uses indicate a gradual decreasing trend. In addition, the ratio for the total world production/nonfood uses indicates a decreasing trend in recent 6 years. On the other hand, part of nonfood production has been directed as feed for aquaculture. In the near future, human direct consumption will increase, and in this case, the production of nonfood uses will be influenced strongly by the amount of human direct consumption.

The purpose of this study is to evaluate fundamental aspects of the relationship between aquaculture and capture fishery production in Japan viewed in relation with global fisheries.

2 Japan Fishery Products in the Last Three Decades

First, focusing on the last three decades[1], between 1979 and 1988, the total production has been fluctuating at $1{,}000\times10^4$ tons level; since then, between 1989 and 1998, production has been dropping down to less than $1{,}000\times10^4$ tons continuously from 1991 reaching 668×10^4 tons in 1998. Since then, fishery production has continued to shrink, and in the third decade between 1991 and 2008, it reached 559×10^4 tons in 2008, and since then, the decreasing trend continued irretrievably, fluctuating between 500×10^4 tons and 400×10^4 tons during the last 4 years (Fig. 1). Generally speaking, Japanese fishery production is basically on a downward trend since 1988, and especially, in the last 5 years, the mean annual fishery production is 457×10^4 tons which correspond to 37 % of the peak level in 1984 ($1{,}282\times10^4$ tons).

Here, we will just make a short mention about inland capture fishery and aquaculture production because of its small volume of production. In 2012, inland capture fishery production was 3.3×10^4 tons and aquaculture production was 3.4×10^4 tons. Since 1979, both productions showed a decreasing trend with capture fishery having slightly less average rate of decrease compared with aquaculture.

In Japan, although inland capture fishery and aquaculture account for only a handful of fishery productions (66,902 tons in 2012), they are important traditional regional activities in each region.

Large amounts of fishery products are also imported from many foreign countries reaching 459×10^4 tons in 2012, in which 335×10^4 tons were utilized as food and 124×10^4 tons as nonfood use in Japan. The amount of imported fishery products is almost equivalent to domestic production.

3 Global Fishery Product Trend

According to preliminary data for 2011 (SOFIA 2012), global capture fishery and aquaculture production supplied 154 million tons of fish in 2011 of which about 131 million tons were destined as food for human consumption (excluding aquatic plants) (Fig. 2). For the past 5 years, marine capture fisheries have been gradually decreasing from 80.2 to 77.4×10^6 tons, while on the other hand, aquaculture shows a gradual increasing trend from 16.0 to 19.3×10^6 tons. From the point of view of human consumption, the demand for aquatic products has been increasing from 17.4 to 18.6 kg per capita food fish supply. As pointed out in SOFIA 2012, fish and fishery products represent a very valuable source of protein and essential micronutrients for balanced nutrition and good health. In the future, it is anticipated that human consumption of fishery products will increase worldwide.

[1]Refer for information about Year books of annual statistics on capture and aquaculture production during 1980 to 2013. Statistics and information department, Ministry of Agriculture, Forestry and Fisheries, Tokyo (in Japanese)

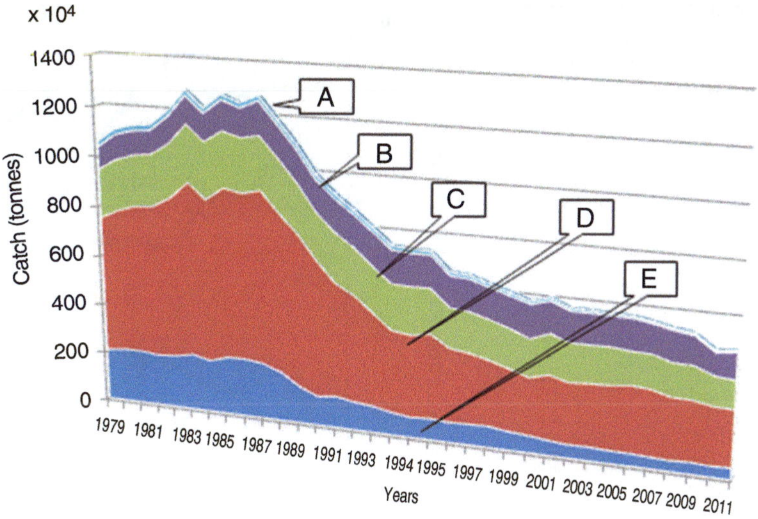

Fig. 1 Yearly variations in the total Japanese fishery products between 1979 and 2012. *A*, inland capture fishery and aquaculture; *B*, aquaculture; *C*, costal fishery; *D*, offshore fishery; *E*, far sea fishery

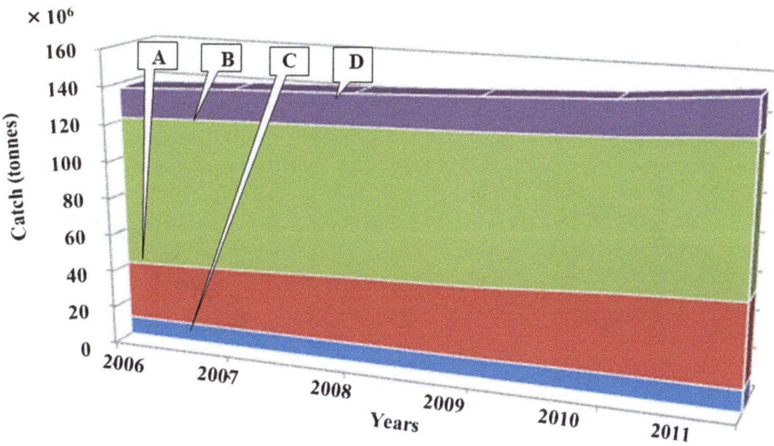

Fig. 2 Yearly variations for the last 5 years in global fishery products (the result for 2011 estimated values by FAO). *A*, marine aquaculture; *B*, marine capture fishery; *C*, inland aquaculture; *D*, inland capture fishery

On the other hand, according to a recent assessment of global trends in the state of world marine fish stocks since 1974, the mass of all fish standing stocks has been fully exploited (Fig. 3) in the FAO statistical areas (16 areas in the ocean).

The result of the assessment based on the available regional data and information for the statistical areas indicated that nearly 60 % of fish standing stocks were already fully exploited, and over 30 % were overly exploited; non-fully exploited stocks were no more than about 10 % since the last half of the 1990s. The biomass of all fish standing stocks has been decreasing continuously, and since then, the mass of non-fully exploited stock showed a steep decline.

4 Japan and Global Trends in the State of Aquaculture

According to the Statistics and Information Department, Ministry of Agriculture, Forestry and Fisheries (2014), in Japan, recent aquaculture production indicates a gradual decreasing trend (Fig. 4). Production rapidly dropped to 0.87×10^6 tons in 2011, regaining a little in 2012 at 1.0×10^6 tons level. This drop was caused by the lack of fishery statistical data in the Tohoku region of Japan.

Aquaculture products reached a peak in 2002 (1.3×10^6 tons); since then, it has been fluctuating at 1.0×10^6 tons level except for

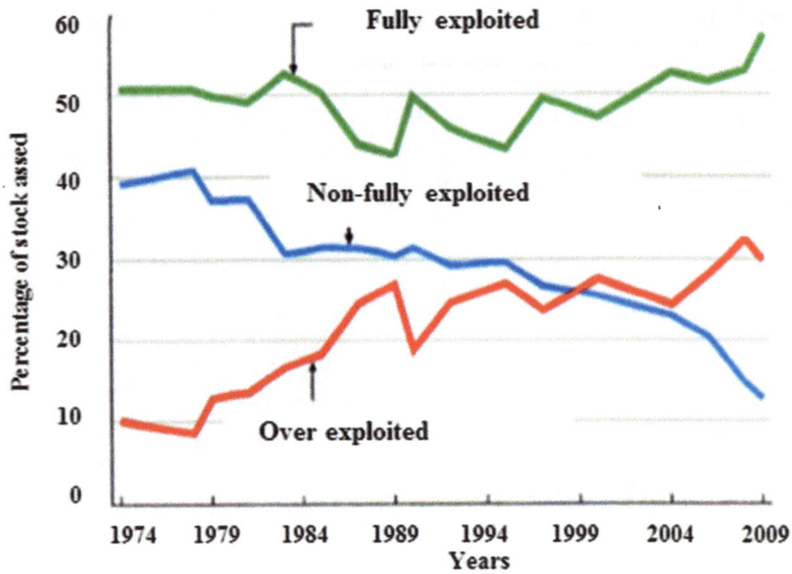

Fig. 3 Global trends in the state of world marine fish stocks since 1974 (SOFIA 2012)

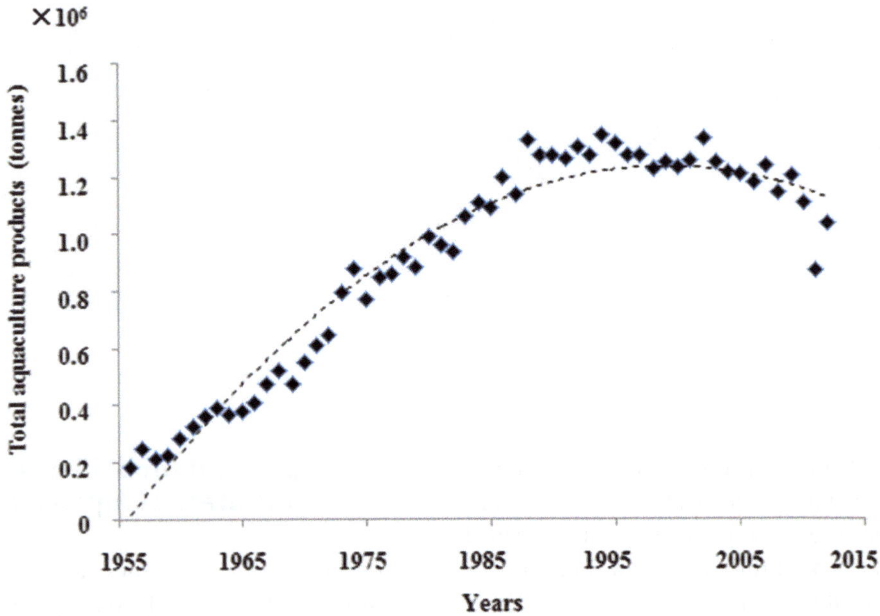

Fig. 4 Yearly variations in the total aquaculture products; regression equation for the results, $p = -660.5y^2 + 58146y - 43567$; $r^2 = 0.95$; *p* products; *y* year

2011. However, it is assumed that aquaculture production is basically in a decreasing trend. Furthermore, it is reasonable to consider that the decreasing trend in aquaculture would continue in the future, and it is very difficult to expect a large increase in production.

In order to clarify the production structure of aquaculture products, we first focus on the production of each aquaculture species (Fig. 5). Seaweed supplied 42 % of the total production, 33 % by shellfish, and the remaining 24 % from fishes. So, in the case of Japan, shellfish and

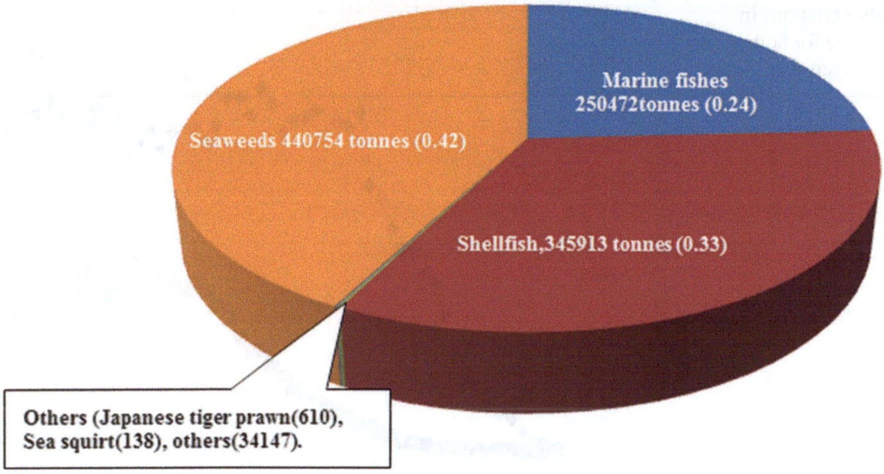

Fig. 5 Aquaculture products (total products, 1,039,483 tons) in 2012 (Source, annual statistics on fisheries and aquaculture production; statistics and information department, Ministry of Agriculture, Forestry and Fisheries (2014))

seaweed make up 75 % of aquaculture production. The quantity supplied by farmed fish (0.25×10^6 tons) is also less than one-fourth of the total coastal capture fishery production (1.0×10^6 tons).

The statistics concerning farmed species mostly constitute (1) coho salmon, (2) yellowtails, (3) Japanese jack mackerel, (4) crevalle jack, (5) red sea bream, (6) bastard halibut, (7) pufferfish, (8) pacific bluefin tuna, and (9) others. Considering the composition ratio, both yellowtails (0.64) and red sea bream (0.23) are the majority, with other species contributing only 0.13. In addition to yellowtail and red sea bream, pufferfish are also of historical importance because their official statistical data have been recorded since 1961, while other species have been described in the 1970s. From a historical point of view, these three species were precursors of fish farming in Japan.

When focusing on the historical supply trend of farmed fish, here, yellowtails and red sea bream are considered as representative species (Fig. 6). Since yellowtails, red sea bream, and pufferfish have been recorded in fishery statistics from 1961, it is reasonable to consider that Japan marine fish aquaculture started first with these three species.

In the case of pufferfish, the quantity supplied was between 103 and 4,179 tons in 2012; the contribution ratio is only 0.017. During the same period, the quantity supplied by yellowtail and red sea bream changed between 1,921 and 160,215 tons and between 2 and 56,653 tons, with contribution ratios 0.64 and 0.23, respectively. The quantity supplied by these two species is distinctly important from other farmed fishes. However, focusing on the last decade (Fig. 6), though the quantity supplied by yellowtails has been variably shifting, it fluctuated around 1.5×10^5 level, showing the same trend or a slight decline (inclination of linear regression equation; −677 (contribution rate, $r^2 = 0.089$)). On the other hand, the quantity supplied by red sea bream has been showing a consistent declining trend during the last decade (inclination of linear regression equation, −2,506; (contribution rate, $r^2 = 0.86$)). The point of irrevocable decline could occur if the quantity reached the level in the 1970s for the former and the 1990s for the latter. In addition, it is assumed that red sea bream would not foresee a recovery from its declining state.

5 Global Trends in Aquaculture

According to SOFIA 2012, the total global aquaculture production reached around 60×10^6 tons (Fig. 7); in 2010, the quantity supplied increased 23 times compared with 1970. When focusing on

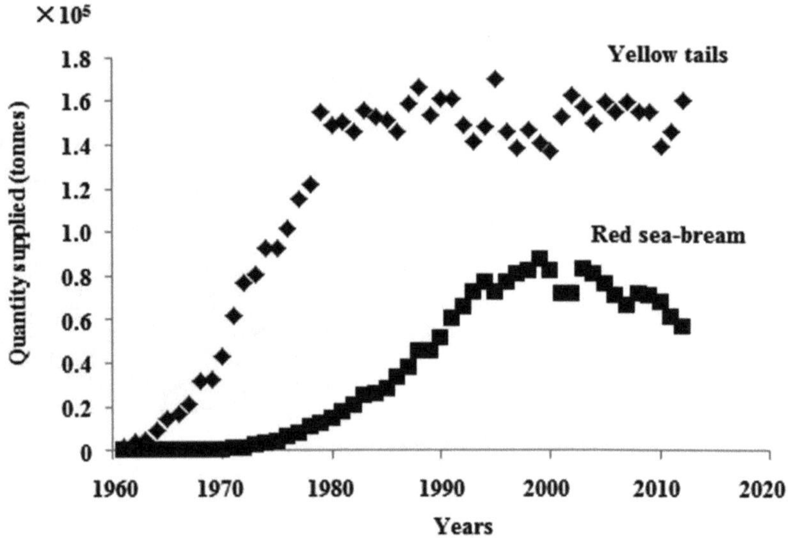

Fig. 6 Yearly variations in quantity supplied for both yellowtails and red sea bream

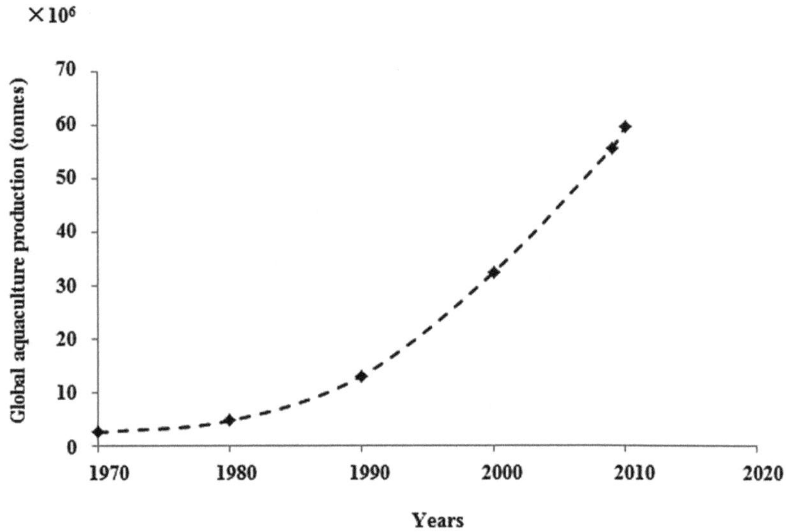

Fig. 7 World aquaculture productions (From SOFIA 2012)

the decade-over-decade growth rate between 1970 and 2010, the growth rate has fluctuated between 2.7 and 1.8, showing a declining trend during the last decade. In addition, FAO estimates the global aquaculture increasing ratio to decrease 2.4 % between 2012 and 2021.

In 2010, in terms of quantity, the percentage of production from freshwater indicates almost 60 %, and the share of marine aquaculture indicates a decline to just 30 %. Brackish water aquaculture yields only 7.9 %. It is reasonable to consider that in global aquaculture production, freshwater aquaculture exceeds other culture environments. In Japan, the growth rate slightly decreased (see Fig. 4), and decade-over-decade indicates 1.8 in 2010, indicating an extremely different situation with the global trend. In the case of Japan, the production from aquaculture is on a declining trend (Fig. 4). Furthermore, the percentage of production from freshwater is only 11 % (33,957 tons); the production of freshwater-farmed fish is extremely small and shows a decreasing trend, while world aquaculture production has been increasing continuously. The composition of world aquaculture production has been dominated by carps (71.9 %, 24.2×10^6

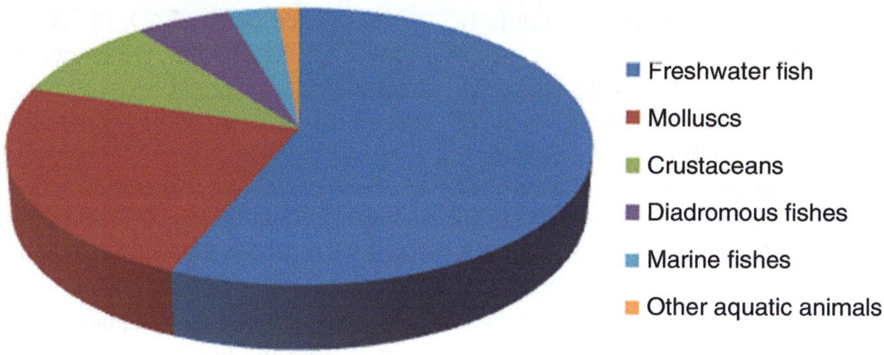

Fig. 8 Species supplied from aquaculture in 2010

tons). In other words, carps have been spreading into the mainstream of farmed fish production in global aquaculture which is extremely different with the trend in Japan (Fig. 8).

When focusing on the farmed fish, almost all species are carnivorous species in Japan, and control of feeding is absolutely necessary, that is, farmed aquaculture has been supported by huge amount of feed. Therefore, carnivorous farmed species have not contributed anything to the net food supply. However, it seems to be a favorite food for consumers under the present economic regime. On the other hand, in world aquaculture production, freshwater species have been contributing to the net food production, specifically, farmed carps (33.7×10^6 tons; 56 %). Among carps, 27.7 % are non-fed filter feeders and the rest are fed with low-protein feeds. In order of quantity produced, farmed freshwater species in the world are (1) silver and bighead carps, (2) Indian major carps, (3) grass carp, (4) tilapias and other cichlids, and (5) common carp.

These freshwater species have been contributing enormously to the net food production for human. In contrast, Japan farmed carp supply reached a peak in 1977 (297,295 tons) and then have been decreasing, reaching 2,964 tons in 2012, which is around one-tenth of the peak quantity in 1977. The contribution rate for total freshwater aquaculture production of carp was 0.35 in 1977 and drastically decreased to only 0.087 in 2012. According to the above historical events, it is assumable that it would be very difficult to recover from the 2012 level.

In terms of quantity, the percentage of production from freshwater is only 11 % (33,957 tons) in 2012. In contrast, marine-farmed species, especially yellowtails and red sea bream, are dominant species with huge feed dosage during breeding period for aquaculture.

From a point of view of quantity supplied, the contribution of freshwater aquaculture is of extremely low value. On the other hand, 69 % of the total world aquaculture production has been supplied from freshwater-farmed species. In contrast, the current state in Japan is extremely different with the global trends.

In Japan, marine fish aquaculture started from the 1950s; since then, through rapid change in supplied products in the 1980s, it increased rapidly in 1990 (fishes, 25×10^4 tons; shellfish, 34×10^4 tons; prawn, 1,596 tons; mollusks, 693 tons; seaweed, 44×10^4 tons).

For marine aquaculture products, fish occupied about 24 % of the total aquaculture production (Fig. 5), and in the case of marine-farmed fishes, the composition is dominated by the carnivorous species yellowtails and red sea bream (87 %).

In contrast, according to SOFIA 2012, marine fishes identified as among the top major species are (1) jacks, pompanos, and mackerels, (2) drums and croakers, (3) gilthead sea breams, and (4) other sea breams. These species supply less than 20×10^4 tons of the global total supply of marine-farmed fishes equal to 1.8×10^6 tons (3.1 % of the total amount).

FAO estimated that the proportion of non-fully exploited stocks has decreased since 1974, while the percentage of overexploited stocks increased. Since the feed of marine-farmed species is being supplied by huge amount of baitfish, carnivorous farmed species production is basically controlled by the amount of baitfish production. Baitfish include the nonfood species. However, the nonfood species have been supplied not only as baitfish but also for other intended purposes such as fish meal, oil, feed for others, etc.

In addition, FAO calculated that the fully exploited stocks produce catches that are at or very close to their maximum sustainable production with no possibility for further expansion. In terms of quantity supplied, the production of carnivorous farmed species is closely related with the quantity supply of baitfish stocks. However, the state of marine stocks might be at or very close to their maximum sustainable production. For reference, the last 6-year trend in the ratio of nonfood uses to human consumption is shown in Fig. 9. Nonfood uses show a gradual decreasing trend. Under the recent state of global production, it would be extremely difficult to supply feed for aquaculture as competition with human consumption increases.

As pointed out by FAO, the declining global catch over the last few years together with the increased percentage of overexploited fish stocks and the decreased proportion of non-fully exploited species around the world convey a strong message – the state of world marine fisheries is worsening and had a negative impact on fishery production. If this is correct, it is time to rethink the responsible fishery production system.

In Japan, marine fish aquaculture production started from the last half of the 1950s, and the main species till now are yellowtails and red sea bream which are both high-priced species in Japan. In 2012, the production of yellowtail was 16.0×10^4 tons and red sea bream 5.6×10^4 tons. These two species compose a large part of marine-farmed fish (87 % in 2012).

In addition, the former is appreciated as "promotion" fish (the fish name changes with growth), and the latter is commonly seen in traditional celebrations such as weddings and other congratulatory events. Recently, sea bream is very rarely used as a congratulatory gift. From the last half of the 1960s to the first half of the 1990s, the fish aquaculture production rapidly increased (Fig. 6). This period corresponds to an increase in offshore production, in particular, the

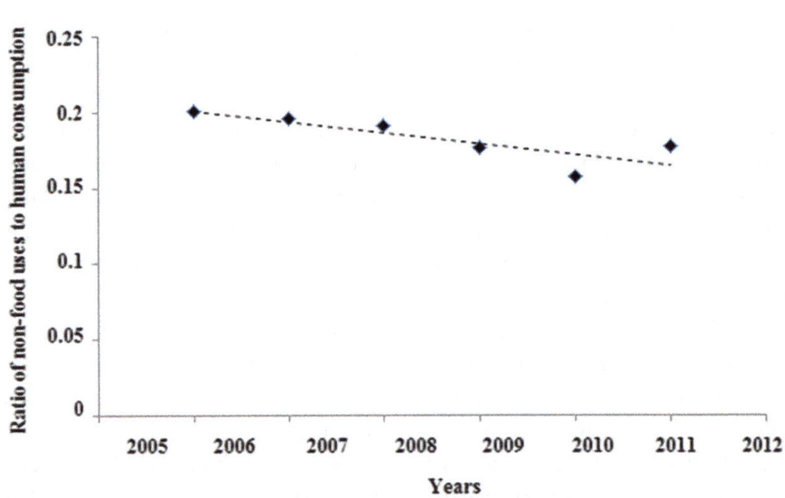

Fig. 9 Yearly changes in ratio of nonfood uses to human consumption (estimated relationship: ratio $=-7.1\times10^{3}\times$years$+14.5$; contribution ratio, $r^2=0.68$)

production of sardines which was maintained at a high level (from the first half of the 1980s to the first half of the 1990s) (Year book of Annual statistics on Fisheries and Aquaculture production 2013). The huge amount of sardines had been used as feed fish for both these two carnivorous species. The market price of sardines has been very low. The price of feed is a significant problem for aquaculture production because carnivorous fish requires much feed fish. Although depending on the species and aquaculture conditions such as the number of fish in a unit volume, feed requirement of both fishes ranged from 6 to 8 times the amount of production as wet weight (Furuya 1999). In order to estimate energy conversion efficiency of typical aquaculture species, such as yellowtail and sea bream in Japan, a preliminary calculation was conducted based on food calorie database. In the calculation, common baitfish feed used in Japan aquaculture, that is, sardines, mackerels, pacific saury and Japanese sand lance, etc., was selected. Actually, the mean value (197 kcal per 100 g weight; converted energy $197 \times 10^3 \times 4.184$ J) obtained for the aforementioned baitfish was used, and the value of feed requirement was assumed as 6 times the amount of production as wet weight. The estimated energy conversion efficiency for yellowtails is 0.21 and for red sea bream 0.16. Here, these values are underestimated because actually almost all feeds used are made from fish meal, and therefore, there is a production loss in the course of processing. In the calculation, production loss was written off. Both species productions have been maintained at around 2.0×10^5 tons level in the recent two decades. When focusing on the recent 6 years from 2006 to 2011, a difference in the degree of decrease can be seen, with yellowtail production indicating a gradual decrease (see Fig. 6) and red sea bream showing rapid reduction.

Recently, carnivore fish aquaculture production was badly affected by increased feed price due to reduced baitfish production (Demura 2010). In addition, as pointed by FAO, 80 % of natural resources as human food are already overexploited, indicating that considering the natural resource condition, there is no place for sufficient allowance to pass on to roundabout production such as carnivore fish aquaculture production. Of course, new feed has been developing without animal protein, but it is still unpredictable and uncertain elements still remain.

There is a need to address the urgent task of reconsidering optimum fishery production system and about the balance between capture fishery and aquaculture production for maintaining and developing the fishery. To avoid the depletion of a finite natural fishery resource, people have to stop and reconsider about the amount of optimum production process and the system of carnivore fish aquaculture. Socioeconomic activities, such as production of popular foodstuffs, should be considered within the bounds of nature. In the case of aquaculture, we have been getting huge amounts of living organisms as feed for carnivore fish from the sea. And even in the processing stage of feed, there is release of and output of garbage. The roundabout method of production such as aquaculture of high-priced fish exists in combination with risks involved. The aquaculture production throughout its production process needs a large amount of natural resources such as living organisms. On the other hand, natural resources are finite, and we could not overuse these organisms' sustainable self-recovery.

For example, imported nonfood use was 1.15×10^6 tons in 2011; on the other hand, fish meal was imported from Chili (34,642 tons), Peru (101,494 tons), Russia (585 tons), and the USA (8,018 tons) reaching a total of 247,045 tons (Year book of Annual statistics on foreign trade statistics 2011).

Domestic production of nonfood use was 4.8×10^5 tons, and although the specific items are unknown, almost of all them might be directed to feed of farmed fish.

In February 2014, the Japanese government instituted a guideline for optimum production of carnivorous species such as red sea bream and yellowtails for supplying adequate aquaculture production to meet demand (production volume is 7.2×10^4 tons and 14×10^4 tons, respectively) in the 2014 fishing period.

The carnivore species aquaculture is separate from basic food production, particularly in animal protein food production because this production method reduces the amount of food. The global production of carnivorous species are going to continue; however, as pointed by FAO, over 80 % of natural resources as human food are already overexploited. Baitfish resources for aquaculture are also included in overexploited natural resources. In addition, as pointed out previously (Yamane 2010), nonfood use consumption indicates a decreasing trend in the recent 10 years, and it is assumed that the amount of human direct consumption is one of the major factors. When considering the actual state in the global fishery production, it is very important to improve the roundabout production system of carnivore fish aquaculture system. To make optimum food production system for humans, a potential of bias on both quantity and quality of fishery production that reflect on the state of natural stocks should be considered. Natural resources are finite and we could not overuse these organisms' sustainable self-recovery.

For example, a simple production model is proposed under several limitations. As a first step, a logistic model was used to describe the relationship between quantity of baitfish production and total capture fishery production.

As a representation of the problem, only the relationship between the quantity of fish production of capture fishery and those of aquaculture is described. To obtain the best capacity of fishery products, the percentage for each production unit as optimum production was figured out. The relationships between total fishery production, total capture fishery products, and those for aquaculture can be defined as (Kadota et al. 2012)

$$T = (1-\alpha)y_f + f\left(\alpha y_f\right) \tag{1}$$

where T is the total fishery products, y_f is the total product from capture fishery, $f(\alpha y_f)$ is total aquaculture products, and α is a parameter indicating percentage of baitfish supplied for aquaculture feed.

Since aquaculture production may be subjected to a restriction such as quantity of the permitted area, environmental conditions, etc., it is assumed that the production would hit a peak somewhere. Here, as a model, Eq. (2) was chosen representing the relationships between total fishery products, total capture fishery products, and aquaculture products:

$$f\left(\alpha y_f\right) = A\left(1-\mathrm{e}^{-\beta\alpha y_f}\right) \tag{2}$$

where A and β are constant coefficients concerning capture fishery products. Developing Eq. (2), the following equation is obtained to represent total fishery products and other production units:

$$T = (1-\alpha)y_f + A\left(1-\mathrm{e}^{-\beta\alpha y_f}\right) \tag{3}$$

$$\alpha = \frac{\ln(A\beta)}{\beta y_f} \tag{4}$$

Since actual data of baitfish used for aquaculture could not be identified, the α values were calculated with the known value A as 40 % of total capture products and 50 % of nonfood use allotted to aquaculture baitfish (Fig. 10). The data applied for calculation is the 2006 FAO data.

According to the estimated value to obtain the optimum production level from given stocks, 17 % of capture products allotted to aquaculture baitfish would result in peak total fishery production. Though under limited conditions, the results would represent the actual state.

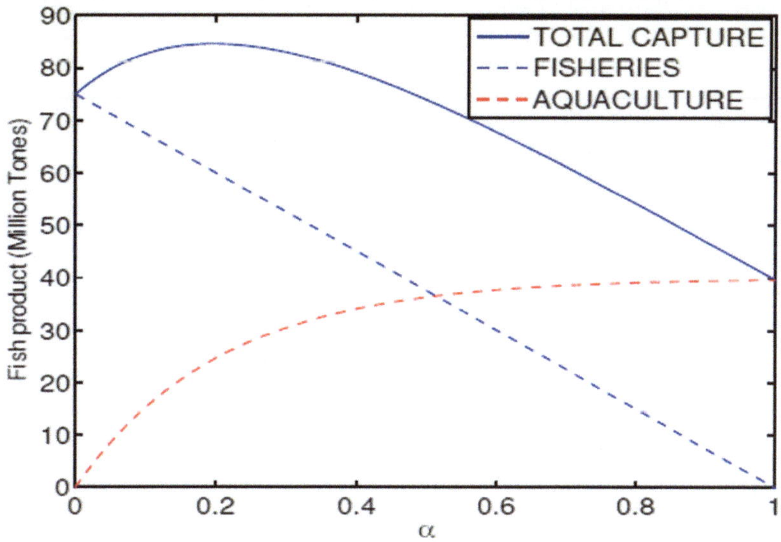

Fig. 10 Relationship between total fishery products and α (percentage of baitfish supplied for aquaculture feeding)

References

Demura M (2010) Price trends of fish meal and influence on aquaculture. Mon Rev Agric For Fish Financ 63(10):45–49. (in Japanese)

Fisheries white paper (2013) Cabinet Office, Government of Japan, 2014 (in Japanese)

Furuya K (1999) Toward to establishment of sustainable fish aquaculture production system. In: Matsumiya M (ed) Governance and conservation of fisheries resources. Kaiyo Publishing Company, Tokyo, pp 204–209. (in Japanese)

Kadota M, Torisawa S, Yamane T, Takagi T (2012) Constructing a mathematical model of a fishery system to estimate ideal aquaculture and capture fishery production. Fish Eng 48(3):197–203

The State of World Fisheries and Aquaculture (SOFIA) (2012) FAO Fisheries and Aquaculture Department, Food and Agriculture Organization of the United Nations, Rome, Italy

Yamane T (2010) Set up the smart relationship between capture fishery and aquaculture: technology collaboration in fisheries, Report of round table meeting for fishing technology 56:37–44. (in Japanese)

Year Book of Annual Statistics on Foreign Trade Statistics (2011) Ministry Finance, Tokyo, 2014 (in Japanese)

Year Book of Annual Statistics on Capture Fisheries and Aquaculture Production (2013) Statistics and Information Department, Ministry of Agriculture, Forestry and Fisheries, Tokyo, 2014 (in Japanese)

Sato-umi: An Integrated Approach for Sustainable Use of Coastal Waters, Lessons from Human-Nature Interactions During the Edo Period of Eighteenth-Century Japan

Teruhisa Komatsu and Tetsuo Yanagi

Abstract
The early modern period in Japan is called the Edo Period. During this time, the Tokugawa shogunate maintained a policy of isolation that required Japan to remain self-sufficient from 1620 to 1850. From the beginning of the eighteenth century to the mid-nineteenth century, the Japanese population remained stable, with approximately 30 million people depending on the sustainable use of agriculture and fisheries. The manner in which the Japanese people utilised natural resources during this period, especially in coastal areas, may inform the sustainable use of coastal waters today. This article reviews the life of coastal villages in the Edo Period under Sato-umi, which is the human use and management of coastal seas for high productivity while maintaining high biodiversity. Traditional practices from the Edo Period, such as fish fences constructed of stones and harvesting of seagrass, could increase habitat diversity and fish production. Uo-tsuki-rin is a practice dating from the Edo Period or earlier, in which local people managed common or feudal lords' coastal forests. After the Edo Period, privatisation of forests and timber harvesting led to the deterioration of coastal ecosystems. Lessons from the Sato-umi concept of the Edo Period combined with Uo-tsuki-rin could contribute to modern integrated coastal management.

1 Introduction

During the early modern period in Japan, known as the Edo Period, Japan isolated itself from the outside world in a policy known as "Sakoku", which means to break off relations with other countries. This policy of isolation by the Tokugawa shogunate resulted in Japanese

T. Komatsu (✉)
Atmosphere and Ocean Research Institute (AORI), University of Tokyo, President of S.F.J.O. (*) of Japan, 5-1-5, Kashiwanoha, 277-8564 Kashiwa-shi, Japan
e-mail: komatsu@aori.u-tokyo.ac.jp

T. Yanagi
Research Division, International EMECS Center, 1-5-2 Wakihama Kaigandori, Chuoku, Kobe 651-0073, Japan

H.-J. Ceccaldi et al. (eds.), *Marine Productivity: Perturbations and Resilience of Socio-ecosystems*, DOI 10.1007/978-3-319-13878-7_30,

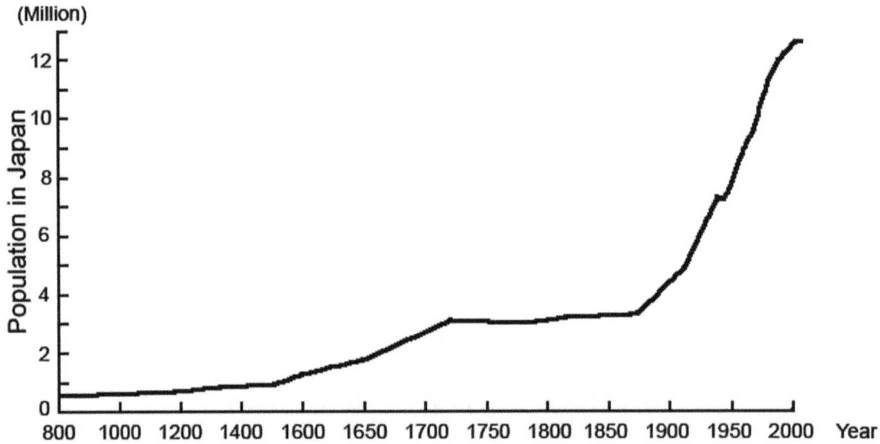

Fig. 1 Changes in population size of Japan from 800 A.D. to 2000 A.D. based on data from the Land Planning Bureau of the Ministry of Land, Infrastructure, Transport and Tourism (2014)

self-sufficiency from 1620 to 1850. From the beginning of the eighteenth century to the mid-nineteenth century, the Japanese population remained stable, at approximately 30 million people, and this population was supported entirely by local agriculture and fisheries (Fig. 1) (Land Planning Bureau of Ministry of Land, Infrastructure, Transport and Tourism 2014). The stable population size during this period suggests that resources were used sustainably. Thus, an investigation of the practices during the Edo Period could inform present sustainable use of coastal waters.

The Japanese word for a village is "sato", and for a mountain it is "yama". "Sato-yama" refers to a village near a mountain where the forest is used for firewood, charcoal and construction, and grasses, located relatively far away from the village, is used as fertiliser (Ogushi 2009). Fertiliser made from grasses was very important for use in rice fields because livestock, such as horses, cows, sheep and goats, had not been depastured in Japan. Human interactions produced secondary forests that attracted insects and small animals to pollinate plants and disperse seeds. Mature forests tend to be dominated by a few climax species and, thus, have lower biodiversity than secondary forests, which include various stages of succession (Ogushi 2009). Under the sato-yama system, villagers maintained self-sufficiency by managing the mountains near villages for high biodiversity and productivity.

The Japanese word for sea is "umi", and Yanagi (2005) proposed the concept of "sato-umi" for human management of coastal seas for high productivity while maintaining high biodiversity. Under sato-umi, coastal waters are maintained in a successional, rather than a climax, stage to maximise biodiversity and productivity.

Areas of sato-yama are located between high mountains and towns on flat land, such as alluvial fans or basins. Similarly, sato-umi are located between towns and seas (Fig. 2). Thus, the concepts of sato-yama and sato-umi both represent the sustainable management of ecotones by humans, including subsistence activities to obtain biological resources. In this article, we examine the practices of human interaction with coastal waters to maintain sato-umi during the Edo Period.

2 Production of Habitat by Fishing Facilities

Modern fishing facilities are constructed of artificial materials such as synthetic fibres and metal, whereas natural materials, such as gravel and wood, were historically used. In the intertidal or upper subtidal zones, fish fences were

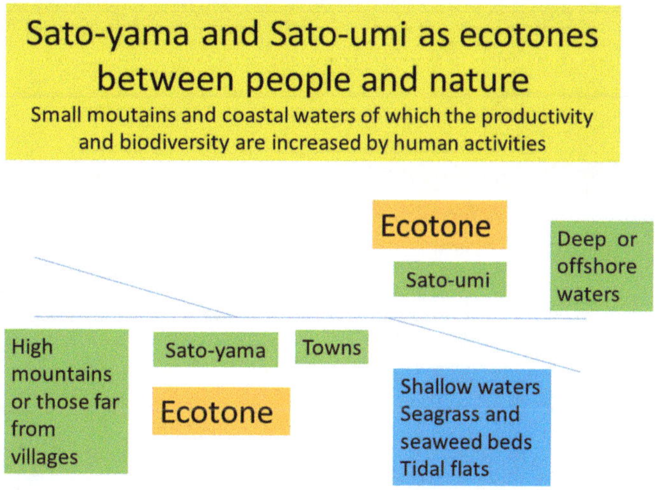

Fig. 2 Diagram of Sato-yama and Sato-umi as ecotones between towns on flat land, such as an alluvial fan or a basin, and high mountains and between towns and deep seas

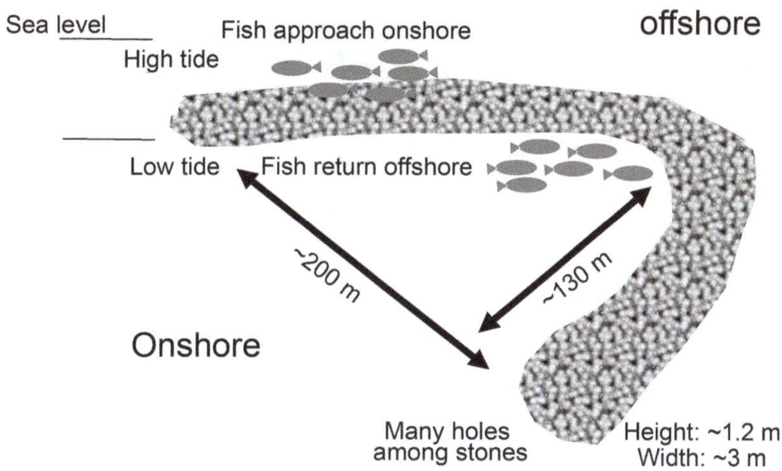

Fig. 3 Schematic diagram of a fish fence consisting of stones in Shiraho, Ishigaki Island, based on information from the Shiraho Conservation Committee of Fish Prolific Seas (http://www.sa-bu.com/what/kachi_4.html accessed on 1 May 2014). Fish move onshore during high tide and return offshore during low tide. This causes fish to concentrate at the centre of the fence. The many holes and surfaces among the stones produce new habitat for marine organisms

constructed of rocks deployed in a triangular or semi-circular shape with the trapping part at the deepest point. These traps utilised tidal changes to trap fish (Fig. 3).

When the tide lowers, fish move away from the shore. Some of them swim offshore within the fence, which guides fish to the trap. On Ishigaki Island in the Ryukyu Archipelago, many fences were used to catch fish (Fig. 4), but they fell into ruin by the late 1980s due to lack of maintenance by the local people (Shiraho Conservation Committee of Fish Prolific Sea 2014).

However, one of these fences was restored in 2006, at Sobari in Shiraho, Ishigaki Island (Fig. 4), by the Shiraho Conservation Committee of Prolific Seas (2014). A survey on and near the fences before and after restoration showed that sessile animals and fish were more numerous after the restoration (Kamimura 2011) (Fig. 5). These increases are due to the stereoscopic structure of the rocks and holes in the stone walls that provide habitat for both fish and sessile animals. As a result, this type of fishing leads to increased biodiversity and

Fig. 4 Map showing the historical spatial distribution of fish fences in Shiraho, Ishigaki Island

allows small juvenile fish to escape from the fence (Kamimura 2011). Because the length of the fish fences was longer than that of the coastline in front of the fences (Fig. 4), fish fences had a big influence on marine environments around Ishigaki Island. This same type of trap has also been constructed of bamboo or wood, which also provides a habitat for attached organisms and can enhance the productivity of lower trophic levels in coastal waters.

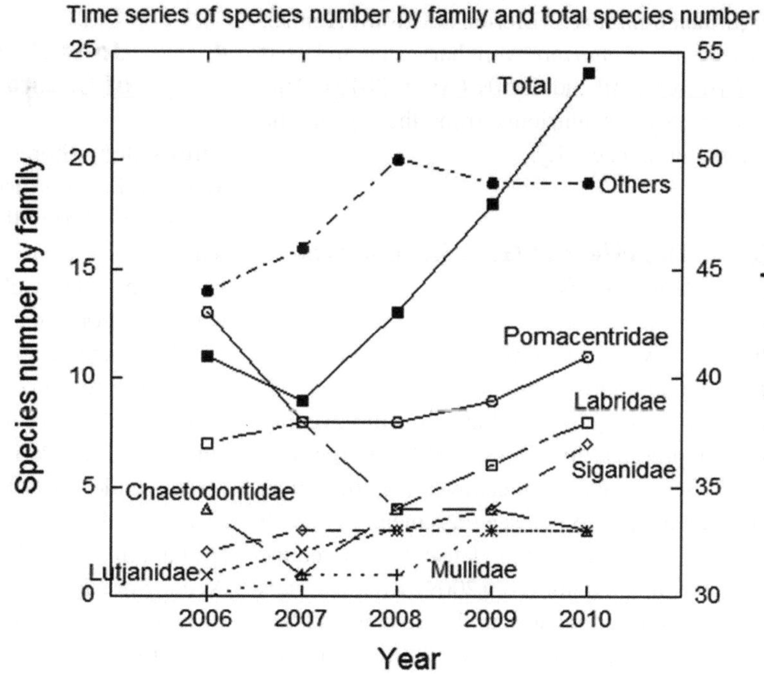

Fig. 5 The number of families and species of fish observed during 30 min of snorkelling in spring from 2006 to 2010 based on the figure in Kamimura (2011). Restoration of a fish fence was conducted in autumn of 2006

3 Harvesting Seagrass for Fertiliser

3.1 Harvest of Seagrass in the Seto Inland Sea

Seagrass meadows and seaweed forests are very important habitats in coastal waters. They influence light (Komatsu 1989), dissolved oxygen (Komatsu 1989), pH (Komatsu and Kawai 1986), water temperature (Komatsu 1985; Komatsu et al. 1982, 1994) and water motion (Komatsu 1996; Komatsu and Murakami 1994). Costanza et al. (1997) classified their social value as very high based on ecological services such as nutrient cycling, food supply, etc. However, economic development in Japan since the 1960s has destroyed many seagrass beds through reclamation and eutrophication (Komatsu 1997). From the Edo Period through to the 1960s, seagrasses and seaweeds were preserved and used as fertiliser for rice, vegetable and fruit fields. During this period, coastal villages were divided into villages of fishermen and villages of farmers who fished during the off-seasons of farming. The latter collected sardines and seagrasses for fertiliser (Hamaguchi 2007).

The prize-winning film "A Naked Island" by famed Japanese film-maker Kaneto Shindo depicts the ordinary life of a family on a small island in the Seto Inland Sea just after World War II. In this film, the family harvests seagrass, *Zostera marina* L., using two bamboo poles per person on a boat during the rainy season. The rainy season produces luxuriant growth of *Z. marina* in June in the Seto Inland Sea. Only the above-ground parts of *Z. marina* are harvested. *Z. marina* has both sexual and asexual reproduction. In a well-developed *Z. marina* bed, seeds do not germinate, due to low light caused by the canopy of leaves. In this situation, the plant reproduces asexually by producing new shoots from the root system. Thus, harvest of only the above-ground parts does not impact reproduction in well-developed seagrass meadows. Harvested leaves of seagrass and *Sargassum* spp., which wash up on shore, are also used as fertiliser on farm fields. In the brackish Lake Nakaumi in Shimane Prefecture, Japan, red algae, *Gracilaria vermiculophylla* (Ohmi) Papenfuss; brown algae,

Sargassum thunbergii (Mertens ex Roth) Kuntze; and seagrass, *Z. marina*, were harvested and used for fertiliser until the 1950s (Anon 2012). This practice recycles nutrients from the sea to the land via human activity.

3.2 The Effect of Gaps in Seagrass Beds on Fish

Climax ecosystems tend to have low biodiversity due to monotonous habitats, while optimal successional ecosystems produce rich biodiversity and high productivity (Ogushi 2009). In the Seto Inland Sea, seagrass meadows are the most important habitat. When seagrass meadows reach a climax community, the dense leaves reduce space for fish. Tanimoto (2009) created gaps inside a seagrass meadow in the Seto Inland Sea and measured fish numbers by setting trammel nets inside the dense seagrass meadow, in the artificially produced gaps and along the border of the seagrass meadow. The results showed that the number of fish and fish species were more abundant in the artificially produced gaps than inside the seagrass meadow (Fig. 6). This suggests that harvesting seagrass leaves could produce fish habitat inside a dense seagrass meadow and that optimal human interaction augments biodiversity and productivity of fish in seagrass meadows.

4 Uo-tsuki-rin Management of Coastal Forests

Forests along shores or rivers play important ecological roles in protecting marine and freshwater organisms. During the Edo Period fishermen and feudal lords protected forests under a policy known as "Uo-tsuki-rin" (Hamaguchi 2007), which is a combination of the Japanese words for "fish", "gather", and "forest". Feudal lords promoted afforestation by local people. Most of these forests have been preserved by villagers since the Edo Period, and some are now preserved under Article 25, the Forest Protection and Consolidation Act. Uo-tsuki-rin areas covered about 54,000 ha in Japan in 2005 (Fisheries Agency of Japan 2014).

During the Edo Period, in Kumamoto Prefecture, the rules of Uo-tsuki-rin required people to plant two seedlings per family on the first day of the year (Bureau of Fisheries 1934). Each seedling was marked with a plate that named the planter. This order applied not only to common forests but also to private ones. Timber harvesting was managed by dividing forests into inner and outer areas and allowing harvesting alternately in each area no more than every 30 years. These duties provided unpaid forced labour (corvée) for local people.

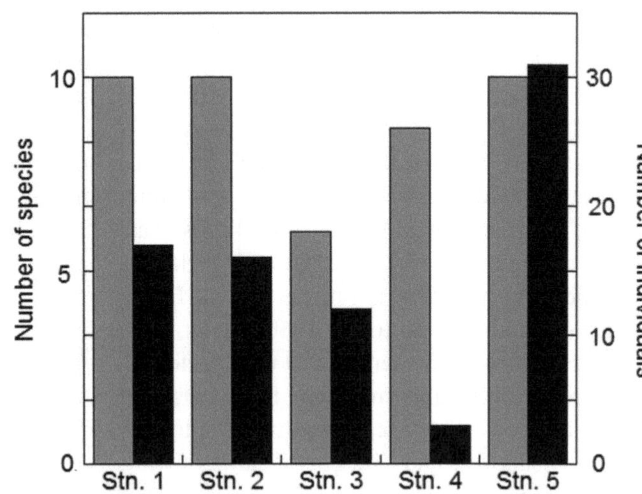

Fig. 6 Numbers of species (*grey bar*) and individuals (*black bar*) caught in trammel nets on the border of seagrass beds (Stations 1 and 2), inside dense seagrass beds (Stations 3 and 4) and in the hole inside dense seagrass beds (Station 5), based on data from Tanimoto (2009)

In the Tokushima Prefecture, Uo-tsuki-rin were protected during the Edo Period as commons or feudal lords' property. In the Meiji period of the nineteenth century, after the Edo Period, Uo-tsuki-rin were privatised and devastated due to lack of management by local people (Bureau of Agriculture and Commerce 1894). Water source forests were also devastated during the Meiji period. Akkeshi Lake, in the Hokkaido Island, is one example. Forests along the Bekanbeushi and Ohoro Rivers flowing into Akkeshi Lake were cut and burned. Mud from the rivers flooded the lake, causing massive death of oysters (Inukai and Nishio 1937). Cutting forests in Uo-tsuki-rin and remote areas in the Meiji Period also damaged seaweed meadows of Konbu, *Saccharina* spp. (Hamaguchi 2007), and herring that spawn on macrophytes in coastal waters (Miura 1971). The concept of Uo-tsuki-rin forests involves a form of integrated coastal management that includes land use that influences water quality of coastal waters. This practice established the sustainable use of coastal waters.

5 Conclusion

Some ecologists insist that intact marine protected areas are necessary for the preservation of biodiversity. However, it is impossible to prohibit people from entering and using marine protected areas along the Japanese coast, where very dense populations live. Lessons from the interaction of local people with nature in the Edo Period demonstrate how optimal human use of coastal waters increases the diversity of habitats, species and productivity. Sato-umi practices of local people in the Edo Period led to increases in biodiversity and productivity, and the Uo-tsuki-rin concept expands these practices to integrated coastal management. Thus, the Sato-umi and Uo-tsuki-rin concepts could be used to optimise human activities in coastal waters and forests.

Acknowledgements This study is part of the Environmental Research and Technology Development Fund (S13) of the Japanese Ministry of the Environment. The authors thank the French-Japanese Oceanographic Society for inviting the first author to discuss the Sato-umi concept at the 15th Colloquium of French-Japanese Oceanography in France.

References

Anon. (2012) Harvesting seaweeds and seagrasses. Report on Eco-activities in Lake Nakaumi 10:1–4. (in Japanese)

Bureau of Agriculture and Commerce (1894) Special research on fisheries related subjects. Bureau of Agriculture and Commerce, Ministry of Agriculture and Commerce, Shueisha, 731pp. (in Japanese)

Bureau of Fisheries (1934) Kumamoto Prefecture. In: Bureau of Fisheries, Ministry of Agriculture and Forestry (ed) Documents of surveys on fisheries systems in old Shogunate Period. Nogyo-to-suisan-sha, Tokyo, pp 327–334, 1116pp. (in Japanese)

Costanza R, d'Arge R, de Groot R, Farber S, Grasso M, Hannon B, Limburg K, Naeem S, O'Neill R, Paruelo J, Raskin RG, Sutton P, van den Belt M (1997) The value of the world's ecosystem services and natural capital. Nature 387, 6630:253–260

Fisheries Agency (2014) Environment beautification and conservation of coastal zone. http://www.jfa.maff.go.jp/j/kikaku/tamenteki/kaisetu/engan/index.html. Accessed on 1 May 2014. (in Japanese)

Hamaguchi H (2007) The formative process of the uotsukirin system on the coastal areas: from Hansei period to Meiji period. Shisen 106:1–18. (in Japanese)

Inukai T, Nishio S (1937) A limnological study of Akkeshi Lake with special reference to propagation of the oyster. J Fac Agric Hokkaido Imp Univ 40:1–33

Kamimura M (2011) Possibility of biodiversity conservation using "Satoumi" as a keyword. Report of Shiraho's international stone tidal weir summit. Reg Stud 8:17–28. (in Japanese with English abstract)

Komatsu T (1985) Temporal fluctuations of water temperature in a *Sargassum* forest. J Oceanogr Soc Jpn 41:235–243

Komatsu T (1989) Day-night reversion in the horizontal distributions of dissolved oxygen content and pH in a *Sargassum* forest. J Oceanogr Soc Jpn 45:106–115

Komatsu T (1996) Influence of a *Zostera* bed on the spatial distribution of water flow over a broad geographical area. In: Kuo J, Phillips RC, Walker DI, Kirkman H (eds) Seagrass biology: proceedings of an international workshop. The University of Western Australia, Nedlands, pp 111–116

Komatsu T (1997) Long-term changes in the *Zostera* bed area in the Seto Inland Sea (Japan), especially along the coast of the Okayama Prefecture. Oceanol Acta 20:209–216

Komatsu T, Kawai H (1986) Diurnal changes of pH distributions and the cascading of shore water in a *Sargassum* forest. J Oceanogr Soc Jpn 42:447–458

Komatsu T, Murakami S (1994) Influence of a *Sargassum* forest on the spatial distribution of water flow. Fish Oceanogr 3:256–266

Komatsu T, Ariyama H, Nakahara H, Sakamoto W (1982) Spatial and temporal distributions of water temperature in a *Sargassum* forest. J Oceanogr Soc Jpn 38:63–72

Komatsu T, Murakami S, Kawai H (1994) Some features of jump of water temperature in a *Sargassum* forest. J Oceanogr 52:109–124

Land Planning Bureau of Ministry of Land, Infrastructure, Transport and Tourism (2014) Land planning on the web. Land Planning Bureau of Ministry of Land, Infrastructure, Transport and Tourism. http://www.kokudokeikaku.go.jp/plan/problem.html. Accessed on 10 Jan 2014

Miura M (1971) Disappearance of spring spawning herring population in Hokkaido and its recovery – consequence of exploitative forest cutting, Suisankai (Fisheries World), February Issue, p 20. (in Japanese)

Ogushi R (2009) Problems of sato-yama (2). Kahokugata Lake Sci 12:37–44. (in Japanese)

Shiraho Conservation Committee of Fish Prolific Sea (2014) Restoration of fish fence at Sobari, Shiraho, Ishigaki. http://www.sa-bu.com/what/photo.html. Accessed on 1 May 2014

Tanimoto T (2009) Interannual change of seagrass beds in the Seto Inland Sea. In: Proceeding of 8th symposium on marine environments, organisms and restoration techniques of coastal environments. Executive Committee of 8th Symposium on Marine Environments, Mie University, Tsu, pp 55–60. (in Japanese)

Yanagi T (2005) "Sato-umi"; a new concept for the coastal sea management. Rep Res Inst Appl Phys Kyushu Univ 129:109–111

Reviving the Seto Inland Sea, Japan: Applying the Principles of Satoumi for Marine Ranching Project in Okayama

Takehiro Tanaka and Yoshitaka Ota

Abstract

This contribution introduces two Japanese marine ranching projects implemented in the western Seto Inland Sea off Okayama, Japan. These two marine ranches, costing a total of USD 40 million, cover 300–500 ha and are comprised of existing marine reefs (both artificial and natural) and sea grass beds. In providing organized coastal environments with habitats optimal for nurturing the development of commercially exploited fish species, these artificially enhanced areas are intended to increase the levels of primary production and stock recruitment. Additionally, parts of these zones are protected as no-take zones or areas where bottom trawling is prohibited; measures were voluntarily implemented by local fishers' organizations. From the perspective of marine ecosystem recovery, these sites have been created to ensure sustainable marine communities, nurturing not only commercially valuable stocks but also the larger surrounding environment in this region. Moreover, we hope to ultimately recreate a hydrological cycle that mirrors natural nutrient flow by connecting these networks to local freshwater systems. These modifications to the coastal ecosystem are based on the Japanese concept of Satoumi – coastal, community-based resource management that combines both scientific knowledge and a traditional understanding of the local environment.

1 Introduction

This contribution introduces two Japanese marine ranching projects implemented in the western Seto Inland Sea off Okayama, Japan. These two marine ranches, costing a total of USD 40 million, cover 300–500 ha and are comprised of existing marine reefs (both artificial and natural) and sea grass beds. In providing organized coastal environments with habitats optimal for nurturing

T. Tanaka (✉)
Satoumi Research Institute, Okayama, Japan
e-mail: takebom7@gmail.com

Y. Ota
NF-UBC Nereus Program, Fisheries Centre,
University of British Columbia, Canada
e-mail: y.ota@fisheries.ubc.ca

H.-J. Ceccaldi et al. (eds.), *Marine Productivity: Perturbations and Resilience of Socio-ecosystems*,
DOI 10.1007/978-3-319-13878-7_31, © Springer International Publishing Switzerland 2015

the development of commercially exploited fish species, these artificially enhanced areas are intended to increase the levels of primary production and stock recruitment. Additionally, parts of these zones are protected as no-take zones or areas where bottom trawling is prohibited; measures were voluntarily implemented by local fishers organizations. From the perspective of marine ecosystem recovery, these sites have been created to ensure sustainable marine communities, nurturing not only commercially valuable fish stocks but also the larger surrounding environment in this region. These efforts have been successful in contributing to the enhanced harvest of fish stocks and brought stability of marine ecosystem in the areas.

Regarding the socio-ecological context of those projects, it shall be noted that those projects are designed and managed on the basis of the Satoumi concept, that is, "High productivity and biodiversity in the coastal sea area with human interaction" (Yanagi 2008). And both implementations of marine ranching were not simply the technological solution to aggregate local productivity of stock but also facilitated by local fishers who are engaged with community-based activities to restore nursery habitats, such as sea grass beds. Thus, those projects contribute to the greater challenge that has been amounted to arguing the continued relevance of the "community" effort of coastal management. We argue that the orchestrated attempt to recover and recreate coastal ecosystem with strong initiative of fishers communities is critical in the face of increasing environmental challenges, such as climate change and loss of biodiversity, to recompose the unity and resilience of coastal living and environment.

2 Okayama Coastal Environment and Fisheries

The two marine ranching projects introduced here are both deployed in Okayama Prefecture, facing Japan's largest enclosed body of ocean, the Seto Inland Sea. The area has rich marine heritage with 20 commercial fishing towns where fishers are engaged with small-scale fisheries and low-impact aquaculture including seaweed and oyster farming, whose total production covers approximately 25 % of the entire fisheries production in the Japanese coast.

Ecologically, this high productivity originated from the unique oceanographic environment of the inland sea. The sea has two distinctive hydrographic features, including numerous narrow channels (seto) between more than 3,000 scattered islets and wider body of open sea with gradual water circulation (nada). The sea accumulates micronutrients released from river basins, while the channel itself conveys the water slowly to resuspend the sediment and maintain healthy turbulence.

However, there has been constant decline of fish catch and little recovery of the stock due to the rapid coastal development which initially started in the 1950s (lost 60 % of costal natural habitat by 1980) and the continuous fishing pressure, though moderate but constant for some species. This has called for the urgent stock recovery plan, which considered the use of ecosystem-based approach and the involvement of fishing communities to be appropriate for the unique biophysical characteristics of the area and socioeconomic context, especially its role to sustain the ownership and interest in the conservation of fish stock as well as marine ecosystem among the stakeholders, especially the fishing communities.

3 Historical Degradation of Marine Habitats in Okayama

Up to the 1940s, 10 % of the marine area was covered by mud flat and sea grass bed, providing spawning and nurturing habitat for various marine species. After the 1980s the sea area was changed by large-scale reclamation and landfill, almost 90 % of those mud flat and sea grass bed have disappeared and were also polluted by various land and sea activities, such as 35 years of aggregate extraction, large-scale oil spill, herbicide from inland agriculture, and mercury from local industrial sectors. This continued up to the 1980s.

Fisheries of Okayama survived through these changes, and the marine ecosystem continued to provide limited spawning and nurturing ground for more than 60 species that are targeted fish stocks. During the same period, the number of fishers engaged with low-impact aquaculture for oyster and seaweed has increased, but numbers of trawling and netting have been decreased; the number of fishers has decreased into half during the last two decades. To respond to both ecological degradation and decline of small-scale fishing industry, from the 1970s, the Okayama government, together with other prefecture governments in the Seto Inland Sea region, has engaged with the maintenance of the fish stock and recovery of the coastal environment.

4 The Design of Marine Ranching Projects

Marine ranching is a system to implement to enhance the fish stock by creating ideal habitat for fish artificially. There are several procedures and also various environmental factors that we need to take into account, particularly to recreate healthy nutrient circulation and ecological continuity of marine ecosystem serving to accommodate the habitats, which fish require to mature and live.

To implement marine ranching, we first assessed the state of marine ecosystem in terms of its habitats and biodiversity, if the area is suitable for nurturing targeted species and identified the need for supplemental assistant to the marine ecosystem, including deploying artificial reef in order to orchestrate the total package of different marine habitats for different stages of fish life history. After this initial assessment, we studied fish behaviors, including their mobility pattern and swimming ability into the consideration as we design the area of marine ranching. This is necessary because for nurturing area for juveniles in particular, we need to design a contrived spatial structure with mechanistic feeding system within the current marine ecosystem by putting the sea grass bed and cultivating small prawn and crabs for their food.

The first selected area for this project was in the west part of Okayama, the sea area of 350 ha between two small islands, Takashima and Shiraishi-jima. Then, the second was in the east part, called Hinase. For both areas, we selected seven species including the Japanese red sea bream, gilthead, rockfish, marbled rockfish, sea bass, red-spotted grouper, and fat greenling. So far, the first ranching has doubled the catch amount of targeted species and marked approximately 15 % increase of the general fish catch. However, it has also created a conflict between commercial and recreational fishers, who are attracted by the ample fish stock.

It is important to note that the success of first ranching owes to local fishing communities whose member conducted the day-to-day maintenance of the artificial reef and feeding system and took an initiative to designate volunteer non-take zones. The member of the fishing community in the second area is now utilizing the marine spatial planning and integrated coastal management to consider ecosystem-based measures such as MPA in the area. In this community, there have already been volunteer activities by fishers to reconstruct sea grass bed by replanting the seed for the last 30 years (Fig. 1).

5 Conclusion: Satoumi and Marine Ranching

The sustainability of Japanese coastal fisheries has been at stake for social and ecological pressures accumulated by various anthropogenic activities, including fisheries and coastal pollutions for several decades. Then, how can coastal communities protect their environment while maintaining economic activities that use marine resources sustainably?

This question can be addressed by investigating either national or regional approaches in marine policy that successfully promote the sustainable use of coastal ecosystem services. It can also be answered through the analysis of local coastal attempt, which we described in this paper.

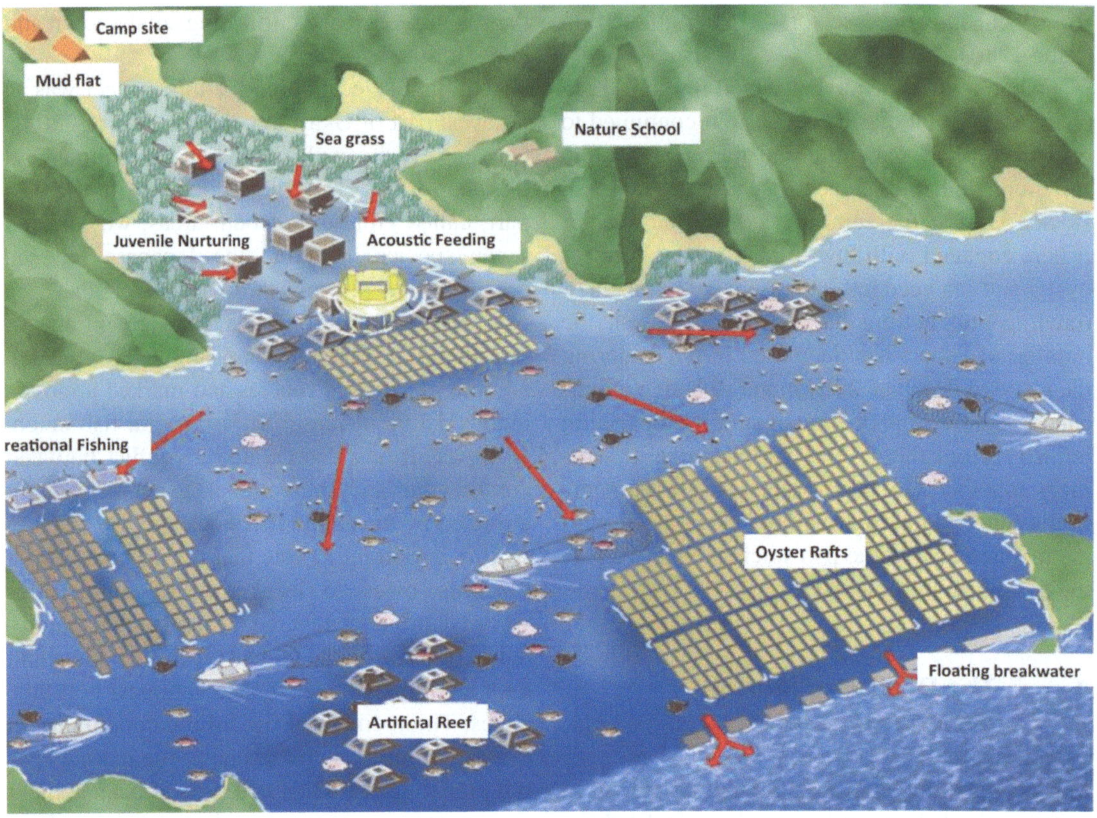

Fig. 1 Planning image of Hinase Marine Ranching. *Arrows* indicate the direction of fish movement as they grow through different life stages

Berque and Matsuda (2013) argues that the hands-on approaches of coastal communities are the key to conserve or restore their marine environment, but technological support and scientific knowledge are necessary for the local-scale innovation. The implementation of marine ranching in Okayama remains to be a challenge for both its cost-efficiency and fisheries governance because the cost can only be recovered by the increase of fisheries production of the enhanced marine ecosystem. Marine geo-engineering can undermine the idea of conserving biodiversity and sustainable use of coastal and marine resources to the work of production sectors, mainly fishers communities, if the social capital of those fishing communities in Okayama is not considered. Thus, the benefits of the Satoumi approach in comparison to other contemporary coastal management schemes rest on its appeal to the fishers themselves.

Technically, marine ranching is set to recreate a hydrological cycle that mirrors natural nutrient flow by connecting these networks to local freshwater systems. Simultaneously, the modifications to the coastal ecosystem based on the Japanese concept of Satoumi coastal, community-based resource management provide both scientific knowledge and a traditional understanding of the local environment, juxtaposed with technical understanding of the marine ecosystem is vital for both implementation and governance.

References

Berque J, Matsuda O (2013) Coastal biodiversity management in Japanese satoumi. Mar Policy 39:191–200

Yanagi T (2008) Sato-Umi – a new concept for sustainable fisheries. In Tsukamoto T et al (eds) Fisheries for global welfare and environment. TerraPub, Tokyo, pp 351–358

Mediterranean Prud'homies

Christian Decugis

Abstract

Fishermen on the French Mediterranean coasts are assembled in old and original collectives called prud'homies and follow their own rules. The individuals in charge of these areas, the prud'hommes, manage coastal fishing areas, act as auxiliary police, settle disputes between members, participate in the regulation of the use of coastal areas, and have an important social influence. They take into account the evolution of available short- and medium-term resources and set rules that are accepted by all fishermen in the collective.

1 Introduction

The Mediterranean fishing communities periodically elect representatives called 'prud'hommes' to organize and sustain fishing in the territory in which they practice their profession.

A 'prud'homme' is a person with extensive experience in and knowledge of the fishing techniques and the people who practice this activity. He is a wise man with impeccable morals who is elected by his peers. The 'prud'homme' is selected from the older generation: he must have been a professional fisherman for at least 10 years and a skipper for at least 5 years. He must take an oath before the administrative authorities, which reflects an individual commitment rather than an act of a higher maritime authority. There are no written laws: the oral tradition prevails.

The 'Prud'homies' system has spanned the centuries, as attested by the contents of old archives: patent letters of King René and kings of France since the fifteenth century; the Decree of the Constituent Assembly in 1792 recognized the merits of prud'homies of Marseille and Cassis and authorized others to create prud'homies ports on this model; the 1859 Decree that still governs the operation of such meetings. It is likely that, throughout this long history (Payan d'Augery 1873; Poujade 1936), we can see realistic adaptations of the organization in pursuit of fishing activities, and, more recently, aquaculture and some coastal management of the different littoral areas (Alegret 1999).

C. Decugis (✉)
9 avenue de la Fontaine, 83700 Saint-Raphaël, France

Prud'homie des Pêcheurs de Saint-Raphaël,
14 avenue Commandant Guilbaud,
83700 Saint-Raphaël, France
e-mail: chris.decugis@gmail.com

H.-J. Ceccaldi et al. (eds.), *Marine Productivity: Perturbations and Resilience of Socio-ecosystems*,
DOI 10.1007/978-3-319-13878-7_32, © Springer International Publishing Switzerland 2015

2 Role of Prud'homies in Fisheries

Various marine species are exploited during their seasonal concentrations at the coastal border or further offshore (Féral 1977). Generally, fishing areas are cramped or inaccessible, the ocean bed falls very rapidly to 800–1,000 m, and the coasts are frequently windy. Some fisheries are conducted in coastal ponds, and most species are scarce. Multipurpose and various small-level activities constitute the fishermen's lives.

These daily environmental variations require almost immediate responses from prud'hommes: to draw lots for fishing stations; organize their rotation between fishermen; expresses it arouses strong interest; resolve conflict between different trades; take precautions following heavy catches; disseminate new technological innovations over the untimely excesses of certain personalities, etc. (Bonzon 2000). This local management justifies these various structures: 33 prud'homies share the shoreline of Port-Vendres in Menton, including Corsica. By itself, the local Fisheries Committee Var covers eight prud'homies and six prud'homies sections, totaling 200 fishermen spread over 432 km of coastline.

3 Roles of Prud'homies

French Mediterranean coasts, with their strong residential and tourist appeal, are subject to intense navigation, constant coastal development, and significant risks of pollution. Throughout history, prud'hommes have mobilized to preserve the quality and extent of their fishing grounds and have therefore played – and continue to play – a role in preserving coastal environments. It is this longstanding history of land management and the ability to adapt to contemporary problems that may appear on the coastline that gives the institution its modernity. Further, the very small scales on which they operate over the entire coastline gives it a certain efficiency.

Prud'homies are usually of modest means (Marceille 1919). Internal management is based on EU principles and are the fruit of lengthy experience and understanding of a common maritime field and resource. It is very important indeed to establish and maintain harmonious competition rules within the fishing communities of the same resource. In practice, they must educate and discipline fishermen who are generally poorly educated (in the school sense), often temperamental, and working in a space where no limit is realized. Land management sees the prud'hommes involved with local authorities, organizations in the maritime field, or associations interested in the marine environment, which employ prud'hommes in various projects. Prud'hommes sometimes instigate some of these projects. Their strength lies in their knowledge of the maritime field and their constant presence on the water.

Currently, and in recent years, the future of fishing-related professions and the use of natural marine resources, as well as the institutions themselves, have been subject to hardships for the sake of economic and political order.

With the industrialization of fisheries in the Mediterranean, which began in the 1960s, the small-scale sector has managed to adapt by turning to the capture and commercialization of extreme-specific products (domestic sales of diverse fresh products, hake, eel, etc.), moving away from the less marketable species (sardines, anchovies, etc.) that previously represented the bulk of the landings. This shift related to the relative modernization of small fleets and fishing gear, and was accompanied by a decrease in the number of fishermen and boats, further increased by the residential and tourist specialization of coastal areas.

The second crisis experienced by this sector since the 1980s relates to the constitution of the whole EU and the economic and political measures that accompanied it, at both European and national levels.

At the national level, the mining sector of coastal marine resources is assessed for general measures such as license Commissioning Operation (Permis de mise en Exploitation

[PMEs]) and Reduction Plans fleets, under guidance programs Pluri-annual (POP) of the Common Fisheries Policy (CFP). For a plethora of fisheries, these heavy devices endanger the renewal of the profession in many ports where the population of fishermen has already reached a minimum threshold.

Other measures, such as those regarding vocational training, employment conditions, and safety at sea continue to hamper the regeneration of the fishing profession in these small coastal trades. Aware that this profession is changing due to the specialization of the regions in Europe, fishermen are willing to consider its redefinition. Devices to set up must be in line with the preservation and enhancement of coastal ecosystems, the regional tourist and residential development (e.g. in the reduction of engine power, with reference to trawlers), or the introduction of diplomas or safety rules defined in relation to ocean navigation.

Chaired by professional fishermen, Prud'homies are essential for consultation with a panel of partners. This partnership, a central link in the operation of local and regional structures fully integrates partners in the fisheries sector alongside the economic, institutional, and territorial development players.

Networking representatives of users of the area, including users of the sea, can exchange and discuss actions to be taken to preserve and enhance the marine environment and those that depend on the economic activities.

4 Other Types of Cooperatives

In Europe, the Mediterranean fisheries differ from those in Spain, Italy, or Greece in terms of the limited number of fishermen spread over the entire coastline, and the fact that this profession has maintained a long tradition of local management. While some countries, like Greece, are plaintiffs in regulatory matters, France has accumulated too many measures and the current problem faced by the profession is to integrate the 'prud'homal' mode of management within European and national systems. In this regard, it should be noted that the Commission, in a technical context, emphasizes management by sector of activity and not as a global management territory, which seems to better reflect the territoriality of the small fishing industry.

This specificity at the European level of Mediterranean coastal fishing, and the harmonization of its operation by its specific institutions imbued with lengthy historical experience could be used for other fishing communities and even constitute a kind of 'laboratory'. Several reasons support this view. The French Mediterranean coast is characterized by a variety of geological settings: exposure to wind and prevailing currents is found all around the Mediterranean basin; the continental shelf is often not extended, there exist steep coastal cliffs, ponds, canals, deep and rough lagoons, deltas, contributory rivers, islands, peninsulas, sandy or rocky areas, seagrass, etc. Each of these situations requires specific operating methods and fishery management.

5 Prud'homies and Society

Fishing faces many competing activities: industrial areas, shellfish, aquaculture, tourism, agriculture, wildlife farms, urban, port, military, and so on. Numerous activities are concentrated in Port-Vendres and Menton. Conflicts have to be resolved with the cooperation of fishermen.

Historically, various forms of similar fishing prud'homies have existed throughout the Mediterranean. The French prud'homies are currently the most developed, and some countries have observed the spirit of these social structures, born of long experience (Giovannoni 1998). This is, for example, the case in Tunisia and Morocco. One can imagine that the cultural background behind the creation and organization of prud'homies is not so different from what might work well elsewhere. Prud'homies have, at different stages of their past progressively inserts several waves of immigration without any major problems on their specificity.

So, if small fishing is given a chance and helped to integrate into the regional economy, a beneficial effect on the preservation of marine ecosystems will be seen (Tempier 1986). Because it is based on management areas, the new righteousness (i.e.: ecological and scientific probity) of the rules of exploitation by fisheries is a significant issue in our regional economies. Finally, the original experience of fisheries management that embodies the prud'homie is included in highly topical debates led by scientists and policy makers at an international level.

6 Conclusion

Decision-making prud'homies are at the intersection of formal law, regional performance, the authority of local administrative structures, economic pressures, recent scientific achievements, and common sense born of professional experience who can manage the natural resources in which they live, outside the rules established by distant jurists (Féral 1987).

Such social structures, born of continuous adaptation to the surrounding environment, also exist in other countries. Japan has a long history of harmonious relationships, via traditional cooperatives, between human communities and the natural environments in which they live.

References

Alegret J-L (1999) Espace, ressources et historicité: la dimension sociale de la pêche en Méditerranée nord occidentale. In: Séminaire en sciences sociales européennes des pêches maritimes, 4th atelier Hermopolis, Syros. Mai 1998. Gestion des pêcheries dans les mers du sud ed. David Symes, 1999

Bonzon A (2000) Development of economic and social indicators for the management of Mediterranean fisheries. Mar Freshw Resour 51:493–500, Sydney, Australia

Féral F (1977) L'évolution de la pêche artisanale en Bas-Languedoc. Econ Mérid 97:79–90, Montpellier, France

Féral F (1987) Un hiatus dans l'administration et la politique des pêches maritimes: les prud'homies de pêcheurs de Méditerranée. Norois, Tome 133 (fascicule 133–135):355–378, Persée édit

Giovannoni V (1998) Des citoyens extraordinaires: les marins-pêcheurs de la Méditerranée française, L'anthropologie maritime comme champ de l'anthropologie sociale et culturelle. Les cahiers de la DRAC PACA "La construction des sources, anthropologie maritimes", Marseille, pp 26–36

Marceille G (1919) Les prud'hommes pêcheurs de la Méditerranée. Thèse Sciences politiques et économiques, Toulouse

Payan d'Augery C (1873) Les prud'hommes pêcheurs de Marseille et leurs archives. Imprimerie J. Nicot, Aix-en-Provence, France

Poujade J (1936) Les prud'homies des pêcheurs de la Méditerranée. Thèse droit, Paris

Tempier E (1986) Prud'homie et régulation de l'effort de pêche. CRPEE, Econ Mérid Tome 1–2 (fascicule 133–134): 41–50

Changes, Adaptations, and Resilience: The Case of French Oyster Farming

Catherine Mariojouls and Jean Prou

Abstract

The history of French oyster farming is marked by threats and difficulties, thus subsequent adaptations. There have been some significant changes in the oyster practices, the economic operation of enterprises, and the place of oyster farming in coastal areas, whereas the persistence of the oyster farming sector facing significant threats has periodically been questioned. Environmental changes have played an important role in this activity closely related to the environment: pollution from various sources, increasing episodes of contamination by phycotoxins, extreme weather events, and climate change. Recently, the high-mortality syndrome questions the issue of the conditions for the sustainability of the activity. Taking into account the conflicting uses in the coastal zone and the importance of relationships with watershed activities, the development policies must also analyze the adaptation capacity of oyster farming socio-ecosystems in these environments highly dynamically, spatially, and temporally. Can we talk about the resilience of oyster culture? We introduce the concept of socio-ecosystem applied to oyster farming, as appropriate for the analysis of resilience. What are the questions lying behind the oyster farming resilience and which method can be used for its study? Using the case of French oyster farming, this communication proposes to discuss some answers to these questions, for different aspects.

1 Introduction

The French oyster farming since its inception in the nineteenth century has been regularly obliged to evolve in order to adapt to new situations, so that the history of oyster farming offers a rich range of difficulties or threats, and subsequent evolutions, for an analysis. We shall first present chronologically the main trends in production,

C. Mariojouls (✉)
AgroParisTech, 16 rue Claude Bernard,
75231 Paris Cedex 05, France
e-mail: catherine.mariojouls@agroparistech.fr

J. Prou
IFREMER, Ronce-les-Bains,
17390 La Tremblade, France
e-mail: jean.prou@ifremer.fr

H.-J. Ceccaldi et al. (eds.), *Marine Productivity: Perturbations and Resilience of Socio-ecosystems*,
DOI 10.1007/978-3-319-13878-7_33,

the main events, and changes which the sector went through. We shall use the available information to set up a typology of the threats and corresponding adaptations developed by the sector. We shall finally introduce a new framework, the socio-ecosystem of oyster farming, as a more pertinent concept for analyzing the capacity of adaptation for oyster farming. In the last part, we shall come to the notion of resilience and present some proposals regarding how it can apply to oyster farming and why some elements would require further research.

2 History: Main Trends and Events in the Oyster Farming Production Over a Century

Historically, in France oyster farming has been developing in the 1850s when new facilities like rail transportation (bringing oysters to cities and routing first tourists to the coast) created a growing consumer demand that could not be met anymore by harvesting market-size oysters from the wild. The starting point was the invention of spat collector for native flat oyster *Ostrea edulis*, by Davaine (1853) and Coste (1866) in the first marine biology laboratories.

The history of French oyster appears as a succession of dramatic development of production and collapses, which causes have been identified or not, and that led to several changes of species chosen and cultured. Figure 1 describes the variations in production volumes, the main biological causes identified, and the adaptations that have been put in place to allow the continuation of the activity despite several crises.

For the native flat oyster *Ostrea edulis*, the production curve shows fluctuations and points out the two mass mortality periods: one starting on 1920 and one starting in the 1970s, due to parasitic diseases marteiliosis and bonamiosis that led almost to an extinction of farming of flat

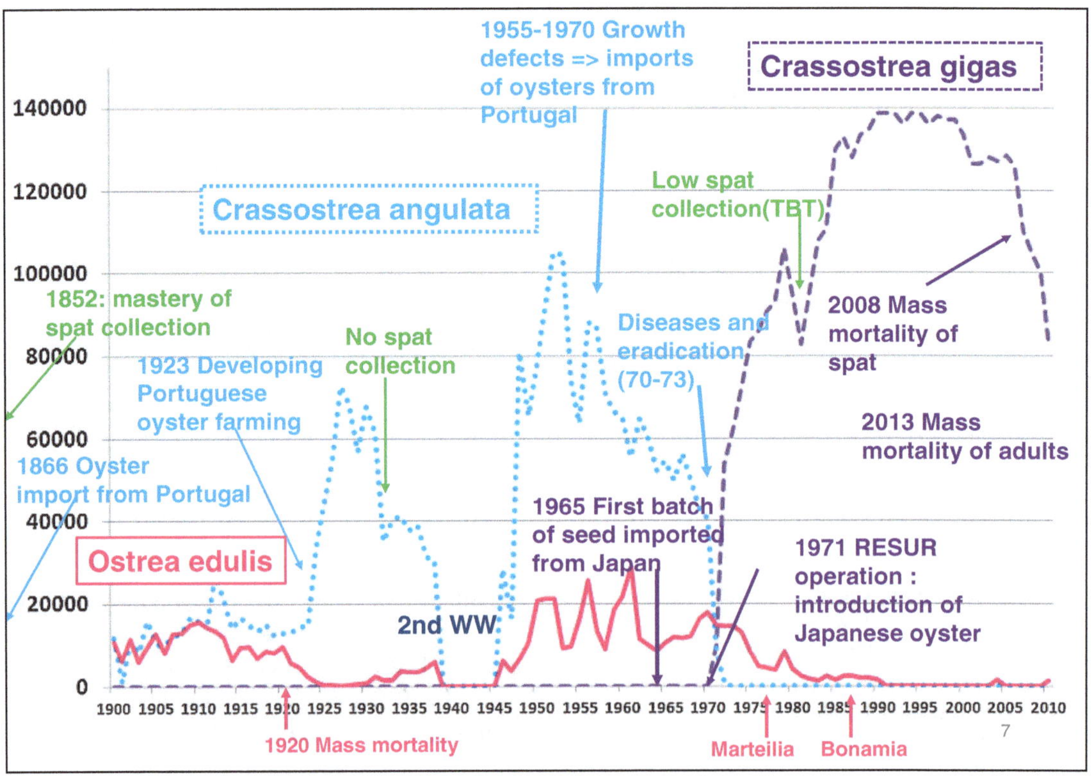

Fig. 1 Production of oyster culture in France (tonnes) and main changes (After IFREMER data)

oyster in the 1980s (Anonyme, 1923). Noticeable are the events before 1900: spat collection was developed from its mastery in 1852, and the first import of Portuguese oysters *Crassostrea angulata* from Portugal was in 1866. But the real development of farming of that species was in 1923, at the time of mass mortality of *O. edulis*. The production of *C. angulata* developed, except during World War II, until mass mortality due to iridovirus in the early 1970s led to a collapse of this species.

The French oyster farming recovered thanks to the introduction of *Crassostrea gigas* from Japan and America in the 1970s and has grown to a high level, up to 140,000T in the 1990s. In the late 1990s were already observed summer overmortalities, but the last decade, in 2000s, is marked by mass mortalities: first for spat from 2008 due to the herpes virus OsHV-1 microvar, leading to a decrease of production level as low as 80,000 T in 2012, and, very recently, in 2013, for adults, due to *Vibrio aesturianus*. Today, the French oyster farming sector is facing a serious crisis.

3 An Attempt to Analyze the Major Threats and Changes and Adaptations

Table 1 attempts to identify major changes and disruptions that the oyster industry has faced and faces now, classified into four broad categories: anthropogenic, ecological, biological, and technical major events. Listed are also the main adaptations implemented in the oyster industry, adaptations which are due to either oyster companies or institutional actors including research and government.

3.1 General Aspects

In oyster farms, there are both technical changes and new economic strategies that have been implemented to adapt. In addition to these immediate coping strategies, it is noticeable also that the oyster business developed proactive strategies, like diversifying their activities (farm visits, trading shellfish), but these fall outside the scope of this communication. As for research, in case of disruption, a set of programs are developed to understand new phenomena and to propose adaptive changes in the systems of oyster culture and in the institutional framework governing the production and the marine environment. Regarding government and public organizations, in addition to alerting devices and crisis management if violent disruption occurs in coastal marine environment, the adaptation to changes and disruptions takes the way of regulation, by creating new legal tools or adapting existing tools.

This presentation provides a basis for the analysis that we want to lead but is necessarily simplified and does not pretend to be exhaustive. Two important parameters, scale and time, are not considered here, while they would be of importance for a more detailed analysis: some disturbances are occurring within a short time (big storms, pollution accident, etc.) and in a specific area, appealing for quick and simple reaction, while a number of changes and threats occur on a long period of time and at a large scale, allowing progressive adaptations on one main pattern or on regional patterns but sometimes questioning the persistence of the oyster production.

We cannot perform here a detailed analysis of this set, but we choose to bring out the analysis of the adaptations developed by oyster farmers facing recurring mortalities over a period of time close to a century. This will allow going further in the analysis of adaptation and resilience in the last part of the communication.

3.2 The Case of Mortalities

Mass mortalities of oysters are considered abnormal and affect the economy of oyster farms. They mark the history of oyster farming and have sometimes been dominating events in the evolution of the industry. They are mostly related to pathogens, identified and characterized by scientists. Bonamiosis and marteiliosis have been described to explain the mortality of flat oyster *Ostrea edulis*. Similarly, an iridovirus

Table 1 Main threats and changes for French oyster farming and adaptations

Threats and changes		Adaptations in the oyster industry
Anthropogenic	Various pollutions	Development of scientific monitoring, changes in controls and regulations
	Microbiological pollution	Regulation, more stringent; self-controls and purification plants
		Involvement of oyster farmers organizations in the management of watersheds
		Relocation of oyster culture
	Increasing use of coastal zone by other activities	New planning and management tools in public policy, with consultation with local stakeholders, and involvement of oyster farmers organizations
Ecological	Increase in frequency of toxic phytoplankton blooms	Development of monitoring, changes in controls and regulations; farmers lobbying for modification of control procedures, considered as too stringent
		Temporary interruptions of sales
	Changes in geomorphology	Oyster farming both a cause and victim of changes in geomorphology
		Creation of "schemes of structure": regulatory constraints for the maintenance of oyster farming parks
Biological and technical	Mortalities due to epidemics during the twentieth century	Change in species: *O. edulis*, then *C. angulata*, and then *C. gigas*
		Imports of oysters for supplying the trading
	Interannual variability of recruitment	Adaptation of sale price of oysters, in the possible range for market
	Development of hatcheries, use of triploids	Wide adoption of hatchery spat, but among farmers, differences in opinion about hatcheries and triploids and socioeconomic cleavages
	Summer over-mortalities of spat, mass mortality since 2008	Changes in farming techniques; restocking with "resistant" oysters
		Restart of shellfishing on wild stocks
		Development of direct sales, prices rising, diversification of activities
		Disappearance of the weakest companies. Public support to oyster companies
	Mass mortality of adults, since 2013	Early sales of oysters, shellfishing on wild stocks/import of oysters for trade
Météorological and climatic	Major storms	
	(a) Martin (1999)	(a) Reconstruction of parks and restart activity in affected areas; public supports
	(b) Xynthia (2010) (sea rise)	(a) Sales closing because of *Pseudo-nitzschia* bloom; farming devices replacement (insurance)
	Global change: temperature rise	Opening of new sites for spat collection (Brest Bay)
	Global change: ocean acidification	Not yet a noticeable effect in France

was responsible for the high mortality of oyster *Crassostrea angulata* called "Portuguese." Finally, concerning the oyster called "Japanese" *Crassostrea gigas*, a herpes virus OsHV-1 microvar and a vibrio, *Vibrio aesturianus*, are identified as the main causes of the recent and current mortality. When mortalities affect the economy of oyster farms, oyster growers react by introducing various changes in their farming and further in their commercialization.

3.2.1 Relation to Wild Stocks: Foundation of Oyster Farming, Solution in Case of Crisis

As mentioned before, controlling the collection of *O. edulis* oyster larvae on a collector was the foundation of oyster farming. The farming industry of flat oyster could develop in several zones of the Atlantic coast: Arcachon, Marennes-Oléron, and Brittany. Later, for cupped oyster species, *C. angulata* and *C. gigas*, the same pattern occurred with the difference that as these species are not breeding on the entire Atlantic coast, Arcachon and Marennes-Oléron became the spat collection centers providing other farming zones, effectively Brittany, Normandy, and the Mediterranean sites.

Despite little known, until the 1980s the collection of spat on collectors has always been accompanied by fishing oysters on natural stocks (oysters of small size for further farming or bigger for direct selling). But in the 1990s, the use of oyster fishing on wild stocks was no longer valid since the controlled reproduction of oyster in hatchery helped to provide farmers with spat easier to work and not subject to the interannual variations of spat collection in the natural environment.

From 2008, to offset the high mortality of young oysters due to OsHV-1, two adaptations appeared: (1) an increase of spat stocked for farming, through increasing the number of collectors installed and also using a larger proportion of hatchery spat, and (2) a return to the additional practice of fishing wild oysters, here at commercial size, to compensate for the loss of volume driven by mortality.

3.2.2 Imports of Oysters

Imports of oysters were often used to compensate for mortality or for supplying the market. Thus, in 1866 the lack of oysters resulted in the importation of Portuguese oysters *Crassostrea angulata* from the Tagus Estuary. At the same time, an accidental cargo release of *C. angulata* into the estuary of the Gironde had allowed the introduction of the species on the French coast, followed by a rapid expansion south of the Loire estuary, but that species had been rejected by production centers breeding the traditional flat oyster, i.e., in Brittany. It is at the end of the First World War that consumer demand and high mortality (1921–1923) on flat oyster forced professionals to adopt the Portuguese oyster. Similarly, after World War II, restarting consumption led to the importation of oysters from Portugal and other countries from 1955 (Anonyme, 1955). In February 1966, a producer of La Tremblade (Charente Maritime), P. Jarno, imported from Japan the first batch of spat of oysters *Crassostrea gigas*. Its rapid growth and tolerance for iridoviruses which decimated the Portuguese oyster (1970–1973) led to its adoption by professionals at individual level (imports of seed from Japan) and at collective level (reseeding from natural deposits by adult oysters from British Columbia, USA).

4 The Social-ecological System, a New Concept for Analyzing Adaptations and Resilience

It is obvious that changes in ecosystem affect the development of oysters and thus the production of farmed oysters. How oyster farmers have adapted to disturbances or simple changes, with implications for oyster production, and why? Is it only the fact of oyster farmers or a set of actors involved in oyster industry, including research and public authorities? The analysis of disturbances and changes in ecosystem and of the adaptations set in response, in place, shows the multiplicity of actors involved and the plurality of adaptive registers that are mobilized. The concept of social-ecological system (SES), allowing a holistic analysis, seems a suitable tool.

The question of adapting to changes in social-ecological systems (SESs) is central in contemporary research work. Holling in 1973 introduced for the first time the notion of resilience of ecological systems within a reflection on their stability against fluctuations and so opened a wide field of ecological research about the resilience of ecosystems. The development of the concept of social-ecological system, and questions for the definition and implementation of sustainable development, led to the study of the adaptability and the resilience of SESs. Taking into account that the numerous works about resilience have led to various interpretations and some confusion, Walker et al. (2004) provided an interpretation and analysis of the concepts of resilience, adaptability, and transformability in SESs. They consider that the paths taken by a SES depend on three attributes: resilience, adaptability, and transformability.

While agroecosystems are subject to rather important research regarding their health, their resilience, and their stability (Xu and Mage 2001; Malézieux 2012), we found little research results about the resilience of marine farming ecosystems.

The resilience after coastal disasters has been analyzed by Adger et al. (2005) in an approach linking explicitly ecosystems and human societies in a social-ecological system. Perry et al. (2010) propose a social-ecological approach which recognizes the interdependence between biophysical and human social components. They point the variety of drivers for changes in marine systems, both for biophysical aspects and for fishing-dependent human communities and their interactions. They demonstrate the need to develop, notably for policies, approaches which maintain the capacities of fish and fishing communities to adapt to the impacts of globalization and environmental changes, but they do not discuss the resilience of the social-ecological system.

The concept of social-ecological system is very pertinent for coastal ecosystems including oyster farming but to our knowledge has not been used for analyzing their resilience and adaptability. The practice of shellfish aquaculture viewed as an ecological disturbance has indeed been studied and the resilience of the ecosystem been questioned by Dumbauld and Ruesink (2008) in the case of Chesapeake estuaries and more widely in the case of West Coast estuaries (USA) by Dumbauld et al. (2009) in an extended review on the ecological role of bivalve shellfish culture in the estuarine environment, but without using the concept of SES. The SPICOSA program uses modelling applied to coastal SES zones including shellfish farming, for research in ICZM, not centered on oyster farming nor its resilience (Ballé-Béganton et al. 2010).

The concepts developed by Perry et al. (2010) for marine systems, with strong interdependence between biophysical and human societies, could be applied to oyster farming in coastal ecosystems. We identified (cf. Table 1) the main drivers for changes in the biophysical system and the human societies and the main responses by human communities. We can consider, as underlined by the authors, that the human communities' responses to marine ecosystem variability can ameliorate or exacerbate these changes.

For French oyster farming, we propose a design of socio-ecosystem (Fig. 2) and its evolution along time, characterized by the succession of several species in the ecosystem, the replacement of one species being done when the conditions for the life of the previous, or a profitable farming, are not sufficient any more. This SES has two main components: (1) the ecosystem, home for the farmed species, in the adequate conditions, with limits for each species and (2) the social and technical system, in which the main stakeholders are the farmers but also the researchers and public authorities acting through public policies. All contribute to changes and adaptations. The farming itself takes place in the intersection of the two, through the farming system (including farming practices) associated successively to the three farmed species, while some important evolution in the spat origin takes place, from natural spat collection to hatchery-produced spat of increasing importance today. Also it must be underlined that farming practices establish the framework of the relationships between the farmed oysters and the ecosystem and thus play a central role to set the conditions for the results of the farming as an economic production. The farming practices themselves show changes over time. Within the period of time where each species is farmed, it has generally been observed that the efficiency of farming first increased and reached a peak before decreasing.

Perry et al. (2010) insist on the importance of different scale approaches for marine social-ecological system, using mainly the example of fisheries. For oyster farming, while the local "natural scale" (bay, watershed, etc.) is indispensible for management, it is also necessary to consider wider scales for governance of some aspects: animal health, contaminants, and sustainability for social and economic aspects.

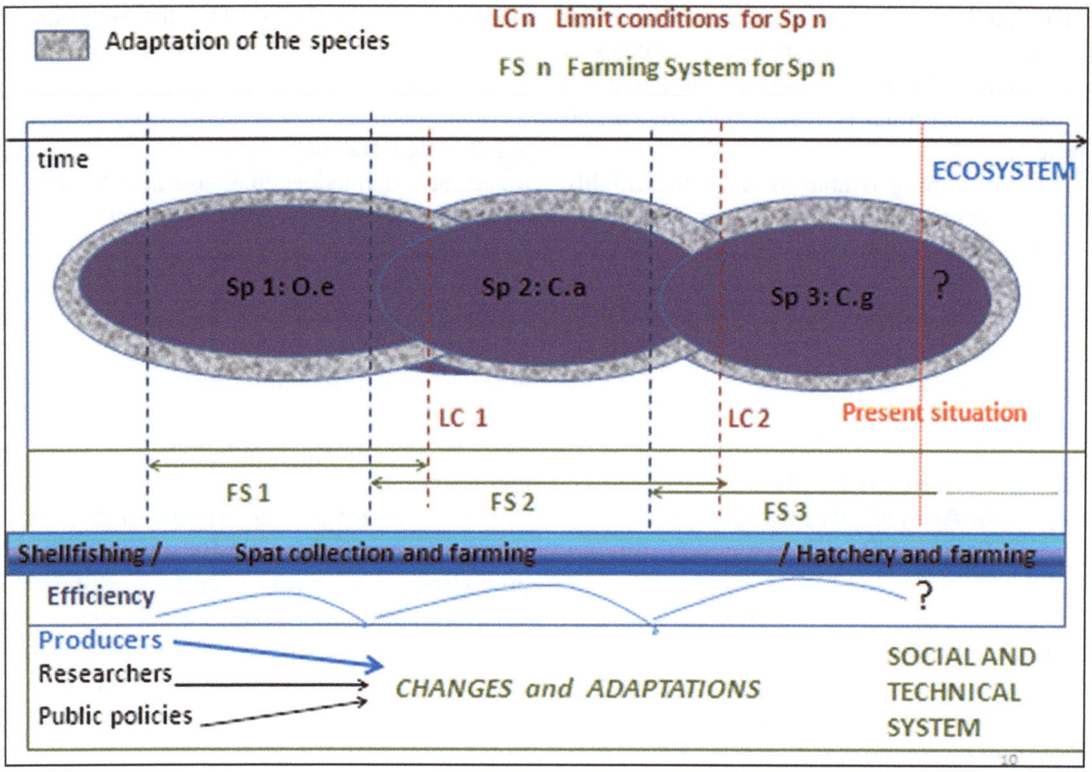

Fig. 2 Attempt to represent the French oyster farming as a dynamic social-ecological system

5 Is the French Oyster Farming Resilient?

5.1 Questions About Resilience

5.1.1 The Resilience of Ecosystems That Produce Shellfish and Support Shellfish Farming

Shellfish farming is an activity practiced in highly dynamic ecosystems, at all time scales: millenary (last glaciation, sea level rise, siltation of estuaries, strong currents, sediment movements, etc.), annual-seasonal (winter storms, etc.), monthly (tide coefficient), and daily-semidiurnal (tides). These geomorphological changes (Bertin and Chaumillon 2006; Kervella 2009) have shaped ecosystems, providing habitat for shellfish.

Analyzing the resilience of oyster farming leads to consider many questions: (1) about the resilience of these ecosystems; (2) conversely, about the impacts of oyster farming on the ecosystem and its resilience; (3) and some about the necessary management of natural oyster beds as reservoirs of genetic biodiversity used by oyster farming.

5.1.2 Disturbances of Shellfish Farming and Ability to Adapt

The ability of adaptation of oyster farming to disturbances is linked to several facts: the farming devices may be affected but are flexible enough to avoid a real destruction; the farm buildings on seaside are often able to accept accidental sea rise because they are already built and used in conditions of level variations due to tides; and the oyster, biologically, is flexible, and, for instance, able to overcome some pollution events, through self-purification as a filter feeder.

Animal health appears as a special issue: disease outbreaks occur often, but a frequent situation is the presence of pathogens with no important losses. It raises questions about the role played by the ecosystem and its changes in the regulation of animal diseases and about the impact of farming practices on the farmed oysters.

Oyster farming is able to adapt in a highly dynamic ecosystem, because it has been organized to have the necessary flexibility. Then, the question is to understand how and why the adaptive process is altered: because of extreme disturbances or because of inadequate farming operations or too strict rules.

5.2 Some First Elements of Analysis of French Oyster Farming Resilience

We consider it is possible to analyze the adaptability and resilience of oyster farming as a social-ecological system facing fluctuations, and we propose some elements for a first analysis, using the case of the French oyster farming. Walker et al. (2004) give the following definition of resilience "Resilience is the capacity of a system to absorb disturbance and reorganize while undergoing change so as to still retain essentially the same function, structure, identity, and feedbacks" and consider that there are four crucial aspects of resilience: latitude, resistance, precariousness, and panarchy. For the oyster farming social-ecological system, these aspects would deserve an in-depth study, but we can here propose some hypothesis.

5.2.1 Latitude

Latitude is defined as "the maximum amount a system can be changed before losing its ability to recover (before crossing a threshold which, if breached, makes recovery difficult or impossible)." The French oyster culture has developed a wide range of adaptations allowing successive recoveries and can be qualified as having a large latitude. In particular, the succession of species (*O. edulis*, *C. angulata*, *C. gigas*) can indeed lead to conclude that the "latitude" of the SES is wide. Although the adoption of a new species can be long (37 years between the biological implantation of *C. angulata* –1867- and the start of farming –1923-), it is done by giving up the previous, more or less clearly. In 1970, the adoption of *C. gigas* was rapid as *C. angulata* was eradicated by the disease.

5.2.2 Resistance

Resistance, defined as "the ease or difficulty of changing the system; how "resistant" it is to being changed," is highly questionable according to components of the SES: ecologically, there is indeed an ability of the farmed oysters to adapt to threats and changes in environment and techniques, and, socially, there are many situations, for instance, differences among oyster farmers. Some producer subgroups can be resistant to changes, for example, the adoption of a technical innovation like triploid oysters proposed by research. While easily adopted by some producers, it has been rejected by some farmers because of not only technical aspects but also a wide set of arguments: environmental, economical, and philosophical (Boulanger and Mariojouls 2012).

5.2.3 Precariousness

It is looking to "how close the current state of the system is to a limit or "threshold"." This concept deserves to be acutely looked at, as the present situation of French oyster farming social-ecological system effectively questions the existence of thresholds for this system: for biophysical characteristics in relation with the adaptability of the animals and for human communities and oyster farming companies.

Researchers, significant players in the SES, play regularly a role as they establish some elements of "latitude" and "precariousness." Concerning oyster diseases, a vaccination is difficult to develop because of a limited immune system of oysters, and breeding in open environment makes difficult the control of pathogens. The only "recovery" solution appears to be genetic selection. Interestingly, a resistant strain introduces a new space of "latitude" and decreases the "precariousness" on biological and economical viewpoints. But on the social viewpoint, it reduces the "latitude" as it establishes an obligation for farmers to buy selected spat from hatchery.

About precariousness, an in-depth analysis about historical data would be required, but at present, the French oyster farming is in a high precariousness.

5.2.4 Panarchy

Panarchy is taking into account that "the resilience of a system (...) will depend on the influences from states and dynamics at scales above and below." This notion opens a wide range of questions to be studied and discussed. Indeed, the analysis of panarchy for oyster faming should include several items: (1) the state and dynamics of watershed, as perturbations on watersheds generate impacts on coastal environments and shellfish; (2) the state and dynamics of seawater at the regional level, at the appropriate scale; and (3) the state and dynamics of ocean, including the effects of climate change on coastal environment and of ocean acidification on shellfish.

6 Conclusion

French oyster farming appears as a complex and highly dynamic system, which has been able to adapt to numerous threats and changes both in the natural environment and economic conditions. The resilience of the oyster farming social-ecological system is important in its social and economic components. As for the biological and ecological components, resilience does exist with regard to changes in the environment but is very small in case of epidemics, and this phenomenon would require an in-depth analysis as it is the key point for the sustainability of the activity. New paradigms that rely on ecosystem services provided by biodiversity add a new dimension and make necessary a renewed vision of relationship between shellfish farming and environment. The question is how to support preventively the development and management of coastal areas in order to make oyster farming a sustainable activity. Issues like monoculture, cultural practices, and management of natural stocks in relation to uses can be revisited in terms of resilience provided by biodiversity.

References

Adger WN et al (2005) Social-ecological resistance to coastal disasters. Science 309(5737):1036–1039

Anonyme (1923) Lecture pour tous –1923. La crise de l'huître plate, Paris

Anonyme (1955) Journal des débats parlementaires – 1955. Sénat. Séance du 30 Juillet 1955 sur les Importations d'huîtres

Ballé-Béganton J, Lample M, Bacher C, Fiandrino A, Guillard V, Laugier T, Mongruel R, Pérez Agúndez JA (2010) A modelling platform for complex socio-ecosystems: an application to freshwater management in coastal zones. In: Swayne DA, Yang W, Voinov AA, Rizzoli A, Filatova T (eds) International Environment Modelling and Software Society (iEMSs), Fifth Biennial Meeting, Ottawa, Canada

Bertin X, Chaumillon E (2006) The implication of oyster farming in increasing sedimentation rates in a macrotidal bay: the Marennes-Oléron Bay, France. Cah Biol Mar 47:19–22

Boulanger M, Mariojouls C (2012) Evolutions des pratiques ostréicoles et divergences d'opinion chez les professionnels: quels fondements, quels arguments? Baies et Rias 43:4–5

Coste V (1866) Voyage d'exploration sur le littoral de la France et de l'Italie. Deuxième édition suivie de nouveaux documents sur les pêches fluviales et marines. Imprimerie Impériale, Paris

Davaine C (1853) Recherche sur la génération des huîtres. Thunot et Cie, Paris

Dumbauld BR, Ruesink J (2008) Oyster culture in Willapa Bay Washington: a case study and framework for examining resilience of coastal estuaries to disturbance. J Shellfish Res 27(4):1004

Dumbauld BR, Ruesink J, Rumrill S (2009) The ecological role of bivalve shellfish aquaculture in the estuarine environment: a review with application to oyster and clam culture in West Coast (USA) estuaries. Aquaculture 290:196–223

Holling CS (1973) Resilience and stability of ecological systems. Annu Rev Ecol Syst 4:1–23

Kervella S (2009) Dynamique des sédiments fins et mixtes des zones intertidales de la baie de Marennes-Oléron. Thèse de doctorat, Université de La Rochelle

Malézieux E (2012) Designing cropping systems from nature. Agron Sustain Dev 32:15–29

Perry RI, Barange M, Ommer RE (2010) Global changes in marine systems: a socio-ecological approach. Prog Oceanogr 87:331–337

Walker B, Holling CS, Carpenter SR, Kinzig A (2004) Resilience, adaptability and transformability in social–ecological systems. Ecol Soc 9(2):5

Xu W, Mage JA (2001) A review of concepts and criteria for assessing agroecosystem health including a preliminary case study of southern Ontario. Agric Ecosyst Environ 83:215–233

New Monitoring Method to Assess the Marine Algae Distribution and Fish School in Marine Ecosystems: The Hachiri-ga-se Hill (Off Mishima, Hagi, Japan) Case Study

Akira Hamano, Hideaki Tanoue, Takahito Fujiwara, and Teruhisa Komatsu

Abstract

The importance of seamounts (<1,000 m height) outside the continental shelf as hot spots for marine ecosystems (from primary producers such as phytoplankton to top predators like tuna) is currently being globally discussed. On the other hand, as sea-hills (>500 m height) situated on the continental shelves are smaller than seamounts, even though they contribute significantly to coastal ecosystems by enhancing the supply of prey and providing resting grounds for numerous fish populations, they have not been fully recognized. Therefore, lately, the importance of sea-hills as a spawning and nursery ground and migration area for marine resources has been pointed out.

The Hachiri-ga-se hill off Mishima, Japan, is located in a part of the 100 m isobath in the southwestern part of the continental shelf and is an important spawning and nursery ground and supports coastal fisheries such as pole and lines. However, there have been fears about the devastation of the fishing ground due to unplanned fishing activities and change in marine environments.

A. Hamano
Department of Fisheries Science and Technology, National Fisheries University, Shinomoseki, 2-7-1, Nagata-Honmachi 759-6595, Japan
e-mail: ahamano@fish-u.ac.jp

H. Tanoue
Department of Fisheries Science and Technology, National Fisheries University, Shinomoseki, 2-7-1, Nagata-Honmachi 759-6595, Japan

Aix-Marseille University, Mediterranean Institute of Oceanography (MIO), Université du Sud Toulon-Var, CNRS/INSU, IRD, UM 110, Campus universitaire de Luminy, case 901, 13288 Marseille cedex 09, France

T. Fujiwara
National Fisheries University, Shinomoseki, 2-7-1, Nagata-Honmachi 759-6595, Japan

T. Komatsu
Atmosphere and Ocean Research Institute (AORI), University of Tokyo, President of S.F.J.O. (*) of Japan, 5-1-5, Kashiwanoha, 277-8564 Kashiwa-shi, Japan

H.-J. Ceccaldi et al. (eds.), *Marine Productivity: Perturbations and Resilience of Socio-ecosystems*, DOI 10.1007/978-3-319-13878-7_34,

Therefore, how we should protect the fishing grounds and effectively manage the coastal fisheries for sustainable harvest is a very important and urgent issue. In this study, we proposed an acoustic monitoring method to assess the distribution of marine algae vegetation and fish schools. These results show that the acoustic monitoring method would be effective to provide an example of solution for protection and management of these offshore sea-hill fishing grounds.

1 Importance of the Protection and Management of Offshore Sea-Hill Fishing Grounds

Sea-hills, which we propose in this paper to be included in the areas for the protection and management of fishing grounds, are the hills on the seafloor 500 m high or less and large-scale natural reefs interspersed on the continental shelves. With regard to the mountains on the seabed 1,000 m high or more found in the ocean, i.e., seamounts, a number of studies have been conducted on the management of their resources (Menard 1964; IHO-IOC 2001). Because seamounts are supplied with nutrients from the depths of the sea by the upwelling generated by the seabed topography and flows, primary production is very active around the seamounts; thus, many fish, the consumers of these nutrients, live there (Rogers 1994). Seamounts are mostly distributed in the open sea out of the exclusive economic zone (EEZ), and thus there have been fears about the depletion of resources as a result of, among others, unplanned fishing activities; therefore, international surveys, including those by the UNEP, have been carried out on these seamounts (e.g., IUCN 2013). On the other hand, the areas around sea-hills interspersed on the continental shelves, which are the subject of this paper, have been used as fishing spots more familiar to local fishers, such as those of pole and line and gill net fishing (Hamano et al. 2001). However, it has begun to be feared that due to the global-scale depletion of marine resources in recent years, as well as the increases in the price of crude oil, large-scale fishing fleets using trawl nets and purse seine, whose main fishing areas have been in the open sea, may turn their eyes to sea-hills situated nearer to their ports, and thus surveys on the management of resources around sea-hills were started.

2 An Example of Surveys on the Fishing Ground Environment at "Hachiri-ga-se" in the Southwestern Part of the Japan Sea (East Sea)

The continental shelf off the San'in Coast in the southwestern region of the Japan Sea (East Sea) is used as the bases of various fishing methods, such as large- and medium-type purse seine fishery, offshore trawl fishery, angling, long-line fishery, and stake net fishery (e.g., Ogawa et al. 1977). In 1999, the survey of the fishing ground environment of the Hachiri-ga-se area, one of the natural offshore hills existing on the continental shelves, was mainly started by the National Fisheries University, universities in the district, and the fisheries experiment station of Yamaguchi Prefecture (Hamano and Nomura 2002) (Fig. 1). Hachiri-ga-se is an important fishing area used daily by local fishers and is also famous as a place for recreational fishing, such as the pole and line fishing of bluefin tuna (*Thunnus orientalis*). As for the area north of Hachiri-ga-se, partly because the area is close to the borders of Korea and China, it has often been reported that overseas fishing fleets operate there without permission (e.g., Hamano and Uchida 2000).

In the above-mentioned survey, in an effort to obtain information necessary for "responsible fishing" by Japanese fishers, such devices as acoustic survey equipment, ROVs, and larva and

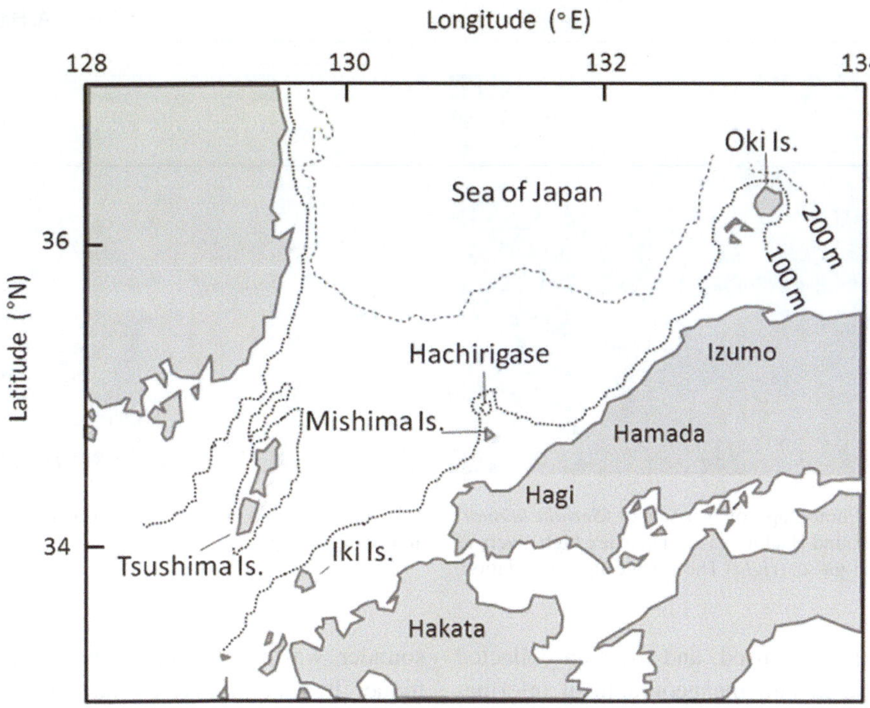

Fig. 1 "Hachiri-ga-se," one of the most important natural reef fishing grounds on the continental shelf in offshore waters southwest of the Sea of Japan

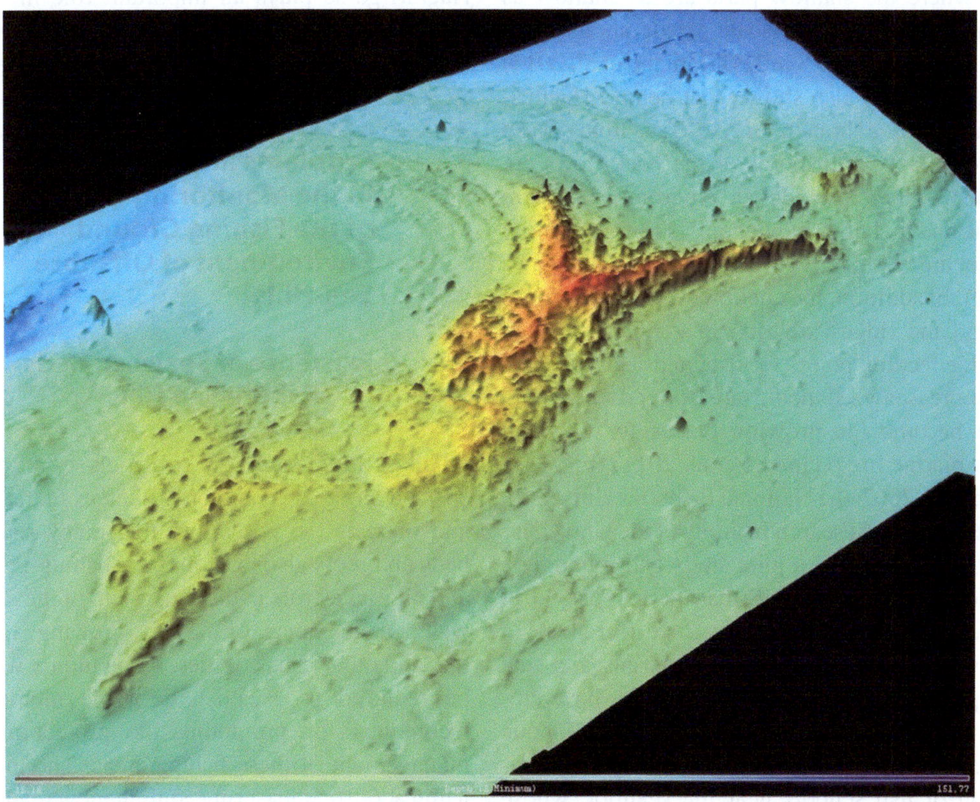

Fig. 2 Three-dimensional image of the bottom topography obtained by multibeam echosounder. *Color bar* indicates depth

Fig. 3 (*left*) Photograph of fish school (*Seriola lalandi*) assembling around Hachiri-ga-se. There are high divers of fish in Hachiri-ga-se. (*right*) The marine algae vegetation, *Ecklonia cava*, was observed at the Hachiri-ga-se by ROV and scuba diving

juvenile nets were used, and the data collected were integrated into the geographical information system (GIS) and visualized by two- or three-dimensional images (Tanoue et al. 2007). As a result, it was found that Hachiri-ga-se is an area where the shallow parts are as shallow as 18 m and where large-scale rapids about 90 m high spread for over 8 miles (Hamano et al. 2011) (Fig. 2). It was also shown that behind Hachiri-ga-se where a branch of the Tsushima Warm Current flows, whirlpools of the back flows are formed, which provide an environment where larvae and juveniles having only a low swimming ability are liable to stay (Hamano and Shinagawa 2010). In addition, the shallow water depth zones around the ridges are rich in zooplankton and provide feeding points for fish, and seaweed beds (offshore seaweed beds) are formed there and have become the growing places for fish and other marine life (Hamano et al. 2011) (Fig. 3).

In June 2007, the National Fisheries University and the Ocean Research Institute of the University of Tokyo conducted a joint study and found by a space analysis of the GIS that the fish caught in these areas differ according to differences in the water depth and bottom materials and that complicated differences in topography have created a distribution of various fish species (Tanoue et al. 2008). A technique for estimating the distribution of fish schools near the seafloor, whose measurement was difficult by a quantitative echo sounder, was developed, and by using this technique, it was discovered that threeline grunt (*Parapristipoma trilineatum*), a coastal fish, in the spawning season is distributed at a high density around the reefs which suggested that Hachiri-ga-se plays an important role in reproducing the fishery resources in the coastal waters (Tanoue 2009).

3 Problems and Solutions in the Monitoring of the Fishing Ground Environment of Offshore Sea-Hills

3.1 Existence of Seaweed Beds in Offshore Sea-Hills

Offshore hills, such as Hachiri-ga-se, are interspersed along the continental shelves in the EEZ, and there is the possibility that many seaweed beds and reefs also exist in these fishing grounds and play a key role in the ecosystem of the coastal waters as the areas for the stay, growing, spawning, and other fish activities. In the future, we will be required to establish desirable monitoring techniques for promoting the protection and management of these fishing areas. The method using acoustic measuring equipment was adopted in the Hachiri-ga-se study, and one of these monitoring techniques can investigate large sea areas quickly and thus is an effective method; this technique was also used for the monitoring of seamounts (Kloser 1996; Komatsu et al. 2002b). However, seaweed beds may have been formed around the sea-hills in the shallow sea areas, such as offshore natural reefs, and for these sea-hills, there is the need to develop new techniques taking into account the estimation of larvae and juveniles distributed in and around the seaweed beds. Therefore, in this study, we summarized the problems facing the development of these new methods to provide an example of the solutions.

3.2 Measuring Technique Proposal for Offshore Seaweed Beds

Seaweed beds play a very important part in the coastal ecosystem because they provide larvae and juveniles with places for growing and hiding and fish and seaweed-eating animals with feeding points and also have such functions as CO_2 fixation and the purification of water quality (e.g., Mann 1973; Komatsu et al. 2005). The main methods for studying the distribution of seaweed beds in coastal shallow water areas include visual inspection from a vessel, diving inspection by quadrate survey, and image analysis using aerial photos or satellite images (e.g., Norris et al. 1997; Komatsu et al. 2002a, c). However, these methods are not appropriate because much time and labor are required, and there will be effects of waves and ocean conditions in offshore hills far away from land. One of the possible techniques for estimating the distribution and quantity of seaweed beds in these areas is an acoustic method using a quantitative echo sounder. An acoustic method is an underwater remote sensing method by which an ultrasonic wave is sent to an object in the sea and information about the sea can be obtained from the reflected intensity of the wave returned from the object (MacLennan and Simmonds 1992).

Because seaweeds are distributed along the seabed, there are some problems when they are used as the objects of an acoustic study. First of all, there will be cases where the echo reaction of the seabed and that of seaweeds is overlaid, and thus these echo reactions must be distinguished from each other. Next, it should be noted that a large number of fish, especially larvae, juveniles, and plankton, exist in and around seaweeds. Therefore, in this study, we introduce a method to separate and distinguish the seaweed bed by a three-stage method. In the first stage, seafloor discrimination will be made by obtaining the standard values for seafloor discrimination using the discriminatory analysis method, which is used in the process of binarized image processing. In the second stage, the adult fish and larvae and juveniles swimming in and around the seaweed bed will then be separated and distinguished by the frequency difference method using multiple frequencies. In addition, the examination of the method for identifying the seaweed bed from acoustic data based on the optimum combinations obtained will be conducted. The outcome will be used to make space interpolation by kriging in order to estimate the distribution of seaweeds. Examples of the application of the measuring technique at Hachiri-ga-se in 2012 and 2013 will be described below:

First stage: seafloor discrimination using the binarized discriminatory analysis method

To avoid the overestimation of echoes from the seafloor based on the sound intensity obtained by a quantitative echo sounder, there is a need to separate and discriminate the echoes from the seabed by fixing a threshold value for sound intensity in the post-processing. In the past, this threshold value was empirically fixed on the basis of an echogram (echographic data). Thus in the Hachiri-ga-se survey, we adopted a mathematical method involving the discriminatory analysis by binarized image processing, which was developed by Hamano et al. (2005) for sound intensity, thereby excluding empirical judgments at the time of conducting the separation and discrimination (see Fig. 4: example results in Hachiri-ga-se).

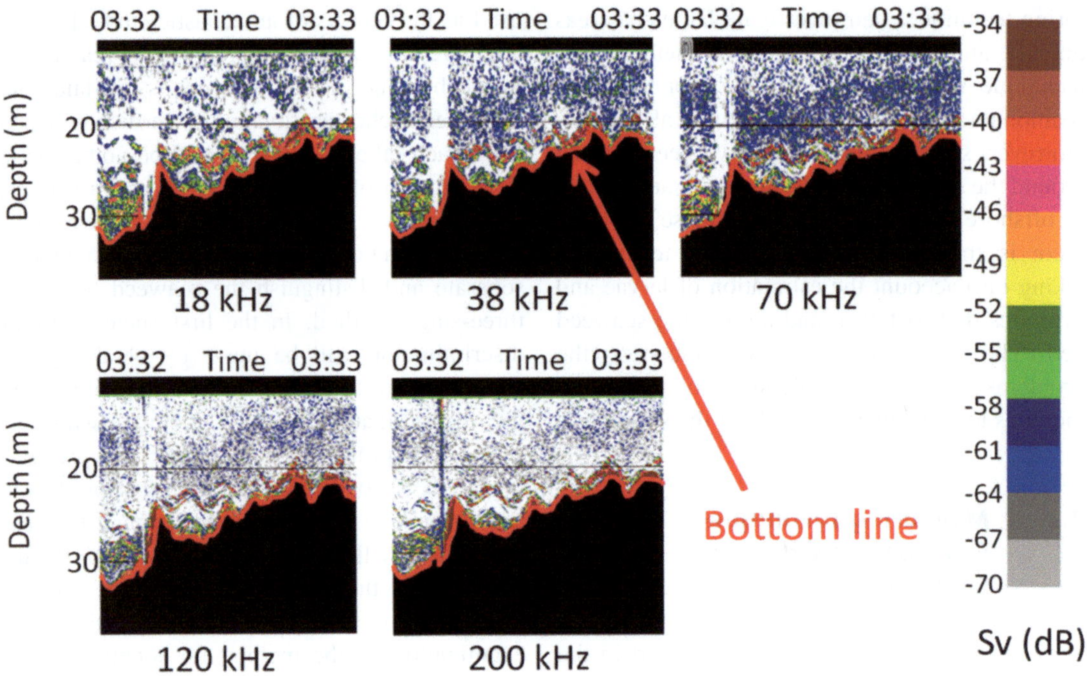

Fig. 4 Examples of echograms with the drawn bottom lines (*Red lines*) determined by the binarization method

Fig. 5 Relationship between frequency and acoustical intensity of main marine algae vegetation *Ecklonia cava* in Hachiri-ga-se (*error bars* represent the standard deviation of the mean). Acoustical intensity of *Ecklonia cava* with 200 kHz is higher than 18 kHz and 38 kHz

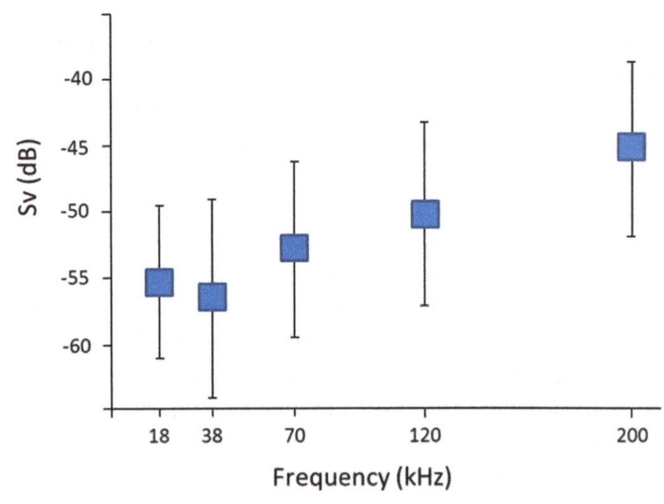

Second stage: separation and discrimination of larvae, juveniles, and seaweeds by the frequency difference method

The acoustic-reflected intensity depends on the frequency of the ultrasonic wave (see Fig. 5: example of results in seaweed bed). The frequency difference method is a technique for separating and discriminating fish and zooplankton by taking advantage of this property of reflected intensity (Greenlaw and Johnson 1983). Seventy percent of the fish ($n=273$) collected in the Hachiri-ga-se study were the juveniles of rabbitfish (*Siganus fuscescens*), and their body length was 7.4 mm on average. The data from collected fish were used to obtain the theoretical values (Furusawa's the-

Fig. 6 Relationship between body length of fish and ⊿TS obtained by ⊿Sv difference method in each of two frequencies (18, 38, 70, 120, 200 kHz). Each mark indicates theoretical ⊿TS of *Siganus fuscescens* estimated by the bladder resonance model

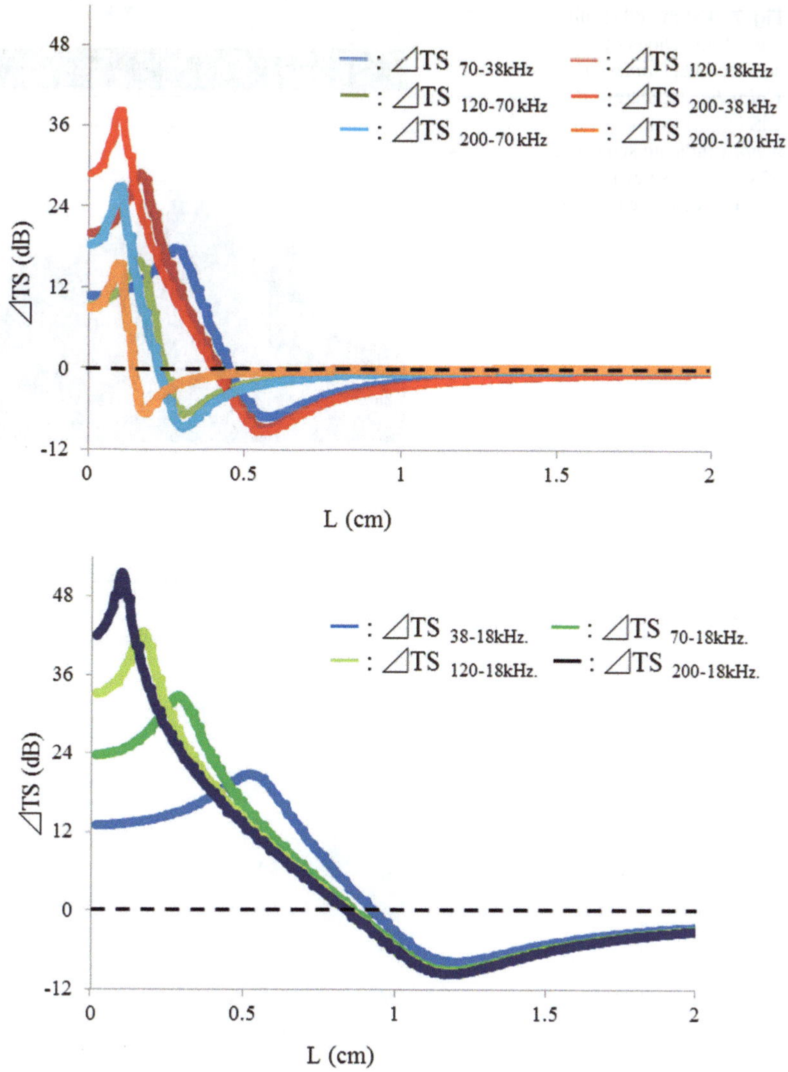

oretical bladder resonance model which was adapted by Iida et al. 1999) for five frequencies (18, 38, 70, 120, and 200 kHz) by the frequency difference method, and it was found as a result that two combinations of frequencies, i.e., 18 kHz and 200 kHz ($⊿Sv_{200\text{-}18}$) and 38 kHz and 200 kHz ($⊿Sv_{200\text{-}38}$), were effective in separating and discriminating the larvae and juveniles (see Fig. 6: example of results for rabbit fish). Thus we were able to identify the reactions of the larvae and juveniles whose ⊿Sv values of $⊿Sv_{200\text{-}18}$ and $⊿Sv_{200\text{-}38}$, which are continuous ⊿Sv on the seafloor, were negative. We then excluded the reactions away from the seabed, because it was evident that there were no reactions due to the seaweeds. We dealt with only the reactions obtained that were continuously distributed on the seafloor as those from seaweeds (see Fig. 7: example of results for Hachiri-ga-se).

Third stage: estimation of distribution by the space interpolation method

Based on the information obtained in the second stage, we estimated the distribution of the seaweed bed by kriging using the space interpolation method (Fig. 8). Kriging is used in many fields, including the estimation of ore reserves, and is the technique for objectively

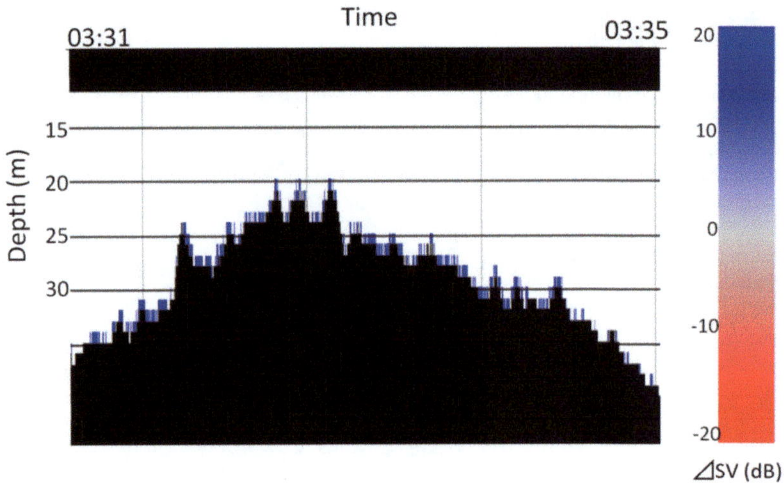

Fig. 7 Estimated result regarding echoes of seaweed (*Ecklonia cava*). *Color bar* indicates $\varDelta Sv_{200\text{-}38kHz}$ value. *Blue echoes* indicate seaweed: $\varDelta Sv_{200\text{-}38kH}$ is negative in the seaweed (see Fig. 5)

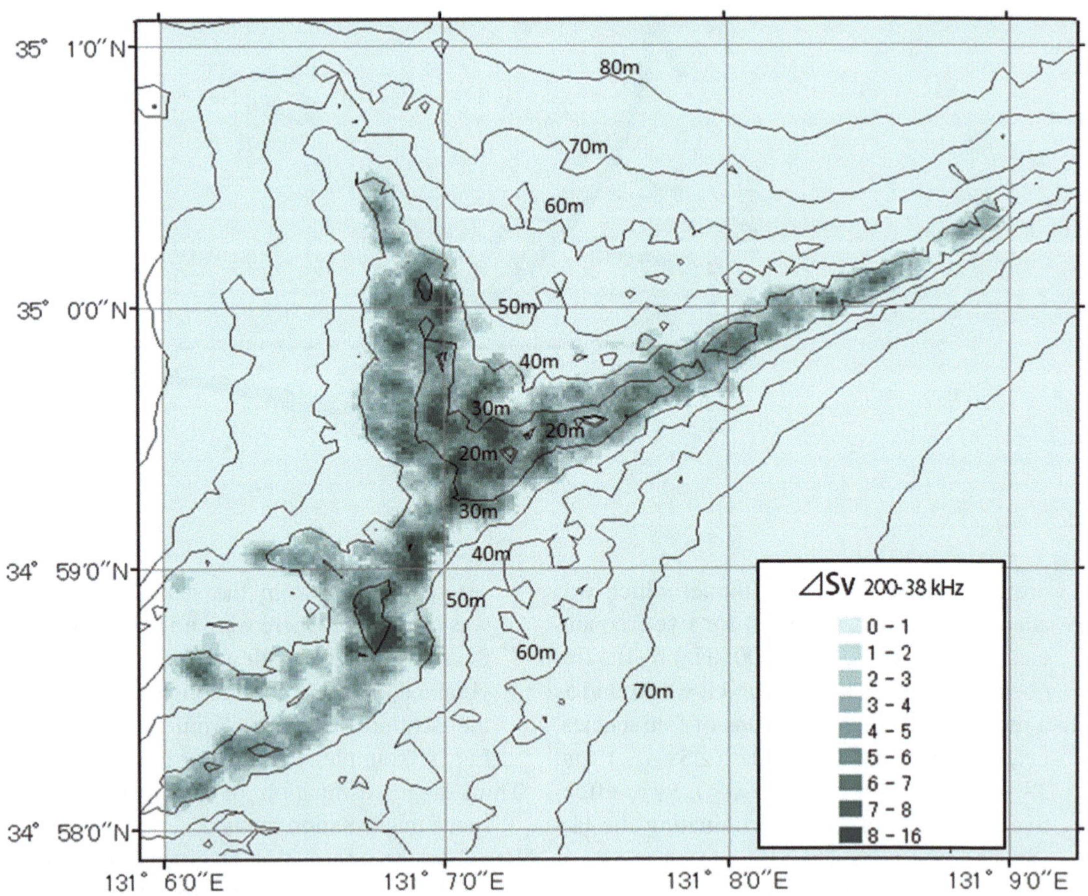

Fig. 8 Distribution of main marine algae vegetation (*Ecklonia cava*) estimated by kriging method

determining the scope of the observation points, which are used to give weight to estimate the observation data, in order to obtain an estimated value at an optional point in space (Zeiler 1999). Kriging has been applied by Tanoue et al. (2008) and Minami et al. (2012) to the space interpolation of fish schools and seaweed beds (Tanoue et al. 2008; Minami et al. 2012). The total area of seaweed beds at Hachiri-ga-se has been estimated at 2.95 km^2, and 94.4 % of the observation results by underwater TV cameras at the 54 locations agreed with the outcome obtained in this study.

4 Conclusion

As the problems of marine protected areas (MPAs), perturbations in oceans, and resilience have been internationally discussed, the importance not only of seamounts in the open sea but also of hills in offshore zones as the habitats of fish has begun to be pointed out. Hachiri-ga-se, situated off Mishima Island, Hagi, Yamaguchi Prefecture, in the southwestern part of the Japan Sea (East Sea), is one of the representative offshore sea-hills in Japan. The most important characteristic of Hachiri-ga-se is its great biodiversity. It has been shown that due to the complicated seabed topography of Hachiri-ga-se and the direct effects of the Tsushima Warm Current, extensive seaweed communities have been formed there, where a wide variety of fish, larvae, and juveniles are distributed, making it an ideal place for the stay, feeding, and rearing of marine life. However, the loss of seaweed beds has been increasing as a result of "isoyake" (withering of seaweeds) in recent years, and it has become more important than ever to check the situation of the maintenance and preservation of seaweed beds. Here, we proposed a new monitoring method to estimate the area on the hills in the offshore zone where seaweed beds are distributed by adopting the frequency difference method using multiple frequencies, the technique for identifying plankton, larvae, and juveniles, instead of the conventional acoustic survey using a single frequency by applying a survey method with a quantitative echo sounder. As a result, the distribution of seaweed beds could be mapped from georeferenced data gathered by acoustic sonar and optical camera surveys. Based on this result, it was shown that our proposed acoustic method would be effective as a technique for estimating the distribution of seaweed beds and juveniles living in the offshore sea-hills.

The protection and management of offshore sea-hill fishing grounds have been regarded as important both internationally and in Japan. As "responsible fisheries" and the management of marine resources have been significantly required, it is considered important to continue the monitoring surveys and to follow their perturbations and their resilience in the coming years in order to promote a sustainable fishery production in harmony with the environment and ecosystems and maintain and develop various types of coastal fisheries. The future adaptations of societies living from the marine resources will be established under the best conditions.

Acknowledgments We express our thanks to Dr. N. Murase, Dr. Y. Kajikawa, Mr. T. Nakamura, and Mr. K. Sadamitsu for their kind help in the fieldwork at Hachiri-ga-se and Dr. S. Ruitton and Dr. M. Mohri for their useful comments. We are grateful to Dr. H.-J. Ceccaldi, Dr. M. H. Tusseau-Vuillemin, Dr Y. Henocque, and Dr. G. Stora for editing our manuscript. This work was partially supported by the city of Hagi.

References

Greenlaw CF, Johnson RK (1983) Multiple-frequency acoustical estimation. Biol Oceanogr 2:227–252

Hamano A, Nomura K (2002) A preliminary survey of marine living resources around Hachiri-ga-se off Mishima, Hagi, Japan. In: Proceedings of Techno-Ocean 2002, Techno-Ocean Network Kobe, Japan. CD-ROM. Daishin Planning, Kobe, 4 p

Hamano A, Shinagawa S (2010) Meanings of the protection and management of offshore sea-hill fishing grounds, collection of the lectures at the symposium on the conservation of Hachiri-ga-se fishing grounds, Mishima Island. Hagi city, 3 p (in Japanese)

Hamano A, Uchida K (2000) Report of the survey on fishing grounds at Hachiri-ga-se, Mishima Island, Hagi, in 1999. Hagi city, 15 p (in Japanese)

Hamano A, Kobayashi T, Shinagawa S (2001) Report of the survey on fishing grounds at Hachiri-ga-se, Mishima Island, Hagi. Hagi city, 159 p (in Japanese)

Hamano A, Tanoue H, Kieser R (2005) The three-dimensional GIS using sonar information for costal fisheries. ICES CM. 2005/U:19, 12 p

Hamano A, Mohri M, Takikawa T, Murase N (2011) Report of the survey on fishing grounds at Hachiri-ga-se, Mishima Island, Hagi city, Hagi, in 2010. Hagi city, 48 p (in Japanese)

IHO-IOC (2001) Standardization of undersea feature names. Guidelines proposal form terminology. International Hydrographic Bureau, Monaco, p 30

Iida K, Mukai T, Mori H (1999) Frequency characteristics of acoustic volume backscattering strength of sound scattering layer and estimation of target strengths of constituent organisms. Bull Jpn Soc Sci Fish 65:66–72 (in Japanese)

IUCN (2013) Seamounts project: an ecosystem approach to management of seamounts in the southern Indian Ocean. Gland, 60 p

Kloser RJ (1996) Improved precision of acoustic surveys of benthopelagic fish by means of a deep-towed transducer. ICES J Mar Sci 53:407–413

Komatsu T, Igarashi C, Tatsukawa K, Nakaoka M, Hiraishi T, Taira A (2002a) Mapping of seagrass and *Sargassum* bed using hydro-acoustic methods. Fish Sci 68:580–583

Komatsu T, Mitani I, Ishida K, Inagaki T, Hamano A (2002b) Assessment of *Beryx splendens* biomass around the Nakanba seamount in Izu Archipelago, Japan, by echosounder. Fish Sci 68:429–430

Komatsu T, Takahashi M, Ishida K, Suzuki T, Hiraishi T, Tameishi H (2002c) Mapping aquaculture facilities in Yamada Bay in Sanriku coast, Japan, by IKONOS satellite imagery. Fish Sci 68(Suppl I):584–587

Komatsu T, Mikami A, Sagawa T, Tanoue H (2005) Distribution of seaweed beds – to know its changes. Kiyo Mon 37:575–579 (in Japanese)

MacLennan DN, Simmonds EJ (1992) Fisheries acoustics. Chapman & Hall, London, p 325

Mann KH (1973) Seaweeds: their productivity and strategy for growth. Science 182:975–981

Menard HW (1964) Marine geology of the Pacific, International series in the earth sciences. McGraw-Hill, New York, 271 p

Minami K, Hamano A, Tojo N, Nakamura T, Yasuma H, Miyashita K (2012) Estimation of the spatial distribution of a *Sargassum* bed in Kuruminose, Yamaguchi, Japan, using an acoustic method. Bull Jpn Soc Sci Fish 78:171–179 (in Japanese)

Norris JG, Sandy WE, Mumford T, Bailey A, Turner T (1997) Estimating basal area coverage of subtidal seagrass bed using underwater videography. Aquat Bot 58:269–287

Ogawa Y, Nakahara Y, Tanaka R (1977) Hydrographic nature of coastal fishing grounds in the southwestern Japan Sea. Bull Seikai Reg Fish Res Lab 321:53–106

Rogers AD (1994) The biology of seamounts. Adv Mar Biol 30:305–350

Tanoue H (2009) Estimation of fish abundance in the sea hill by conjointly using quantitative echosounder and geographical information system. Doctor thesis, Graduate School of Agriculture and Life Sciences, The University of Tokyo, 118 p (in Japanese)

Tanoue H, Hamano A, Komatsu T, Boisnier E (2007) The efficient coupling of GIS and acoustic survey in mapping fish abundance around a hill: the Hachiri-ga-se case study. In: Nishida T, Kailola PJ, Hollingworth CE (eds) GIS/Spat Ana Fish Aquatic Science 3. Fishery Research Group, Kawagoe, pp 115–128

Tanoue H, Hamano A, Komatsu T, Boisnier E (2008) Assessing bottom structure influence on fish abundance in a marine hill by using conjointly acoustic survey and geographic information system (GIS). Fish Sci 74:469–478

Zeiler M (1999) Modeling our world. Environmental Systems Research Institute, Copenhagen, 297 p

Reduction and Aging of the Fishery Work Population in Japan and the Possibility of Its Recovery from an Ergonomic Perspective

Hideyuki Takahashi

Abstract

The increasing age and decreasing number of fishermen are a serious problem in Japan. Principal causes may include the decline of fishery catches (probably caused by the decrease in fisheries resources), the stagnations of fish prices (probably related to economic conditions), and the lack of improvement in work environments for many years. In other words, the Japanese fishery is losing its appeal as an occupation because the work is hard and risky and the profitability is poor. This study focused on an investigation of the work environment of the Japanese fishery and on the KAIZEN (voluntary improvement by the person(s) concerned) of the fishery from an ergonomic perspective. Several cases of the small trawl fishery, a representative of the coastal fishery in Japan, were investigated to understand and evaluate their work. All tasks on the boats were video recorded, and the times required and physical burdens of the main tasks were evaluated. Through the analyses, a task, namely, fish sorting, was considered to need preferential improvement, and possible solutions were indicated. Finally, problems and prospects in applying the KAIZEN principle to the fishery are discussed.

1 Introduction

The decreasing number and the increasing age of fishermen represent a serious problem in Japan. There were approximately 500,000 fishermen in the 1980s; however, that number has declined to approximately 200,000 now. According to the Food and Agriculture Organization, the number of fishermen is increasing or leveling off in major countries other than Japan (Food and Agricultural Organization of the United Nations 2012). Although the overall mean age of the Japanese population is increasing (Ministry of Internal Affair and Communications 2012), that among fishermen is increasing faster. The portion of aged fishermen (≥65 years old) has exceeded 30 % since 2003 (Ministry of Agriculture, Forestry and Fisheries 1975–2010). So why are fishermen in

H. Takahashi (✉)
National Research Institute of Fisheries Engineering, Fisheries Research Agency, 7620-7 Hasaki, Kamisu-shi, Ibaraki, Japan
e-mail: hideyuki@affrc.go.jp

H.-J. Ceccaldi et al. (eds.), *Marine Productivity: Perturbations and Resilience of Socio-ecosystems*,
DOI 10.1007/978-3-319-13878-7_35,

decline in Japan? The author considers that there are three main factors: reduced fish catches, slumping fish prices, and poor work environments. Until the 1980s, the fishery was a worthwhile occupation in Japan, despite the hard work, because the fishermen were able to catch enough fish and to sell them at acceptable prices. However, both the catches and the prices have declined since the 1990s. Although the causes are unclear, these declines may be related to the decrease in fisheries resources due to overfishing and to economic recessions. Moreover, the work environment has remained poor. Even if fishermen engage in hard work, they cannot catch enough fish and sell them at acceptable prices; thus, the occupation has become less appealing in Japan.

To revive the fishery in Japan as a healthy occupation under the current situation, one in which fishermen's incomes will not increase in the short term, the work environments must be improved by means of ergonomic labor-saving techniques. Namely, it is important to restore fishermen's work environments to reasonable and healthy conditions by means of the so-called KAIZEN principle. However, fishermen cannot be expected to use systematic methods to evaluate and improve their work environments by themselves in the small fishery management bodies that occupy the majority of the fishery of Japan. Nonetheless, third parties can support fishermen by conducting objective investigations of their work environments and then suggesting effective improvement policies.

Such improvement of the fishery work environment can occur at various levels. For instance, there are infrequent but fatal or injurious accidents such as a ship capsizing or a fisherman falling from the ship or getting caught in machines, and there are other ongoing physical symptoms in daily work caused by other factors. There are also emotional problems in workplaces, such as those that occur in the context of human relationships. The goal of this research was to address the causes of musculoskeletal symptoms in daily work because little is known about the specific influence of fishery work on fishermen's bodies in relation to the various fishing methods in Japan. In other words, specific fishery work has been scarcely explored scientifically. As various fishing boats and methods are accompanied by differing work environments and methods, the needs of each and the plans for improvement will vary accordingly. Consequently, we should first understand what happens at the sites of fishery work through the accumulation of case studies.

In the current study, several cases of small trawl fishery, a representative coastal fishery in Japan, were investigated. The characteristics of the work sites were defined quantitatively, and the results are discussed with respect to possible approaches to improving the work environment.

2 Material and Method

The small trawl fishery is that in which relatively small fishing boats with gross tonnages of less than 15 tonnes are used. Several types of fishing gear are used in the small trawl fishery: Danish seine, beam trawl, dredge, and otter trawl. The fish catch per year by the small trawl fishery in Japan remained at around 0.4–0.5 million tonnes during the last decade, corresponding to 7–9 % of the total annual fish catch in Japan (Ministry of Agriculture, Forestry and Fisheries 2014). The small trawl fishery is distributed widely over most coastal prefectures. In this study, the author investigated the actual work of four small trawl fishing boats from across Japan, operating in Aichi, Chiba, Hiroshima, and Osaka prefectures (the individual case studies are described in Takahashi 2009, 2013a, b; Takahashi et al. 2012).

In each investigation, the fishing operation was recorded during a fishing trip using waterproof video cameras attached to the boat to understand and quantify the characteristics of fishery work. It is crucial to identify time-consuming tasks and physically excessive tasks in order to investigate how to improve them. Accordingly, two analyses were carried out using the recorded video images to quantify the time required for and physical burden of each representative task.

First, the time series of the tasks performed on each boat during a fishing trip were arranged, and then the proportion of the whole navigation time required for each representative task was determined. Second, the physical burden of the fish-sorting task, judged as the most time-consuming task by the time study, in each boat was estimated based on the crew's work postures recorded on video images. The Ovako working posture analysing system (OWAS) method, one of the simplest and best-known methods, was used in this study (Karhu et al. 1977). In the OWAS method, a work posture is determined categorizing posture based on three portions of the body (upper body, upper limbs, and lower limbs) and the weight of objects handled. When a work posture is identified precisely, the physical burden is ranked in one of four action categories (AC); these serve as an index of the degree of demand for improvement. That is, the number associated with the AC reflects the harmfulness of the work posture on the subject's body (AC1 being the best posture and AC4 being the most harmful). When a harmful posture occurs with high frequency, preference should be given to improve the work.

3 Results

The results are summarized in Fig. 1. The typical workflow on a small trawl fishing boat entails casting the net(s), towing the net(s), hauling the

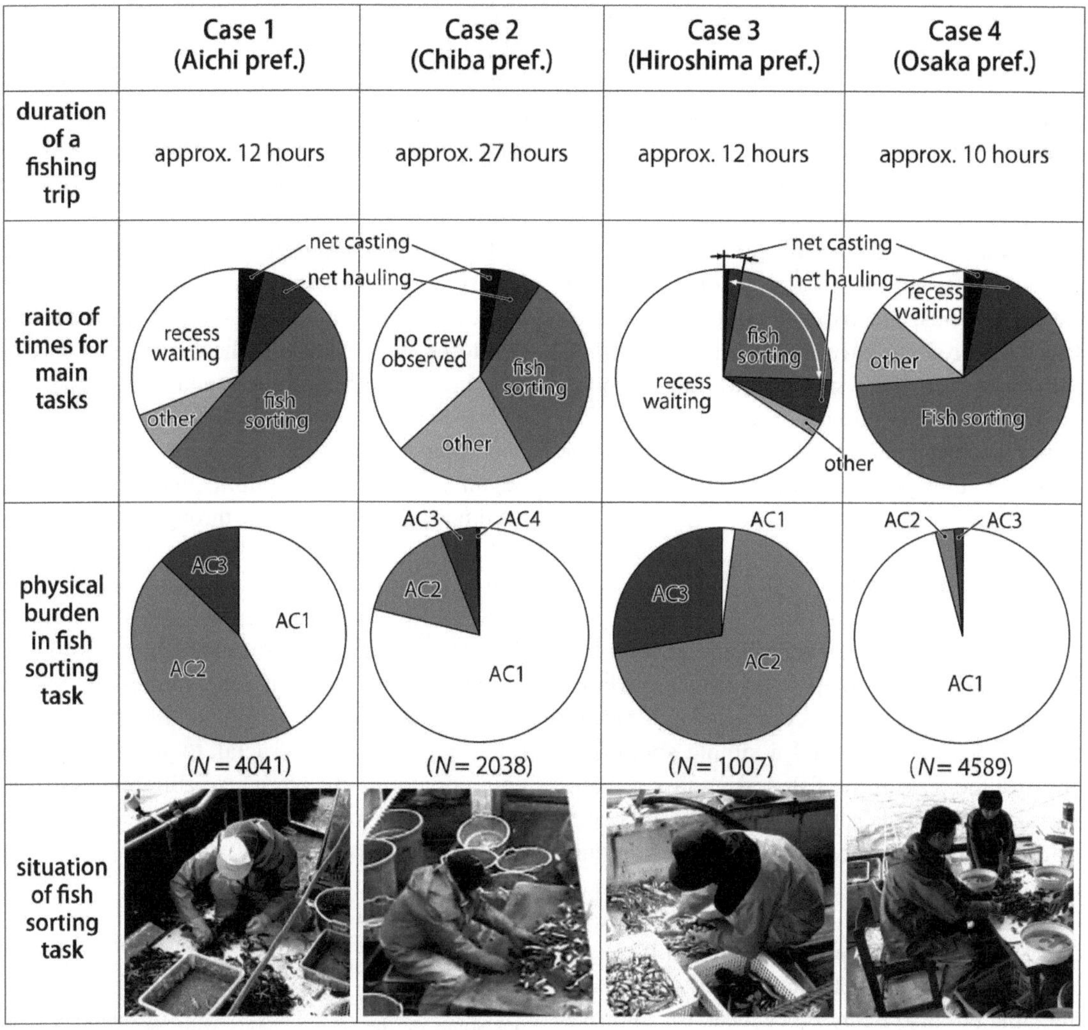

Fig. 1 The proportion of the navigation time required for each task (*third row*), the physical burden associated with the fish-sorting task as estimated by the OWAS method (*fourth row*), and the type of fish sorting performed (*fifth row*) on each of the four fishing boats investigated

net(s), unloading fishes from the net(s), sorting the fishes, and storing the sorted fishes in the fishhold(s). Fish sorting and storing are usually performed during net towing. The portion of the whole navigation time required for each representative task (Fig. 1, circle graphs in the third row) varied in each case; however, fish sorting was the most time-consuming task on all of the boats studied. The AC (the index indicating the degree of need for improvement in the OWAS method) for the fish-sorting task in each boat (Fig. 1, circle graphs in the fourth row) could be classified as either the type of work where the crew sorts fishes on the deck floor in a squatting posture (Aichi and Hiroshima prefectures) or the type where the crew sorts fishes on a table while seated (Chiba and Osaka prefectures). The latter posture appeared to represent the ergonomically better work posture, and the results demonstrated this quantitatively.

Summarizing these results, the fish-sorting task was the most time consuming in all four cases studied, and the physical burden was undesirable when the fishermen sorted fishes on the deck. Thus, the fish-sorting task should be improved preferentially. According to the comparison among four small trawl fishing boats from across Japan, the use of a table during fish sorting appears to be a simple solution to improve the fishermen's physical burden.

4 Discussion

The results of this study identified the most significant task in the small trawl fishery in Japan through the quantitative comparison of several case studies and proposed a simple solution to improve the task. If the fish-sorting task remains unchanged, the fishermen will have an increased risk of developing serious low back symptoms. However, no table for fish sorting has been introduced in many small trawl fishing boats even though many fishermen seem to readily recognize its value. It is likely that not only the idea of the table but also many other ideas for improvement will not be accepted easily. The author considers that there are two main reasons why fishermen do not readily implement the suggestions for improvement.

The first factor is the limited capacity of the fishing boat. The volume of Japanese fishing boats is regulated by gross tonnage as a means of controlling their catch, and fishermen typically aim to achieve the full permissible volume to maximize their fishing capacity. Consequently, no space is set aside for safe and comfortable work, and the fishermen are forced to work in uneasy postures and sometimes in an unsafe manner.

The second factor involves fishermen's way of thinking and knowledge. Many fishermen are proud that they work in the environment, namely, the sea, where death is so near. Such prides often bring about the wrong way of thinking in that they take it for granted that they work in a dangerous environment. Moreover, veteran fishermen are hesitant to modify a way of working that has been established over a long time. Therefore, it is not easy for them to implement a new method, even if it is clearly reasonable.

Given this situation, it is difficult to improve the fishery work environment dramatically in the short term. However, in the long term, it will be effective to promote reform in the fishermen's way of thinking through awareness and suggestions for effective improvements. In Japan, a seminar for fishermen concerning safe work environments began last year as a public undertaking. The accumulation of knowledge through investigations of fishery work will also provide useful information for fishermen through such seminars.

KAIZEN solutions are occasionally obtained through comparative studies of many case studies. Expanding the range of case studies to foreign countries will be beneficial to the fisheries of other regions. It is crucial to understand the common terms, to recognize differences based on culture for each country, and to disseminate knowledge to each country.

Acknowledgments I am grateful for all the fishermen who kindly cooperated with my researches and for the people who helped the implementations of the investigations. I am also thankful with my colleagues who carried out some of the investigations with me. I am greatly indebted to Professor Hubert-Jean Ceccaldi for his kind advice and encouragement.

References

Food and Agricultural Organization of the United Nations (2012) The state of world fisheries and aquaculture 2012. Food and Agricultural Organization of the United Nations, Rome, p 209

Karhu O, Kansi P, Kuorinka I (1977) Correcting work posture in industry: a practical method for analysis. Appl Ergon 8(4):199–201

Ministry of Agriculture, Forestry and Fisheries (1975–2010) The 5-12th census of fisheries. http://www.maff.go.jp/j/tokei/census/fc/index.html (in Japanese)

Ministry of Agriculture, Forestry and Fisheries (2014) Survey of sea surface fishery production 2012. http://www.maff.go.jp/j/tokei/kouhyou/kaimen_gyosei/index.html (in Japanese)

Ministry of Internal Affair and Communications (2012) The labour force survey 2012. http://www.stat.go.jp/data/roudou/ (in Japanese)

Takahashi H (2009) Work analysis on a small trawl fishery at Choshi city, Chiba prefecture. Fish Eng 46(1):1–8 (in Japanese)

Takahashi H (2013a) Work condition of the small trawl fishery in Hatsukaichi city, Hiroshima prefecture. J Ergon Occup Saf Health 15s:120–123 (in Japanese)

Takahashi H (2013b) Work analysis of a small trawl fishery (dredge net fishery) at Kishiwada city, Osaka prefecture. J Fish Technol 6(1):89–98 (in Japanese)

Takahashi H, Saeki K, Watanabe K (2012) Work analysis of a small trawl fishery at Minami-chita district, Aichi prefecture. Fish Eng 49(2):133–140 (in Japanese)

Part VI

Communications: Status and Future of Marine Protected Areas

Regulation and Management of Marine Protected Areas in Japan

Naoki Amako

Abstract

In response to the global movement for establishing MPAs and MPA networks, such as the CBD COP decisions, Japan defined the MPAs and decided to promote their establishment. The Ministry of the Environment formulated its "Marine Biodiversity Conservation Strategy" in 2011, which specified the definition of MPAs in Japan, along with the existing systems that can be regarded as MPAs. The purposes of the existing systems can be categorized into three groups: (1) protection of natural scenery, (2) protection of natural environment or habitats, and (3) protection and cultivation of fishery resources. The regulations imposed in each area depend on the purpose, but they regulate either development activities, taking designated species (e.g., endangered species), or fishery activities. One distinctive character of in Japan that MPAs regulate fishery activities is that the fishery is managed by the resource users themselves. Accordingly, local fishers are engaged not only in fishery operations but also in resource management (e.g., resource assessment and setting individual catch quota) and ecosystem conservation (e.g., maintaining seaweed beds and coral reefs). There is a controversy as to whether some MPAs can really be regarded as MPAs. As stated in the "National Biodiversity Strategy of Japan 2012–2020," it is important to consider standards and methods for evaluating the effects of marine protected areas from the biodiversity standpoint.

1 Background

The movement for conservation of marine biodiversity has become larger in recent years. The World Summit on Sustainable Development (WSSD) and the Conference of the Parties to the Convention of Biological Diversity (CBD COP)

N. Amako (✉)
Ministry of the Environment, 1-2-2 Kasumigaseki, Chiyoda-ku, Tokyo 100-8975, Japan
e-mail: naoki_amako@env.go.jp

H.-J. Ceccaldi et al. (eds.), *Marine Productivity: Perturbations and Resilience of Socio-ecosystems*, DOI 10.1007/978-3-319-13878-7_36, © Springer International Publishing Switzerland 2015

both set a target to establish marine protected areas (MPAs) and MPA networks, and the CBD COP10 adopted "Aichi Biodiversity Targets" which include the target to conserve at least 10 % of coastal and marine areas.

Surrounded by the sea of high productivity, Japan has used and conserved its marine areas since the ancient times. One of the major characteristics is that the marine resources have been managed by local users themselves, not by the governments. As early as 757 AD, the government rules called "Yoro Code" stipulated that "Mountains, rivers, scrubland, bogs and coastal areas are for common use and to be managed by local users themselves." "Goseibai Shikimoku," which was the legal code of the Kamakura shogunate, reiterated the above in 1232 AD, and "Ritsuryo principles" of Tokugawa shogunate in 1721 AD stipulated that "Coastal areas should be used and managed by the local communities, while offshore areas are for common use." Accordingly, local fishers have been engaged not only in fishery operations but also in resource management and ecosystem conservation (Makino 2011). The activities of local fishers may include, but not limited to, resource assessment, setting individual catch quota, and maintaining seaweed beds or coral reefs, some of which can be regarded as the elements of MPA management.

In the modern society of Japan, fishery resource is managed not only by resource users themselves but also by the governments. Also, there are protection measures on natural scenery and wild fauna and flora including endangered species and their habitats. These measures include area-based conservation, but the term "marine protected areas" was not used until recently.

2 Process of Defining "Marine Protected Areas" in Japan

In response to the global movement to promote the establishment of MPAs, the first thing Japan did was define the MPAs.

2.1 Marine Biodiversity Conservation Strategy

The Ministry of the Environment, with the support from the panel of experts in the field of marine ecology, law of the sea, fisheries, marine industry, etc., formulated its "Marine Biodiversity Conservation Strategy" in 2011.

2.1.1 Definition of MPAs

The "Marine Biodiversity Conservation Strategy" contained the definition of MPAs as follows: "Marine areas designated and managed by law or other effective means, in consideration of use modalities, aimed at the conservation of marine biodiversity supporting the sound structure and function of marine ecosystems and ensuring the sustainable use of marine ecosystem services." In formulating this definition, the definition of protected areas by IUCN and the definition of marine protected areas by CBD were used as references. One of the characteristics of Japan's definition is that it aimed not only at conservation but also at ensuring the sustainable use. This is based on the idea that sustainable use is dependent on healthy ecosystem, so that maintenance of a healthy ecosystem is an imperative part of sustainable use.

2.1.2 Areas Meeting the Definition of MPAs

The "Marine Biodiversity Conservation Strategy" also indicated the list of existing systems that meet the above definition of MPAs (Table 1). These existing systems were selected based on the purpose of designation and the regulations that apply. The purposes of the existing systems can be categorized into three groups: (1) protection of natural scenery, (2) protection of natural environment or habitats, and (3) protection and cultivation of fishery resources. The regulations imposed in each area depend on the purpose, but they regulate either development activities, taking designated species (e.g., endangered species), or fishery activities. These MPAs can overlap, so that various purposes could be fulfilled by combining different regulations.

Table 1 Areas meeting the definitions of MPAs

Area (system)	Purpose of designation	Description of major regulations
(1) *Protection of natural scenery*		
Natural park (Natural Parks Law)	Protection of outstanding natural scenery and promotion of its use	Mainly regulation on developments, such as landfills (ordinary zone: notification required; marine park zone: permission required; prohibition on taking organisms on some areas). Special areas (permission required) may be set in brackish water zones
Natural coastal protected zone (Act on Special Measures Concerning Conservation of the Environment of the Seto Inland Sea)	To maintain the state of nature, so that seashores and ponds could be used for bathing, shellfish gathering, and so forth in the future	Regulation on developments, such as construction of new structures, transformation of land properties, mining of minerals, and earth and rock quarrying (concerned prefectures must be notified)
(2) *Protection of natural environment or habitats for organisms*		
Nature conservation area (Nature Conservation Law)	Conservation of the outstanding natural environment requiring particular conservation	Regulation mainly on developments, such as land transformation (ordinary zone: notification required; marine special zone: permission required; prohibition on taking organisms on some areas)
Wildlife protection area (Wildlife Protection and Proper Hunting Law)	Protection of wildlife	Regulation on hunting. Developments, such as the construction of structures, are also regulated in special protection zones, and the use of power-driven vessels is additionally regulated in special protection designated zones
Natural habitat conservation area (Law for the Conservation of Endangered Species of Wild Fauna and Flora)	Conservation of nationally endangered species of wild fauna and flora	Regulation on development in monitored zones (notification required). In controlled zones, the harvest of designated species and the use of power-driven vessels are regulated in addition to regulation on development (permission required). Additionally, access is restricted for restricted entry zones
Natural monument (Law for the Protection of Cultural Properties)	Protection of animals, plants, geographic features, and minerals of high scientific value	Requires permission on activities that alter the current state or adversely affect its conservation
(3) *Protection, cultivation, etc. of aquatic animals and plants*		
Protected water surface (Act on the Protection of Fishery Resources)	Protection and cultivation of aquatic animals and plants	Development, such as landfill and dredging (license system), and harvest of designated aquatic animals and plants are regulated for water surfaces suitable for egg laying and growth of juvenile fish
Coastline marine resource development area, designated sea area (Marine Fishery Resources Development Promotion Law)	To promote the streamlining of the development and use of marine fishery resources through measures to promote stock enhancement and aquaculture of aquatic animals and plants systematically	Development, such as seabed transformation and digging, is regulated (it must be notified to the governor or the Minister of Agriculture, Forestry and Fisheries). Prefectures must formulate a "coastline marine resource development plan"

(continued)

Table 1 (continued)

Area (system)	Purpose of designation	Description of major regulations
Area designated by prefecture, fishery operator group, etc.	To protect and cultivate aquatic animals and plants and to ensure sustainable use	Regulation on harvest of specified aquatic animals and plants, etc.
(Underlying systems) harvest regulation zone (Fishery Act and Act on the Protection of Fishery Resources), water surfaces covered by the resource management regulations, and autonomous efforts by fishery cooperatives (Fishery Cooperative Act)		
Common fishery right area (Fishery Act)	To enhance fisheries productivity (protecting and cultivating aquatic animals and plants and ensuring sustainable use), etc.	The harvest of aquatic animals and plants (area, period, fishing method, number of vessels, etc.) is regulated by the rules about the exercise of fishery rights (approved by the governor). A right to petition based on real rights, a right to claim compensation or damages, and, at the same time, a charge of the infringement on fishery rights will apply to infringement by any third party

2.2 Adoption on the Cabinet Level

Two months after the formulation of the "Marine Biodiversity Conservation Strategy" in 2011, the Headquarters for Ocean Policy, which is comprised of cabinet members, adopted the definition of MPAs and the areas that meet this definition, along with the policy to establish MPAs. The area covered by the known existing MPAs is about 8.3 % of Japan's territorial waters and EEZ, excluding the overlaps between different types of MPAs.

In 2012, the cabinet endorsed "National Biodiversity Strategy of Japan 2012–2020," which was revised to reflect the "Aichi Biodiversity Targets." It set a national target to appropriately conserve and manage at least 10 % of coastal areas and ocean areas by 2020.

At the same time, not only the area covered by MPAs but also the level of management is important, and the enhancement of management, such as reviewing the location and the type of MPAs in accordance with the changes in natural and social environment, should be pursued as necessary.

2.3 Management of MPAs

The above mentioned MPAs are managed by different agencies or entities – i.e., the Ministry of the Environment, the Agency for Cultural Affairs, the Fisheries Agency, the prefectural governments, the fishery cooperatives, etc.

The level of management also differs from one type of MPA to another, according to the purpose of the MPA. The MPAs autonomously managed by local fishers are patrolled almost everyday, whereas the national parks are managed by rangers who do not visit the site as often as they wish. In what is called "Sato-umi," the fishers may try to make their fishing ground richer by maintaining seaweed/sea grass beds or coral reefs or by installing artificial reefs, etc.

As many of the MPAs managed by the Ministry of the Environment are overlapping with other MPAs, the Ministry manages their MPAs collaboratively with local governments, local fishers, scientists, NGOs, etc.

3 Criticism and the Possible Way Forward

There is a controversy as to whether some MPAs can really be regarded as MPAs. The effectiveness of regulations applied on these MPAs is questioned. As stated in the "National Biodiversity Strategy of Japan 2012–2020," it is important to consider standards and methods for evaluating the effects of marine protected areas from the biodiversity standpoint. Recognizing that management effectiveness is an element of CBD PoWPA (Programme of Work on Protected Areas) and Aichi Target 11, the Ministry of the Environment held a regional capacity-building workshop on MPA management effectiveness in September 2013.

Also, the Ministry of the Environment is currently conducting selection of important marine areas for biodiversity conservation in Japan's territorial waters and EEZ, basically using the EBSA (ecologically or biologically significant marine area) criteria which were adopted at CBD COP9. The result of this exercise would serve as a basis for promoting the enhancement of MPAs and networking thereof.

Reference

Makino M (2011) Fisheries management in Japan: its institutional features and case studies. Fish & Fisheries Series 34. doi:10.1007/978-94-007-1777-0_2. Springer Science+Business Media B.V

Limits of the Concepts of Marine Protected Areas: Adaptation of Human Populations and Their Professions in the Different Types of Marine Protected Areas

Hubert-Jean Ceccaldi

Abstract

Marine protected areas (MPAs) were created in reaction to large-scale disturbance that humankind has caused to the equilibrium, productivity, and production of marine ecology. This chapter considers some specific characteristics of different types of MPAs. It attempts to assess how the protection of these areas allows them to return to close to their original state after disturbance. It also attempts to discern the ways in which societies whose existence or activities depend on these protected areas may adapt their total conservation or use the environment more or less aggressively.

The creation, existence, and maintenance of these new marine areas induces greater control of their condition, and new coastal environment laws have led to the creation of new activities and professions.

Human societies are beginning to implement new technical and legal methods of control. They must realize the maintenance of ecosystem quality through new approaches. Finally, they must consider that these new activities must be integrated into the functioning of natural marine ecosystems and that additional resources should be allocated to these new missions.

1 Introduction

In modern times, people have often lost touch with nature, despite that our entire evolution has taken place within it. Our imprint on nature will not cease; decisions must be made to protect it, including limiting access rights to the natural environment and limiting excessive direct exploitation of marine resources.

Although the term 'Marine Protected Area' (or MPA) crosses several different concepts,

H.-J. Ceccaldi (✉)
Directeur d'études (h) Ecole Pratique des Hautes Etudes, Institut Méditerranéen d'Océanologie (M.I.O.), Oceanomed, Bâtiment Méditerranée, 13288 Marseille cedex 09, France

Président of S.F.J.O. (*) of France, Marseille cedex 09, France

Société franco-japonaise d'Océanographie, 27 rue Rocca, 13009 Marseille, France
e-mail: ceccaldi.hubert@orange.fr

H.-J. Ceccaldi et al. (eds.), *Marine Productivity: Perturbations and Resilience of Socio-ecosystems*, DOI 10.1007/978-3-319-13878-7_37, © Springer International Publishing Switzerland 2015

fundamentally, it is a clearly defined geographical marine area to which access rights are prohibited or severely restricted and the protection of which is recognized by all surrounding countries.

During the Renaissance period, the relationship between humankind and nature was one of survival for peasants, and one of aesthetics for aristocrats. The bourgeois humanism of the eighteenth and nineteenth centuries was based on the massive use of natural resources, the transformations and trade of which gradually become global. We are now forced to consider other relationships between humanity and nature because the facts are binding on us; we are in a new era of very limited size and resources worldwide. As early as 1931, in *Perspectives on the World Today*, Paul Valéry wrote, "The time of the finite world begins." This new reality for humanity is now unavoidably upon us.

How shall we now realize the creation of MPAs in different countries and cultures – Western, African, Muslim, South American, Chinese, and Siberian – or, more clearly, in Europe and Asia and, finally, in terms of this conference, between France and Japan.

This is all the more necessary in our country, given that France has a presence in the four great oceans (Atlantic, Indian, Pacific, and Southern). With waters covering about 11 million square kilometers under jurisdiction, France is the second largest maritime nation, behind only the USA. These territories occur in all climates, and many are in tropical or equatorial zones. Their ecosystems are rich and ecologically diverse: coral reefs and lagoons, atolls, faults, estuaries, deltas, mangroves, etc. The very important marine territories of Japan cover about 4.5 million square kilometers.

Recall the 15th Franco-Japanese Symposium of Oceanography theme: 'Marine productivity: disturbance and resilience of socio-ecosystems'.

We believe that overall productivity of the marine environment has led naturally to the generation of wealth and has gained economic value in human societies. The intensive exploitation of natural products, harvested mainly by fishing, is the primary disturbance of the marine environment. The creation of MPAs aims to return, at least partially, the disturbed natural environments to their previous state. So ideas are needed to induce and develop resilience in natural environments.

This trend must continue innovating other objectives and new jobs based on scientific research and techniques currently underway or that will appear in the future.

2 Different Types of Marine Protected Areas

Several national or regional international organizations have attempted to establish a classification system for MPAs, according to the degree of protection applied. Common rules will apply to countries with different histories and cultures in the face of natural processes independent of the rules and artificial boundaries of human societies.

The establishment of new MPAs changes or disrupts the behavior of corporations that operate ancestral or, more recently, more technical or industrial marine environments. Although significant progress has been made in this direction recently, new efforts will be needed.

- First type I (or Ia)

Protected category Ia areas are set aside to conserve biodiversity or some unique geological features. Human impact is minimized, and visits or audits verify their proper condition.

- Second type (or Ib)

In this category, the areas are generally large, intact, or slightly modified. Their character and natural influence is maintained without permanent human habitation or are sparsely populated. This preserves their natural ecology, and natural processes predominate for their conservation for future generations.

- Third type (or II)

Protected areas in this category are large near-natural areas set aside to protect large-scale ecological processes, as well as cash and characteristics of regional ecosystems. People may conduct non-disruptive educational, recreational, scientific, or spiritual activities while taking into account local traditions.

- Fourth type (or III)

The purpose of these protected areas is to allow the protection of specific natural structures, such as a topographic feature, a special coastal relief, an underwater cave, a unique feature, or nearby islands with special vegetation. They are quite small, but visitors attach great importance to the area. Their existence also protects the habitats of local marine organisms and biodiversity.

- Fifth Type (or IV)

Protected areas in this category are designed to protect particular species or habitats. They are managed through regular and active interventions to meet the requirements of specific species or to maintain habitats.

- Sixth type (or V)

These are areas of particular character, with significant ecological, biological, cultural, or landscape features. The protection maintains historical practices and customary management.

- Seventh type (or VI)

These MPAs, large in size, are designed to preserve ecosystems and natural habitats. Controlled management of traditional natural resources is undertaken in which people intervene in a modest and balanced manner to participate in their maintenance.

Japanese authorities have adopted a six-level classification system that closely resembles the previous (FAO 2011) and that of the National Agency for Marine Protected Areas, established in France in 2006 (Anonymous 2012a, b).

3 Limits of These Definitions

This useful classification provides a general framework when considering the degree of utilization of the natural environment by man. However, it is rather theoretical and qualitative because boundaries between different types of MPAs are not (and cannot) be clearly defined.

MPAs usually involve coastal areas. We tend to think that these areas should be returned to the condition they were in before exploitation by, and even the existence of, humankind.

For now, this classification is difficult to adapt to countries, jurisdictions or policies with a very low environmental culture that largely disregard the needs of ecology, the evolution of natural ecosystems, or their operation, let alone this classification.

Previously, only a portion of those that have been classified into one category or another have been correctly classified. Designation often relied on geographical, topographical, or administrative requirements, such as 'national park', 'sanctuary', 'regional park', 'prefectural', 'county', 'Park Blue Coast', 'Kushiro National Park Shitsugen', 'National Park Rikuchūkaigan', etc.

Moreover, other limitations of this classification are clear. It does not integrate protected areas that are further offshore, which are poorly protected and often exploited by factory ships. It does not integrate the deep ecosystems that are slow to recover once disturbed and are often exploited by destructive trawling.

4 Operation and Effects of Marine Protected Areas

4.1 Macro-function

At a macro-functional level, MPAs can be considered as areas of protection for particular species, including the production of eggs and larvae, leading to the protected reconstitution of broodstock for these areas, as well as the export of juveniles to neighboring areas, where they can replenish populations, rebalance ecosystems, and restore fishery areas. The main interest is to build networks of MPAs (Anonymous 2008) in order to establish fishing areas that should, in their turn, be managed, as they are not inexhaustible.

Marine areas receiving water bodies carrying juveniles born in the MPAs should have the characteristics necessary to achieve high productivity in terms of nutrients and photosynthesis. It should be evaluated on a regular basis to ascertain the potential of organic production areas considered. Examples include fine 12-year studies of fish populations in the marine park of Côte Bleue, near Marseille (Le Direach et al. 2010) or the nature reserve of Banyuls (Lenfant et al. 2012).

It would be pointless to create new MPAs in marine zones with naturally low productivity. Following the example of the exemplary Marine National Park of Port Cros (Doussan et al. 2005), the Mediterranean MedPan network was created to improve the efficiency of management of MPAs in the Mediterranean. It is a model that includes 58 members (mostly management agencies throughout the Mediterranean basin) and 33 other partners that manage more than 80 MPAs in 18 Mediterranean countries (Gabrié et al. 2012).

In principle, the International Union for Conservation of Nature (IUCN) categories for protected areas would be applicable to all types of MPAs. However, as the number of terrestrial protected areas is lower and newer, they do not have the benefit of the same level of experience and knowledge.

Their effects should be better known in a concrete way, which involves many objective and *in situ* experiments and observations by competent staff.

In most cases, the category has been determined, but not the management objectives. Finally, we often forget that many MPAs extend to both terrestrial and marine areas, and that the properties of both types of protected areas depend on the mutual influence of one on the other.

4.2 Perturbations of Marine Protected Areas

Proper management of MPAs requires a good knowledge of natural, cyclical, or unexpected structure and functioning of ecosystems existing in areas considered variations.

Periodic, seasonal, or even circadian changes are not always taken into account, and they are much more important than one might think, especially in quantitative evaluations to be performed at different times of the day or night (Molinero and Flos 1992; Marchand and Masson 1988; Power et al. 2000).

One must also reflect on typhoons and tsunamis, which can have dramatic effects and destroy large portions of the seabed, as happens frequently along the coast of Japan and has done since time immemorial. The return to 'normal' operating conditions of MPAs – resilience – can happen quickly or take less time depending on the importance and depth of each disturbance.

To some extent, these large changes, periodical or not, weather or seismic, we consider the human scale can be considered 'normal' in the global long-term functioning of the ecosystems considered (Salvat 2014).

4.3 Integrity of Marine Protected Areas

The natural environments of the entire planet are no longer in their original state. We must, in this view, differentiate between the nature of the waste that can be recycled in ecosystems forever and the waste from human activities that the areas are often not able to recycle (Ceccaldi 2004). They are already affected by human influence, and this is particularly the case for marine pollution and fisheries.

Plastic debris are everywhere. Not only do they accumulate in the centers of the great oceans to occupy huge areas dubbed 'the plastic continents' by journalists, but also micro- and nano-debris from their mechanical, physical, chemical, or biological destruction inconspicuously invade water bodies and enter the food chain, the consequences of which are still poorly measured.

The situation for soluble drug products from medical or pharmaceutical chemical businesses is similar; these can pass through wastewater treatment plants and retain their original power.

Sand and gravel samples indicate that the integrity of MPAs is also strongly modified, even from a great distance, as the equilibrium profiles and composition of the sea floor is affected.

The effects of airborne particles such as ash emitted by volcanoes or earthworks at sea are also worth mentioning. Their composition can be neutral but often toxic, as pesticides come from distant sources. They can also impede marine filter feeders.

5 Role of Marine Protected Areas

5.1 Biodiversity, Production, or Exploitation?

Although the need for the creation of MPAs has been expressed formally for years (UNESCO 1996), it should be made clear whether the MPAs are intended only to protect biodiversity – with a conservation project – or whether their role is to promote species to be used in adjacent areas or even within the seventh or sixth type of MPA, with moderate catches that do not impair the ecological balance in the zone (Berque and Matsuda 2013). In this area, the traditions established in the fishing cooperatives in Japanese coastal villages – sato-umi – (Yanagi 2007, 2008) lead to improvements and friendly management of natural environments.

Strong political trends exist towards spatial planning of sea areas engaged in economic activities in shared marine space, e.g. tourism, transport, various extractions, offshore wind turbines, etc. As part of the International Marine Protected Areas Congress (IMPAC 3), Environment Ministers from 20 countries wanted to prioritize the objectives of nature conservation.

This choice is important because it is essential to know the flow of living species (adults, juveniles, or larvae) and dissolved or particulate matter that are driven by ocean currents in MPAs.

It is equally important to also assess the flow coming out of MPAs, living and/or non-living so they form the basis of populations then removed by fishermen (Parrish 1999; White and Kendal 2007).

Although the need for rational management will exist for several years (Salm et al. 2000), the ratings are less well known than in terrestrial environments where observations are much easier (Larrieu and Gonin 2009). Future observations will require means that are only beginning to exist, in terms of sensors, observation devices, meters, integrators, and means of calculation and modeling.

5.2 Are Marine Protected Areas the Best Solution to Preserve Biodiversity?

The basic idea is that, usually, mankind destroys species for its use, thereby changing the ecosystem by removing food or other species. Unfortunately, people most often impoverish the environments in which they operate; MPAs are intended to restore the initial conditions, which is deemed to be the best way to proceed.

However, we envisage a very different approach to act positively in MPAs: the installation of passive structures, rich in nutrients, to move towards the surface sea waters of the bottom to help increase the productivity of coastal waters and biological production of the area (Yanagi and Nakajima 1991; Yanagi 2007).

Moreover, advances in the technology of artificial reefs that copy and enhance this 'engineering of nature' creates natural habitats that encourage marine organisms and provide more opportunities for survival and development.

We think we can make better use of the increase in natural production in the volume of water of the area and its sea floors, and the amount and characteristics of marine habitats, thereby increasing the number of filter particles in the photic zone. It is also possible to establish a large number of shelters for larval stages and juveniles usually present in the ecological conditions of the areas considered.

In the nineteenth century, French zoologists such as Antoine Fortuné Marion proposed the establishment of rocks on sandy–muddy coastal floors near Marseille to increase the number of rockfish. A further recent development has been achieved in Marseille with the outstanding operation 'Reefs Prado' (Récifs Prado), part of which is available for artisan fishing; another part is not and in fact plays the role of an MPA. We could give many other examples. This concept can be multiplied, and the countless fields of various artificial reefs settled along the coast of Japan is one of the finest examples in the world.

6 Towards Artificial Marine Protected Areas

It was first thought that the best way to simply preserve natural ecosystems was to deny humankind, deemed to be the main destroyer, parts of the marine field and return to nature its free operation. The most common assumption is that these natural conditions are the most productive.

However, progress in the drawings and achievements of artificial reefs has shown that specific artificial structures favor the presence and development of certain species (Ceccaldi 1998, 1999). The production of usable organic matter created in the photic layer is much better used on these supports on the seabed, where bacteria destroy this matter. Much progress remains to be made in this area to guide the composition and functioning of coastal ecosystems.

Moreover, in some cases, one can deny access to marine areas carefully managed by people, and these areas behave as MPAs.

This is the case with some of the artificial islands built in Japan in front of the city of Kobe. The last island built to create the city's airport is surrounded by experimental sub-marine amenities for the implantation of algae forming the basis of new ecosystems. Fishing is prohibited and this has created, in effect, an artificial MPA. Similar facilities should be considered in all modern port areas (Ceccaldi 2006).

Other models to protect migratory species may be taken in the case of bird migrations that take place each season between Scandinavia and Africa. Several protected artificial fields facilitating such migration areas were established in France, such as Lake Chantecoq (in the Marne prefecture) classified as a national reserve, the Marais d'Orx or Arjuzanx reserve on a former mining site (in the Landes), the artificial lake Puydarrieux in the Pyrenean mountains, and the creation and maintenance of wetlands in the Camargue, in the Rhone delta, etc.

Artificial arrangements to protect wildlife are a reality. While some marine species do migrate, which is sometimes very important, most move with the currents, which ensures distribution, especially in the larval stage, in their planktonic phase, and juvenile forms, when their swimming abilities have not yet been acquired. Other marine species migrations are performed through a natural process of transport on natural packages of seaweed torn from the sea floors by storms (Senta 1966; Anraku and Azeta 1967; Safran 1990).

Nowadays, many species live in association with various floating objects, mostly plastic (Barnes 2002; Barnes et al. 2009; OSPAR 2009; Galgani 2014) due to the industrial and commercial activities of our developed societies. But other debris exist, such as that from the destruction wrought by the tsunami on the coast of Tohoku in northern Japan. These debris are estimated at 35 million cubic meters on the ground, and a similar volume is estimated to have been driven into the sea. Much of the debris has disrupted natural parks, especially the coastal park Rikuchu Kaigan. Whether the presence of these materials – wood, cement, metals, plastics, etc. – will be harmful or beneficial in terms of new submarine habitats is unknown as yet.

7 Services Provided by Ecosystems Versus Socio-ecosystems

Governments, economists, political decision makers, and the public have begun to understand that natural ecosystems provide immense benefit to human communities, and this benefit has even begun to be quantified in purely economic terms. This is a positive initial approach. However, 'service rendered' usually means that the profits go in one direction: from natural resources to humans, who are there to be served. We use these resources, often without taking into account the negative consequences of that use. Only now are decisions beginning to reflect that humans are part of the ecosystem and not simply there to be served – we are integrated into the overall operation.

This integration differs widely between cultures living in direct contact with nature, particularly in marine zones.

However, we must also ensure the presence of breeding species, constituting regional ecosystems, which provide eggs and are essential for the permanent recovery of larvae in neighboring ecosystems. We can also consider exports as larval feed, and even as multiple streams DNA willing to give new life forms that would not have existed if the spawners had been destroyed by various types of fishing. If the overall marine productivity (nutrients, light, and photosynthesis) and benthic plankton is ensured, whatever the level of protection, it is not the same for the production, which is very dependent on these habitats and survival, in good conditions, species considered each ecosystem component.

8 The Need for Further Research

MPAs give irreplaceable opportunities to understand in detail the functioning of ecosystems in undisturbed conditions, such conditions support the metamorphosis of larvae of benthic species (Delort et al. 2000).

Some areas are specifically preserved when they are home to rare species (Le-Saout et al. 2013). This is the case at sea for horseshoe crabs (Shuster and Sekiguchi 2009), some species of sea turtles, and other species whose existence is provided for in specific areas.

Moreover, the continued progress of sensors and computer technology will lead us to develop observation and transmitter networks that enable modeling real-time character stands and particle flux, eggs, and larvae MPAs in their environment.

To the extent that we can consider MPAs as export sites for eggs and larvae, these particular environments will enable the monitoring of different levels of biodiversity at the ecosystem, species, genetic, and molecular levels.

In other domains, we need to understand the process of creation, of present and future evolutions of MPAs.

Historians have been primarily concerned with major events, famous people, conflicts, societal institutions, or just the daily lives of citizens. However, the evolution of society did not take place in a vacuum, but in a context of constant physical economic, social, and cultural interference. An understanding of these factors is essential to encourage countries without a conservation tradition to create MPAs.

9 Information and Training

The knowledge of local people and amateur divers is often very important but also often neglected (Hamilton et al. 2012; Leleu et al. 2012). To delineate and effectively manage MPAs, great efforts to inform decision makers, training specialists, and control and defense will be absolutely essential. Its most important task is the general education of the public and the policy makers, as well as specialist training to manage and monitor the MPAs.

For the public who do not have the opportunity to take part in SCUBA diving, one of the most effective ways to learn about the realities of underwater life is to visit aquariums in which food webs and functional ecosystem models are represented: Nausicaa in Boulogne-sur-Mer, France; Aquarium La Rochelle in La Rochelle, France; the Oceanographic Museum of Monaco; Aquarium des Lagons in Nouméa, New Caledonia; Océanopolis in Brest, France; Okinawa Churaumi Aquarium in Okinawa, Japan; Georgia Aquarium in Georgia, USA; Dubai Aquarium and Underwater Zoo in Dubai, UAE; the Oceanografic Valencia in Valencia, Spain; Shanghai Ocean Aquarium in Shanghai, China; the Aquarium of Genoa in Genoa, Italy, etc.

Very few MPAs have met previously detailed knowledge on their operation, their effects and their specific objectives. They have not developed precise proposals for local decision makers and scientific experts (Salm et al. 2000).

Educating anglers who make withdrawals clearly defined in MPAs points can play a big role (Smallwood and Beckley 2012).

10 Preservation and Protection of Marine Protected Areas

The establishment of an efficient international, or at least regional, MPA policing authority is essential. However, the proliferation of MPAs, and increases in their surface areas makes them more expensive to control and police. New technologies will enable relay controls from ships to monitor protected surfaces. Modern satellite performance; the use of new, more efficient, and improved drones (e.g. Shadow View Foundation [2013] in Italy, Namibia, and the Philippines) can be the basis of new integrated systems for continuous monitoring at any time of day or night as well as effective interventions to chase intruders. 'MPA police' will be necessary in the future.

Nobody wants marine species to suffer the same fate as the African rhinoceros and elephants, the populations of which have been decimated over past decades and gamekeepers of which are killed by poachers.

It is obviously not appropriate to compare the behavior of elephant killers (whose objective is to sell tusks to mainly Asian finishers) with that of small-scale fishermen who work hard.

To whom do wild elephants and other animals living in the savannah belong? Similarly, at sea, who owns the fish and has the right to take them from the natural environment? Do they not constitute a common good of humanity? How do we manage their lives and/or their disappearance? Who owns those designated areas, the artificial reefs, animals from hatchery stock and released into MPAs? (Ceccaldi and Nakagawa 2003).

We can also consider the creation of MPAs as a new approach to appropriate more directly the marine environment and its resources areas – protecting it better to use it. Some countries are clearly aware; however, many others, especially developing countries, probably because they have other emergency priorities, are unaware of the problem and are, at best, at the beginning of their approach to managing their coastal environment (Agardy et al. 2003).

Furthermore, the management of MPAs will vary widely between countries.

In any case, many more resources are required for MPA networks to be established and functioning and these means are insufficient in several European countries.

In France, substantial resources have been allocated to this mission, including through the creation of a national agency and/or associations such as the MedPAN (Network of Manager of Marine Protected Areas in the Mediterranean) network, which is developing the network of Mediterranean MPAs. But these means should be increased, because the marine territory for which France has a responsibility to the planet is the second largest in the world. Moreover, as in other countries, many organizations and agencies have similar missions and very close missions. These should be included to make better use of current and future resources in the development of this sector.

The origins of the dualism between people and natural marine environments have deep and often invisible roots; we need to understand these to eliminate ethnocentric barriers (Descola 2005; Descola and Larrère 2011).

They are better understood when we explore the traditional, historical, sociological, cultural, anthropological (Lévi-Strauss 2011), or religious motivations that guide development decisions, pedagogies of education controls sites, or different types of MPA exploitation. To further explore these points is outside the scope of this chapter. Certainly, investigating these themes further can lead to discoveries.

11 Conclusion

Facts are stubborn things: growth in our developed societies will necessarily be limited rather than infinite, and will gradually reduce due to population growth, increased average longevity, and improved standard of living.

In response to human interference in the marine environment, the creation of MPAs is actually a new form of human influence on natural environments that is not yet managed. Protected areas currently only account for 2 % of marine areas, and they are expected to reach

10 % in 2020. Some of these protected areas are only beginning to be observed. However, the decision has not yet been made to establish MPAs to conserve existing biodiversity or whether this biodiversity can be enhanced and improved by the addition of artificial habitats, what proportion and with what techniques such protected areas can reasonably be exploited for the benefit of humankind through careful environmental management of samples in these semi-artificial environments.

More generally, and despite significant efforts in recent years to create and develop comprehensive approaches to preserve marine areas, have not yet been developed to clearly understand past relationships, current and future relationships of humankind with the marine environment, they gradually take possession

In other words: mankind take progressively possession of sea and oceans, without knowing the future relationships between man and the sea.

References

Agardy T, Bridgewater P, Crosby MP, Day J, Dayton PK, Kenchingyon R, Laffoley D, McConney P, Murray PA, Parks JE, Peau L (2003) Dangerous targets? Unresolved issues and ideological clashes around marine protected areas. Aquat Conserv Mar Freshw Ecosyst 13:353–367

Anonymous (2008) Establishing marine protected area networks – making it happen. IUCN World Commission on Protected Areas (IUCN-WCPA). National Oceanic and Atmospheric Administration and The Nature Conservancy, Washington, DC, 118 p

Anonymous (2012a) Stratégie nationale pour la création et la gestion des aires marines protégées. Ministère de l'Écologie, du Développement durable, des Transports et du Logement, Agence des Aires marines protégées, (Mars 2012)

Anonymous (2012b). Guide méthodologique pour la création et la gestion concertée d'une aire marine protégée. Ministère de l'Écologie, du Développement durable, des Transports et du Logement, Agence des aires marines protégées (Mars 2012)

Anraku M, Azeta M (1967) The feeding activities of Yellowtail larvae *Seriola quinqueradiata* Temmink et Schlegel, associated with floating seaweeds. Bull Seikai Reg Fish Res Lab 35:41–50

Barnes DKA (2002) Invasions by marine life on plastic debris. Nature 416:808–809

Barnes DKA, Galgani F, Thompson RC, Barlaz M (2009) Accumulation and fragmentation of plastic debris in global environments. Phil Trans R Soc B 364:1985–1998. doi:10.1098/rstb.2008.0205. Published 15 June 2009

Berque J, Matsuda O (2013) Coastal biodiversity management in Japanese satoumi. Mar Policy 39:191–200

Ceccaldi HJ (1998) Some examples of Japanese artificial reefs and some recent progress. In: Abstr. Symp. European Artificial Reef Research Network, Southampton, 8–10 janvier 1998, 4 p

Ceccaldi HJ (1999) "Artificial reefs" versus "Underwater structures to enhance ecology and fisheries": an attempt of classification and future ways of research. In: Proceedings of the conference, 7th C.A.R.A.H. Sanremo, 7–11 Oct 1999, p 448

Ceccaldi HJ (2004) Déchets produits par la nature et déchets produits par l'homme: une révision des concepts, p. 25. Rés.V ème Conf. internat. "Droits de Propriété, Économie et Environnement: les déchets. Aix-en-Provence, 23–25 juin 2002

Ceccaldi HJ (2006) Futur des extensions portuaires et protection du milieu marin. Ann Inst Médit Transports Marit 21: 279–319

Ceccaldi HJ, Nakagawa (2003) Le progrès technologique au service de l'exploitation durable du littoral. IV ème Conf. internat. "Droits de Propriété, Économie et Environnement: le Littoral. Aix-en-Provence, 26–28 juin 2002: 189–195, Bruylant édit. Bruxelles

Delort E, Watanabe N, Etoh H, Sakata K, Ceccaldi HJ (2000) Analysis of initial fouling process in coastal environment: effects of settlement, attachment, and metamorphosis promoters. Mar Biotechnol 2(3):224–230

Descola P (2005) Par delà nature et culture. Gallimard NRF, 640 p

Descola P et Larrère R (2011) L'écologie des autres. L'anthropologie et la question de la nature. Paris, Éditions Quae, coll. Sciences en questions, Paris, 110 p

Doussan I et al (2005) Guide de création d'un espace marin protégé en Méditerranée : aspects juridiques, biologiques et écologiques. Les Actes. Parc National de Port-Cros, publ., 48p

FAO (2011) Fisheries management. 4. Marine protected areas and fisheries. FAO Technical Guidelines for Responsible Fisheries (Compiled by Jessica S. Sanders, Dominique Gréboval and Antonia Hjort No. 4, Suppl. 4. Rome. 198 pp

Gabrié C, Lagabrielle E, Bissery C, Crochelet E, Meola B, Webster C, Claudet J, Chassanite A, Marinesque S, Robert P, Goutx M, Quod C (2012) The status of marine protected areas in the Mediterranean Sea. MedPAN & CAR/ASP. Ed: MedPAN Collection. 260 pp

Galgani F (2014) Marine litter within the European marine strategy: framework directive. In: Proc. 15th Fr.-Japon Oceanogr (in press)

Hamilton R, Sadovy de Mitcheson Y, Aguilar-Perera A (2012) The role of local ecological knowledge in the conservation and management of reef fish spawning aggregations. In: Sadovy de Mitcheson Y, Colin PL (eds) Reef fish spawning aggregations: biology,

fisheries and management, vol 35, Fish and fisheries series. Springer, Dordrecht, pp 331–369
Larrieu L, Gonin P (2009) L'Indice de Biodiversité Potentielle (IBP) : une méthode simple et rapide pour évaluer la biodiversité potentielle des peuplements forestiers. Revue Forestière Française, 6–2008 - pp 727–748
Le Direach L et al (2010) Suivi des peuplements de poissons de la Réserve marine du Cap Couronne (Parc marin de la côte bleue) – Bilan 1995–2007. GIS Posidonie – Parc marin de la Côte Bleue, 1–108
Leleu K, Alband F, Pelletier D, Charbonnel E, Letourneur Y, Boudouresque CF (2012) "Fishers" perceptions as indicators of the performance of marine protected areas (MPAs). Mar Policy 36(2):414–422
Lenfant P et al. (2012) Suivi temporel du peuplement ichtyque au sein et à proximité de la RNN Cerbère-Banyuls, année 1,2 et 3 été-automne 2007, 2009 et 2011. CEFREM UMR 5110 CNRS-UPVD, publ., 60p.
Le-Saout S, Hoffmann M, Shi Y, Hughes A, Bernard C, Brooks TM, Bertzky B, Stuart H, Butchart M, Stuart SN, Badman T, Rodrigues ASL (2013) Protected areas and effective biodiversity conservation. Science 342(6160):803–805
Lévi-Strauss C (2011) L'Anthropologie face aux problèmes du monde moderne. Seuil édit. coll. La Librairie du XXIe siècle, 160 p
Marchand J, Masson G (1988) Inshore migration, distribution, behaviour and feeding activity of sole, *Solea solea* (L.), postlarvae in the Vilaine estuary, France. J Fish Biol 33:227–228
Molinero A, Flos R (1992) Influence of season on the feeding habits of the common sole *Solea solea*. Mar Biol 113:499–507
OSPAR (2009) Marine litter in the North-East Atlantic Region: assessment and priorities for response. London, 127 pp
Parrish R (1999) Marine reserves for fisheries management: why not. Calif Cooper Ocean Fish Investig Rep 40:77–86
Power M, Attrill MJ, Thomas RM (2000) Environmental factors and interactions affecting the temporal abundance of juvenile flatfish in the Thame estuary. J Sea Res 43:135–149
Safran P (1990) Drifting seaweeds and associated ichtyofauna: floating nursery in the Tohoku waters. La Mer (Tokyo) 28:225–239
Salm RV, Clark J, Siirila E (2000) Marine and coastal protected areas: a guide for planners and managers. IUCN, Washington, DC, 390 p
Salvat B (2014) Health and degradation of coral reefs: assessment and the future. In: Proc. 15th Fr.-Japon Oceanogr (in press)
Senta T (1966) Experimental study on the significance of drifting seaweeds for juvenile fishes. 1. Experiments with artificial drifting seaweeds. Bull Jpn Soc Sci Fish 32(8):639–641
Shadow View Foundation (2013) Vasteland 78, 3011 BN Rotterdam. info@shadowview.org
Shuster CN, Sekiguchi K (2009) Basic habitat requirements of the extant species of horseshoe crabs (Limulacea). In: Tanacredi JT, Botton ML, Smith DR (eds) Biology and conservation of horseshoe crab. Springer, Dordrecht/Heidelberg/London/New York, pp 115–129. ISBN 978-0-387-89958-9
Smallwood N, Beckley LE (2012) Spatial distribution and zoning compliance of recreational fishing in Ningaloo Marine Park, north-western Australia. Fish Res 125–126(2012):40–50
UNESCO (1996) Biosphere reserves: the Seville strategy and the statutory framework of the world network. UNESCO, Paris, p 19pp
Valéry P (1931) Regards sur le monde actuel, Paris, Librairie Stock, Delamain et Boutelleau: 216 pages
White C, Kendal BE (2007) A reassessment of equivalence in yield from marine reserves and traditional fisheries management. Oikos 116:2039–2043
Yanagi T (2007) Sato-Umi – a new concept for coastal sea management. TERRAPUB, Tokyo, 86 pp
Yanagi T (2008) "Sato-Umi" – a new concept for sustainable fisheries. In: Tsukamoto K, Kawamura T, Takeuchi TD, Beard D, Jr., Kaiser MJ (eds) Fisheries for global welfare and environment. 5th world fisheries congress. Research Institute for Applied Mechanics, Kyushu University, Kasuga, pp 351–358
Yanagi T, Nakajima M (1991) Change of oceanic condition by the man-made structure for upwelling. Mar Pollut Bull 23:131–135

Health and Degradation of Coral Reefs: Assessment and Future

Bernard Salvat

Abstract

Coral reefs are in a crisis. We wonder about their survival in the face of human demographic pressures and anthropogenic degradation as well as considering the predictions of climate change. In short, what will happen to coral reefs under the influence of global change?

The main characteristics of the coral reef ecosystem are recalled related to corals and their association with algae, as builders due to calcification, on their distribution, biodiversity and biogeography. Services to human are many with the existence of islands, protection of coastal zones, fisheries and food security, sources of material, development of tourism, pearl culture, etc., all services which have been economically evaluated. Coral reefs suffer from natural events of which they recover since they exist (cyclones, bleaching and mortality, outbreaks of predators, diseases) except if these events succeed too frequently. Anthropogenic causes of degradation increased since half of the last century under the pressure of demography and development (filling, digging, sedimentation, wastewaters with nutrients and pollutants, overfishing, tourism substructure). Nevertheless, coral reef health status is variable according to regions. Factors of climate change will be either beneficial for coral reefs (sea level rise), probably with no more impact than today (cyclones), or much worrying (sea temperature increase, acidification).

We comment predictions on the future of coral reefs which have been published. Along with time, the threats for the future of coral reefs changed

B. Salvat (✉)
CRIOBE Moorea Polynésie Française, USR 3278
EPHE-CNRS-UPVD, Labex Corail, Université de
Perpignan Via Domitia, Perpignan, France
e-mail: bsalvat@univ-perp.fr

H.-J. Ceccaldi et al. (eds.), *Marine Productivity: Perturbations and Resilience of Socio-ecosystems*,
DOI 10.1007/978-3-319-13878-7_38,

in the scientific community due to science progress. Acidification is a good example appearing less than two decades ago and presently the most worrying factor, as well as adaptation of corals to temperature which was unsuspected one decade ago. These facts lead us to consider that published predictions will be modified by new discoveries and incline us to be cautious. We comment on the catastrophic ambiance which characterises the scientific production on the future of coral reefs amplified by the media in our society.

1 Introduction

Coral reefs are the richest marine ecosystem in terms of biodiversity, equivalent to the rainforests on land. According to the latest report of the IPCC (2014), this is the most threatened ecosystem on the planet. Although its surface in tropical coastal areas, reefs and lagoons, is of the order of half a million square kilometres, the coral ecosystem brings humanity considerable services (services rendus, voir plus haut) to over half a billion people in nearly 80 countries. Natural and anthropogenic degradation has resulted in a decline in health over the past century and especially for the past 50 years. A global health status needs to be considered regionally. Predictions of global change, demography, development and climate change lead to pessimism for the near future argued by many scientists. It should put these items with a view on what scientists said since half a century and in a broader temporal context, over millions of years, where we have to distinguish regarding humanity and the planet.

2 Coral Reefs

Coral reefs are the largest ever built construction on the surface of the planet by living beings and that persisted over time. Corals behind these buildings are primitive colonial invertebrates whose basic unit is the polyp, a kind of small anemone. Due to their association with symbiotic algae, zooxanthellae per billion in cells of polyps, corals have the ability to build a limestone skeleton. But this association limits corals living in shallow waters for photosynthesis which is essential for their symbiotic algae. Corals also require warm waters. All these constraints limit their distribution to tropical and subtropical coastlines.

This association is symbiotic and both partners have metabolic benefits. The ability of calcification is prodigious; a square metre of corals can produce up to 10 kg of calcium carbonate per year, and dead coral skeletons accumulate on top of each other meanwhile undergoing eustatic or tectonic movements. There are some 650 species of these symbiotic corals, hermatypic and reef builders, while other species, ahermatypic and without zooxanthellae, are living in the depths of the ocean.

Coral reefs are distributed in two biogeographic provinces: the first is the Indo-Pacific from the African east coast to the Pacific without really being drawn to west American shores and the second is the Caribbean. Species of reef complexes of these two provinces are not the same. The reefs cover about 280,000 km^2 as built construction (Spalding et al. 2001), but with sandy lagoons and associated ecosystems such as seagrass beds and mangroves, the total area exceeds 600,000 km^2. The Great Barrier Reef alone, which stretches over 2,000 km along the eastern coast of Australia, covers more than 350,000 km^2. After Australia and Indonesia, France has the third largest area of coral reef ecosystem mainly with New Caledonia (35,000 km^2) and French

Polynesia (15,000 km^2), but second for its exclusive economic zone (EEZ) of 11 million km^2, after the United States.

More than a third of marine species in the world live in coral reefs, about 95,000 species, and there are 32 of the 34 phyla that exist on our planet (Reaka-Kudla 1997; Spalding et al. 2001; Bouchet 2006). If vertebrates are relatively few, some invertebrate groups have tens of thousands of species such as molluscs and crustaceans. It is in the 'Coral Triangle' (Indonesia, the Philippines, Solomon) that the maximum species richness is found.

3 Services Provided by Coral Reefs to Mankind

Once human populations are located on coral coasts, they will develop close links with their coral reef inhabitants and their habitats whether it was exploitation of resources or various uses. This was the case in the Caribbean, as on the coast of East Africa in the Indian Ocean, in the Southeast Asia and in the Pacific as migration developed from west to east over the past two millennia. Cultural ties between these human populations and the sea are still very much alive especially in communities where the subsistence economy is dominant. We are witnessing in cultural demands with the goal of reviving the traditional ways of resource management along with political claiming on identity which validity is recognised by the Convention on Biological Diversity and special agreements on the recognition of Nagoya traditional knowledge and benefit sharing (CBD, Nagoya 2010).

Beyond these cultural aspects, ecosystem services provided by coral reefs to humanity are increasingly put forward with their monetary value for the ecosystem to be considered in what it brings to the economy.

Reef phenomenon saved oceanic islands of their disappearance; coral formations growing near the water surface while the system subside that affects and has removed the original volcano often covered by hundred meters of corals accumulated over millennia (over 2,000 m at Bikini). On some 420 atolls in the world, mainly in the Pacific (84 in French Polynesia alone), well over a hundred are uninhabited and pristine, while other host populations are living mainly on reef resources. Several atolls have paid a heavy price for particular human activities (nuclear testing sites, Pacific war, etc.).

The protection service of the coast by coral reefs is capital in many oceanic islands and continental coasts (Hoegh-Guldberg et al. 2007) where human activities are growing protected from swell and tempered from the damage during cyclones and tsunamis (Wilkinson et al. 2006 during the 2004 tsunami in Indonesia and swept the Indian Ocean).

Fishing is of course the most immediate service to the populations of several million residents in coastal areas with coral reefs (Whittingham et al. 2003; Cinner et al. 2008), especially in low coral islands (Burke et al. 2011). The fisheries from the lagoon reef environment are valued at 4 million tonnes, less than 5 % of global fisheries, but it is important as subsistence and for local marketing.

Reefs are also the place of supply of materials for some traditional buildings as in the Maldives but also elsewhere for the construction of infrastructure (roads, ports, airports) given the fact that it is cheaper to dig the fringing reef than to establish careers ashore. This is a perfect demonstration of the tragedy of the commons (Hardin 1968; Wilkinson and Salvat 2012).

The development of mass tourism since the end of World War II participates in the economies of many tropical countries including small island states whose resources are limited and whose isolation makes exports less competitive (Salvat and Paihle 2002). The emblematic vision of paradise reef scenery with sun, coconut, white coral sand beach and colourful reef fish is associated with water and underwater sports such as diving with renowned destinations to evolve with sharks. The Great Barrier Reef receives over 2.7 million visitors each year (Caillaud et al. 2012).

Pearl culture, in which several Pacific countries have embarked as French Polynesia with a maximum of 11 tonnes of pearls exported in 2000, the first value of exported income for the country, is recent.

To demonstrate the benefits to humans and the economy, studies of evaluation of services of coral reefs have developed in the late 1990s (Cesar 1996 – Costanza et al. 1997). They are tools for decision in support for the conservation of this ecosystem. A worldwide value of coral reefs has been estimated at $ 380 billion, that being 1 % of the value of all biomes on the planet (Costanza et al. 1997). Benefits of the coral reef ecosystem in the world economy were estimated at $ 29 billion, including 11 for tourism, 11 for coastal protection and 7 for fisheries. However, these estimates do not take into account or underestimate the cultural aspects as well as political stability in some countries between urban and rural areas.

4 Pressures on Coral Reef Health

If the health of coral reefs has particularly deteriorated over the past century, this is mainly due to the pressure of human activities due to demography and development. There is a synergy between the natural degradation factors, which have always existed and whose reefs accommodate themselves well, with growing impacts from human activities (Fig. 1). To better understand this increasing pressure of human demography on reef resources and habitats is to mention that reefs are not now larger and more productive than two millennia ago. The planet had less than 400 million people in 1000, it crossed 2.5 billion in 1950 and we are at a time when there are more than 7 billion people...productivity of coral reefs is not 20 times it was a millennium ago!!

Natural disturbances affecting the reefs are earthquakes (rare), tsunamis, cyclones, population explosions of predators and periods of thermal anomalies of seawater, to mention only the most important. Tsunamis cause more damage to reefs when flooded waters are withdrawing to the sea carrying lots of materials from land than damage caused by the energy of breaking waves (Wilkinson et al. 2006). Cyclones can completely destroy a reef that will need a dozen or 15 years to recover their initial state provided that no other major disturbances occur later (Harmelin-Vivien 1994; Salvat et al. 2008; Williams et al. 2008; Salvat and Wilkinson 2011). Outbreaks of the coral-eating crown-of-thorns starfish (*Acanthaster planci*) that decimated hundreds of square kilometres of reefs in certain periods in different parts of the Pacific in the 1970s are still unexplained, but we interpret them as natural phenomena despite the evocation of anthropogenic causes that

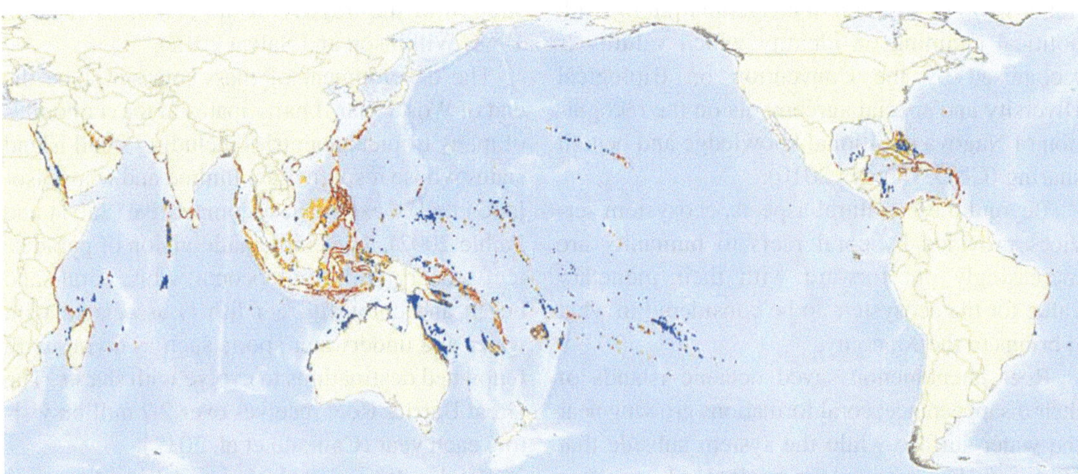

Fig. 1 Areas of coral reefs in *red* or *orange* are the most threatened; those in *blue* are the less impacted (After Burke et al. 2011)

have been made. These outbreaks occurred from time to time when the coral cover of the reefs is optimum but without any defined cycle. Such was the case on the Great Barrier Reef in the 1970s and 1980s or in French Polynesia in 1978–1982 and 2006–2007 (Salvat et al. 2008; Adjeroud et al. 2009), and recently, they provoked 42 % of coral mortality (De'ath et al. 2012).

The phenomena of coral bleaching correspond to the break of the association between the coral and its symbiotic algae. This is generally a response to heat stress especially at positive temperature anomaly of ocean waters in the summer. These phenomena may lead to colony mortality if stress is too important depending upon the temperature differential and the duration of the anomaly. These bleaching events affecting mortality of large coral reef areas have increased over the last 50 years with the warming of ocean waters. The 1980s saw the development of this bleaching mortality to several large areas of the Indo-Pacific and Caribbean (Brown 1997; Wilkinson and Souter 2008; Eakin et al. 2010; Salvat et al. 2008; Burge et al. 2014). These events are consistent with the IPCC report in 2013 indicating that the surface ocean temperature has risen by 0.85 °C since 1880 and a tenth of a degree per decade since 1971, 0.4 °.

Diseases and epidemics affecting reef animals over large geographic regions are also the cause of major degradations whose causes are not clearly elucidated (Sutherland et al. 2004). These disturbances have been very important in the Caribbean and have changed the structure of reef communities (Gardner et al. 2003; Weil and Cróquer 2009.) such as loss of *Acropora* or mortality of the sea urchin *Diadema*. These deceases are often linked to bleaching coral mortality (Willis et al. 2004) or other factors such as wastewater (Sutherland et al. 2011) or abundance of macro-algae (Nugues et al. 2004).

Anthropogenic disturbances have developed since the mid-twentieth century (Salvat 1987). Most anthropogenic coral reef degradations occurred in the early 1960s with the destruction of corals or annihilation by sedimentation caused by land mismanagement of the watershed or the massive releases of untreated wastewater with their nutrients: smothering reefs, fill or excavation on fringing reef areas for construction of airports or ports, expansion of urban areas and hotel complexes, organic pollution profiting to algae to the detriment of corals, etc. (Ryther and Dunstan 1971; Banner 1974; Endean 1976; Salvat 1978, 1987 to name but precursors papers).

The pesticide pollution of reef communities seems general but in varying concentrations according to region and their uses in urban and agricultural areas (Ramade and Roche 2006). It has a few drawbacks to public health through the consumption of fish or invertebrates. The presence of several organochlorine insecticides (HCH, aldrin, DDT) was reported in organisms of the Great Barrier Reef in Australia (Shaw et al. 2010) and even in the centre of the Pacific, French Polynesia, with chlordecone or kepone (Roche et al. 2011; Salvat et al. 2012). The presence of the herbicides, group of triazine and substituted urea (diuron) and chloroacetic acid derivatives both in Australia (Lewis et al. 2009) and in French Polynesia (Salvat et al. 2012) is a concern for coral reef health given the concentrations found and considering the considerable sensitivity of symbiotic algae of corals to these molecules that inhibit photosynthesis (Ramade and Roche 2006).

Overfishing is responsible for a very significant degradation of reef communities not only by the plunder of resources by a population more and more numerous, especially in Southeast Asia and in areas where the subsistence economy is still valid, but also because of the progress of technology in place for fishermen since the 1970s: aluminium boats, engines and net-nylon diving equipment. Moreover, the development of the economy, often related to tourism, induces a local commercial fishery which requires more exploitation. Fisheries are threatened by a scarcity of resources in some regions (Newton et al. 2007), and more than half of fisheries in the South Pacific are considered unsustainable (Bell et al. 2009). Such a situation is the same in the Indian Ocean (McClanahan et al. 2011) and in the Caribbean (Paddack et al. 2009). Many species of fish reproduce by spawning aggregations at certain times of the year in some places; these aggregations are declining in both the Indo-Pacific and the Caribbean (Russell et al. (2014).

5 The Condition of Reefs from Global to Regional

The scientific coral reef community was asked in the early 1990s by intergovernmental organisations (United Nations Environment Program, Intergovernmental Oceanographic Commission, World Meteorological Organisation, International Union for Conservation of Nature) to launch a long-term monitoring system on coral reefs related to climate change. At that time, scientists believed that the most important cause of future degradation of coral reefs would be the demographic pressure and development which was increasing considerably since 50 years. Meetings from 1990 to 1993 (IOC reports) and in Panama in 1996 led to the constitution of the Global Coral Reef Monitoring Network (GCRMN) headed by Clive Wilkinson who published the first 'Status of Coral Reefs of the World' in 1998, and others have been regularly published. These reports contributed largely to the sensibilisation of politicians and correlatively through the International Coral Reef Initiative to the development of coral reef science all over the world. About 20 % of the world's reefs are considered too severely destroyed or disturbed to regenerate, and 35 % are at risk of substantial deterioration in the years to come (Wilkinson 2008). Nothing better than a map can give a general picture of the health of the coral reef ecosystem in the world. The map (Fig. 1) from the World Research Institute (Burke et al. 2011) indicates coral areas where risks of degradation are impending: Southeast Asia, the Caribbean and East Africa with a total of more than 60 % of reefs. This is mainly due to demographic pressure and development. It is noteworthy that in central Pacific, far from continents and with less population pressure, the reefs are less threatened and relatively healthy (Salvat et al. 2008; Burke et al. 2011; Chin et al. 2011).

6 Climate Change and Coral Reefs

We will not address the causes of climate change and will retain its main consequences and predictions for the coral reef ecosystem.

Cyclones would not be more frequent than in the previous decades, but their intensity will be greater, levels 4 and 5 on the Saffir-Simpson scale in part due to a higher temperature of ocean waters. This phenomenon has already been observed during the recent decades (Hoyos et al. 2006; Elsner et al. 2008) and would continue with perhaps more intensity (Knutson et al. 2010). However, these predictions vary according to regions. It is clear that the occurrence of more violent and destructive cyclones disrupts reefs especially with the synergy that can be caused by other disrupting events of natural or anthropogenic origins (Salvat and Wilkinson 2011).

Rising sea level is already underway with an average of just less than 2 mm per year since 1960. It was predicted at + 18–59 cm by 2100 according to the IPCC report in 2007. It is much revised upwards in the report of 2013 and is expected to be a minimum of 0.5–1 m. The rate of accretion of reefs since the last Ice Age and during the sea level rise was about 2 cm/year and sometimes greater (Montaggioni 2005; Camoin et al. 2012) which is compatible with what would occur with the prediction of sea level increase related to climate change worst scenario. We must thus distinguish what is beneficial or harmful to coral reefs on the one hand and humanity on the other. For coral reefs, a sea level rise is rather good news because they will grow upward and they will expand after the flooding of intertidal reef flats and colonisation of coastal areas, but some negative factors would also occur such as sedimentation and turbidity (Hopley and Kinsey 1988; Storlazzi et al. 2011). For people in coastal and low-lying islands and their infrastructure, it is quite different, and the consequences will be disastrous (IPCC 2013, 2014). Land loss and saltwater intrusion would be catastrophic in atolls, physically threatened in their existence beyond the problems of climate refugees of their populations (Maldives, Bahamas, Kiribati). Some argue on the persistence of atolls with reefs growing up and rising in tandem with the sea (Pala 2014). Nevertheless, there will be a delay between the two phenomena, and reef accretion during Holocene reveals less rates (Perry et al. 2013; Hubbard et al. 2014).

The rise in ocean temperature will cause bleaching events and coral mortality during summer. Hoegh-Guldberg in 1999 predicted that bleaching mortality would be very frequent in 2015 and each year in 2040. Since, our knowledge on the relationship between corals and their zooxanthellae and characteristics of the zooxanthellae has made much progress with somme adaptability (Logan et al. 2013; Palumbi et al. 2014). It appears that some coral species are less sensitive to limited temperature anomalies than others with also geographic variations (McClanahan et al. 2007), and other species are capable of adaptation or acclimatisation (Guest et al. 2012; Logan et al. 2013; Palumbi et al. 2014). The latest report from the IPCC predicts a rise in temperature, mean values, between 1.8 and 4 °C in 2100. A too rapid speed of temperature rise would not allow corals to adapt according to Donner et al. 2005, and most of the scientific papers indicate that there will be almost yearly mortalities for the mid-twenty-first century (Donner, 2009; Burke et al. 2011). Changes in sea temperature and currents are expected to redistribute the geographic distribution of fish available for catch with a decrease in the tropics which tend to be the most socioeconomically vulnerable areas (Cheung et al. 2010).

Ocean acidification is already evidenced by a decrease of one-tenth of a pH unit over the past century. It could not be stopped and would decrease from 8.2 to 7.8 in 2100 representing a decrease of 100 times faster than in the last million years and a final value that the ocean has not been known for the last 20 million years. The challenge is significant because experimental studies in laboratory or in natural habitats where pH is less basic show much less calcification of benthic and pelagic organisms with calcareous tests (Gattuso et al. 1998; Langdon and Atkinson 2005; De'ath et al. 2009; Fabricius et al. 2011). Most studies agree on an announcement published 15 years ago (Kleypas et al. 1999) with the prediction that the calcification of organisms would be reduced by a third in 2050.

The fight against the consequences of climate change can only be taken at the worldwide level for emissions of greenhouse gas. There is no way for a country to fight whatever against the consequences of more intense cyclones, rising sea temperatures and ocean acidification, all phenomena that lead to degradation and affect coral reef resilience. Only the rise in sea level can be thwarted in some countries able to deploy extremely expensive works that do not own the majority of small island technology.

7 Predictions of Climate Change and Catastrophic Ambiance

The future of coral reefs is disquieting in the context of global change with demographic pressure which will increase, especially in coastal areas, and predictions of climate change at global scale although variable according to regions. This is the reef ecosystem across the whole planet that might be achieved. We would like to comment on two aspects: predictions and science progress and the ambiance of catastrophism.

7.1 Predictions and Science Progress

How do we consider with objectivity scientific statements and predictions which are published over time and what credibility should we give to present scientific explanations and prediction on the future of coral reefs?

When scientific information, either facts or predictions, are published, they consider of course the level of knowledge at the moment. We do think that our knowledge of the complexity we try to capture and seize is incomplete and that one scientific conclusion at a moment can be false later. It is more than that for prediction. We give two examples: one general and the other limited.

The general example refers to the scientific community considerations over time when considering factors of climate change and their consequences on the future of the coral reef ecosystem. In the early 1990s, the scientific community was considering that the reefs were threatened in their very existence for reasons of population pressure in developing countries. The issue of climate change was announced to debate at end of the 1980s with mainly two factors: the temperature and sea level rise. It is only later that the issue of acidification of marine waters and its impact on calcifying organisms with calcareous tests appeared. The first publication on the relationship between aragonite saturation and pH dates only from 1992 (Smith and Buddemeier), and that was the beginning of the acidification problem for coral reefs with the first work of Gattuso et al. 1998. As revealed by two surveys conducted in 1993 and 2004 (Kleypas and Eakin 2007), coral reef scientists were concerned and published on usual reef damage, ranking as 37th the acidification problem which was just emerging. However, this factor is now considered the first threat to coral reefs and for the future and as important enough to create work on its economic impact (Brander et al. 2009). The advanced and pioneer papers (Smith and Buddemeier 1992; Gattuso et al. 1998) are led to completely revise the main future causes of degradations of coral reefs and prediction. One cannot forget that science in progress may bring on the desk other factors, new or not, with or without interference with known or unknown factors, in our ability to handle the complexity of nature.

Another example of prediction, more than a decade ago, was on what was mentioned to happen for coral reefs when considering the effect of the rise of sea surface temperature. Due to ecological observations at that time, some papers concluded that bleaching and mortality of corals will be very common in 2015 (Hoegh-Guldberg 1999). This prediction 15 years ago is wrong. Moreover, since that time, the scientific knowledge is improving on the adaptation or acclimatization of coral species to a rise of sea level temperature (Guest et al. 2012; Logan et al. 2013; Palumbi et al. 2014). The present knowledge does not allow the simple prediction 15 years ago considering that elevated temperature induces coral mortality. This is not so simply the case.

These examples show that we must not forget that predictions are only predictions with current knowledge and are delivered with some uncertainty to be confirmed, amplified or denied later according to the advancement of knowledge showing all the unexpected complexity of organisms and ecosystems. These predictions must be considered with caution before concluding that the reefs will survive or will exist no more in 2100. These predictions cannot take into account future advances in knowledge that all go towards more complexity of interacting factors causing change. Scientists cannot imagine and foresee the potential of nature to find solutions or the ingenuity of man to find technological parades faced to degrading factors.

7.2 Catastrophism

At this stage of thought, it is necessary to consider prediction in an objective analysis considering what we know, which is far more incomplete, and without engaging ourselves in catastrophism. Catastrophe and disaster are synonyms (Ramade 2006) but express an attitude to assume the worst. This ambiance of catastrophism that joins many myths and legends was back in fashion and strength in 1980 when it was popularized that the meteorite killed the dinosaurs, explaining the massive biological crisis in the late Cretaceous (Babin 2007). We are informed by the media and by all recent means of communication (the Internet) of all catastrophic events all over the world. Such a situation, globalisation, reinforced the 'catastrophic atmosphere' in which we are considering global events as well as regional or local ones. Catastrophism is inherent in our society and science and media have their responsibility on this atmosphere. Related to the future of coral reefs, the catastrophic predictions have been a debate among scientisits after some of them published worst-case scenarios at the expense of truth (Grigg 1992; Maynard et al. 2008; Hoegh-Guldberg 2009). Beyond publishing results of their research, scientists entitled their papers with a catastrophic resonance for different reasons. One of them is for authors to catch the attention of their colleagues and of decision makers on the importance of their research which needs more support. The example of the outbreaks of the crown-of-thorns starfish on the Great Barrier Reef in Australia in the 1960s is demonstrative. The consequences of these outbreaks, which are natural events, caused important mortality of the corals. To fight against what would be a disastrous situation ecologically for the barrier reef and economically for the country (due to tourism), sensational papers were published in science periodicals as well as in newspapers and amplified by the media with subsequent political reaction. It was the start point of the important development of Australian research on coral reef (Wilkinson 1990).

8 Conclusion

It is clear as we wrote before that 'the future of coral reefs is disquieting in the context of global change with demographic pressure which will increase, especially in coastal areas, and predictions of climate change at a global scale'. Nevertheless, considering the past evolution of our knowledge on factors influencing the future of coral reefs and predictions that have been made and overall taking note of all changes that occurred in knowledge and predictions over time, we cannot conclude that the coral reef ecosystem will collapse before the end of this century.

Such an issue will be a ruin and a disaster for a million of people in a disordered world, probably with many other modifications. This is the challenge for humanity. For the planet, it will be one more catastrophic mass extinction according to Veron (2008) since we consider that life appeared on our planet. Veron listed the causes of past five extinctions and observed that the primary causes were linked to the carbon cycle and concluded that the 'ocean acidification has the potential to trigger a sixth mass extinction'. It would take only some million years for a gap on the reef phenomenon on our planet, waiting for other new extinctions and other gaps until the end of its existence – the challenge of life and reefs on our planet.

Acknowledgements We thank and greatly appreciate comments from two colleagues who reviewed the manuscript: Hubert Ceccaldi who organised the 15th French-Japanese Symposium on Oceanography in Marseille and François Ramade. Thanks to Patrick Prouzet for the layout recommendations and help.

References

Adjeroud M, Michonneau F, Edmunds PJ, Chancerelle Y, Lison de Loma T, Penin L, Thibaut L, Vidal-Dupiol J, Salvat B, Galzin R (2009) Recurrent disturbances, recovery trajectories and resilience of coral assemblages on a South Central Pacific reef. Coral Reefs 28:775–780

Babin C (2007) Autour du catastrophisme. Vuibert, Paris, pp 1–169

Banner AH (1974) Kaneohe Bay, Hawaii: urban pollution and a coral reef ecosystem. Proc Second Coral Reef Symp Brisb 2:685–702

Bell JD, Kronen M, Vunisea A, Nash WJ, Keeble G, Demmke A, Pontifex S, Andrefouet S (2009) Planning the use of fish for food security in the Pacific. Mar Policy 33:64–76

Bouchet P (2006) The magnitude of marine biodiversity. In: Duarte CM (ed) The exploration of marine biodiversity, scientific and technological challenges. Fundacion BBVA, Bilbao, Espana, pp 33–64

Brander LM, Rehdanz K, Tol RSJ, Van Beukering PJH (2009) The economic impact of ocean acidification on coral reefs. ESRI Work Pap 282:1–33

Brown BE (1997) Disturbances to reefs in recent times. In: Birkeland C (ed) Life and death of coral reefs. Chapman and Hall, New York, pp 354–378

Burge CA, Eakin CM, Friedman CS, Froelich B, Hershberger PK, Hofmann EE, Petes LE, Prager KC, Weil E, Willis BL, Ford SE, Harvell CD (2014) Climate change influences on marine infectious diseases: implications for management and society. Annu Rev Mar Sci 6:249–77

Burke L, Reytar K, Spalding M, Perry A (2011) Reefs at risk revisited. World Resources Institute, Washington, DC, pp 1–114

Caillaud A, Damiens F, Salvat B, Wilkinson C (2012) Preventing Coral Grief: a comparison of Australian and French coral reef protection strategies in a changing climate. Sustain Dev Law Policy XII(2): 26–31, 63–64

Camoin G, Seard C, Deschamps P, Webster JM, Abbey E, Braga JC, Iryu Y, Durand N, Bard E, Harmelin B, Yokoyama Y, Thomas AL, Henderson GM, Dussoullez P (2012) Reef response to sea-level and environmental changes during the last deglaciation: Integrated Ocean Drilling Program Expedition 310, Tahiti Sea Level. Geology 40:643–646

CBD (Convention on Biological Diversity) Nagoya (2010) www.biodiv.org/programmes/areas/marine/protected.asp

Cesar HSJ (1996) Economic analysis of Indonesian coral reefs. Working Paper Series "Work in progress", World Bank, Washington DC, pp 1–97

Cheung WWL, Larm VWY, Sarmiento JL, Kearney K, Warson R, Zeller F, Pauly D (2010) Large-scale redistribution of maximum fisheries catch potential in the global ocean under climate change. Global Change Biol 16(1): 24–35

Chin A, Lison de Loma T, Reytar K, Planes S, Gerhardt K, Clua E, Burke, Wilkinson C (2011) Status of coral reefs of the Pacific and outlook: 2011. Publishers Global Coral Reef Network, pp 1–260

Cinner JE, McClanahan TR, Graham NAJ, Daw TM, Maina J, Stead SM, Wamukota A, Brown K, Bodin A (2008) Vulnerability of coastal communities to key impacts of climate change on coral reef fisheries. Global Environ Change (0). doi:10.1016/j.gloenvcha.2011.09.018. http://www.sciencedirect.com/science/article/pii/S0959378011001579

Costanza R, d'Arge R, de Groot R, Farber S, Grasso M, Hannon B, Limburg K, Naeem S, O'Neill RV, Paruelo J, Raskin RG, Sutton P (1997) The value of the world's ecosystem services and natural capital. Nature 387: 253–260

De'ath G, Lough JM, Fabricius KE (2009) Declining coral calcification on the Great Barrier Reef. Science 323:116–119

De'ath G, Fabricius KE, Sweatman H, Puotinen M (2012) The 27-year decline of coral cover on the Great Barrier Reef and its causes. PNAS 109(44):17995–17999

Donner SD, Skirving WJ, Little CM, Oppenheimer M, Hoegh-Guldberg O (2005) Global assessment of coral bleaching and required rates of adaptation under climate change. Glob Chang Biol 11:2251–2265

Donner SD (2009) Coping with commitment: projected thermal stress on coral reefs under different future scenarios. PLoS One 4(6):e5712. doi:10.1371/journal.pone.0005712

Eakin CM, Morgan JA, Heron SF, Smith TB, Liu G et al. (2010) Caribbean coral in crisis: record thermal stress, bleaching, and mortality in 2005. PLoS One 5: el13969

Elsner JB, Kossin JP, Jagger TH (2008) The increasing intensity of the strongest tropical cyclones. Nature 455:92–95

Endean R (1976) Destruction and recovery of coral reef communities. In: Jones OA, Endean R (eds) Biology and geology of coral reefs, II, Biology 2, Academic Press, NewYork, pp 215–253

Fabricius KE, Langdon C, Uthicke S, Humphrey C, Noonan S, De'ath G, Okazaki R, Muehllehner N, Glas MS. Lough JM (2011). Losers and winners in coral reefs acclimatized to elevated carbon dioxide concentrations. Nat Clim Change 1: 165–169. doi:10.1038/nclimate1122

Gardner TA, Côté IM, Gill JA, Grant A, Watkinson AR (2003) Long-term region-wide declines in Caribbean corals. Science 301:958–960

Gattuso JP, Frankignoulle M, Bourge I, Romaine S, Buddemeier RW (1998) Effect of calcium carbonate saturation of seawater on coral calcification. Glob Planet Change 18:37–46

Grigg RW (1992) Coral reef environmental science: truth versus the Cassandra syndrome. Coral Reefs 11:183–186

Guest JR, Baird AH, Maynard JA, Muttaqin E, Edwards AJ, Campbell SJ, Yendall K, Affendi YA, Chou LM (2012) Contrasting patterns of coral bleaching susceptibility in 2010 suggest an adaptive response to thermal stress. PLoS One 7(3):1–8

Hardin G (1968) The tragedy of the commons. Science 162:1243–1248

Harmelin-Vivien ML (1994) The effects of storms and cyclones on coral reefs: a review. J Coast Res (Special issue n° 12 on Coastal Hazards): 211–231

Hoegh-Guldberg O (1999) Climate change, coral bleaching and the future of the world's coral reefs. Mar Freshw Res 50:839–866

Hoegh-Guldberg O (2009) Climate change and coral reefs: Trojan horse or false prophecy? Coral Reefs 28:569–575

Hoegh-Guldberg O, Mumby PJ, Hooten AJ, Steneck RS, Greenfield P, Gomez E, Harvell CD, Sale PF, Edwards AJ, Caldeira K, Knowlton N, Eakin CM, Iglesias-Prieto R, Muthiga N, Bradbury RH, Dubi A, Hatziolos ME (2007) Coral reefs under rapid climate change and ocean acidification. Science 318:1737–1742

Hopley D, Kinsey DW (1988) The effects of rapid short term sea level rise on the Great Barrier Reef. In: Pearman GI (ed) Greenhouse: Planning for Climatic Change. CSIRO, Brill, Leiden, pp 189–201

Hoyos CD, Agudelo PA, Webster PJ, Curry JA (2006) Deconvolution of the factors contributing to the increase in global hurricane intensity. Science 312:94–97

Hubbard D, Guschler E, Davies P, Montaggioni L, Camoin G, Dullo WC, Storlazzi C, Field M, Flechter C, Grossman E, Sheppard C, Lescinsky H, Fenner D, McManus J, Scheffers S (2014) Island outlook: warm and swampy. Science 345:1461

IPCC (2013) Summary for policymakers. In: Stocker TF, Qin D, Plattner GK, Tignor M, Allen SK, Boschung J,

Nauels A, Xia Y, Bex V, Midgley PM (eds) Climate change 2013: the physical science basis. Contribution of Working Group I to the fifth assessment report of the Intergovernmental Panel on Climate Change. Cambridge University Press, Cambridge/New York
IPCC (2014) Climate change 2014: impacts, adaptation, and vulnerability. In: Field, CB, Barros VR, Dokken DJ, Mach KJ, Mastrandrea MD, Bilir TE, Chatterjee M, Ebi KL, Estrada YO, Genova RC, Girma B, Kissel ES, Levy AN, MacCracken S, Mastrandrea PR, White LL (eds) Contribution of Working Group II to the fifth assessment report of the Intergovernmental Panel on Climate Change. Cambridge University Press, Cambridge/New York
Kleypas JA, Eakin CM (2007) Scientists' perception of threats to coral reefs: results of a survey of coral reef researchers. Bull Mar Sci 80:419–436
Kleypas JA, Buddemeier RW, Archer D, Gattuso JP, Langdon C, Opdyke BN (1999) Geochemical consequences of increased atmospheric carbon dioxide on coral reefs. Science 284:118–120
Knutson TR, McBride JL, Chan J, Emanuel K, Holland G, Lansea I, Kossin JP, Srivas-Tava AK, Sugi M (2010) Tropical cyclones and climate change. Nat Geosci 3:157–163
Langdon C, Atkinson MJ (2005) Effect of elevated pCO2 on photosynthesis and calcification of corals and interactions with seasonal change in temperature/irradiance and nutrients enrichment. J Geophys Res 110(C9): C09S07
Lewis SE, Brodie JE, Bainbridge C et al (2009) Herbicides: a new threat to the Great Barrier Reef. Environ Pollut 157:2470–2484
Logan CA, Dunne JP, Eakin CM, Donner SD (2013) Incorporating adaptive responses into future projections of coral bleaching. Glob Chang Biol. doi:10.1111/gcb.12390
Maynard J, Baird A, Pratchett M (2008) Revisiting the Cassandra syndrome; the changing climate of coral reef research. Coral Reefs 27:745–749
McClanahan TR, Ateweberhan M, Muhando CA, Maina J, Mohammed MS (2007) Effects of climate and seawater temperature variation on coral bleaching and mortality. Ecol Monogr 77:503–525
McClanahan TR, Graham NAJ, MacNeil MA, Muthiga NA, Cinner JE, Bruggemann JH, Wilson SK (2011) Critical thresholds and tangible targets for ecosystem-based management of coral reef fisheries. PNAS 108:17230–17233
Montaggioni LF (2005) History of Indo-Pacific coral reef systems since the last glaciation: development patterns and controlling factors. Earth Sci Rev 71:1–75
Newton NK, Cote IM, Pilling GG, Jennings S, Dulvy NK (2007) Current and future sustainability of island coral reef fisheries. Curr Biol 17:655–658
Nugues M, Smith G, van Hooidonk R, Seabra M, Bak R (2004) Algal contact as a trigger for coral disease. Ecol Lett 7:919–923
Paddack MJ, Reynolds JD, Aguilar C, Appeldoorn RS, Jim Beets J et al (2009) Recent region-wide declines in Caribbean reef fish abundance. Curr Biol 19:1–6
Pala C (2014) Studies suggest that atoll islands will rise in step with a rising sea. Science 345:486–497
Palumbi SR et al (2014) Mechanisms of reef coral resistance to future climate change. Science 344:895
Perry CT, Smithers SG, Gulliver P (2013) Rapid vertical accretion on a 'young' shore-detached turbid zone reef: Offshore Paluma Shoals, central Great Barrier Reef, Australia. Coral Reefs 32(4): 1143–1148
Ramade F (2006) Des catastrophes naturelles. Dunod Paris 1–258
Ramade F, Roche H (2006) Effets des polluants sur les écosystèmes récifaux. Rev Ecol 61:3–33
Reaka-Kudla ML (1997) The global biodiversity of coral reefs: a comparison with rain forests. In: Reada-Kudla ML, Wilson DE (eds) Biodiversity II: understanding and protecting our biological resources. Joseph Henry Press, Washington, DC
Roche H, Salvat B, Ramade F (2011) Assessment of the pesticides pollution of coral reefs communities from French Polynesia. Rev Ecol 66(1): 3–10
Russell MW, Sadovy de Mitcheson Y, Erisman BE, Hamilton RJ, Luckhurst BE, Nemeth RS (2014) Status report world's fish agregation 2014. Report by Science and Conservation of Fish Agregations (SCFRA) in collaboration with the ICRI Ad Hoc Committee for Reef Associated Fisheries, 13pp4. Report by Science and Conservation of Fish Agregations (SCFRA) in collaboration with the ICRI Ad Hoc Committee for Reef Associated Fisheries, 13pp
Ryther JR, Dunstan WM (1971) Nitrogen, phosphorus and eutrophication in the coastal marine environment. Science 171:1000–1013
Salvat B (1978) Agressions humaines en milieu corallien. Parcs 3(2):1–4
Salvat B (1987) Human impacts on coral reefs: facts and recommendations. Antenne Museum EPHE, French Polynesia, ISBN 2.905630.06.X, pp 1–253
Salvat B, Paihle C (2002) Islands and coral reefs, population and culture, economy and tourism: world view and a case study of French Polynesian (chapter 14). In: di Castri F, Balaji V (eds) Tourism, biodiversity and information, Backhyus Pub., Leiden, pp 213–231
Salvat B, Wilkinson C (2011) Cyclones and climate change in the South Pacific. Rev Écol 66:105–115
Salvat B, Aubanel A, Adjeroud M, Bouisset P, Calmet D, Chancerelle Y, Cochennec N, Davies N, Fougerouses A, Galzin R, Lagouy E, Lo E, Monier C, Ponsonnet C, Remoissenet G, Schneider D, Stein A, Tatarata M, Villiers L (2008) Le suivi de l'état de santé des récifs coralliens de Polynésie française et leur récente évolution. Rev Ecol 62:145–177
Salvat B, Roche H, Berny P, Ramade F (2012) Recherches sur la contamination par les pesticides d'organismes marins des réseaux trophiques récifaux de Polynésie française. Rev Ecol 67:129–148
Shaw M, Furnas MJ, Fabricius K et al (2010) Monitoring pesticides in the Great Barrier Reef. Mar Pollut Bull 60:113–122
Smith SV, Buddemeier RW (1992) Global change and coral reef ecosystem. Annu Rev Ecol Syst 23:89–118

Spalding MD, Ravilious C, Green EP (2001) United Nations Environment Programme, World Conservation Monitoring Centre. World Atlas of Coral Reefs. University of California Press, Berkeley, pp 1–416

Storlazzi CD, Elias E, Field ME, Presto MK (2011) Numerical modeling of the impact of sea-level rise on fringing coral reef hydrodynamics and sediment transport. Coral Reefs 30(1):83–96

Sutherland KP, Porter JW, Torres C (2004) Disease and immunity in Caribbean and Indo-Pacific zooxanthellate corals. Mar Ecol Prog Ser 266:273–302

Sutherland K, Shaban S, Joyner J, Porter J, Lipp E (2011) Human pathogen shown to cause disease in the threatened Elkhorn Coral Acropora palmata. PLoS One 6(8):e23468. doi:10.1371/journal.pone.0023468

Veron JEN (2008) Mass extinctions and ocean acidification: biological constraints on geological dilemmas. Coral Reefs 27:459–472

Weil E, Cróquer A (2009) Local and geographic variability in distribution and prevalence of coral and octocoral diseases in the Caribbean I: community-level analysis. Dis Aquat Org 83:195–208

Whittingham E, Campbell J, Townsley P (2003) Poverty and reefs. DFID-IMM-IOC-UNESCO, 1–260

Wilkinson C (1990) Preface in special issue of Coral reefs on Acanthaster planci. Coral Reefs 9:93–94

Wilkinson C (2008) Status of coral reefs of the world: 2008. Global Coral Reef Monitoring Network and Reef and Rainforest Research Centre, Townsville, pp 1–296

Wilkinson C, Salvat B (2012) Coastal resource degradation in the tropics: does the tragedy of the commons apply for coral reefs, mangrove forests and seagrass beds? Mar Pollut Bull 64:1096–1105

Wilkinson C, Souter D (2008) Status of Caribbean coral reefs after bleaching and hurricanes in 2005. Global Coral Reef Monitoring Network, and Reef and Rainforest Research Centre, Townsville, pp 1–148

Wilkinson C, Souter D, Goldberg J (2006) Status of coral reefs in tsunami affected countries: 2005. Australian Institute of Marine Science, Townsville, pp 1–154

Williams DE, Miller MW, Kramer KL (2008) Recruitment failure in Florida keys Acropora palmata, a threatened Caribbean coral. Coral Reefs 27:697–705

Willis BL, Page CA, Dinsdale EA (2004) Coral disease on the Great Barrier Reef. In: Rosenberga E, Loya Y (eds) Coral health and disease. Springer, Berlin, pp 69–104

Part VII

Round Table: Sato-umi, The Wealth of the Commons

French and Japanese speakers at the round table- From the *left* to the *right*: C. Decugis, Dr S. Pioch, Prof. C. Mariojouls, Dr Y. Hénocque, M. T. Tanaka, Prof. T. Komatsu, Prof. Y. Koike, Prof. S. Seino and Dr H. Takahashi.

Sato-umi, the Wealth of the Commons: Moderator's Summary

Yves Hénocque

The round table was held as the last and concluding event of the symposium second session in Marseille. The whole session was covered by Japan's national TV channel, NHK, and part of it was used in the making of a special program on sato-umi nationally broadcasted in March 2014. It took place with the participation of a special guest, Mr. Gérard Romiti, the President of the National Committee of Maritime Fisheries and Marine Fish Farms (CNPMEM). Mr. Romiti starts with a presentation of the role of the Committee.

The CNPMEM (National Committee of Maritime Fisheries and Marine Fish Farms) is a private body with public service responsibilities, which represents all professionals in the production sector of maritime fisheries and fish farming. Apart from the CNPMEM headquarters in Paris, this professional body is now organized along the coast of mainland France and the overseas territories in 14 regional committees and 12 departmental or interdepartmental committees.

More and more of the work of the CNPMEM is devoted to ensuring the responsible and balanced exploitation of the marine resources and achieving recognition of the value of the jobs and the men and women working in the sector. It is responsible for regulating the maritime fisheries and actively participates in drawing up French and European regulations for the industry.

All of the matters dealt with by the CNPMEM pursue the same objective: to gain recognition at all levels of French, European, and international decision making for the commitment and responsibility of the professionals of the sector, especially as an example of a "responsible fishery and sustainable development" where fishermen, rather than mere "predators," appear as the "sustainers" of marine ecosystems functioning. In that sense, French coastal fishermen are much interested in developing exchanges and learn from the Japanese fishers' unique experience in marine ecosystem management lastly transcribed into the concept of sato-umi.

Mr. Romiti's overall message was then more particularly illustrated through (1) Mr. Decugis' presentation of varied initiatives made by the local fisheries association (APAM) from the Var department and (2) Prof. Mariojouls' presentation of new practices in oyster farming in France. These changes are also concerning the fishers' onboard working conditions for which a good

With the participation of Yves Hénocque (moderator), Christian Decugis, Takehiro Tanaka, Yasuyuki Koike, Teruhisa Komatsu, Catherine Mariojouls, Sylvain Pioch, Hideyuki Takahashi, Satako Seino, and Gerard Romiti (special guest).

Y. Hénocque (✉)
Institut Français pour l'Exploitation de la Mer (Ifremer), Conseiller Principal Politique maritime et Gouvernance, General-Secretary of S.F.J.O. (*) of France, Correspondant Asie-Pacifique (JAMSTEC), 155 rue Jean-Jacques Rousseau, 92138 Issy-les-Moulineaux cedex, France
e-mail: yves.henocque@ifremer.fr

H.-J. Ceccaldi et al. (eds.), *Marine Productivity: Perturbations and Resilience of Socio-ecosystems*,
DOI 10.1007/978-3-319-13878-7_39, © Springer International Publishing Switzerland 2015

deal of research is now carried out as explained by Dr. Takahashi.

Literally, "sato" in Japanese means a local community or village where people live their life, and "umi" is the common word for the sea so that the first meaning of sato-umi is the sea associated with a local community or village, in other words, a symbiotic relationship between human activities at the seashore and nature in coastal areas. Therefore, sato-umi aims at well-balanced sustainable management of coastal seas not only in water quality but also in both ecosystem and ecosystem services, while the implementation of the sato-umi concept has been promoted in many ways and many places in Japan.

The concept is derived from a much older one that applies to rural land management under the term of satoyama. Indeed, both approaches are totally interlinked as underlined by Prof. Komatsu who presented the traditional and sustainable use of the marine eelgrass as a fertilizer for agriculture in the Seto Inland Sea. The land is giving the nutrients to the sea and in return, the sea is giving the fertilizers for cultivating the land. Everywhere, farmers and fishers have learned to live in harmony with nature. In France, several typical agricultural landscapes may be considered as a kind of satoyama and a salt pond marshland as a kind of sato-umi. This has been somewhat lost and should be regained through local and scientific knowledge.

Sato-umi is deeply rooted in the traditional way of coastal management where community efforts were undertaken in a comprehensive and integrated manner from upland forest and rivers to coastal seas. In that regard, sato-umi may be regarded as based on traditional methods of sustainable community-based management of the coast and the sea.

However, the period of high economic growth has induced important social and natural condition changes that lead to the degradation and loss of natural capital (environment, landscape, and resources). In reaction, the term sato-umi became the mark of a kind of "rebirth" to restore once lost rich and healthy coastal seas using community-based participatory activities as a concept incorporated into official institutional systems of national policies such as the Basic Ocean Plan (2008) based on the National Ocean Policy (Basic Ocean Act, 2007) and other environmental strategies. Among others, it is worth mentioning the Act for the Promotion of Nature Restoration (2003) based on the unprecedented idea of initiating nature restoration projects involving governments and municipalities based on the initiatives of local organizations rather than governments or municipalities' regular plans.

At the local level, municipalities from the Seto Inland Sea also introduced the concept of sato-umi as an official policy. One of the participants to the round table, Mr. Tanaka, introduced one successful example in the municipality of Hinase, Okayama prefecture.

There are about 200 fishing families who make their living from coastal fisheries. Like in many other places in the Seto Inland Sea, industrialization and land reclamation severely damaged the coastal ecosystem. Many species of marine life disappeared as swaths of natural habitat, including mud flats and eelgrass beds, were mostly destroyed and as the water quality highly degraded by river runoff polluted by household sewage and toxic chemicals from industrial plants. As a result, fishers have seen their fish stock more and more depleting, shifting to oyster farming for a good number of them. In the 1960s and onward, a governmental project to enhance local fish stocks ("from catching fish to cultivating fish") was implemented in the Seto Inland Sea. It involved releasing cultured juvenile fish and shellfish stock (sea bream, shrimps, crabs, etc.) into the coastal enclosed sea in order to enhance stocks of commercially valuable fish.

But after several attempts to increase fish stocks that way, the fishing community (about 200 families) in Hinase recognized that it would not be possible to recover fish stocks simply by artificially increasing the number of juveniles. Something was missing, in particular, one of the main habitats, i.e., the sea grass beds or amamo-ba, the area of which declined drastically from 590 ha in the 1940s to just 15 ha in 1985.

In 1985, the Hinase Fisheries Cooperative Association (FCA) initiated the project to restore the eelgrass beds. Through observation from fishers and with the help of local fishery officials including Mr. Tanaka, they found that in some of the areas where the eelgrass beds were before, water salinity and seabed conditions had been deeply changed. The local fishery officials then helped in artificially adjusting the seabed conditions and salinity by carpeting the seabed floor with grinded oyster shells and mitigating the wave movements through the use of floating wave-dissipating devices.

As a consequence of a long-term effort, the amamo-ba in Hinase increased from just 12 ha in 1985 to almost 100 ha in 2009. The recovery is still in progress, but so far, the application of both local and scientific knowledge has been highly positive in recreating the habitat and in orchestrating cooperation by the government sector (prefecture), scientific experts, and the fishing community.

Still today, as it is the case in some highly urbanized areas in France, connectivity between the land and the sea has been lost. In that regard, one may say that the recent, devastating earthquake and tsunami have restored the Tohoku area to the state it was in before the era of Japan's high economic growth. But the government's basic principle is not only to rebuild the once existing seawalls but to build a continuous wall along the entire Tohoku coast in Fukushima, Miyagi, and Iwate prefectures. In Kesennuma, this seawall will be a giant structure (almost 15 m!), over three times as high as the previous one, and it will be sloped (i.e., 91 m wide concrete structure as its base).

Clearly for Dr. Seino, such a structure standing right at the boundary between land and sea will destroy the wetlands and tidal flats, making it impossible for them to function again. Unfortunately and in spite of local opposition, it looks like the government has no time for discussion (deadline for construction is the fiscal year 2015, i.e., in between April 2015 and March 2016), making it a matter of disaster reconstruction hence not needing any kind of environmental assessment.

As already highlighted by Prof. Komatsu, the "forest is the lover of the sea" (Mori wa umi no koibito), and this was applied as well after the great Tohoku earthquake in Kesennuma, Moune Bay, with tree-planting activities.

Prof. Koike proposes to expand these activities further to other Tohoku bays (Mangoku-ura Bay, Ishinomaki Bay, Matsushima bay, etc.). After the first 3 years of "repair," the 4th year of "expansion" should lead to the multiplication of sato-umi-related initiatives by local fishery cooperatives and municipalities.

Considering these kinds of environmental degradation and thus losses in natural capital leads to the question of the values underpinned in the satoyama and sato-umi approach. For instance, the ecosystems provide direct use values such as food, fiber, fuelwood, and water among others. At the same time, satoyama and sato-umi also produce a number of indirect values that include flood and water regulation, nutrient recycling and temperature regulation, carbon sink, and transport of larvae through currents or pollination on land. Then, one may consider the option values, which will include the maintenance of the satoyama or sato-umi harmonious arrangement for future generations as a basic element of their cultural heritage.

But Dr. Pioch states that values are not absolute numbers since they differ depending on social groups. For example, local communities (fishers) value many of the direct uses like rice production, fishery production, and water/nutrients regulation much higher than residents who might be able to acquire these services from other sources. Urban residents, on the other hand, might place high values on the indirect uses such as climate regulation and cultural services.

These different values of ecosystem services held by different social groups influence the perceptions and attitudes toward satoyama and sato-umi and their use in preserving biodiversity and enhancing a sustainable supply of different ecosystem services. These differences in perception are true between social groups but also in different cultural context, in different countries. The real and common challenge lies in whether or not satoyama and sato-umi can be scaled up so that they can deliver these economic and human development opportunities to local communities in developed and developing countries.

Overall, the round table discussion clearly showed that in Japan, the sato-umi concept and practice cannot be disconnected from a unique and very original trait of fishing right where local communities and fishermen comanage their coastal and marine resources through fisheries and/or aquaculture (e.g., oyster farming) activities. Successful community-based management or comanagement, when it involves both local communities and local governments, has evolved in many parts of the world including France though in a much less systematic way. If traditional socioeconomic systems governed by customary practices and laws are allowed to determine fishery management plans and policies, some of the environmental damage of large-scale, industrial, mixed-stock fisheries could be avoided.

In conclusion, going back to Gerard Romiti's statement at the opening of this round table, because the protection of marine resources is an integral part of overall environmental conservation including restoration, and because it is a particularly vital issue for fisheries, the very operators who are the fishers must be cast as major players in the environmental conservation struggle.

In other words, fishermen must realize that they are the main components in the marine resource use system and that they therefore must actively participate in ecosystem-based fishery based on mutual agreements through existing or to-be-created governance forms and procedures. Because this issue concerns all the fisheries around the world and because of the unique system of fishing rights in Japan, this is a model that could inspire the French fishery cooperatives while developing sato-umi-like approaches in marine and coastal waters.

Part VIII

Short and Preliminary Communications

Pôle Mer Méditerranée

Guy Herrouin

Abstract

Created in 2005, the "Pôle Mer Méditerranée" is a business and innovation world-class sea cluster located in the South of France. Its ambition is to contribute to the sustainable development of the maritime and coastal economy in the Mediterranean basin, in Europe, and more widely at an international scale. The network of the "Pôle Mer Méditerranée" involves 360 members including major companies, SMEs, research institutes, and academic structures.

1 The Strategic Areas

The "Pôle Mer Méditerranée" stimulates and encourages innovation through collaboration around six strategic business areas (see Fig. 1).

1.1 Key Figures

- 360 members
- ¾ of industry (more than 160 SMEs)
- ¼ of research and training
- More than 200 labeled projects for a total amount of 546 M€
- 200 granted projects

1.2 Our "Ambition"

To contribute to the sustainable development of the maritime and coastal economy.

1.3 Our "Job"

To encourage the innovation between our member-associating companies and research laboratories. Promote the results at the international level.

2 History of the Cluster

- 2004: The French government set a call for tender for the constitution of competitiveness clusters.
- Phase 1:
  2005: Creation of two competitiveness cluster "Mer Bretagne" and "Mer Provence-Alpes-Côte

G. Herrouin (✉)
Pôle Mer Méditerranée, La Seyne-sur-Mer, France
e-mail: herrouin@polemermediterranee.com;
http://www.polemermediterranee.com

H.-J. Ceccaldi et al. (eds.), *Marine Productivity: Perturbations and Resilience of Socio-ecosystems*,
DOI 10.1007/978-3-319-13878-7_40,

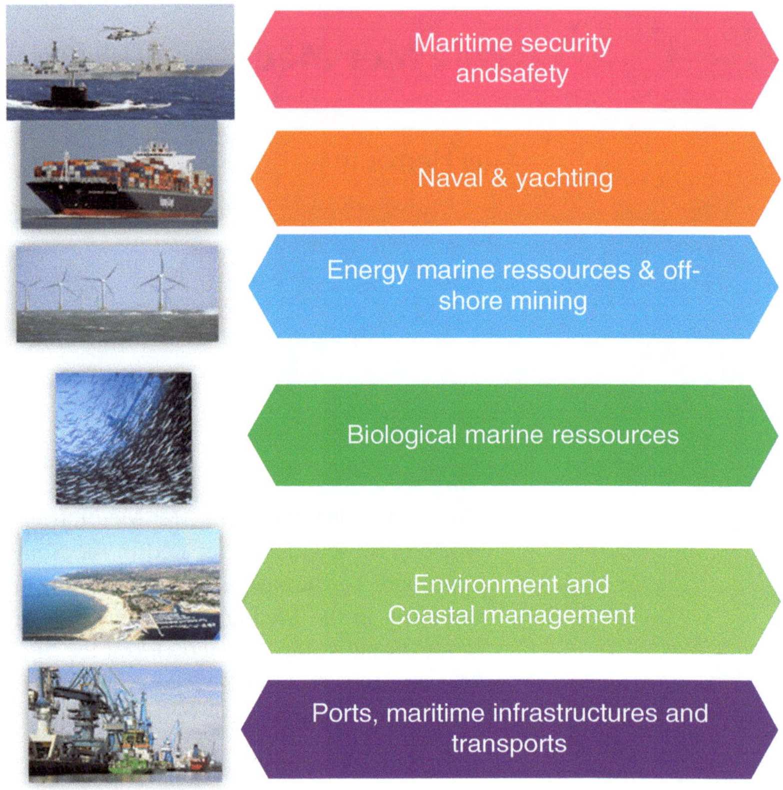

Fig. 1 The six strategic business areas

d'Azur." Awarding of the label competitiveness cluster with world vocation
2008: First evaluation – classified in A category. All objectives of Phase 1 were reached

- Phase 2: Innovation oriented
  2009: Contract of performance for the period 2009–2012
  2012: Second evaluation – classified in A category as highly performant cluster
- Phase 3: Business oriented
  2013: Signature of new contract of performance 2013–2018, extension in Languedoc-Roussillon area, and change of the name "Mer PACA" in "Mer Méditerranée"

3 Focus on Two Market-Oriented Programs

3.1 Sustainable Coastal Development and Ecological Engineering

- R&D on Marine ecological restoration solutions
- Multifunctional, integrated, and eco-designed ports
- Key figures: 13 labeled projects and 14 M€ total budget

3.2 Sea and Coastal Environment Service Providers

- Services for tourist use
  Service information
  Management service environmental data
  Aid services to marine management
  Prevention and management of environmental risks including climate change
- Key figures: 51 labeled projects for a total budget of 92 M€

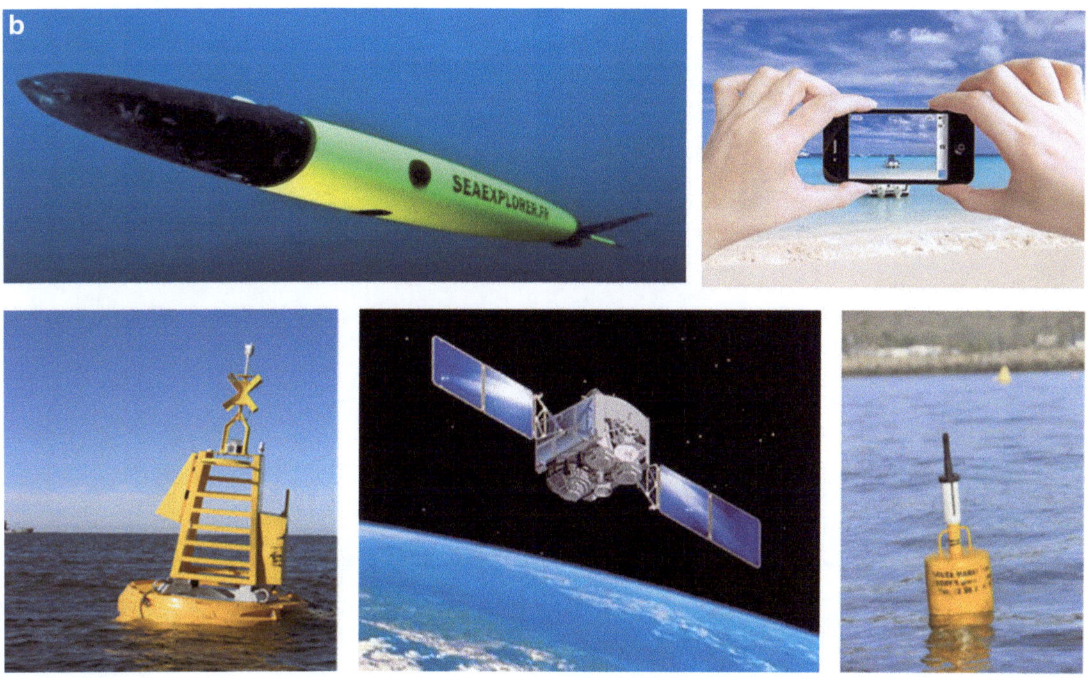

Satellite Tagging of Blue Sharks (*Prionace glauca*) in the Gulf of Lions: Depth Behaviour, Temperature Experience and Movements – Preliminary Results

F. Poisson, Y. Mitsunaga, T. Kojima, S. Torisawa, B. Séret, H. Demarcq, A. Banègue, and J.M. Groul

Abstract

The lack of reliable fishery-dependent data and fundamental understanding of the biology of most shark species causes concern for the sustainable management of shark populations in the Mediterranean Sea. The blue shark (*Prionace glauca*), a wide-ranging shark occurring in all tropical and temperate seas, is also one of the most heavily fished sharks in the world and so in the Mediterranean. In other oceans, its distribution is depicted as complex with spatial and temporal segregation by sex and size. In the Mediterranean Sea, movements and behaviours of this species are poorly known. The study aims at investigating on habitat occupancy, residency times and migratory pathways as well as providing behavioural data on temperature experience and swimming depth of the blue shark. This study strives to identify habitats and regions that are essential for the survival of sharks, while also determining when and where sharks are most

F. Poisson (✉) • H. Demarcq
IFREMER-Unité Halieutique Méditerranée
(HM, UMR MARBEC (MARine Biodiversity
Exploitation and Conservation)), Avenue Jean Monnet,
34203 Sète Cedex, France
e-mail: Francois.poisson@ifremer.fr

Y. Mitsunaga • S. Torisawa
Faculty of Agriculture, Kinki University,
204-3327 Naka-machi, Nara City 631-8505, Japan

T. Kojima
College of Bioresource Sciences, Nihon University,
Japan Kameino, Fujisawa, Kanagawa 2520880, Japan

B. Séret
Département Systématique et Evolution, IRD,
Muséum national d'Histoire naturelle,
CP 51 55 rue Buffon, 75231 Paris Cedex 05, France

A. Banègue
Fédération Française Pêcheurs en Mer (FFPM),
6 Boulevard de la Côte Radieuse,
66140 Canet en Roussillon, France

J.M. Groul
Seaquarium du Grau du Roi,
Avenue du Palais de la Mer,
30240 Grau Du Roi, France

Association Stellaris, Avenue du Palais de la Mer,
30240 Grau Du Roi, France

H.-J. Ceccaldi et al. (eds.), *Marine Productivity: Perturbations and Resilience of Socio-ecosystems*,
DOI 10.1007/978-3-319-13878-7_41,

vulnerable and will assist in the conservation of the species. The use for the first in the Mediterranean Sea of different types of satellite tags is proposed to investigate the ecology of this species. The preliminary results of the "smart tags" deployed on eight individuals are presented.

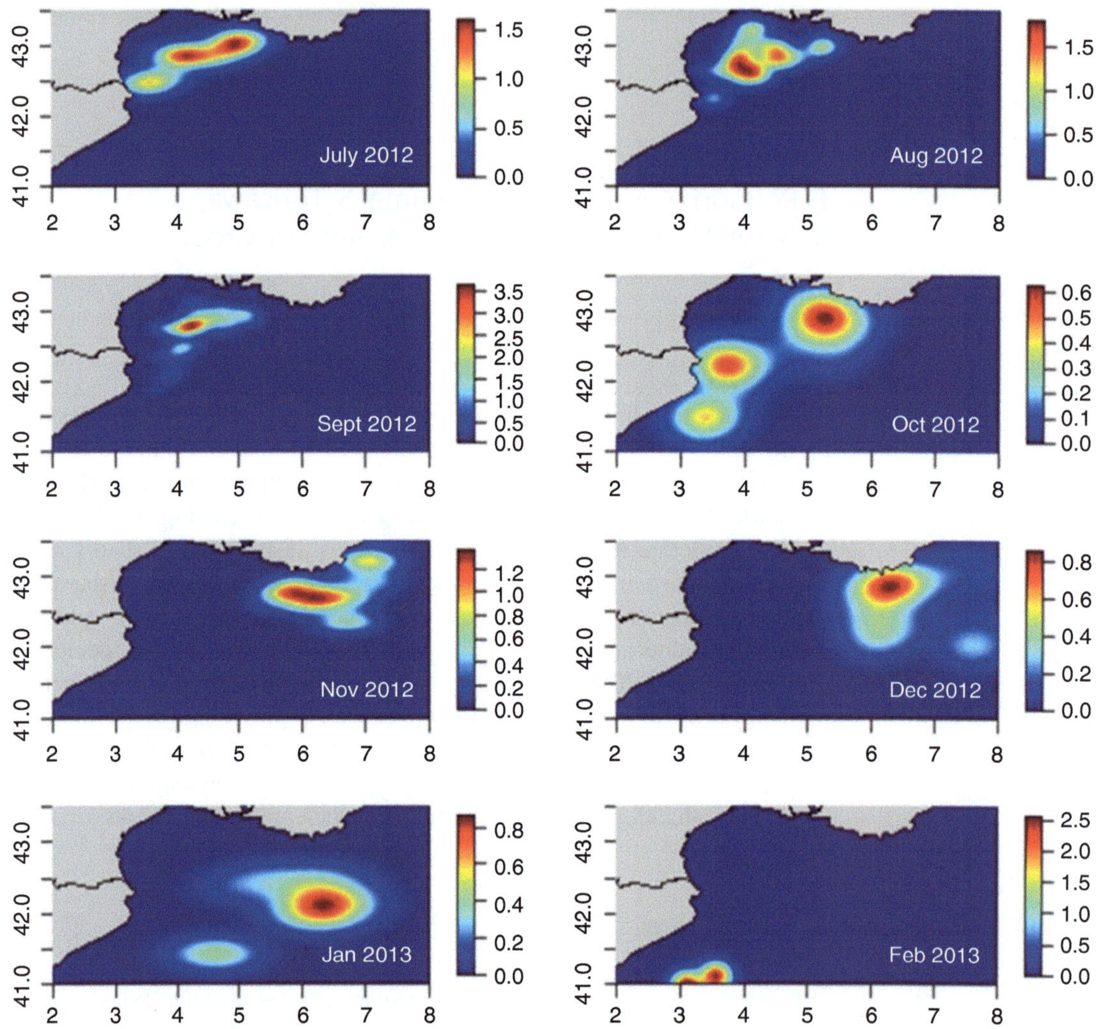

Fig. 1 Kernel density plot showing the major areas of prolonged residency and monthly movement patterns of a female shark (fork length: 130 cm)

Embryological Development of *Pinna nobilis* in Controlled Conditions

S. Trigos, N. Vicente, J.R. García-March, J. Torres, and J. Tena

Abstract

The Mediterranean fan mussel *Pinna nobilis* has been long time used for culinary, decorative, or textile purposes. For this reason their populations have been reduced over a large part of the Mediterranean Sea. *Pinna nobilis* is an iconic species and develops vital ecosystem functions, such as substrate fixation or recovering of suspended particles in the water column. Moreover, it supports a large number of species enhancing biodiversity. This study will present the first experiments tested in laboratory in order to complete the larval cycle of this species. The experiments were carried out in collaboration between the Paul Ricard Oceanographic Institute (France) and the IMEDMAR (Spain).

Around the world we can find lots of different species with high importance in the ecosystem network. However, in every sea or ocean, certain species are considered as relevant or emblematic species. These specimens are part of the tradition and represent the culture of a determinate region. One of these specimens is the bivalve *Pinna nobilis*. This endemic species has inhabited the Mediterranean Sea before hominids begin to walk upright since fossils have been found from the Miocene-Pliocene. Later, their populations have been used by Egyptians and Romans to manufacture clothes using their byssus threads known as "sea silk" due to its high quality and smoothness. The Muslims have used the nacre of their shells in the nineteenth century to fabricate jewelry, clothing buttons, or knife handles, and some Mediterranean regions such as Croatia, Turkey, or Greece continue to eat its meat.

Fortunately, *P. nobilis* is currently protected by the European legislation since 1992. Despite these regulations, lots of different actions such as

S. Trigos (✉) • J.R. García-March • J. Torres • J. Tena
Institute of Environment and Marine Science Research (IMEDMAR), Universidad Católica de Valencia, Avenida del Puerto s/n, 03710 Calpe, Alicante, Spain
e-mail: trigos_37@hotmail.com

N. Vicente
Institut Océanographique Paul Ricard, Ile des Embiez, 83140 Six Fours les Plages, France

Institut Méditerranéen de la Biodiversité et de l'Ecologie marine et continentale (IMBE), Aix-Marseille Université, Aix-en-Provence, France

H.-J. Ceccaldi et al. (eds.), *Marine Productivity: Perturbations and Resilience of Socio-ecosystems*, DOI 10.1007/978-3-319-13878-7_42,

Fig. 1 Adult individual of *Pinna nobilis* collected in the natural environment. Picture: N. Vicente

the use of trawls and anchors by boats; the regression of sea grass meadows, where they usually live; or their capture by scuba divers who commercialize the shells for ornamentation have increased the pressure on their damaged populations (Vicente 1991; Vicente and Moreteau 1991). Thus, it is necessary to develop new strategies which contribute to reinforce damaged populations which are under high risk of disappearance and to restore ecosystems degraded largely by human activities. Therefore, the proposal of this work intends to constrain the methodology to obtain juvenile specimens of *P. nobilis* in laboratory conditions in order to supply harmed environment with new cohorts increasing their population density and consequently favoring the conservation of natural environments. It could suppose the solution to ensure the durability of this emblematic species in the Mediterranean Sea (Garcia-March 2005).

In this work the first data on the hatchery of *P. nobilis* are shown. We have carried out numerous essays to develop the best hatchery protocol for this endangered bivalve. In order to obtain healthy gametes, 22 adult fan mussels collected (Fig. 1) during summer months (July to September) in 2012 and spawning induced by thermal shock (Helm et al. 2006).

A variation of 10 °C in water temperature was enough to easily induce spawning in all individuals >44.5 cm. The egg mass was non-buoyant and larvae started swimming only 1 day after fertilization, which indicates that females could maintain eggs in the body cavity until they are fecundated and larvae start swimming. An average of 725,000 were obtained from adults and some fan mussels were able to spawn repeatedly during several days. Some individuals changed from producing sperm to eggs in a few days. Interestingly, some individuals spawned simultaneously eggs and sperm, inducing self-fertilization. We were able to maintain the larvae alive for 25 days, but the vellum was destroyed, likely by bacteria, and the offspring stopped swimming and feeding and then died. Future experiments will consist firstly in improving the survival of larvae. After solving this question, other subjects such as the selectivity of different substrates for larvae settlement or gonadal maduration in captivity will be also tested.

References

Garcia-March JR (2005) Aportaciones al conocimiento de la biología de *Pinna nobilis* Linneo, 1758 (Mollusca Bivalvia) en el litoral mediterráneo ibérico. Universitat de Valencia, Servei de Publicacions, Valencia. http://www.tesisenxarxa.net/TDX/TDX_UV/TESIS/AVAILABLE/TDX-0628106-132411//garcia.pdf

Helm MM, Bourne N, Lovatelli A (2006) Cultivo de bivalvos en criadero. Un manual practico. FAO Documento Técnico de Pesca. No. 471. Roma, 184 pp

Vicente N (1991) Estudio ecológico y protección del molusco lamelibranquio *Pinna nobilis* L., 1758 en la costa mediterránea. Iberus 9(1–2):269–279

Vicente N, Moreteau JC (1991) Statut de *Pinna nobilis* L. en Méditerranée (Mollusque Eulamellibranche). Gis Posidonie publ., Marseille, pp 159–168

Mangrove Forest Ecosystem Services with Reference to the Transportation of Organic Materials to Coral Reefs: A Preliminary Study in Palau

Makoto Tsuchiya, Izumi Mimura, Yuka Yano, Noelle W. Oldiais, Yimnang Golbuu, Yoko Fujita, and Kaoruko Miyakuni

Abstract

Mangrove forests provide several important ecosystem services including (1) storage of high quantities of organic materials and nutrients; (2) sources of organic material supply to coral reefs; (3) environmental purification functions via feeding behaviors of benthic animals; (4) refuge for young coral reef fish; (5) sites for marine leisure activities such as kayaking; (6) habitat provision to birds, insects, and other animal species; (7) opportunities for food resources; and (8) opportunities for environmental education including ecotours; and others. To clarify these services, a preliminary survey was conducted in Palau, where mangroves connected to coral reefs are dominant along the coast.

Accumulations of organic materials were conspicuous around a river mouth area but were dispersed in coral reef areas. Research into biogeochemical cycles is necessary to clarify the importance of the ecosystem services of mangrove forests, and information on the sustainable use of mangrove and coral reef ecosystems for tourists is required. Ecosystem evaluations should also be conducted through interdisciplinary study.

M. Tsuchiya (✉)
Faculty of Science, University of the Ryukyus, Okinawa, Japan
e-mail: tsuchiya@sci.u-ryukyu.ac.jp

I. Mimura • Y. Yano • N.W. Oldiais
Graduate School of Engineering and Science, University of the Ryukyus, Okinawa, Japan

Y. Golbuu
Palau International Coral Reef Center, Koror, Palau

Y. Fujita
International Institute for Okinawan Studies, University of the Ryukyus, Okinawa, Japan

K. Miyakuni
Faculty of Tourism Sciences and Industrial Management, University of the Ryukyus, Okinawa, Japan

H.-J. Ceccaldi et al. (eds.), *Marine Productivity: Perturbations and Resilience of Socio-ecosystems*, DOI 10.1007/978-3-319-13878-7_43, © Springer International Publishing Switzerland 2015

1 Introduction

Mangrove ecosystems around river mouths or in inner bay areas play important roles in island and coral reef ecosystems, forming a bridge for the transportation of organic materials and nutrients between terrestrial and/or riverine areas and coastal ecosystems such as coral reefs, sea grass beds, and tidal flats. Mangroves also contribute to the maintenance of biological communities in the coastal zone, as animals in these coastal ecosystems use the organic materials transported from terrestrial regions via river systems and plants utilize the nutrients.

"Ecosystem services" is a relatively recent term, proposed by Ehrlich and Ehrlich in 1981, although the concept of nature's providing benefits to humans is long standing. The widely accepted definition of "ecosystem services" is the benefits provided by ecosystems to humans. As part of the current intensive debate about ecosystem services (e.g., Costanza et al. 1997), mangrove ecosystems have been a topic of discussion. Many papers list mangrove ecosystem services with information on their economic valuation (Bennett and Reynolds 1993; Tam and Wong 1999; Sathirathai and Barbier 2001; Gunawardena and Rowen 2005; Hussain and Badola 2008; Albert et al. 2012; Brander et al. 2012). When these services are discussed with reference to those of coral reefs and other ecosystems, the importance of mangroves must be clarified, and the integrated management of island and coral reef ecosystems could be improved by further consideration of ecosystem connectivity.

In the tropical and subtropical regions where mangrove forests occur, people have traditionally collected mangrove crabs and other fishery products and enjoyed leisure activities among mangrove forests. Conversely, mangrove areas worldwide have been targets for development activities such as the construction of shrimp ponds and reclamation. Ecosystem services must be understood if we are to ensure the sustainable use of such environments. This paper examines mangrove ecosystem services in Palau.

2 Study Area and Methods

A case study was conducted in Palau. Babeldaob Island, the biggest in Palau, is surrounded by coral reefs, and mangroves are very abundant around the river mouth area (Fig. 1). Mangroves are found along 80 % of the coastline of the island, and the remaining 20 % comprises beaches and reefs. The island has 4,025 ha of mangroves.

On Babeldaob Island, mangrove forests are distributed over several hundred meters from riverbanks to inland areas 1 km upstream from estuaries. Along the seafront, *Sonneratia alba* and *Rhizophora mucronata* are common, whereas *R. mucronata* and *R. apiculata* dominate estuarine regions. *Xylocarpus granatum* can be observed farther inland (upstream), and *Nypa fruticans* can be found along riverbanks.

To clarify the role of mangroves in the island coral reef ecosystems of Palau, a reference survey of mangrove ecosystem services was conducted. Then, a study of the distribution of organic materials around a river mouth area was made.

Three survey stations, Ngiwal, Ngaremeduu, and Airai, were selected. At each station, we visited the river area, river mouth area, and coastal zone in a small boat and collected suspended materials using 2-liter plastic bottles. The water was filtered through a GF glass filter, and the sediments were rinsed with diluted water, dried, and weighed. Sediments scooped directly from the bottom were kept in 200-ml glass bottles. After the removal of salt, all samples were dried and weighed. Then, 0.1 N HCl treatment was applied to obtain information on the $CaCO_3$ content of each sample.

The organic matter contents of these samples were measured using a CN analyzer, and stable isotope analyses were performed using methods described by Shilla et al. (2011). In this preliminary report, the results at Ngiwal area, the central part of east coast of Babeldaob Island (N7°31′50″, E134°36′42″), are shown (Fig. 1).

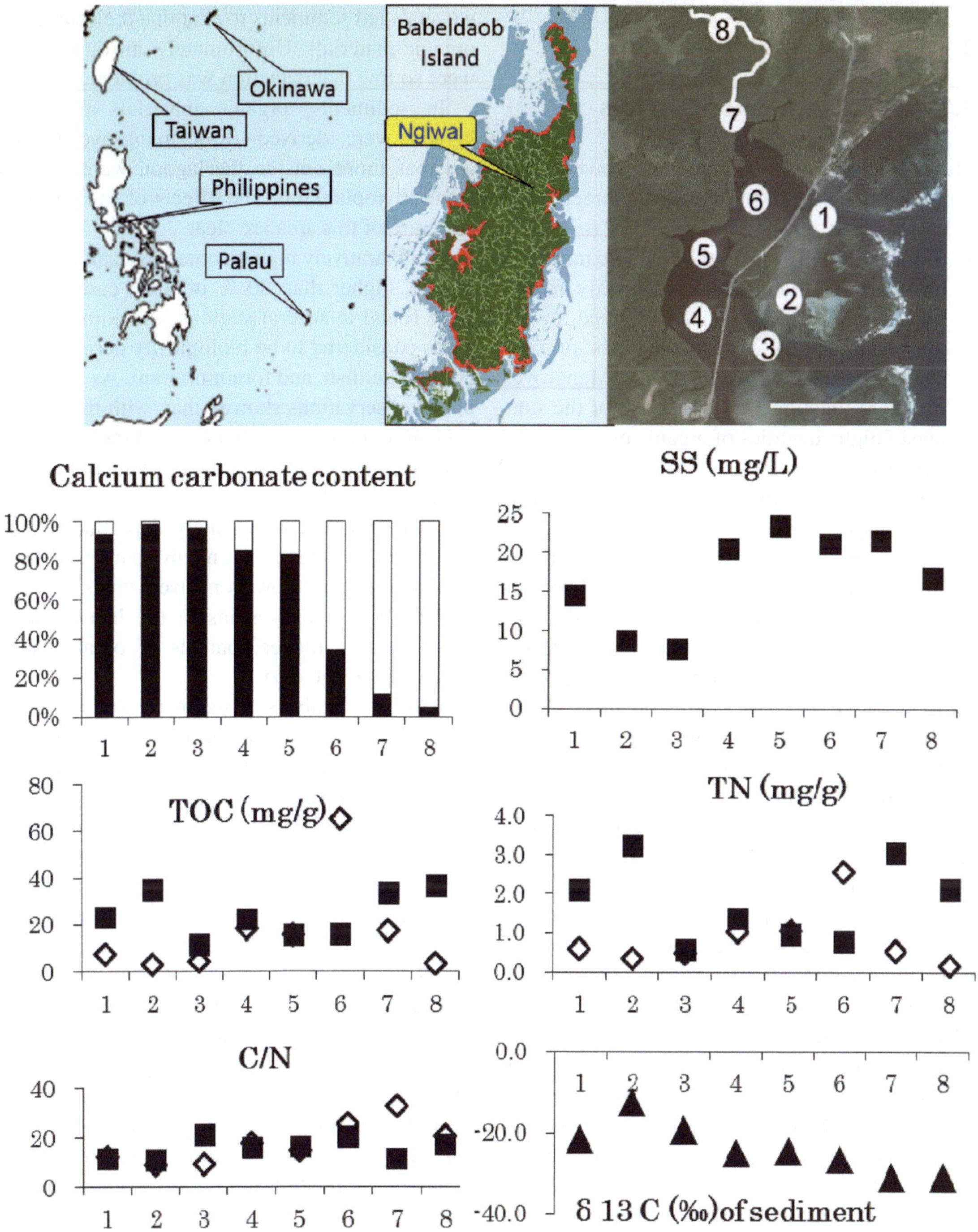

Fig. 1 Study site. Stations (*1–8*) for collecting sediment and suspended materials at Ngiwal, Babeldaob Island, which is surrounded by dense mangrove forests indicated by *red color*. A road with two culverts divides the bay into inside lagoon and outside areas. *1–3* were set outside of the road, *4–6* in the enclosed lagoon, and *7* and *8* in the river. Scale bar, 100 m. Results of analyses at each station were also shown. Content of $CaCO_3$ in sediments is shown in *black* column. ◇and ▲, sediments; ■, suspended materials

3 Results and Discussion

3.1 Biogeochemical Services

Mangrove ecosystems are thought to provide the following services to humans directly or indirectly:

- Storage of high quantities of organic materials and nutrients. Mangroves grow mainly in river mouth regions, environments inclined toward the accumulation of various types of substances that flow from rivers. Mangroves therefore function as warehouses for the storage of high quantities of organic materials and nutrients.
- Sources of organic matter supply. Large amounts of leaves, twigs, and flowers from mangrove plants fall to the mangrove forest floors, eventually breaking down and providing organic material to surrounding tidal flats and coastal areas. This material becomes food for small animals and serves as an important food source for coastal living organisms.

These two services can be clarified by future research into the dynamics of organic materials around mangrove areas.

The topography of this area is unique. A road named Compact Road, which runs around the whole of Babeldaob Island, crosses the bay, forming an enclosed lagoon.

Two small culverts have been used as passages for small boats. Waters in the lagoon are characterized by poor visibility and were observed to contain large amounts of suspended matter. Water flow and currents are mild. However, waters on the opposite side of the road were found to be comparatively clear. This suggests that little water exchange takes place between the lagoon and outer waters.

However, waters surrounding the culverts are constantly exchanged due to tidal movements as well as outflow from the lagoon to the waters beyond. Stable isotope analysis has become a popular method for understanding food webs in aquatic ecosystems because carbon and nitrogen stable isotopes can reveal the organic carbon sources of primary producers. In this research, we analyzed sediments to examine the sources of organic materials. Unfortunately, nitrogen levels were so low that detection was problematic.

In sediments, organic materials from the lagoon were derived from mangrove plants, whereas those outside the lagoon were derived from phytoplankton. The effects of the physical structure of this area are clear.

Comparatively higher contents of calcium carbonate, higher than 80 % in some cases (1–5), were found at several stations. Calcium carbonate is considered to be biologically derived from corals, shellfish, and foraminiferans. As such, the above observations showed that, with the exception of river regions, bottom sediments in the marine areas surveyed had been influenced greatly by coral reefs.

Total organic content analysis produced suggests the formation of an environment where organic matter outflow from rivers tends to accumulate. At some sites inside the lagoon area, comparatively higher contents of organic substances were detected.

C/N ratio analysis of sediment samples taken from sites located between the estuarine region (6) and upper reaches of the river (7 and 8) exhibited higher values.

Commonly, high values for sediments containing substances derived from higher plants with thick cell walls suggest that a survey site, beginning with its mangroves, contains a significant accumulation of matter derived from higher plants.

Early observations of the coral reefs and mangroves of Palau inspired questions about the influences of the vast mangrove systems on coral reefs. It thus seems clear that massive quantities of fallen leaves will affect the quality of surrounding waters and bottom sediments, while generating a relationship with the biota inhabiting river mouths and coastal zones.

Golbuu et al. (2003) investigated sediment behavior in semi-closed environments of river mouths in the Ngerikiil River in Airai State. While watersheds in this area maintained comparatively good environmental conditions up until the 1970s, the construction of the International Airport during 1978–1982 and lowland agriculture, which was particularly popular in the 1990s, aggravated soil runoff, causing sea bottoms in estuarine regions and coastal zones to be covered with fine sediment. They concluded that only 2 % of sediment particles that flow into estuarine regions are transported to the outer seas, with the majority remaining within the bay and contributing continuously to the accumulation of sediment. The amount of sediment accumulated was estimated to reach 20,000 tons in a year.

Golbuu et al. (2011) researched the sedimentation processes of fine particles and turbidity dynamics of watersheds in four terrestrial areas under differing degrees of development on Babeldaob Island. It was found that fine particle sedimentation per 1 km^2 ranged between 29.7 and 216 tons annually, with areas under higher levels of development exhibiting more sedimentation.

For coral reefs, an example illustrating the link between terrestrial areas and coastal zones is the death of corals caused by sediment particle outflow. Some studies have also reported that sediment particle outflow has an effect on coral reef organisms in the Republic of Palau (Fabricius et al. 2007). During periods when the effects of sediment outflow on coral reefs were observed, the quantity of accumulated microscopic particles was found to be 39.6 $mg/cm^2/day$, and the extent of the sediment's effects on coral reefs was reported to depend on the type and species of organism.

This network via biogeochemical cycle between mangroves and coral reefs should be considered in the discussion of coastal management in Palau.

3.2 Environmental Functions

- Environmental purification functions

Mangroves and tidal flats are home to immense numbers of crabs and snails. When these animals feed, they reduce the amount of organic materials in tidal flats through their ability to purify their own habitats.

- Refuges for young coral reef fish

During high tides, it is possible to observe small fish swimming in from the sea to forage among the roots of mangroves. Mangroves are an especially important source of shelter for young juveniles that live on coral reefs.

- Provision of habitats to birds, insects, and other animal species

During low tides, a variety of bird species rest and carry out feeding activities on the tidal flats that extend from mangrove forests. Additionally, birds deposit organic matter taken from other tidal flats in the form of droppings. Such biological processes cannot be ignored when large populations of bird species are present.

The diverse types of birds inhabiting tidal flats are also of interest to bird-watchers. The use of tidal flats by birds is not random; some bird species use shorelines and water edges, and others feed with their feet slightly in the water. Although tidal flats may seem uniform, they should be understood as a collection of many different environments that are utilized by various types of living organisms.

- Opportunities for food resources

To humans, the most important species in Palau mangrove ecosystems is the mangrove crab, and attempts at crab breeding are currently in progress. Land crabs *Cardisoma* spp. are also eaten frequently.

Although some mangrove species, such as mangrove crabs, have been used as food for people in Palau, it is difficult to obtain fish-catch statistical data for mangrove areas. This is also true for information on coral reef fish, although data spanning a 5-year period starting in 2002 have been made public (2006 Statistical Yearbook, Republic of Palau, Bureau of Budget and Planning, Ministry of Finance).

3.3 Socioeconomic Considerations

- Important resources for environmental education

Because mangroves have been widely used in a variety of academic researches, fascinating discoveries have been made, bearing possibilities for new academic progress and improvements to human life. Such information can be used for environmental education, but it is also important to further interpretation of mangrove ecosystems.

- Sites for recreation, such as ecotourism and kayaking

Ecosystem health is vital to the establishment and maintenance of the tourism industry. Dive shops operate because of the presence of healthy coral reefs and the opportunity to view corals and colorful fish. The same is true for tour companies that organize mangrove forest and coral reef boat tours.

In recent years, ecotourism has become increasingly popular, functioning as a source of revenue as well as contributing to the conservation of natural environments. However, injudicious ecotourism can undeniably upset the natural balance and lead to problems. As such, it is important to consider the nature of different types of ecotourism as well as their participant capacities.

Compared with figures for 2002, the number of tourists in Palau is increasing. Tourist visits in 2010 were said to number 90,000. Although many tourists come to Palau to enjoy coral reef seascapes, it is also true that mangrove areas have been used for tourist kayaking activities. Careful management to ensure the coexistence of tourism with nature is thus necessary.

In Okinawa, much debate has surrounded possible ways to increase the number of tourists visiting the prefecture. In 2007, 5.9 million people visited Okinawa, and moves to increase tourist figures to 10 million have been implemented since then.

Palau has a land area of 487 km^2 and a population of 21,000. In comparison, Okinawa has 4.6 times the land area of Palau (2,276 km^2) and a population of 1.4 million people, 67 times of that of Palau. When considering environmental capacities, it is evident that Okinawa has a bigger resident population and hosts a larger number of tourists compared to Palau. While comparisons between Okinawa and Palau should take into consideration differences in their natural environments and the sizes of their ocean areas, it is hoped that future approaches may be formulated bearing in mind their respective present conditions.

It is therefore vital to understand that tourism industries cannot flourish without a healthy natural environment and attempts toward environmental conservation. Of late, discussions on the ways in which tourism and nature may coexist have begun in Okinawa, and attention should to be paid to the ways in which such discussions are converted into practical measures for implementation. It is hoped that such debates will be useful to environmental conservation efforts in Palau. Interdisciplinary discussion from natural and social science perspectives will be useful in studies of the carrying capacities of tourist sites in Palau for their sustainable use.

3.4 Economic Evaluation of Mangrove Ecosystems

To make an economic evaluation of an ecosystem, its value must be divided into two categories: its market value and its nonmarket value. The market value is defined as the value of natural materials and resources in an ecosystem and is evaluated as a straightforward market transaction. In mangrove or coral reef ecosystems, for example, the prices of resources such as fish, shellfish, and crabs are used in this calculation. However, insufficient data are available on such market transactions. For Palau, statistics on fish catches are lacking, and personal transactions between fishermen and consumers are common. Environmental changes also affect fish catches and market prices, but such analyses have not been conducted.

The nonmarket value of mangrove ecosystems is also important. This is the value of ecosystem and/or cultural services. Environmental purification functions, values of recreation sites, and so on are included in this value. To evaluate this nonmarket value, a replacement method must be applied because no prices are assignable for these values.

In order to evaluate the economic values not only of mangrove ecosystems but also of any natural resource, we must quantitatively recognize the environmental situation of the ecosystem in question and its dynamics. Undoubtedly, this is an important interdisciplinary research topic involving several scientific fields, as well as the straight relationships between mankind, natural environment, and exploitation of marine resources.

Acknowledgments We are very grateful to staffs of the Palau International Coral Reef Center for their help. This work was partly supported by special funds from the Sasakawa Peace Foundation, Japan Science and Technology Agency (JST), and Japan International Cooperation Agency (JICA).

References

Albert JA, Warren-Rhodes K, Schwarz AJ, Duke ND (2012) Mangrove ecosystem services & payments for Blue Carbon in Solomon Islands, The World Fish Center, Solomon Islands. AAS-2012-06, 6pp

Bennett EL, Reynolds CJ (1993) The value of a mangrove area in Sarawak. Biodivers Conserv 2:359–375

Brander LM, Wagtendonk AJ, Hussain SS, McVittie AM, Verburg PH, de Groot RS, van der Ploeg S (2012) Ecosystem service values for mangroves in Southeast Asia: a meta-analysis and value transfer application. Ecosyst Ser 1(1):62–69

Costanza R, D'Arge R, De Groot R, Farber S, Grasso M, Hannon B, Linnberg K, Naeema S, O'Neill RV, Parvelo J, Raskin RG, Sutton P, Van den Belt M (1997) The value of the world's ecosystem services and natural capital. Nature 387:253–260

Ehrlich P, Ehrlich A (1981) Extinction: the causes and consequences of the disappearance of species. Ballantine Books, New York, 384pp

Fabricius KE, Golbuu Y, Victor S (2007) Selective mortality in coastal reef organisms from an acute sedimentation event. Coral Reefs 26:69

Golbuu Y, Victor S, Wolanski E, Richmond RH (2003) Trapping of fine sediment in a semi-enclosed bay, Palau, Micronesia. Estuar Coast Shelf Sci 57:941–949

Golbuu Y, van Woesik R, Richmond R, Harrison P, Fabricius KE (2011) River discharge reduces reef coral diversity in Palau. Mar Pollut Bull 62:824–831

Gunawardena M, Rowen JS (2005) Economic valuation of a mangrove ecosystem threatened by shrimp aquaculture in Sri Lanka. Environ Manag 36(4):535–550

Hussain SA, Badola AR (2008) Valuing mangrove ecosystem services: linking nutrient retention function of mangrove forests to enhanced agroecosystem production. Wetl Ecol Manag 16(6):441–450

Sathirathai S, Barbier EB (2001) Valuing mangrove conservation in Southern Thailand. Contemp Econ Policy 19:109–122

Shilla DJ, Tsuchiya M, Shilla DA (2011) Terrigenous nutrient and organic matter in a subtropical river estuary, Okinawa, Japan: origin, distribution and pattern across the estuarine salinity gradient. Chem Ecol 27:523–542

Tam NFY, Wong YS (1999) Mangrove soils in removing pollutants from municipal wastewater of different salinities. J Environ Qual 28(2):556–564

Photographs

Professor Teruhisa Komatsu with Mrs Mireille Hingrez-Céréda, Major of Boulogne-sur-Mer, during the welcome speech

Professor Hubert Ceccaldi discussing with Dr. Yves Reecht during the coffee break at Boulogne-sur-Mer

H.-J. Ceccaldi et al. (eds.), *Marine Productivity: Perturbations and Resilience of Socio-ecosystems*,
DOI 10.1007/978-3-319-13878-7, © Springer International Publishing Switzerland 2015

Drs. Hiroshi Okumura, Hideki Takami, and Hideaki Tanoue discussing during the poster session

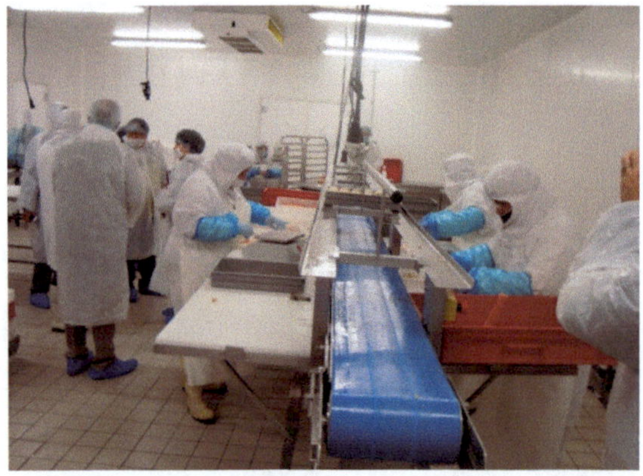

Visit of a unit of seafood product "Ocean Délices" at Boulogne-sur-Mer

A part of the participants at Marseille during the presentation of Mr Takehiro Tanaka

A part of the organization team at the Marseille University with Dr. Georges Stora and Dr. Yasuyuki Koike

Professor Catherine Mariojouls speaking during the round table and from left to right are Christian Decugis, Sylvain Pioch, Catherine Mariojouls, Yves Henocque, Takehiro Tanaka, Teruhisa Komatsu, Yasuyuki Koike, Satoquo Seino, and Hisayuki Takahashi

Professors Hisayuki Arakawa and Yasuyuki Koike with Drs. Yves Hénocque, Yuji Tanaka, and Toshiki Nakano

Drs. Yves Henocque and Patrick Prouzet discussing during the lunch at Marseille

From left to right: Prof. Teruhisa Komatsu, Prof. Yasuyuki Koike, Mr Romiti (Chairman of the National Committee for Fisheries and Aquaculture in France), Dr. Satoquo Seino, and Dr. Hisayuki Takahashi

The 15th Franco-Japanese Symposium of Oceanography "Marine Productivity, Perturbations and Resilience of Socio-Ecosystems," organized by the long-standing partners Société franco-japonaise d'Océanographie de France and Société franco-japonaise d'Océanographie du Japon, reviewed the impacts of natural (storms, typhoons, earthquakes, tsunamis, etc.) and man-made (pollution, buildings in coastal areas, aquaculture, tourism, sports, diving, etc.) perturbations inflicted on natural coastal and marine environments, and examined the resilience of the affected socio-ecosystems and ability of governance responses to face these global/local changes.

The 15th Franco-Japanese Symposium of Oceanography was supported by the two countries' highest authorities.

The book collects 43 selected papers, written by experts from numerous universities and research institutes in both countries. It addresses the needs of marine sciences researchers (natural and social sciences), decision-makers and coastal zone managers, and other stakeholders involved in coastal and marine socio-ecosystems.

Author Index

H.-J. Ceccaldi et al. (eds.), *Marine Productivity: Perturbations and Resilience of Socio-ecosystems*,
DOI 10.1007/978-3-319-13878-7, © Springer International Publishing Switzerland 2015

G

H

I

J

K

MIX
Papier aus verantwortungsvollen Quellen
Paper from responsible sources
FSC® C105338

If you have any concerns about our products,
you can contact us on
ProductSafety@springernature.com

In case Publisher is established outside the EU,
the EU authorized representative is:
Springer Nature Customer Service Center GmbH
Europaplatz 3, 69115 Heidelberg, Germany

Printed by Libri Plureos GmbH
in Hamburg, Germany